III International Scientific-Practical Conference "Graphene and Related Structures: Synthesis, Production, and Application" (GRS-2019)

IOP Conference Series: Materials Science and Engineering
Volume 693

Tambov, Russia
13 - 15 November 2019

Editors:

Anna Godymchuk
Jesus Iniesta Valcarcel
Evgeny Galiunin
Anna Zykova

ISBN: 978-1-7138-0985-2
ISSN: 1757-8981

Printed from e-media with permission by:

Curran Associates, Inc.
57 Morehouse Lane
Red Hook, NY 12571

Some format issues inherent in the e-media version may also appear in this print version.

Institute of Physics
Dirac House, Temple Back
Bristol BS1 6BE UK

Phone: 44 1 17 929 7481
Fax: 44 1 17 920 0979

techtracking@iop.org

Additional copies of this publication are available from:

Curran Associates, Inc.
57 Morehouse Lane
Red Hook, NY 12571 USA
Phone: 845-758-0400
Fax: 845-758-2633
Email: curran@proceedings.com
Web: www.proceedings.com

TABLE OF CONTENTS

IOP Conf. Series: Materials Science and Engineering **693** (2019) 011001 doi:10.1088/1757-899X/693/1/011001

Preface

III International Scientific-Practical Conference "Graphene and Related Structures: Synthesis, Production, and Application" (GRS-2019) was organized by Tambov State Technical University (Russia) in cooperation with Rosnano (Russia), Vernadsky Non-Governmental Environmental Foundation (Russia), Belarusian State University (Republic of Belarus), University of Alicante (Spain), University of Chile (Chile), Technological Institute for Superhard and Novel Carbon Materials (Russia), Russian Academy of Sciences, and Russian-Chinese Research Institute of Graphene in November, 13-15, 2019, Tambov, Russian Federation.

GRS-2019 gathered over 150 engineers, academicians, graduate and postgraduate students, PhD and postdoctoral fellows, researchers and specialists dealing with research and use of carbon nanomaterials (graphene, nanographite, carbon nanotubes and nanofibers, functional and construction materials), as well as metrological maintenance, standardization and quality control of carbon nanotechnology-based products. The Conference provided a platform giving the participants from different disciplines the opportunity toshare and discuss their research findings with a wide audiencein the development of nanotechnologies,elaboration, production and application of graphene-related materials.

The GRS-2019 Proceedings include the papers covering (1) Graphene and nanographite; (2) Carbon nanotubes and nanofibers; (3) Functional and construction nanomaterials; and (4) Metrological maintenance, standardization and quality control of carbon nanotechnology products.

We appreciate the contribution of the Participants, Institutions and Sponsors supporting the event. All paper submitted to the Proceedings were intensively reviewed by prominent experts in various scientific areas from Egypt, China, Poland, Russia, Spain, Iran, UAE, Ukraine, Nigeria, France, Greece, India, USA, Czech Republic, Iraq, Turkey, Brazil, Finland, Pakistan, North Macedonia, Italy, Japan, and Germany. The Editors are very grateful to all Reviewers for their valuable suggestions and comments that permit to improve the papers quality.

Web page of the GRS-2019: http://graphene.tstu.ru

GRS-2019 Organizing Committee:

Prof. Alexey Tkachev, Tambov State Technical University, Russian Federation

Editors of the Special Issue GRS-2019:

Dr. Anna Godymchuk, Tomsk Polytechnic University, Russian Federation

Dr. Jesus Iniesta Valcarcel, University of Alicante, Spain

Dr. Evgeny Galunin, Tambov State Technical University, Russian Federation

Dr. Anna Zykova, Institute of Strength Physics and Materials Science of Siberian Branch of Russian Academy of Sciences, Tomsk, Russian Federation

IOP Conf. Series: Materials Science and Engineering **693** (2019) 011002 doi:10.1088/1757-899X/693/1/011002

Peer review statement

All papers published in this volume of *IOP Conference Series: Materials Science and Engineering* have been peer reviewed through processes administered by the proceedings Editors. Reviews were conducted by expert referees to the professional and scientific standards expected of a proceedings journal published by IOP Publishing.

IOP Conf. Series: Materials Science and Engineering **693** (2019) 012001 doi:10.1088/1757-899X/693/1/012001

An experimental study on the influence of functionalized carbon nanotubes CNT Taunit series on the thermal conductivity enhancement

A J Ali[1, 2] and E N Tugolukov[1]

[1]Technology and Methods of Nano Products Manufacturing, *Tambov State Technical University,* 106 Sovetskaya Street, 392000 Tambov, Russian Federation
[2]Department of Biomedical Engineering, *University of Technology,* Iraq

[1]E-mail: 10678@ uotechnology.edu.iq

Abstract. This paper aims to determine the thermal conductivity for two types of functionalized carbon nanotubes (FCNT "Taunit M") nanofluids. $FCNT_1$ and $FCNT_2$ have been used to enhance the thermal conductivity of nanofluids. Six concentrations of 0.01, 0.02, 0.03, 0.05, 0.07, 0.1 wt.% loading of $FCNT_1$ and three concentrations of 0.01, 0.05, 0.1 wt.% loading of $FCNT_2$ have been used in this experiment. The experiments were achieved at temperatures 40°C. The technique for measuring thermal conductivity was described and manufactured at Tambov State Technical University. As a result of the study, the effect of increasing concentrations of FCNTs in the aqueous medium on thermal conductivity was presented. The relationship was nonlinear in $FCNT_1$. The maximum enhancement thermal conductivity was 11.6 and 7.1% loading $FCNT_1$ and $FCNT_2$ respectively, which achieved at a weight fraction of 0.05 and 0.1 % respectively. Finally, the latest researches in this field were presented for comparison in the results obtained.

1. Introduction

The enhancement of thermal conductivity in conventional heat transfer fluids was an engineering challenge and problem. One of the ways to solve this problem is to take aiming measures to increase the values of thermal conductivity by adding nanoparticles to traditional heat transfer fluids, which leads to increase in the overall heat transfer process. One of the results of this addition to heat transfer fluid is the appearance of nanofluid. Therefore, there are numerous numerical and experimental studies concentrated on nanofluids [1-3]. Nanofluids are engineered by a mixture of the traditional heat transfer fluids like oils, distilled water, fat, Ethylene glycol, and others with nanoparticles 100 nm such as Cu, CuO, Si, MWNT, SWNT, etc. But one of the problems in the nanofluid, which drew the attention of workers in this field is the agglomeration and sedimentation of nanoparticles after various time periods. Therefore, the best improvement which can be obtained is: dispersed uniformly in host-based fluids and suspended stably [4]. Several methods have been used to avoid agglomeration and sedimentation of nanoparticles and to obtain a stable suspension: chemical and physical methods by adding surfactants and using ultrasonic at different frequencies respectively; electrical method by controlling the pH [5, 6]. Carbon nanotubes for nanofluids have been attracted the attention of various researchers due to their advantage such as high thermal conductivity that is higher in comparison with other nanoparticles. It is a result of the aspect ratio which is an important factor in improving thermal conductivity [7, 8]. Carbon nanotubes have thermal conductivity reaching (2000-3500) W/m.K [9]. Carbon nanotube exists as the single type particle such as single (SWCNTs), multi-walled (MWCNTs) and double (DWCNTs) structures or like hybrid nanoparticles recently appeared [10].

In this research thermal conductivity of functionalized carbon nanotubes "Taunit M" ($FCNT_1$ and $FCNT_2$) in the base fluid (distilled water) was measured. The latest researches in this field for the same type of nanoparticles were analyzed for comparison of the results obtained.

2. Experimental part

2.1. Chemicals

There are two types of functionalized carbon nanotubes "Taunit M" ($FCNT_1$ and $FCNT_2$) were produced for the improvement of thermal conductivity in the water. The difference between ($FCNT_1$) and ($FCNT_2$) is the growth at different time periods on the catalyst, where it was 40 minutes for $FCNT_1$ and 5 minutes for $FCNT_2$ in the reactor CVD. Carbon nanostructured materials were manufactured at (NanoTechCenter Ltd) in Tambov Region, Russia [11]. The functionalized CNTs were prepared by using the traditional method of chemical oxidation of CNT with sodium hypochlorite [12, 13].

2.2. Measurement of thermal conductivity of nanofluid and stability

The manufacturers certified measuring equipment that provides high measurement accuracy is complex and expensive. In this regard, the development of experimental methods for determining the thermal conductivity of liquids, including nano-modified, is currently continuing. The transient hot-wire (THW) method is widely used, especially by device KD2 thermal property analyzer [14]. The experimental model of thermal conductivity measurement is manufactured in this research at Tambov state technical university as shown in figure 1. It consists of a cup made of a material that has low thermal conductivity, which removable module is mounted. Also, the module contains a heater and temperature sensors 4 and 5 located along the axis of the cup.

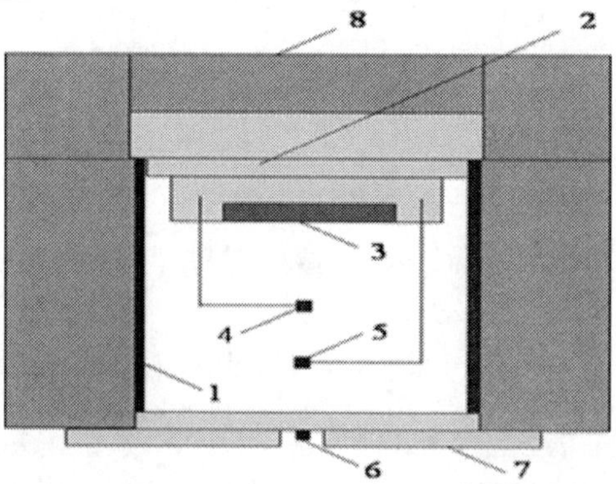

Figure 1. Scheme of the experimental setup:
1- cup; 2- which removable module is mounted; 3- heater; 4, 5 -the temperature sensors; 6-temperature sensor is installed on a cooled metal base; 7- cooled metal base ; 8- heat-insulating coating.

The functionalized carbon nanotubes "Taunit M" nanofluids have high stability in time, as can be seen in figures 2 and 3. In order to ensure accuracy, the measurement was conducted for 3 times for one sample under the same experimental conditions. The rig was calibrated by using distilled water in the considered temperature range before the start of measuring. The error rate did not exceed and has a tolerance accuracy of ±5%. Operates according to principle a somewhat similar is based on transient hot-wire THW. The values of nanofluids effective thermal conductivity at temperature 40 °C have been provided. In order to evaluate the stability of nanofluids, the samples have been maintained under static conditions and standard atmospheric temperature for six months. It was observed neither agglomeration of nanoparticles nor sedimentation during this period. This is a good indicator of nanoparticle stabilization in comparison with other researches.

The main properties and SEM images of the functionalized carbon nanotubes (FCNT "Taunit M") are presented in table 1 and figure 4 respectively.

Figure 2. Samples of $FCNT_1$ and $FCNT_2$ "Taunit M" nanofluids.

Figure 3. Samples of $FCNT_1$ and $FCNT_2$ "Taunit M" nanofluids after six month.

Table 1. Properties of functionalized carbon nanotubes "Taunit M".

Parameter	Value
Inner diameter	4-8 nm
Outer diameter	8-15 nm
Length, µm	2 and more
Bulk density, g / cm	0.03÷0.05
Total amount of impurities %,	up to 5
(after purification)	up to 1
Specific geometrical surface, m^2/g	180-200
Thermal stability (°C)	up to 600
Color	Black

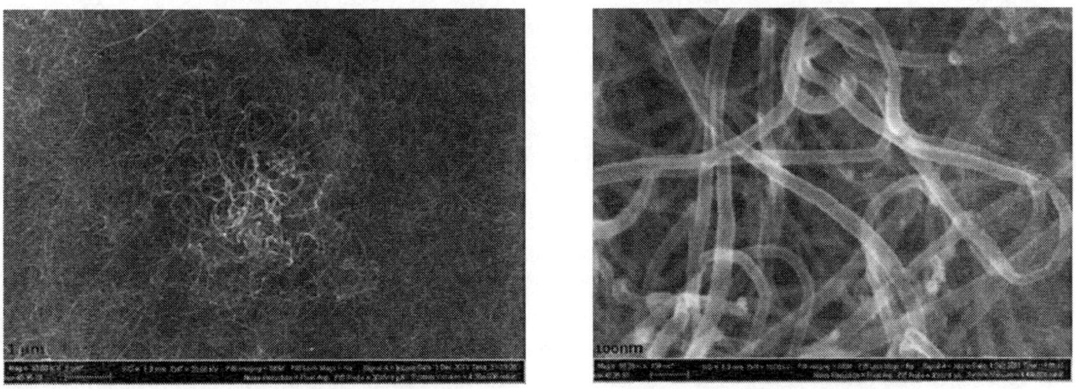

Figure 4. SEM images of the functionalized carbon nanotubes "Taunit M" [15].

Two types of functionalized carbon nanotubes "Taunit M" were evaluated in this experiment. Six concentrations of 0.01, 0.02, 0.03, 0.05, 0.07, 0.1 wt.% loading of $FCNT_1$ and three concentrations of 0.01, 0.05, 0.1 wt.% loading of $FCNT_2$ have been used. The experiment was achieved at temperature 40°C. The maximum improvement of thermal conductivity was 11.6% at 0.05 wt.% loading of $FCNT_1$, that means to achieve an increase in the thermal conductivity of water from 0.625 to 0.71 W/m.K, and for the second type 7.1% enhancement at 0.1 wt.% loading of $FCNT_2$, it means to achieve an increase in the thermal conductivity of water from 0.625 to 0.67 W/m.K. The relationship was nonlinear in $FCNT_1$. The thermal conductivity improvement of $FCNT_2$ remained constant from 0.01 to 0.05 wt.%, then the thermal conductivity rose to 7.1% improvement at 0.1 wt.%. The relationship between thermal conductivity and concentration of the studied FCNT is presented on figures 5 and 6.

It is known that thermal conductivity increase with increasing temperature and concentration. However, this is not always as the higher concentration might affect the dispersion of nanoparticles in the water with the absence of a surfactant. The dispersion / stability of nanoparticles is the most factor importance in enhancing thermal conductivity. In this paper, the improvement of thermal conductivity decreases at 0.07 and 0.1 wt.% loading for $FCNT_1$, due to agglomeration in our solution. The reason was the low volume of solution and due to the low density of CNT, the bulk volume of CNT will be high which will be shrinkage in the solution and losing its stability.

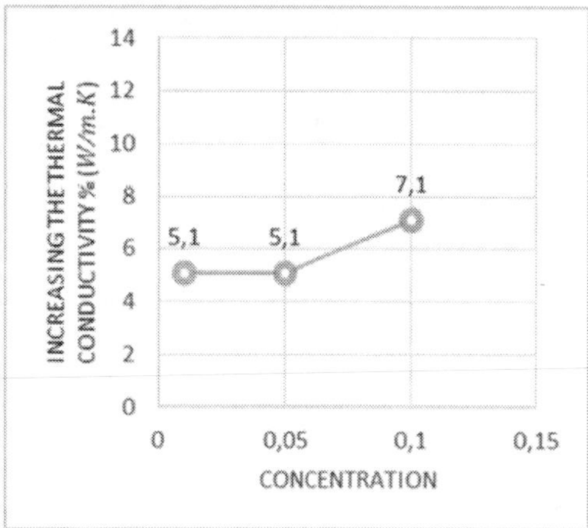

Figure 5. Relationship between increased thermal conductivity VS Concentration $FCNT_2$.

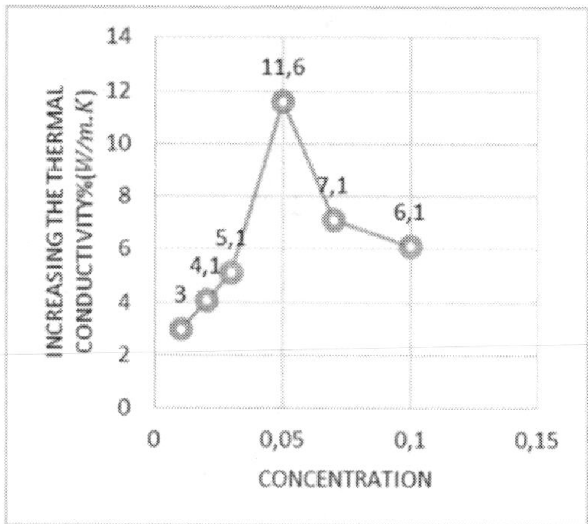

Figure 6. Relationship between increased thermal conductivity Vs Concentration $FCNT_1$.

3. Comparison between $FCNT_1$ nanofluid and other functionalized carbon nanotube nanofluids on enhancement thermal conductivity

The latest researches in carbon nanotube nanofluids were analyzed to compare obtained experimental results for the $FCNT_1$ (table 2). Taunit $FCNT_2$ has been neglected due to the lower enhancement in comparison with $FCNT_1$ type.

According to the literature review of modern studies in various CNT nanofluids Shamaeil et al. [16] have tested 0.02, 0.05, 0.075, 0.1, 0.25 and 0.6 % solid volume fraction double-walled carbon nanotubes (DWCNTs) in ethylene glycol (EG) for thermal conductivity improvement and showed 24.9 % enhancement of 0.6 vol%. Harandi et al. [17] have examined the thermal conductivity of FMWCNTs–Fe_3O_4 in EG at 0 to 2.3 % volume fraction. The maximum improvement of thermal conductivity was 30%, which was achieved at 2.3 vol.%. In the paper of Soltanimehr et al. [18] the highest improvement in thermal conductivity of the MWCNTs/Ethylene glycol EG (40 vol.%) and distilled water (60 vol.%) found to be 34.7% enhancement at 1.0 vol.% loading. Manasrah et al. [19] have reported the effects of pristine carbon nanotubes and functionalization CNTs with polyethylene glycol (MWCNTs-PEG) by using Fischer esterification method to show their dispersion in base fluid with various concentrations of 0.01 wt.%, 0.05 wt.% and 0.1 wt.% without using any surfactants. The maximum improvement of thermal conductivity was 5.77% and 19% at 1 wt.% loading of pure CNT and functionalized MWCNTs respectively. Esfe et al. [20] studied the thermal conductivity of (SiO_2-MWCNT /EG) hybrid nanofluid at 0.025 to 0.86 vol% and showed an enhancement of up to 20.1% more than base fluid thermal conductivity at 0.86 vol %. The thermal conductivity of CuO-SWCNTs in EG40% - water 60% nanofluid with 0.02 to 0.75 vol.% has been investigated by Rostamian et al. [21]. It has been noted the thermal conductivity increase by 36.2% at 0.75% volume fraction. SWCNT–Al_2O_3 in EG (hybrid nanofluid) at volume fractions of 0.04, 0.08, 0.15, 0.3, 0.5, 0.8, 1.5 and 2.5 have been experimentally investigated by Esfe et al. [22]. They observed increase in the thermal conductivity reaches 41.2% suspensions at a volume fraction of 2.5%. In other paper Esfe et al. [23] have investigated the influence of functionalized single walled carbon nanotubes (FSWCNTs) in the base fluid Ethylene Glycol suspension at 0.02, 0.05, 0.075, 0.1, 0.25, 0.5 and 0.75% volume fraction. It was noted maximum enhancement 45% for 0.75 vol%. Mirbagheri et al. [24] have investigated the influence of functionalized multi-walled carbon nanotubes (FMWCNTs) in the base fluid of EG-water (20:80 vol.%) containing concentrations 0.025 to 0.8 vol.% loading. It was observed 27.3% enhancement in thermal conductivity for 0.8 vol.% concentration of FMWCNTs. Omrani et al. [25] investigated MWCNTs with different geometrical characteristics in water based nanofluids containing 0.05 vol.%. The maximum improvement of thermal conductivity reached to 36% at 45 °C in comparison to the water.

Table 2. A literature review on modern studies in various CNT-nanofluids.

Investigator	Particles Hybrid particles	Base fluids	Concentration (%)	Maximum enhancement K(%)	T (°C)	year
our research	FCNTs	water	**0.05 wt.% (0.03 vol.%)**	11.6%	40 °C	2019
Omrani et al.	FMWCNTs	DIW	0.05 vol.%	36%	45 °C	2019
Mirbagheri et al.	FMWCNTs	Water-EG (80-20)	0.8 vol.%	27.3%	50 °C	2018
Esfe et al.	FSWCNTs	EG	0.75 vol.%	45%	50 °C	2018
Manasrah et al	FCNT+PEG	water	1 wt.% (0.6 vol.%)	19%	35 °C	2017
Esfe et al.	SiO_2-MWCNT (30-70%)	EG	0.86 vol.%	20.1%	50 °C	2017
Rostamian et al.	CuO-SWCNTs	EG water (40:60)	0.75 vol.%	36.2%	50 °C	2017
Esfe et al.	WCNT–Al_2O_3 (30 -70%)	EG	2.5 vol.%	41.2%	50 °C	2017
Shamaeil et al.	FDWCNTs	EG	0.6 vol.%	24.9%	52 °C	2016
Harandi et al.	FMWCNTs –Fe_3O_4	EG	2.3 vol.%	30%	50 °C	2016
Soltanimehr et al.	FMWCNTs	EG-water (40:60)	1.0 vol.%	34.7%	50 °C	2016

It remains to be known that stability and the economic side play an equal role in enhancement thermal conductivity nanofluids. In this research it has been obtained an improvement in thermal conductivity taking into account the economic side where the concentration of FCNT was the least in comparison with other researches and the nanofluids have high stability. Because the cost of adding carbon nanotubes in the base fluid is high, so the low concentration is better on the economic side. Finally, we have concluded from this analysis and results that the carbon nanotubes lead to enhance in thermal conductivity. Therefore, the urgent need proposed for future works is to make many experimental studies on the thermal conductivity of CNT nanofluids and Hybrid CNT nanofluids in various basal fluids with the possibility to obtain valuable and new results.

4.Conclusions

The thermal conductivity of functionalized carbon nanotubes "Taunit M" ($FCNT_1$, $FCNT_2$) in the base fluid (distilled water) was measured for these two types. The first type ($FCNT_1$) achieved an improvement of thermal conductivity, especially in the 0.05 wt.%, reaching 11.6%, and the relationship was nonlinear with a concentration of this type, where there was a decrease in improved thermal conductivity reaches to 6.1% at 0.1 wt.%. The maximum enhancement of 7.1% at 0.1 wt.% was shown in the second type $FCNT_2$. The thermal conductivity measuring device was manufactured at Tambov State Technical University. Calibration has been implemented on a device by using distilled water and the error rate ranged from ± 5%. The device can be developed in the future to be more accurate and reliable. The third part of this study was a comparison between the result of this research and other recent researches result on each of CNTs and hybrid CNT nanofluids. It can be concluded that the improvement in thermal conductivity with economic importance in our research result was good. Finally, we need to develop these results.

References

[1] Tawfik M M 2017 Experimental studies of nanofluid thermal conductivity enhancement and applications: a review *J. Renewable and Sustainable Energy Reviews* **75** 1239-1253

[2] Usri N A, Azmi W H, Mamat R, Hamid K A and Najafi G 2015 Thermal conductivity enhancement of Al_2O_3 nanofluid in ethylene glycol and water mixture *Energy Procedia* **79** 397-402

[3] Xing M, Yu J and Wang R 2015 Experimental study on the thermal conductivity enhancement of water based nanofluids using different types of carbon nanotubes *International Journal of Heat and Mass Transfer* **88** 609-616

[4] Yu F et al 2017 Dispersion stability of thermal nanofluids *Progress in natural science: Materials International* **27**(5) 531-542

[5] Hadadian M, Samiee S, Ahmadzadeh H and Goharshadi E K 2013 Nanofluids for heat transfer enhancement - a review *Physical chemistry research* **1**(1) 1-33

[6] Al-Waeli A H, Chaichan M T, Kazem H A and Sopian K 2019 Evaluation and analysis of nanofluid and surfactant impact on photovoltaic-thermal systems *Case Studies in Thermal Engineering* **13** 100392

[7] Evgin T et al 2016 Effect of aspect ratio on thermal conductivity of high density polyethylene/multi-walled carbon nanotubes nanocomposites *Composites Part A: Applied Science and Manufacturing* **82** 208-213

[8] Eatemadi A et al. 2014 Carbon nanotubes: properties, synthesis, purification, and medical applications *Nanoscale research letters* **9**(1) 393

[9] Pop E, Mann D, Wang Q, Goodson K and Dai H 2006 Thermal conductance of an individual single-wall carbon nanotube above room temperature *Nano letters* **6**(1) 96-100

[10] Ncibi M C and Sillanpää M 2015 Optimized removal of antibiotic drugs from aqueous solutions using single, double and multi-walled carbon nanotubes *J. of hazardous materials* **298** 102

[11] Polyakov V A et al 2017 Sorption Activity of "Taunit"-Series Carbon Nanomaterials *International Letters of Chemistry, Physics and Astronomy* **74** 1-8

[12] Abdolkarimi-Mahabadi M and Manteghian M 2015 Chemical oxidation of multi-walled carbon nanotube by sodium hypochlorite for production of graphene oxide nanosheets *J. Fullerenes, Nanotubes and Carbon Nanostructures* **23**(10) 860-864

[13] Yang J C, Yen C H, Wang W J, Horng J J and Tsai Y P 2010 Assessment of adequate sodium hypochlorite concentration for pre-oxidization of multi-walled carbon nanotubes *J. of Chemical Technology & Biotechnology* **85**(5) 699-707

[14] Pryazhnikov M I , Guzei D V , Minakov A V and Lobasov A S 2015 Development and Testing of Experimental Methods Definition of Thermal Conductivity of Nanofluids *J. of Siberian Federal University. Engineering & Technologies* **8**(2) 153

[15] Information on: http://www.nanotc.ru

[16] Shamaeil M , Firouzi, M and Fakhar A 2016 The effects of temperature and volume fraction on the thermal conductivity of functionalized DWCNTs/ethylene glycol nanofluid *J. of Thermal Analysis and Calorimetry* **126**(3) 1455-1462

[17] Harandi S S , Karimipour A, Afrand M , Akbar M and D'Orazio A 2016 An experimental study on thermal conductivity of F-MWCNTs–Fe3O4/EG hybrid nanofluid: effects of temperature and concentration *International Communications in Heat and Mass Transfer* **76** 171-177

[18] Soltanimehr M and Afrand M 2016 Thermal conductivity enhancement of COOH-functionalized MWCNTs/ethylene glycol–water nanofluid for application in heating and cooling systems *Applied Thermal Engineering* **105** 716-723

[19] Manasrah A D , Laoui T , Zaidi S J and Atieh M A 2017 Effect of PEG functionalized carbon nanotubes on the enhancement of thermal and physical properties of nanofluids *Experimental Thermal and Fluid Science* **84** 231-241

[20] Esfe M H , Esfandeh S and Rejvani M 2018 Modeling of thermal conductivity of MWCNT-SiO_2 (30:70%)/EG hybrid nanofluid, sensitivity analyzing and cost performance for industrial applications *J. of Thermal Analysis and Calorimetry* **131**(2) 1437-1447

[21] Rostamian S H, Biglari M, Saedodin S and Esfe M H 2017 An inspection of thermal conductivity of CuO-SWCNTs hybrid nanofluid versus temperature and concentration using experimental data, ANN modeling and new correlation *J. of Molecular Liquids* **231** 364-369

[22] Esfe M H, Rejvani M, Karimpour R and Arani A A A 2017 Estimation of thermal conductivity of ethylene glycol-based nanofluid with hybrid suspensions of SWCNT–Al_2O_3 nanoparticles by correlation and ANN methods using experimental data *J.of Thermal Analysis and Calorimetry* **128**(3) 1359-1371

[23] Esfe M H, Firouzi M and Afrand M 2018 Experimental and theoretical investigation of thermal conductivity of ethylene glycol containing functionalized single walled carbon nanotubes *Physical E: Low-dimensional Systems and Nanostructures* **95** 71-77

[24] Mirbagheri M H, Akbari M and Mehmandoust B 2018 proposing a new experimental correlation for thermal conductivity of nanofluids containing of functionalized multiwalled carbon nanotubes suspended in a binary base fluid *International Communications in Heat and Mass Transfer* **98** 216-222

[25] Omrani A N, Esmaeilzadeh E, Jafari M and Behzadmehr A 2019 Effects of multi walled carbon nanotubes shape and size on thermal conductivity and viscosity of nanofluids *J.Diamond and Related Materials*

Thermal behavior of chitosan as a carbon material precursor under IR radiation

A Vasilev[1,2], M Efimov[1], G Bondarenko[1], V Kozlov[1,2], E Dzidziguri[2] and G Karpacheva[1]

[1]Chemistry of Polyconjugated Systems Laboratory, *A V Topchiev Institute of Petrochemical Synthesis RAS*, 29 Leninsky Avenue, Moscow 119991, Russian Federation

[2]Department of Functional Nanosystems and High-Temperature Materials, *National University of Science and Technology "MISiS"*, 4 Leninsky Avenue, Moscow 119991 Russian Federation

[1,2]E-mail: raver.vasiljev@mail.ru

Abstract. An effective accelerated synthesis procedure of a carbon material based on chitosan via infrared pyrolysis was developed. Infrared radiation heating allows to shorten preparation time significantly. It takes only 2 min of the residence time at 800 °C for complete carbonization of the polymer. The chemical transformations in chitosan during infrared pyrolysis depended on temperature were studied. The formation of C=C–C=N and C=C–C=C conjugated bonds at a low-temperature stage of the carbonization process was demonstrated. It was revealed that the main product of the chitosan pyrolysis at 600-800 °C represents a graphite-like carbon material in the form of carbocyclic structures.

1. Introduction

Carbon supported metal nanoparticles are commonly employed as promising materials for many applications as electrode materials in supercapacitors, gas sensors, carriers for the targeted magnetically-controlled delivery of drugs, and in the designing of new magnetic materials [1-8]. Particularly, carbon supports are perspective in catalysis due to their chemical resistance to aggressive media, thermal stability, electrochemical, catalytic properties, a large specific surface area and a wide variety of carbonaceous precursors.

One of the main methods of obtaining carbon supports is the thermal treatment of carbonaceous materials, for example, woods, seeds, shoots, natural, synthetic polymers, *etc* [9-11]. Chitosan (CS), a linear polysaccharide produced from chitin shells of shrimp and other crustaceans, attracts researchers' attention due to their low cost, environmental friendliness, biocompatibility and water solubility. Since CS exhibits fairly high stabilizing and reducing properties for metal nanoparticles, it can be used as a carbon precursor for the preparation of carbon supported metal nanoparticles [12,13].

The main goal of this work was to investigate the chemical transformation of CS at different stages of infrared (IR) pyrolysis in order to determine the possibilities of applying chitosan as an effective precursor of carbon support of metal nanoparticles.

IOP Conf. Series: Materials Science and Engineering **693** (2019) 012002 doi:10.1088/1757-899X/693/1/012002

2. Materials and methods

CS flakes (MW = 500 kDa, deacetylation degree - 83 %) purchased from "Bioprogress" (Shchyolkovo, Russia) were used in the present research. The carbonization of the polymer samples was carried out via annealing at the desired temperature for 2 min in a nitrogen atmosphere under IR heating conditions. The preparation temperature range was 150-800 °C. The heat treatment was performed in the laboratory device of incoherent IR annealing. The halogen lamps KG-220 (total power of 24 kW) were applied as a radiation source, which had a maximum emission wavelength in the region of 0.9-1.2 μm. The samples are marked as IR-CS-x, where x is the temperature of IR heat treatment.

Powder X-ray diffraction (XRD) analysis was performed using "Difray-401" diffractometer (Scientific Instruments, Saint Petersburg, Russia) with Bragg-Brentano focusing, using $Cr_{K\alpha}$ radiation. Raman dispersion spectra were recorded by "Senterra" microscope (Bruker, Billerica, Massachusetts, USA) with 532 nm laser and 0.2 mW of power. A JEM-2100 microscope (JEOL, Tokyo, Japan) were used to examine the surface morphology of the IR pyrolyzed chitosan through transmission electron microscopy (TEM).

Chemical transformation of chitosan was studied via IR spectroscopy. Infrared spectra were recorded in the reflection mode on a HYPERION-2000 IR microscope attached to an IFS-66 v/s FTIR spectrometer (Bruker; crystal – Ge, resolution - 3 cm^{-1}, and region of recording - 600 – 4000 cm^{-1}). X-ray photoelectron spectroscopy (XPS) was conducted using a PHI5500 Versa Probe II instrument (ULVAC-PHI Inc, Chigasaki, Japan) with a monochromatic Al Kα X-ray source.

Thermogravimetry (TG) measurements were carried out in a nitrogen flow at a rate of 50 mL min^{-1} under non-isothermal conditions from room temperature up to 500 °C at a heating rate of 10 °C min^{-1} on a Discovery TG instrument (TA Instruments, Newcastle, USA). The device furnace was equipped with four halogen lamps to heat a SiC chamber in the infrared range. The sample was loaded in a platinum crucible and placed into the SiC chamber. The weighing accuracy of the device was 0.1 %, the balance sensitivity was ~ 10^{-7} g, and the accuracy of temperature control was 1 °C.

Differential scanning calorimetry (DSC) measurements were carried out in a nitrogen flow at a rate of 50 mL min^{-1} under non-isothermal conditions in the temperature range of 40-500 °C at a heating rate of 10 °C min^{-1} on a Q20 calorimeter (TA Instruments). The calorimetric precision was 0.1 %, and the sensitivity was 1 μW.

3. Results and discussion

Fourier-transform IR spectroscopy is one of the best methods to study the chemical transformations of polymers. Figure 1 shows the results from the FTIR spectroscopy of the initial CS and the heat-treated samples exposed to IR radiation at different temperatures. For clarity, the spectra were shifted relative to each other along the axis of intensity.

As shown in figure 1, the IR spectra of the initial CS exhibit characteristic bands corresponding to CS structural elements: at 1000-1050 cm^{-1}, the stretching vibrations of C-OH and C-O-C bonds; at 1300-1350 cm^{-1}, the stretching vibrations of C-N bonds; at 1631-1650 cm^{-1}, the bending vibrations of NH_2 groups; at 2800-3000 cm^{-1}, a band attributed to the bonds in -CH_2- groups; at 1150 and 1550 cm^{-1}, bands corresponding to amide groups; and at 3200-3500 cm^{-1}, a broad band corresponding to the associated -OH groups [14].

It was found that the CS structure does not change under IR treatment at 150 °C. The band in the region of 664 cm^{-1} indicates the keeping of the polymer crystallinity. At the temperature of 200 °C, the crystallinity band disappears, the intensity of the bands from C-O bonds decreases in the region of 1000-1030 cm^{-1}, and a negligible increase in the intensity of the amide groups can be observed. Increasing the temperature of the IR pyrolysis up to 250 °C leads to the formation of conjugated bonds C=C-C=N (166 cm^{-1}) and the rapid vanishing of the amide groups and C-O bonds from the polymer structure. At the temperature of 300 °C, the CS structure changes significantly. All bands characterize the CS structure absence. The position and intensity of two bands at 862 and 1400 cm^{-1} correspond to the carbonate ion CO_3^{2-}. The new broad band at 1260 cm^{-1} corresponding to C-O-C bonds linked with adjacent conjugated bonds is revealed as the temperature rises up to 400-500 °C. Further increase in the synthesis temperature up to 600-800 °C provides a reduction of the amount of conjugated bonds C=C-C=N(C) and the change of oxidation degree of the carbon support and formation of different types of C-O-X bonds (X – H, N, O). This can be explained by the intensity decrease of the bands corresponding to the conjugated bonds, growth and shift toward long-wave region of the C-O-C bands as well. The IR spectra of the samples obtained at 600-800 °C have a shoulder in the region of 1050-1500 cm^{-1}, which indicates the remaining of some number of the conjugated bonds

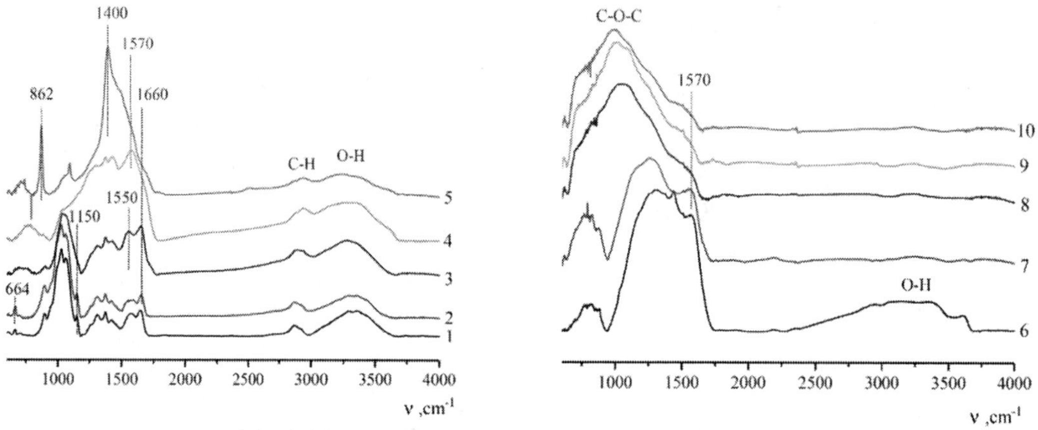

Figure 1. IR spectra of the initial CS (1) sample and the samples IR-pyrolyzed at temperatures of 150 (2), 200 (3), 250 (4), 300 (5), 400 (6), 500 (7), 600 (8), 700 (9), and 800 °C (10).

Figure 2a demonstrates the TG and DSC curves recorded for the CS. The first endothermic peak can be observed between 40-150 °C, with a minimum at 91 °C. A mass loss of 10 % is associated to this effect, and it is attributed to the evaporation of absorbed water in the polymer. The second effect can be observed between 250-330 °C, with an exothermic peak at 300 °C. This effect can also be associated with a mass loss (40 %), which corresponds to the thermal degradation of polymeric chain with vaporization of volatile compounds. The pyrolysis of polysaccharides structure starts by a random split of the glycosidic bonds, followed by a further decomposition forming acetic and butyric acids and a series of lower fatty acids [15-19]. The energy associated with this second effect was found to be 178.8 J g^{-1}. Finally, the consecutive low-intensity exothermic peaks can be observed up to 500 °C, which corresponds to the residual cross-linked degradation of the CS. The mass loss associated with this effect was found to be about 15 %.

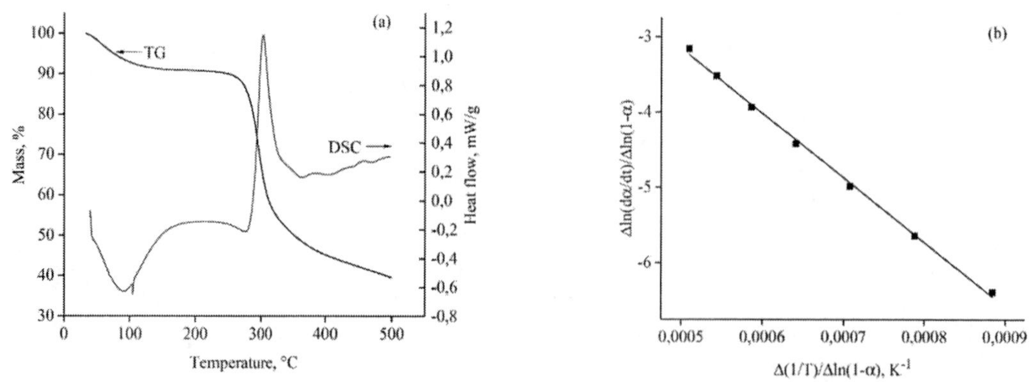

Figure 2. (a) TG and DSC curves of chitosan; (b) Plot for the determination of *E* and *n*.

Kinetics of thermal degradation of CS was evaluated from the thermogravimetric analysis. The TG data were applied to the calculation of the activation energy E and the reaction order n of the main transformation in the temperature range of 250-330 °C by the method of Friedman (1) [20].

$$\ln\left(\frac{\mathrm{d}\alpha}{\mathrm{d}t}\right) = \ln A + n\ln(1-\alpha) - \frac{E}{RT} \tag{1}$$

The activation energy calculated by this method is called effective. The value of the effective activation energy of thermal decomposition is one of the most important indicators that characterizes the resistance of natural and synthetic polymers to high temperatures. Activation energy is also used in the study of the mechanisms of thermal destruction, aging and stabilization of polymeric materials.

Figure 2(b) represents the linearization of the data function $\mathrm{d}\alpha/\mathrm{d}t = f(T)$ and determination of the activation energy and reaction order. The value of Y intercept is the reaction order, and the slope ratio is equal to the value ($-E/R$). According to the results, the activation energy and the reaction order of the main stage of the thermos-oxidative degradation of the CS were found to be 71 kJ mol^{-1} and 1, respectively.

XPS is one of the best tools for the study of the surface chemical modification. The results of the chemical composition and the C1s, N1s and O1s spectra of chitosan pyrolyzed at 800 °C are presented in Table 1 and figure 3, consequently. The main asymmetric peak at 284.5 eV (peak 1) and its π-satellite shifted by 6.0 – 6.1 eV corresponding to carbon with sp^2-hybridization, the symmetric peak at 284.9 eV (peak 2) corresponding to carbon with sp^3-hybridization including the adsorbed hydrocarbons and also two symmetric peaks in the region of higher energies (peaks 3 and 4) can be revealed by the approximation of the C1s spectrum. The position of the third peak corresponds to carbon-nitrogen and carbon-oxygen bonds (C-N, C-O). The fourth peak is located in a wide range of 288.4-290.1 Ev, which contains different groups with double bonds (C=O) including the carboxylic group. However, a large distribution of the low-intensity bond energies of the fourth peak is associated with a higher measurement error. In order to precisely define function groups, the high-resolution N1s and O1s spectra should be considered.

The analysis of the N1s and O1s spectra revealed the peaks corresponding with single bonds at 533.0 eV (C-O) and 400.6 eV (C-N) and with double bonds at 531.3 eV (C=O) and 398.3 eV (C=N) at the edge of the carbon layers. According to the intensity of the single and double bonds of the N1s and O1s spectra, the carbon bonds with sp^3-hybridization are predominant in the pyrolyzed CS.

Table 1. Chemical composition of the CS IR-pyrolyzed at 800 °C, aas determined from the XPS analysis.

Sample	Chemical composition (at. %)		
	C	N	O
IR-CS 800 °C	89.2	5.4	5.1

The XRD analysis showed that the CS IR-pyrolyzed at 400 °C has already a graphite-like structure (figure 4a). XRD patterns 5-7 have graphite reflection peaks located at $2\theta \approx 37°$ and 68°. Since the graphite reflection peaks (002) and (101) of the pyrolyzed CS become sharper with rising temperature, it can be concluded that the carbonized structure is more ordered. Similar to IR spectroscopy, it can be seen that the structure of the CS pyrolyzed at 150 °C does not change and at the temperature of 200 °C, the polymer structure became more defective according to the expansion of the main peak at the XRD pattern that means the decrystallizationprocess. At the temperature of 250 °C the CS structure was changed significantly.

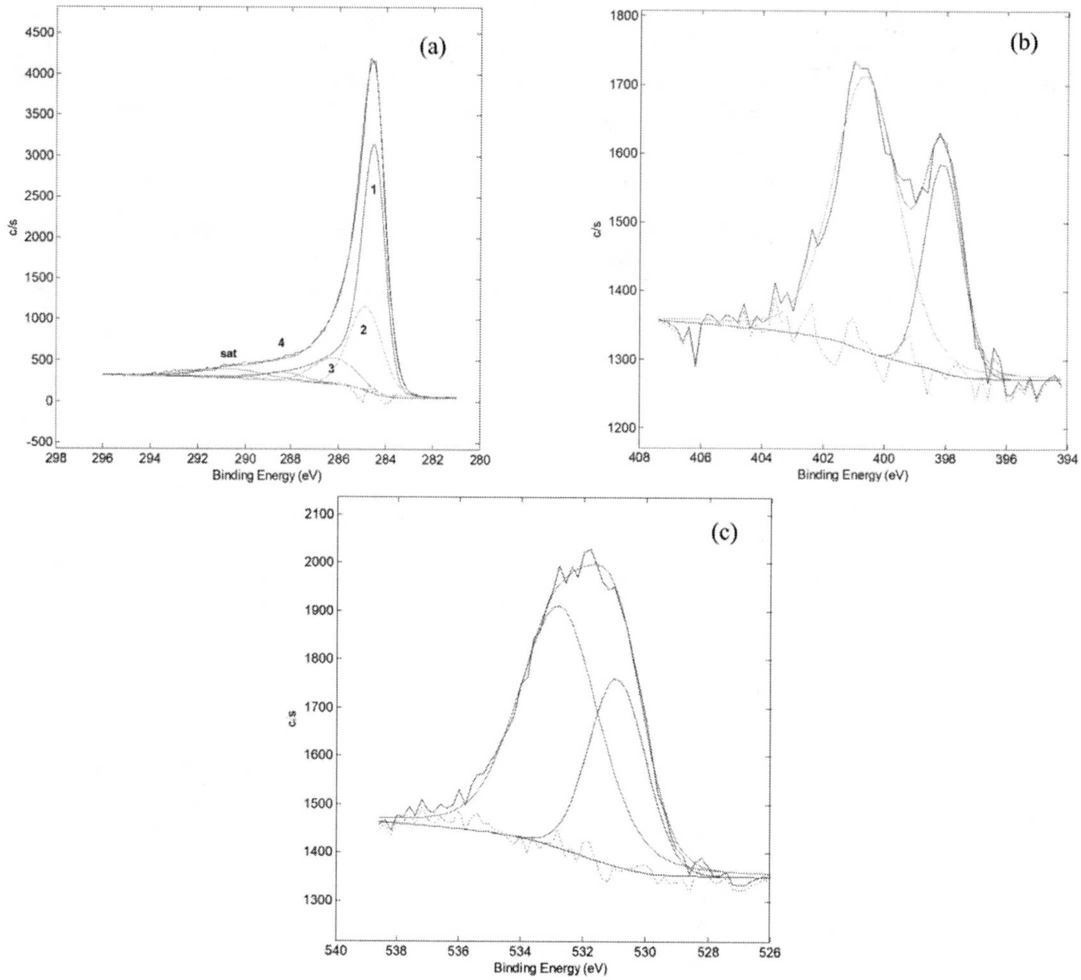

Figure 3. XPS spectra of C1s (a), N1s (b) and O1s (c) for the CS IR-pyrolyzed at 800 °C.

Raman spectra of the IR-pyrolyzed CS samples are characterized by the presence of D- and G-bands, which are characteristic of the presence of the carbon with sp^3- and sp^2-hybridization, accordingly (figure 4b). The Raman spectra of the perfect graphite is characterized by the G-band (1580 cm^{-1}). The appearing of the D-band corresponding to the Raman shift of the diamond at 1350 cm^{-1} indicates the oxidized or substituted carbon atoms formation with the sp^3-hybridization at the edges of hexagonal carbon layers. Since the intensity ratio I_D/I_G was found to increase with rising temperature, it can be concluded that the amount of the carbon atoms with the sp^3-hybridization also increases. This fact was confirmed by the IR-spectroscopy as well. In addition, a wide 2D band can be seen at about 2888 cm^{-1}, which indicates the different location of disordered hexagonal layers of carbon, *i.e.* the amorphous carbon with a low degree of graphitization [21], which was confirmed by the XRD analysis.

The TEM images (figure 5) demonstrate the morphology of the carbonized CS, which is characterized by a homogeneous and layer structure that is typical for most carbon materials. According to the BET-analysis, the specific surface area of the pyrolyzed CS was found to reache approximately 35 m^2/g depending on synthesis conditions.

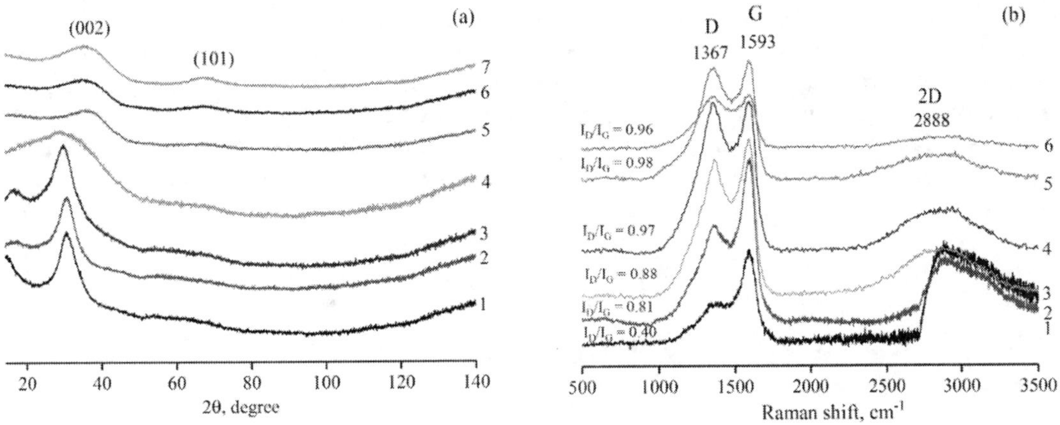

Figure 4. (a) XRD patterns of the initial CS sample (1) and the samples IR-pyrolyzed at 150 (2), 200 (3), 250 (4), 400 (5), 600 (6) and 800 °C (7); (b) Raman spectra of the CS IR-pyrolyzed at 300 (1), 400 (2), 500 (3), 600 (4), 700 (5), and 800 °C (6).

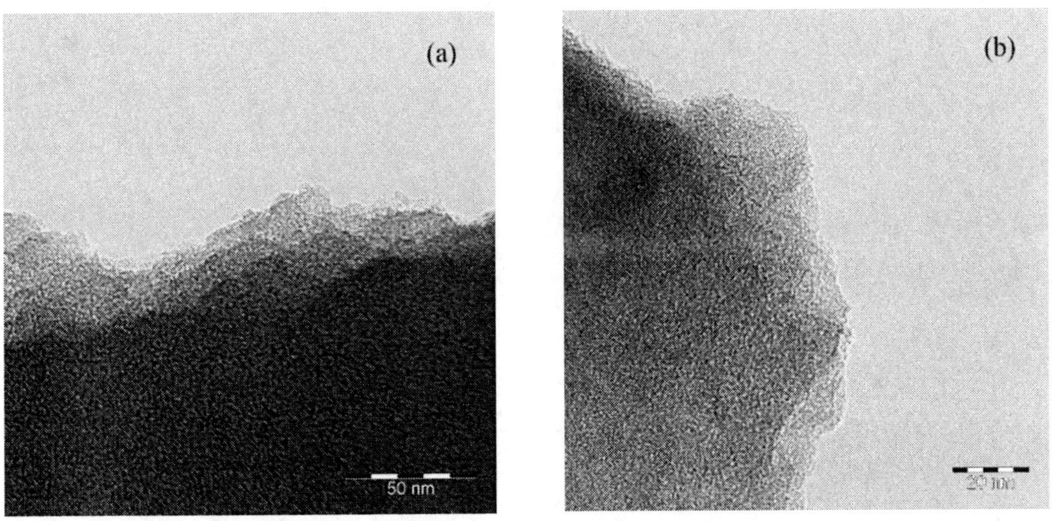

Figure 5. TEM images of the CS IR pyrolyzed at 800 °C.

The previous works have well reported the polysaccharides pyrolysis mechanisms and kinetics [22-25]. It was noted that the main volatile compounds of CS pyrolysis are pyrazines, pyridines, pyrroles and furans. According to the paper [19], the pyrazine compounds are the dominated products among the volatile chemicals of chitosan pyrolysis. However, data about the chemical structure of carbon residue, the main product of chitosan pyrolysis, cannot be observed. Summarizing the research results, the chemical transformation scheme of the CS under IR pyrolysis was proposed (figure 6). At the temperatures of 200-250 °C, the cleavage of the glycosidic bond and the ring-opening from the C_1-O bond breaking (as marked in figure 6) takes place. Intra- and intermolecular dehydrogenation leads to the formation of extended chains of the conjugated bonds C=C-C=N. Then, the scheme can be divided into 3 paths: formation of low hydrocarbons (path 1), formation of pyrazine compounds (path 2), and the formation of carbocyclic structures.

The product composition of polysaccharides pyrolysis is complex. For instance, the pyrolysis of cellulose leads to the formation of more than 100 compounds. Along with carbon residue, the polysaccharides pyrolysis characterized by the formation of water and carbon oxides in abundance [18]. Inspired by the ring-opening reactions during the cellulose pyrolysis, the C_2-C_3 bond in the CS could be further fractured. As a result, the CS is fragmented into aminoacetaldehyde and other low hydrocarbons (path 1 in figure 6). According to path 2, the bimolecular nucleophilic addition reaction occurs leading to the formation of pyrazine compounds. It can be assumed that the carbon residue formation proceeds through dehydration, decarboxylation, aromatization, and intramolecular condensation to form the carbocyclic structures (path 3 in figure 6) [18,26]. The results of the IR-, Raman and X-ray photoelectron spectroscopy showed that the CS pyrolysis at 600-800 °C provides the formation of graphite-like carbon structure. The other above mentioned products of pyrolysis either are volatilize or their amount is very low.

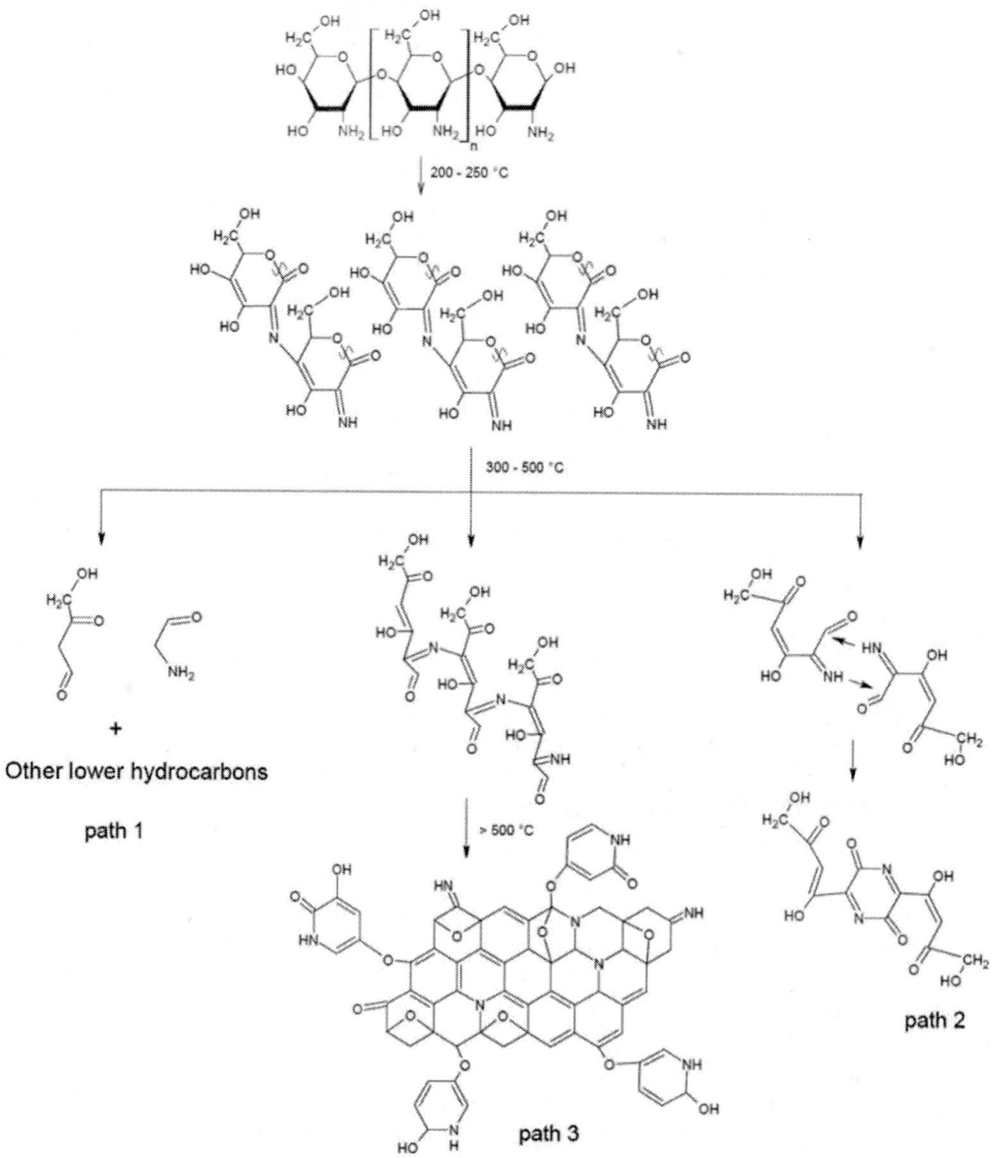

Figure 6. The proposed scheme of the CS decomposition.

4. Conclusion

CS possesses the potential to be a precursor of carbon-supported metal nanoparticles. It was revealed that during the thermal decomposition, especially at 300 °C, a lot of volatile products were released. Some of them (CO, H_2) have reducing properties, thereby making the CS an attractive precursor in terms of the reducing agent. It was shown that during the IR pyrolysis, the CS structure undergoes strong changes up to 800 °C. At relatively low temperatures (200-500 °C), the formation of C=C–C=N and C=C–C=C conjugated bonds was observed. Starting from 500 °C, the length of the polyconjugated system and the number of conjugated bonds was found to decrease due to the carbocyclic structures formation which, probably, begins at 300 – 350 °C. Thus, the CS can be effectively applied as a metal reducer and stabilizing support that hinders the agglomeration of nanoparticles.

5. Acknowledgement

This work was carried out within the State Program of TIPS RAS.

References

[1] Qiu F, Li L, Liu G, Wang Y, Wang Y, An C, Xu Y, Xu C, Wang Y, Jiao L and Yuan H 2013 In situ synthesized Fe-Co/C nano-alloys as catalysts for the hydrolysis of ammonia borane. *Int. J. Hydrogen Energy* **38** 3241 https://doi.org/10.1016/j.ijhydene.2012.12.090

[2] Yu M, Han Y, Li Y, Li J and Wang L 2018 Improving electrochemical activity of activated carbon derived from popcorn by $NiCo_2S_4$ nanoparticle coating. *Appl. Surf. Sci.* **463** 1001 https://doi.org/10.1016/j.apsusc.2018.09.037

[3] He Y and Sun W 2018 Carbon-coated SbCu alloy nanoparticles for high performance lithium storage. *J. Alloys Compd.* **753** 371 https://doi.org/10.1016/j.jallcom.2018.04.183

[4] Quan B, Xu G, Yi H, Yang Z, Xiang J, Chen Y and Ji G 2018 Enhanced electromagnetic wave response of nickel nanoparticles encapsulated in nanoporous carbon. *J. Alloys Compd.* **769** 961 https://doi.org/10.1016/j.jallcom.2018.08.069

[5] Zou T, Wu Y and Li H 2018 Electromagnetic and microwave absorbing properties of carbon-encapsulated cobalt nanoparticles. *Mater. Lett.* **214** 280 https://doi.org/10.1016/j.matlet.2017.12.028

[6] Koutsopoulos S, Barfod R, Eriksen K M and Fehrmann R 2017 Synthesis and characterization of iron-cobalt (FeCo) alloy nanoparticles supported on carbon. *J. Alloys Compd.* **725** 1210 https://doi.org/10.1016/j.jallcom.2017.07.105

[7] Campisi S, Chan-Thaw C E, Wang D, Villa A and Prati L 2016 Metal nanoparticles on carbon based supports: The effect of the protective agent removal. *Catal. Today* **278** 91 https://doi.org/10.1016/j.cattod.2016.04.026

[8] Atsushi F, Dhepe P L 2009 Sustainable green catalysis by supported metal nanoparticles. *Chem. Rec.* **9** 224 https://doi.org/10.1002/tcr.200900004

[9] Chen X, Zhang J, Zhang B, Dong S, Guo X, Mu X, Fei B 2017 A novel hierarchical porous nitrogen-doped carbon derived from bamboo shoot for high performance supercapacitor. *Sci. Rep.* **7** 1 https://doi.org/10.1038/s41598-017-06730-x

[10] Gao L, Lu H, Lin H, Sun X, Xu J, Liu D, Li Y 2014 KOH direct activation for preparing activated carbon fiber from polyacrylonitrile-based pre-oxidized fiber. *Chem. Res. Chinese Univ.* **30** 441 https://doi.org/10.1007/s40242-014-4059-1

[11] Khezami L, Chetouani A, Taouk B, Capart R 2005 Production and characterisation of activated carbon from wood components in powder: Cellulose, lignin, xylan. *Powder Technol.* **157** 48 https://doi.org/10.1016/j.powtec.2005.05.009

[12] Cárdenas-Triviño G, Elgueta C, Vergara L, Ojeda J, Valenzuela A and Cruzat C 2017 Chitosan doped with nanoparticles of copper, nickel and cobalt. *Int. J. Biol. Macromol.* **104** 498 https://doi.org/10.1016/j.ijbiomac.2017.06.040

[13] Adlim M, Bakar M A, Liew K Y and Ismail J 2004 Synthesis of chitosan-stabilized platinum

and palladium nanoparticles and their hydrogenation activity. *J. Mol. Catal. A* **212** 141 https://doi.org/10.1016/j.molcata.2003.08.012
[14] Nakamoto K 1963 *Infrared Spectra of Inorganic and Coordination Compounds* (New York: Wiley) p 328
[15] López F A, Mercê A L R, Alguacil F J, López-Delgado A 2008 A kinetic study on the thermal behaviour of chitosan. *J. Therm. Anal. Calorim.* **91** 633 https://doi.org/10.1007/s10973-007-8321-3
[16] Mucha M, Pawlak A 2002 Complex study on chitosan degradability. *Polimery* **47** 509 https://doi.org/10.1002/pola.1988.080260403
[17] Piskorz J, Radlein D, Scott D C 1986 On the mechanism of rapid pyrolysis of cellulose. *J. Anal. Appl. Pyrol.* **9** 121 https://doi.org/10.1016/0165-2370(86)85003-3
[18] Shen D, Xiao R, Gu S, Zhang H 2013 The overview of thermal decomposition of cellulose in lignocellulosic biomass. *Cellul. – Biomass Convers.* 193 https://doi.org/10.5772/51883
[19] Liu C, Zhang H, Xiao R, Wu S 2017 Value-added organonitrogen chemicals evolution from the pyrolysis of chitin and chitosan. *Carbohydr. Polym.* **156** 118 https://doi.org/10.1016/j.carbpol.2016.09.024
[20] Friedman H L 1964 Kinetics of thermal degradation of char-forming plastics from thermogravimetry. Application to a phenolic plastic. *J. Polym. Sci. C* **6** 183 https://doi.org/10.1002/polc.5070060121
[21] Thorat G M, Jadhav H S, Chung W J, Seo J G 2018 Collective use of deep eutectic solvent for one-pot synthesis of ternary Sn/SnO2@C electrode for supercapacitor. *J. Alloys Compd.***732** 694 https://doi.org/10.1016/j.jallcom.2017.10.176
[22] Alén R, Kuoppala E and Oesch P 1996 Formation of the main degradation compound groups from wood and its components during pyrolysis. *J. Anal. Appl. Pyrol.* **36** 137 https://doi.org/10.1016/0165-2370(96)00932-1
[23] Antal M J and Varhegyi G 1995 Cellulose pyrolysis kinetic: the current state of knowledge. *Ind. Eng. Chem. Res.* **34** 703https://doi.org/10.1021/ie00042a001
[24] Bridgwater A V 1999 Principles and practice of biomass fast pyrolysis processes for liquids. *J. Anal. Appl. Pyrol.***51** 3https://doi.org/10.1016/S0165-2370(99)00005-4
[25] Patwardhan P R, Satrio J A, Brown R C and Shanks B H 2009 Product distribution from fast pyrolysis of glucose-based carbohydrates. *J. Anal. Appl. Pyrol.* **86** 323 https://doi.org/10.1016/j.jaap.2009.08.007
[26] Shen Y, Ma D and Ge X 2017 CO_2-looping in biomass pyrolysis or gasification. *Sustain. Energ. Fuels* **1** 1700 https://doi.org/10.1039/c7se00279c

Electrochemical synthesis of multilayer graphene oxide and its application in composite materials

A Yakovlev[1], E Yakovleva[1], V Tseluikin[2], V Krasnov[2] and A Mostovoy[2]

[1]Phisical Technical Institute, *Yuri Gagarin Saratov State Technical University*, 77 Polytekhnicheskaya Street, Saratov 410054, Russian Federation
[2]Engels Technological Institute (Branch), *Yuri Gagarin Saratov State Technical University*, 17 Svobody Square, Engels Saratov Region 413100, Russian Federation

[1]E-mail: aw_71@mail.ru

Abstract. An electrochemical method for multilayer graphene oxide production by anodic oxidation of graphite dispersed in sulfuric acid is proposed. The possibility of sequential dispersion of graphite during the electrochemical oxidation and hydrolysis is shown. It was established that the carbon nanostructured materials obtained are prone to agglomeration in aqueous dispersions. After sonication, the particle size of the oxidized graphite was found to significantly decrease with the formation of the multilayer graphene oxide. It was elucidated that the structure of the obtained material consists of multigraphene planes with a thickness of 0.01-0.10 μm, and the pore size is 1-10 μm. The possibility of using the graphene oxide to produce polymer composite materials and composite electrochemical coatings is considered. Regarding the kinetics of the epoxy composition curing, it was observed that the graphene oxide affects the structure formation, which was indicated by an increase in the gel formation duration from 45 to 64 min and curing from 53 to 78 min, with an increase in the maximum curing temperature from 105 to 116 °C. Through the method of thermogravimetric analysis, it was established that the introduction of the graphene oxide reduces the thermal stability of the epoxy composites, but provides increased Vica heat resistance from 100 to 125-158 °C. The width of the passive state area for the nickel-graphene oxide composite electrochemical coating was found to be 1.32-1.40 times higher than that for the pristine nickel coating.

1. Introduction

Graphene possesses unique properties, which makes it possible to use for developing supercapacitors [1], superconductors and nanosensors [2-4], composite materials and coatings [5,6]. Currently, applied research employs single- and multilayer graphene materials obtained from graphite oxide, which is mainly synthesized through the Hummers method [7-10]. However, the production of graphite-intercalated compounds and multilayer graphene oxide (GO) is also possible using the electrochemical method in sulfuric or nitric acids [11,12].

Modification of polymers by carbon functional additives (graphite, carbon nanotubes, thermally expanded graphite, fullerene C_{60}, or multilayer graphene) have been widely studied to produce polymer materials with a given set of functional and consumer characteristics [13-15]. Despite the large number of works devoted to the investigation of the effect of various carbon modifiers, at present, there is still no complete information on issues related to studying that effect on the structure formation and the structural and operational characteristics of polymer composite materials.

After co-deposition of metals with particles dispersed from electrolyte suspensions, composite electrochemical coatings (CECs) are formed [6,16]. The functional properties of such coatings are largely determined by the dispersed phase nature. Previous studies have shown that the inclusion of graphite electrochemical oxidation products in the matrix of electroplated coatings leads to an improvement in their tribological and corrosion properties [16]. In this regard, electrochemically synthesized GO appears to be a promising dispersed material for improving the operational characteristics of composite coatings.

Considering the aforementioned, the aim of this work was to study the electrochemical synthesis and structure of multilayer GO, as well as the functional properties of composite materials and coatings modified by it.

2. Materials and methods

Dispersed graphite powders (Standard GB/T 3518-95, Sunshine Resources Holdings Limited, Beijing, China) with a particle size ranging from 160 to 200 μm were employed in the present research. 83-% H_2SO_4 was used as electrolyte. Electrochemical measurements were carried out with an IPC Pro MF potentiostat (Volta, St.-Petersburg, Russia) in a three-electrode cell using a platinum cathode and platinum current wire. A mercury sulfate electrode (MSE) (CHI150, MIRONG, Shanghai, China) was used as a reference electrode. The synthesized compounds were hydrolyzed with subsequent drying at 90-100 °C. Studies on the particle size distribution of oxidized graphite in aqueous suspensions were carried out by an Analysette-22 NanoTec laser analyzer (Fritsch, Idar-Oberstein, Germany). The surface and structure of nanostructured graphite particles were studied using a EXplorer Aspex scanning electron microscope with built-in energy dispersion analysis (Aspex Corporation, Sheffield Drive, Delmont, USA). FTIR spectroscopy of the nanostructured graphite particles was performed on an IRTracer-100 instrument (Shimadzu, Tokyo, Japan). X-ray phase analysis was carried out on an ARL X'TRA X-ray diffractometer (Thermo Fisher Scientific SARL, Ecublens, Switzerland).

To study the possibility of using electrochemically synthesized multilayer GO as a polymer matrix, the previously developed epoxy composition [5], consisting of epoxy resin (ED-20) – 100 mass parts (m.p.), trichloroethyl phosphate (TCEP) – 40 m.p. and polyethylene polyamine (PEPA) as hardener – 15 m.p., was used as CEC. 0.01-0.50 m.p. of the GO were introduced into this epoxy composition. To increase the uniformity of distribution of the GO, its activation and the activation of the binder surface, the composition was sonicated. The parameters of ultrasonic exposure for the dispersed and filled CECs were chosen from our previous paper [17,18]: ultrasound frequency - 22±2 kHz, and duration – 60 min.

Studies on the CECs were carried out as follows: determination of bending stress and modulus of elasticity in bending [according to ISO 178:2010], determination of tensile strength and modulus of elasticity [according to ISO 527-2:2012], determination of compressive strength [according to ISO 604:2002], determination of impact strength [according to ISO 179-1:2010], Brinell hardness test [according to ISO 2039-1:2001], and determination of Vica heat resistance [according to ISO 306:2004]. Changes in mass, their rate and values of thermal effects during sample heating were investigated through the method of thermogravimetric analysis using a Paulik-Paulik-Erdei Q-1500D derivatograph system (MOM, Budapest, Hungary) under the following experimental conditions: sample weight – 100 mg, environment – air, heating interval – to 1000 °C, heating rate – 10 °C/min, and relative error – ≤ 1 %. Thermal conductivity and thermal resistance was determined an ITP-MG4 100 device (Stroypribor, Chelyabinsk, Russia) according to ISO 22007-2: 2015.

For electrochemical studies, the CECs were deposited in electrolytes – $NiSO_4 \cdot 7H_2O$ (220 g/L), $NiCl_2 \cdot 6H_2O$ (40 g/L); CH_3COONa (30 g/L) and graphene oxide (10 g/L) – on a steel base (steel 45) at 25 °C, with the electrolytes being mixed. Electrolytic coatings of pure nickel were obtained from this electrolyte without a dispersed phase. Corrosion tests of the nickel coatings were carried out by recording anode potentiodynamic curves in a 3-% NaCl solution at potential sweep rate V_p = 10 mV/s. The curves were recorded before a sharp rise in current. The corrosion resistance of the coatings was judged by the length of the passive state potential area.

3. Results and discussion

3.1. *Analysis of oxidized graphite*

The electrochemical oxidation was performed in the galvanostatic mode at a current of 400 mA (200 mA/g graphite) with a transmission of electric capacity of 700-1500 mA·h/g and potential control. The increase in the volume of the suspension graphite electrode recorded during the anodic oxidation of the graphite was found to be 180-200 % of the original volume. During the electrochemical oxidation of natural graphite, the carbon matrix undergoes changes accompanied by an increase in interlayer distances from 3.34Å (in the pristine graphite) to 7.78 Å (in the oxidized graphite). This can be explained by the formation of intercalated graphite compounds, as well as the formation of oxygen-containing surface functional groups [9,19-21]. Like in the case of GO synthesized through the Hummers method [9,10], the surface functional groups mainly consist of hydroxyl and phenolic [the peak at 3417 cm^{-1}] and hydroxyl groups between graphene layers [the band between 2800 and 3100 cm^{-1}] (figure 1); the presence of these groups indicates the formation of GO. The sp^2 hybridization of C=C in the graphene structure is confirmed by a peak at 1627 cm^{-1} [9]. The peak at ~2300 cm^{-1} corresponds to absorbed CO_2 molecules [10]. The band between 1106 and 1005 cm^{-1} corresponds to C-O-C (epoxy groups) [10]. The band at 1384 cm^{-1} represents a deformation oscillation in -COOH-groups. As a result of hydrolysis, exfoliation of oxidized graphite particles occurs, resulting in a 10-11-fold increase in the volume.

The distribution range of the oxidized graphite particles in the aqueous suspension is greatly expanded (0.2-1800 µm, modal size 355 µm) (figure 2). The particles are prone to agglomeration, which requires mechanical or ultrasonic action, after which the particle size distribution range becomes equal to 0.2-400 µm, with a modal size of 91 µm.

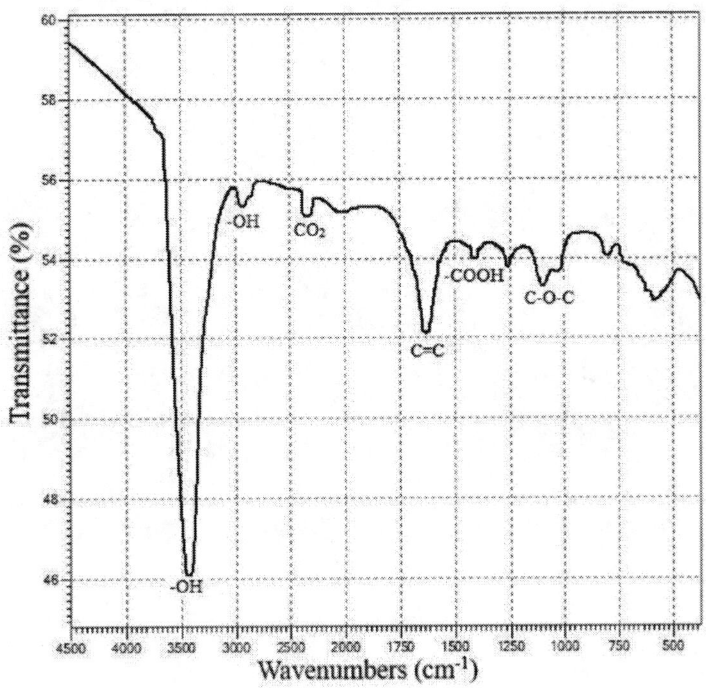

Figure 1. FTIR spectra recorded for the electrochemically oxidized graphite.

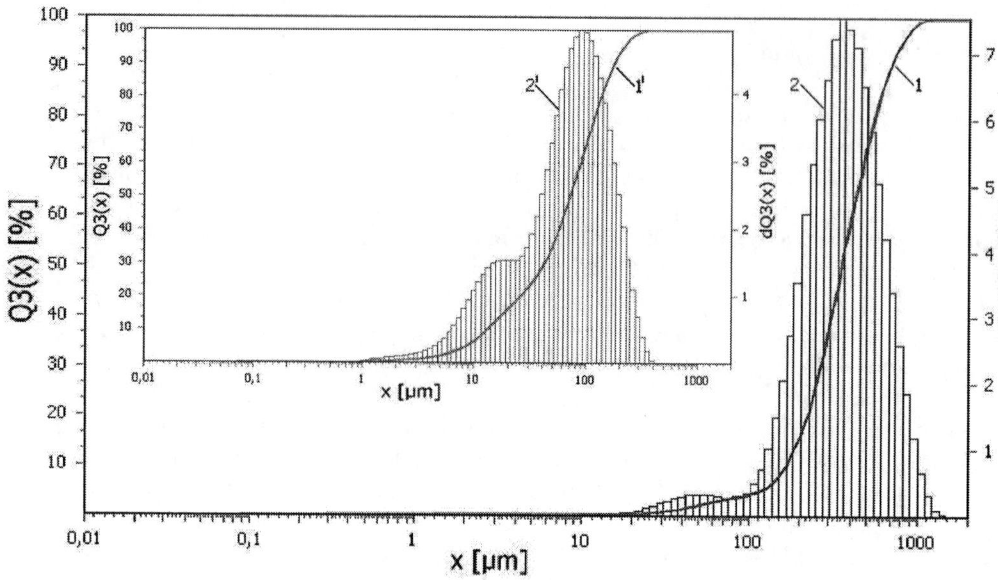

Figure 2. Particle size distribution for the oxidized graphite. 1,1' – integral, 2,2' – differential; 1',2' – when exposed to ultrasound (50 W). dQ3(x) [%] - distribution density; Q3(x) [%] - cumulative distribution; and x[μm] - particle size.

The structure of the oxidized graphite particles represents a sequence of ordered layers, the thickness of individual layers may be < 0.1 μm (figure 3), which can also be noted for the oxidized graphite obtained through the Hummers method [9,10]. The radiograph records the peak characteristic of GO at 2θ = 11.45° (figure 4), which corresponds to the interlayer distance of 7.78 Å and allows to classify the resulting compound as multilayer GO [3,19,20].

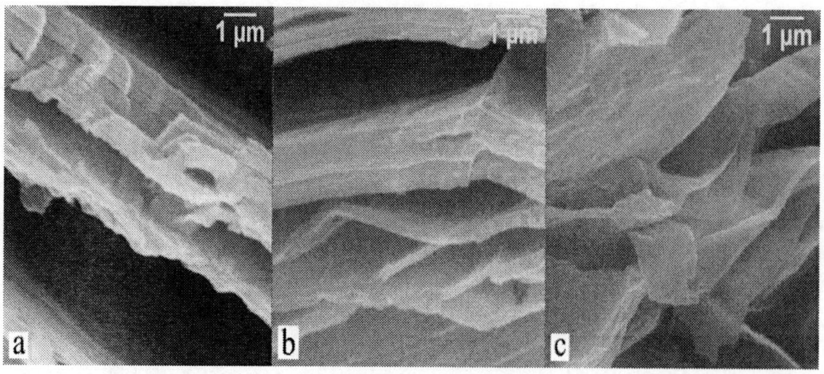

Figure 3. SEM images of the electrochemically oxidized graphite. Electric capacity: 700 **(a)**, 900 **(b)** and 1500 **(c)** mA·h/g.

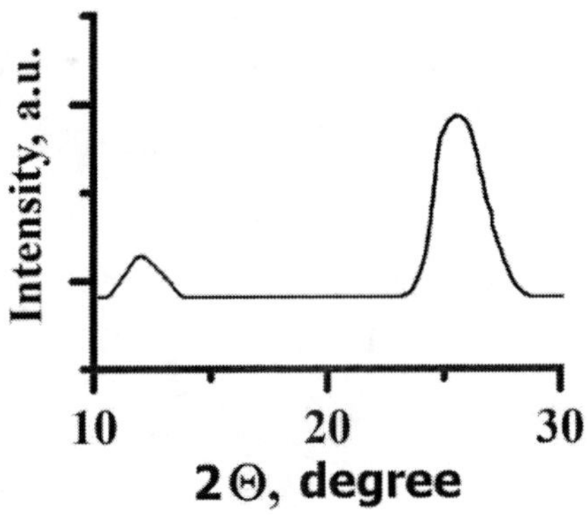

Figure 4. A radiograph of the electrochemically oxidized graphite.

3.2. *Studies of composite materials*

From the results obtained on the epoxy composites presented in Table 1, it can be seen that the optimum GO content is 0.1 m.p., since at this level, the maximum values of physical and mechanical properties were achieved. Relative to the CEC without GO, the parameters of the CEC containing the 0.1 m.p. GO were increased 2.3-fold (bending failure stress and bending elastic modulus), by 57 % (compressive strength), 77 % (tensile strength), 55 % (tensile elastic modulus), 75 % (impact strength), and 50 % (hardness).

Table 1. Properties of the epoxy compositions developed.

Composition (m.p.) + PEPA (15 m.p.)	σ_b, MPa	E_b, MPa	σ_c, MPa	σ_t, MPa	E_t, MPa	a_{imp}, kJ m^{-2}	H_B, MPa
ED-20 (100) +TCEP (40)	53	1750	70	36	1610	8	84
ED-20 (100) +TCEP (40) + GO (0.05)	109	3400	101	56	2114	8	110
ED-20 (100) + TCEP (40) + GO (0.1)	120	4086	110	64	2509	14	126
ED-20 (100) +TCEP (40) + GO (0.5)	93	4282	114	54	2785	10	132

σ_b – bending failure stress; E_b – bending elastic modulus; σ_c – compression strength; σ_t – tensile strength; E_t – tensile elastic modulus; a_{imp} – impact strength; $H_в$ – Brinell hardness; coefficient of variation in properties – 4-7 %.

Regarding the kinetics of the epoxy composition curing, it was observed that the GO affects the structure formation, which is indicated by an increase in the gel formation duration from 45 to 64 min and curing from 53 to 78 min, with an increase in the maximum curing temperature from 105 to 116 °C. Through the method of thermogravimetric analysis, it was established that the introduction of the GO reduces the thermal stability of the epoxy composites (figure 5) but provides increased Vica heat resistance from 100 to 125-158 °C.

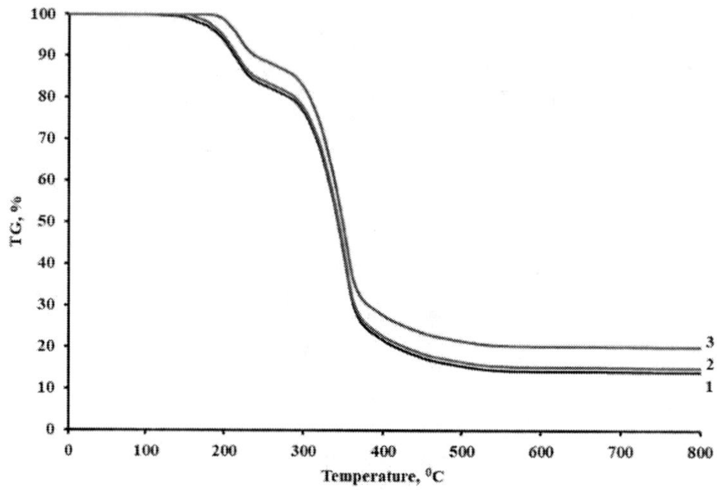

Figure 5. Data of the thermogravimetric (TG – mass loss,%) analysis of the samples:
1 – ED-20 (100 m.p.) + TCEP (40 m.p.) + PEPA (15 m.p.);
2 – ED-20 (100 m.p.) + TCEP (40 m.p.) + GO (0.1 m.p.) + PEPA (15 m.p.);
3 – ED-20 (100 m.p.) + TCEP (40 m.p.)+ GO (1.0 m.p.) + PEPA (15 m.p.).

Due to the fact that the GO has a sufficiently high heat conductivity, the introduction of even its small amounts into the epoxy composition increases the thermal conductivity coefficient by 2.6-3.0 times, and decreases the thermal resistance (Table 2).

Table 2. The GO effect on the thermal conductivity of the epoxy compositions.

Composition (m.p.) + PEPA (15 m.p.)	Heat conductivity coefficient, W (m K)$^{-1}$	Thermal resistance, m^2 K W^{-1}
ED-20 (100) +TCEP (40)	0.105±0.0058	0.086±0.0043
ED-20 (100) + TCEP (40) + GO (0.1)	0.274±0.0090	0.036±0.0015
ED-20 (100) +TCEP (40) + GO (0.5)	0.322±0.0098	0.030±0.0012

The comparison between the optimum CEC developed (GO content – 0.1 m.p.) and the existing analogs showed its competitiveness, as it possesses comparable and even higher (regarding some parameters) physical and mechanical characteristics (Table 3).

Table 3. Comparison between the developed epoxy compositions and analogs.

Composition (m.p)	σ_b, MPa	E_b, MPa	σ_t, MPa	E_t, MPa	a_{imp}, kJ m^{-2}	H_B, MPa
ED-20 (100) + TCEP (40) + GO (0.1) + PEPA (15)	120	4086	64	2509	14	126
Analogs						
ED-20 (100) + Fullerene C_{60} (0.1) + PEPA (10) [4]	75	3100	-	-	9	-
ED-20 (100) + TCEP (42) + WPPT (0.1) + PEPA (15) [7]	97	2614	52	2095	14	120
ED-20 (100) + TCEP (40) +MTW (20) + PEPA (15) [6]	53	-	-	-	7	110

WPPT – whiskers of potassium polytitanates, MTW – millet threshing waste heat-treated at 350 °C.

The electrochemical studies showed that in the presence of the GO, nickel is deposited at less negative potential values (figure 6). During the electrodeposition, in the case of the nickel-GO CEC coating, the current increases compared to the nickel coating without dispersed phase, thereby indicating an increase in the cathode process rate.

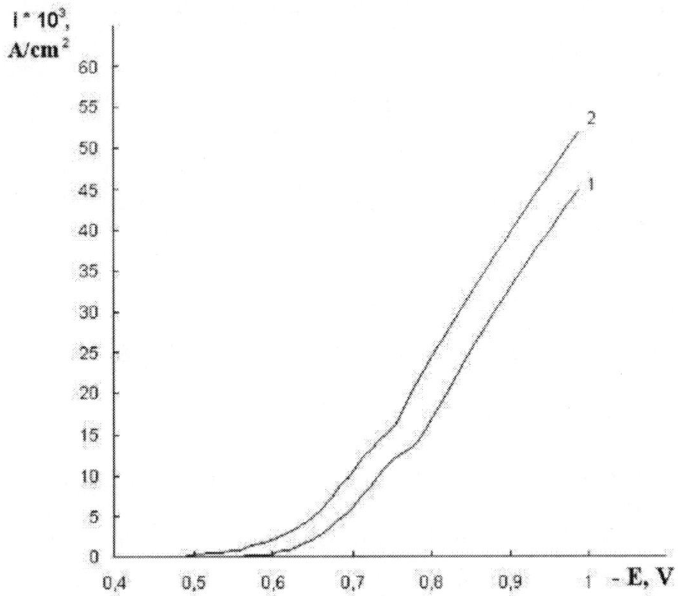

Figure 6. Potentiodynamic polarization curves of the nickel deposition without additive (1) and (2) together with the GO.

Table 4. The width of the passive state area E, V of nickel-based coatings obtained at different cathode current densities i_C.

i_C, A/dm^2	6	7	8	9	10
Nickel	0.60	0.62	0.66	0.68	0.68
Nickel-GO CEC	0.80	0.82	0.90	0.94	0.96

The introduction of dispersed GO particles into the nickel matrix leads to its structural changes, which, in turn, should affect the operational properties of electrolytic coatings. Corrosion resistance is of great practical interest. The studies on the nickel coatings in a 3-% NaCl solution showed that the width of the passive state area for the nickel-GO CEC is 1.32-1.40 times higher than that for the nickel coatings without dispersed phase (Table 4). This effect is due to several reasons. The CEC is more resistant to corrosion the higher area of its surface overlap by particles of dispersed phase because it provides a uniform distribution of corrosion current between the sites that prevent its spreading. When the GO particles are introduced into the nickel matrix, the pores become overlapped. Furthermore, in the composite coatings, the dispersed phase effect on the corrosion is revealed only in the case of formation of particles at phase boundaries or throughout the bulk of compounds more corrosion-resistant than the metal matrix. Otherwise, the occurrence of the corrosion process will not stop but will bypass the particles. Obviously, in the case of the nickel-GO CEC, such compounds are formed.

4. Conclusion

The possibility of obtaining multilayer GO through electrochemical oxidation of natural graphite powder dispersed in H_2SO_4 was shown. It was established that the graphite is consistently exfoliated during anodic oxidation with the formation of graphite oxide and subsequent hydrolysis with the formation of multilayer multigraphene. In comparison with the conventional Hummers method, the main advantage of the technique herein proposed is that the strong oxidizers ($KMnO_4$, $K_2Cr_2O_7$, *etc*) are not used, thereby increasing the environmental safety of the process and the purity of the compounds synthesized.

The possibility of directed control over the performance properties of epoxy composites through using small amounts of finely dispersed multi-layer GO was demonstrated. This provides the development of epoxy composites with high performance properties that meet the requirements of most industries. Thus, these materials can be used for sealing electronic products, as well as for impregnating and filling components in the aviation, shipbuilding and automotive industries.

Besides, it was shown that the introduction of the GO dispersion into the nickel electrolyte facilitates the process of electrodeposition and contributes to the formation of composite coatings. It was found that the GO particles increase the width of the passive state area by 1.32-1.40 times.

Thus, GO-containing CECs can be used as corrosion-resistant materials in machine building and the other industries.

5. Acknowledgment

The present research was funded by the Russian Foundation for Basic Research (RFBR) in the framework of Project No. 18-29-19048.

References

[1] Revo S L, Budzulyak I M, Rachiy B I and Kuzishin M M 2013 Electrode material for supercapacitors based on nanostructured carbon. *Surf. Eng. Appl. Electrochem.* **49** 68 http://dx.doi.org/10.3103/S1068375513010122

[2] Kulkarni G S, Reddy K, Zhong Z and Fan X 2014 Graphene nanoelectronic heterodyne sensor for rapid and sensitive vapour detection. *Nat. Commun.* **5** 4376 http://dx.doi.org/10.1038/ncomms5376

[3] Avouris P and Dimitrakopoulos C 2012 Graphene: synthesis and applications *Mater. Today* **15** 86 http://dx.doi.org/10.1016/S1369-7021(12)70044-5

[4] Fen L, Xue J, Jijun Z and Shengbai Z 2015 Graphene oxide: A promising nanomaterial for energy and environmental applications. *Nano Energy* **16** 488 http://dx.doi.org/10.1016/j.nanoen.2015.07.014

[5] Blokhin A N, Dyachkova T P, Sukhorukov A K, Kobzev D E, Galunin E V, Maksimkin A V, Mostovoi A S and Kharitonov A P 2018 Composite materials using fluorinated graphene nanoplatelets. *Nanosyst. Phys. Chem. Math.* **9** 102 http://dx.doi.org/10.17586/2220-8054-2018-9-1-102-105

[6] Tseluikin V N and Koreshkova A A 2015 Electrochemical deposition and properties of composite coatings consisting of zinc and carbon nanotubes. *Russ. J. Appl. Chem.* **88** 269 http://dx.doi.org/10.1134/S1070427215020135

[7] Hummers W S and Offeman R E 1958 Preparation of graphitic oxide. *J. Am. Chem. Soc.* **80** 1339 https://dx.doi.org/10.1021/ja01539a017

[8] Dreyer D R, Jia H P and Bielawski C W 2010 Graphene oxide: a convenient carbocatalyst for facilitating oxidation and hydration reactions. *Ang Chem. Int. Edit. Engl.* **49** 6813 http://dx.doi.org/10.1002/anie.201002160

[9] Qi L, Xinli G, Yao Z, Weijie Z, Chuang G, Li Z, Xiaojuan W, Hongyi Z, Jian C, Zengmei W and Litao S 2017 Porous graphene paper for supercapacitor applications. *J Mater. Sci. Technol.* **33** 793 http://dx.doi.org/10.1016/j.jmst.2017.03.018

[10] Zaaba N I, Foo K L, Hashima U, Tanb S J, Wei-Wen L and Voon C H 2017 Synthesis of graphene oxide using modified Hummers method: Solvent influence. *P Engin.* **184** 469 http://dx.doi.org/10.1016/j.proeng.2017.04.118

[11] Yakovlev A V, Finaenov A I, Zabud'kov S L and Yakovleva E V 2006 Thermally expanded graphite: Synthesis, properties, and prospects for use. *Russ. J. Appl. Chem.* **79** 1741 http://doi.org/10.1134/S1070427206110012

[12] Yakovlev A V, Yakovleva E V, Zabud'kov S L and Finaenov A I 2010 Electrochemical processes on graphite powder electrodes in HNO3 solutions. *Russ. J. Appl. Chem.* **83** 820 http://dx.doi.org/ 10.1134/S1070427210050113

[13] Xin W, Lei S, Wiwat P, Yuan H and Baljinder K 2013 The effect of graphene presence in flame retarded epoxy resin matrix on the mechanical and flammability properties of glass fiber-reinforced composites. *Composites A* **53** 88 http://dx.doi.org/ 10.1016/j.compositesa.2013.05.017

[14] Kharitonov A P, Tkachev A G, Blohin A N, Dyachkova T P, Kobzev D E, Maksimkin A V, Mostovoi A S and Alekseiko L N 2016 Reinforcement of Bisphenol-F epoxy resin composites with fluorinated carbon nanotubes. *Compos. Sci. Technol.* **134** 161 http://dx.doi.org/ 10.1016/j.compscitech.2016.08.017

[15] Buketov A V, Sapronov A A, Buketova N N, Brailo M V, Marushak P O, Panin S V and Amelin M Y 2018 Impact toughness of nanocomposite materials filled with fullerene C60 particles. *Compos. Mech. Comp. Appl.* **9** 141 http://dx.doi.org/ 10.1615/CompMechComputApplIntJ.v9.i2.30

[16] Tseluikin V N 2019 Preparation and properties of graphite nitrate modified composite electrochemical coatings based on a nickel-chromium alloy. *Inorg. Mater.* **92** 579 http://dx.doi.org/10.1134/S0020168519070215

[17] Mostovoi A S, Yakovlev E A, Burmistrov I N and Panova L G 2015 Use of modified nanoparticles of potassium polytitanate and physical methods of modification of epoxy compositions for Improving their operational properties. *Russ J Appl Chem.* **88** 129 http://dx.doi.org/ 10.1134/S107042721501019X

[18] Mostovoy A S, Kadykova Y A, Bekeshev Y A and Tastanova L K 2018 Epoxy composites modified with microfibers of potassium polytitanates. *J. Appl. Polym. Sci.* **135** 1 http://dx.doi.org/ 10.1002/app.46651

[19] Edwards R S and Coleman K S 2013 Graphene synthesis: relationship to applications. *Nanoscale.* **5** 38 http://dx.doi.org/ 10.1039/c2nr32629a

[20] Wang J, Salihi E C and Šiller L 2017 Green reduction of graphene oxide using alanine. *Mater. Sci. Eng. C.* **72** 1 http://dx.doi.org/ 10.1016/j.msec.2016.11.017

[21] Johnson D W, Dobson B P and Coleman K S 2015 A manufacturing perspective on graphene dispersions. *Curr. Opin. Colloid Interface Sci.* **20** 367 http://dx.doi.org/ 10.1016/j.cocis.2015.11.004

Method of modeling the heat exchange process in the synthesis of carbon nanostructures in arc discharge plasma

G V Abramov[1,*], I A Avcinov[2], A N Gavrilov[2] and N V Sukhanova[2]

[1]Department of Mathematical Support of Electronic Computers, *Voronezh State University*, 1 University square, Voronezh, 394018, Russian Federation
[2]Department of Information and Control Systems, *Voronezh State University of Engineering Technologies*, 19 Revolution Avenue, Voronezh 394036, Russian Federation

[*]E-mail: agwl@yandex.ru

Abstract. The method for calculating the temperature field during the synthesis of carbon nanostructures by plasma of electric arc discharge is proposed. The calculation scheme of heat exchange process is given. An algorithm for constructing a discrete model of the numerical solution of the heat transfer problem by the finite-element method in the environment of multiphysical modeling COMSOL Multiphysics is considered. The research of influence of parameters of different types of finite element grids on the convergence rate, calculations time and simulation accuracy is carried out. The rational parameters of the discretization of the computational domain of the considered system are determined. The results of correspondence of the constructed model of heat exchange process to the physical process are presented.

1. Introduction

Carbon nanostructures (CNS) with unique physicochemical properties are of significant interest for various branches of modern industry [1]. However, their widespread use is constrained by the high cost and low productivity of existing synthesis methods, which is due to the lack of study of the features of the processes of formation of the final product.

Different variations of plasma sublimation of graphite raw materials with subsequent deposition of the synthesized product on the cooled surface is one of the directions which allows to obtain high-quality nanostructured carbon material in sufficiently large volumes. The most famous variety of such technologies is the electric arc synthesis of CNS in an inert gas medium [2]. This method is different by a high passing speed, a small synthesis zone and the ability of obtaining various carbon nanostructures (fullerenes, nanotubes, nanofibers).

The processes occurring in the electric arc synthesis of CNS can be attributed to high-temperature technological processes. Plasma temperature (4500 - 5500 K), heating and cooling of graphite electrodes have a direct impact on the quality of the final product [3]. Therefore, the aim of this work is to develop a methodic of modeling the heat transfer process in the anode-plasma-cathode system in the electric arc synthesis of CNS, which allows to take into account the influence of temperature fields on the evaporation of the anode and the formation of final carbon fractions in sediment deposition and on the walls of the operating chamber. The proposed mathematical model of heat transfer is distinguished by taking into account the total heat flux power in the heat equation of the buffer medium and the mobile boundaries of the considered system.

IOP Conf. Series: Materials Science and Engineering **693** (2019) 012004 doi:10.1088/1757-899X/693/1/012004

2. Theoretical analysis

When constructing a model of the heat exchange process of the considered system, a number of the following assumptions are accepted: the thermophysical parameters of the system components are constant; the plasma of the interelectrode space is considered to be optically thin and the interaction of radiation with the plasma is not considered due to a slight effect on the process [4]; the mechanism of heat transfer in the anode, cathode, and sediment – is convective heat conductivity, in the buffer gas – thermal conductivity, in low-temperature plasma – convective heat conductivity with a volumetric heat source; electrodes in the synthesis process retain uniformity and the constancy of the density. Helium is considered as a buffer medium.

In the course synthesis, there is a change in the configuration of the working zone caused by the destruction of the initial graphite raw material and the growth of sediment, which must be taken into account in the model. Taking into account of the moving boundaries of the electrodes for the process of electric arc synthesis of CNS in a helium medium, the design scheme of heat-exchange will be of the form shown in figure 1.

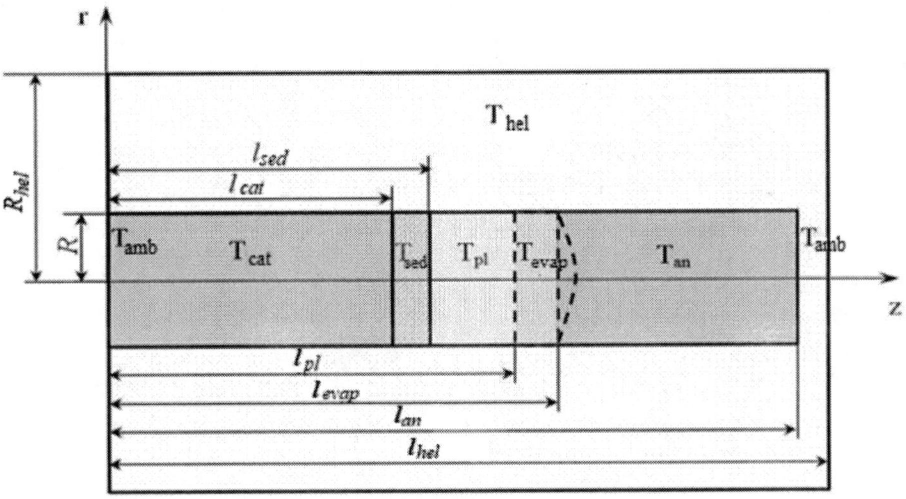

Figure 1. Design scheme of process of heat transfer: T_{an}, T_{cat}, T_{sed}, T_{pl}, T_{evap}, T_{amb} – temperatures, respectively, of the anode, cathode, sediment, plasma, evaporation zone, and ambient; l_{cat}, l_{sed}, l_{pl}, l_{evap}, l_{an} – distance from the beginning of coordinate to the right border of the cathode, sediment, plasma, evaporation, anode, respectively; R – radius of the electrodes; R_{hel} – helium border.

A mathematical model of the heat transfer process, taking into account the accept assumptions in the cylindrical coordinate system has the form:

$$\begin{aligned}
&\frac{\partial T_{an}}{\partial t}=\frac{k_{an}}{\rho_{an}\cdot C_{an}}\cdot\left(\frac{\partial^2 T_{an}}{\partial r^2}+(1/r)\cdot\frac{\partial T_{an}}{\partial r}+\frac{\partial^2 T_{an}}{\partial z^2}\right),\\
&\frac{\partial T_{cat}}{\partial t}=\frac{k_{cat}}{\rho_{cat}\cdot C_{cat}}\cdot\left(\frac{\partial^2 T_{cat}}{\partial r^2}+(1/r)\cdot\frac{\partial T_{cat}}{\partial r}+\frac{\partial^2 T_{cat}}{\partial z^2}\right),\\
&\frac{\partial T_{sed}}{\partial t}=\frac{k_{sed}}{\rho_{sed}\cdot C_{sed}}\cdot\left(\frac{\partial^2 T_{sed}}{\partial r^2}+(1/r)\cdot\frac{\partial T_{sed}}{\partial r}+\frac{\partial^2 T_{sed}}{\partial z^2}\right),\\
&\frac{\partial T_{pl}}{\partial t}=\frac{k_{pl}}{\rho_{pl}\cdot C_{pl}}\cdot\left(\frac{\partial^2 T_{pl}}{\partial r^2}+(1/r)\cdot\frac{\partial T_{pl}}{\partial r}+\frac{\partial^2 T_{pl}}{\partial z^2}+\frac{Q}{k_{pl}\cdot V_{pl}}\right)-u_{pl_r}\frac{\partial T_{pl}}{\partial r}-u_{pl_z}\frac{\partial T_{pl}}{\partial z},\\
&\frac{\partial T_{hel}}{\partial t}=\frac{1}{\rho_{hel}\cdot C_{hel}\cdot V_{hel}}\cdot Q_{\Sigma},
\end{aligned} \quad (1)$$

where k_{an}, k_{cat}, k_{sed}, k_{pl} – the thermal conductivity of the anode, the cathode, the sediment and plasma, ρ_{hel}, ρ_{an}, ρ_{cat}, ρ_{sed}, ρ_{pl} – density of helium, anode, cathode, sediment and plasma, C_{hel}, C_{an}, C_{cat}, C_{sed}, C_{pl} – specific heat of helium, anode, cathode, sediment and plasma; u – the velocity field; V_{hel}, V_{pl} – is the volume of plasma and helium in the synthesis chamber; Q – the power of the heat source, Q_{Σ} – the power of the heat flux in helium.

The total capacity of the heat flow in helium:

$$Q_{\Sigma}=Q_{catt}+Q_{catb}+Q_{sedb}+Q_{plb}+Q_{anb}+Q_{ant}-Q_{loss}. \quad (2)$$

$$Q_{catt}=2\pi\cdot\alpha_{hel}\int_0^R r\cdot(T_{cat}\big|_{z=0}-T_{hel})dr;\quad Q_{catb}=2\pi\cdot R\cdot\alpha_{hel}\int_{l_{cat}}^{l_{sed}}(T_{cat}\big|_{r=R}-T_{hel})dz;$$

$$Q_{sedb}=2\pi\cdot R\cdot\alpha_{hel}\int_{l_{cat}}^{l_{sed}}(T_{sed}\big|_{r=R}-T_{hel})dz;\quad Q_{plb}=2\pi\cdot R\cdot\alpha_{hel}\int_{l_{sed}}^{l_{evap}}\sigma\cdot(T_{pl}^4\big|_{r=R}-T_{hel}^4)dz;$$

$$Q_{anb}=2\pi\cdot R\cdot\alpha_{hel}\int_{l_{evap}}^{l_{an}}(T_{an}\big|_{r=R}-T_{hel})dz;\quad Q_{ant}=2\pi\cdot\alpha_{hel}\int_0^R r\cdot(T_{an}\big|_{z=l_{an}}-T_{hel})dr;$$

$$Q_{loss}=\alpha_{ht}(T_{hel}-T_{amb})\cdot S_{cham}.$$

where α_{ht} – coefficient of heat transfer to the environment; Q – power of the loss; S_{cham} – surface area of the operating chamber.

The boundary and initial conditions for the elements of the developed model have the form:

$0 \le r \le R:$

$$z=0: \quad k_{cat}\frac{\partial T_{cat}}{\partial z}=\alpha_{hel}(T_{cat}-T_{amb});$$
$$z=l_{cat}: \quad T_{cat}=T_{sed};$$
$$k_{cat}\frac{\partial T_{cat}}{\partial z}=k_{sed}\frac{\partial T_{sed}}{\partial z};$$
$$z=l_{sed}(t): T_{sed}=T_{pl};$$
$$k_{sed}\frac{\partial T_{sed}}{\partial z}=k_{pl}\frac{\partial T_{pl}}{\partial z};$$
$$z=l_{evap}(t): T_{pl}=T_{an}=T_{evap};$$
$$-k_{pl}\frac{\partial T_{pl}}{\partial z}=-k_{an}\frac{\partial T_{an}}{\partial z}+\rho_{an}U_{evap}L;$$
$$z=l_{an}: \quad -k_{an}\frac{\partial T_{an}}{\partial z}=\alpha_{hel}(T_{an}-T_{amb});$$

$$r=0: \quad 0\le z\le l_{cat}, \quad \frac{\partial T_{cat}}{\partial r}=0;$$
$$l_{cat}\le z\le l_{sed}, \quad \frac{\partial T_{sed}}{\partial r}=0;$$
$$l_{sed}\le z\le l_{evap}, \quad \frac{\partial T_{pl}}{\partial r}=0; \qquad (3)$$
$$l_{evap}\le z\le l_{an}, \quad \frac{\partial T_{an}}{\partial r}=0;$$

$$r=R: \quad 0\le z\le l_{cat}, \quad -k_{cat}\frac{\partial T_{cat}}{\partial r}=\alpha_{hel}(T_{cat}-T_{amb});$$
$$l_{cat}\le z\le l_{sed}, \quad -k_{sed}\frac{\partial T_{sed}}{\partial r}=\alpha_{hel}(T_{cat}-T_{hel});$$
$$l_{sed}\le z\le l_{evap}, \quad -k_{pl}\frac{\partial T_{pl}}{\partial r}=\sigma(T_{pl}^4-T_{hel}^4);$$
$$l_{evap}\le z\le l_{an}, \quad -k_{an}\frac{\partial T_{an}}{\partial r}=\alpha_{hel}(T_{an}-T_{amb});$$

The functions obtained by approximating of experimental data are used as expressions which describes the moving boundaries of electrodes. For the boundary of the evaporation anode - plasma:

$$l_{evap}(t)=l_{evap}^{init}+(a_1\cdot T^2+a_2\cdot T+a_3)\cdot t, \qquad (4)$$

where a_1, a_2,a_3 – proportionality coefficients, are determined on the basis of experimental data. For the boundary plasma - sediment on the cathode:

$$l_{evap}^{init}=const;\ l_{sed}^{init}=const;\ l_{sed}(t)=l_{sed}^{init}+\frac{\rho_{an}}{\rho_{sed}}\cdot u_{evop}\cdot t. \qquad (5)$$

At the initial moment $t = 0$:

$$T_{cat}(0)=T_{sed}(0)=T_{pl}(0)=T_{an}(0)=T_{hel}(0)=T_{amb}=T_{hel}^{init},\ T_{amb}=const. \qquad (6)$$

The constructed mathematical model of heat-exchange (1-3) for the system under consideration differs by taking into account the total heat flux power in the equation of thermal conductivity of the buffer medium and the mobile boundaries of the anode-plasma-cathode system.

The system of non-stationary differential equations (1-2) describing the heat-exchange process in the system under consideration, taking into account the boundary conditions (3) and the variable boundaries of the system (5-6), is a complex mathematical problem, the finding an accurate analytical solution of which is extremely difficult. Therefore, to obtain a solution, the finite-element method was used [5].

The use of the finite-element method involves the partition of the region in which the solution of differential equations is sought into a number of subfields (finite elements), in each of which an unknown quantity (moving and temperature of body points) has a simple analytical expression. Outside its element, this function is zero. Finite elements have common nodal points where they are related between themselves. The function values in the nodes are the solution to the task. The number of equations at use of the finite-element method is equal to the number of unknown values in the nodes on which the solution of the original system is sought, and is directly proportional to the number of elements. Each of the elements is associated with a limited number of neighbors, therefore so the system of linear algebraic equations has a sparse view, and this greatly simplifies its numerical solution.

The algorithm for constructing a discrete model when using the finite-element method of a continuous magnitude is the following:

1. The phase plane is divided into a finite number of nodal points.

2. The value of a continuous magnitude at each nodal point is considered a variable parameter, which must be found.

3. The whole area of calculation of a continuous quantity is divided by a finite number of elements, which have common nodal points and, in sum, approximates the shape of the entire region.

4. The analogue quantity is approximated on each element by a polynomial found on the basis of nodal values. Each finite element is set by form and function. These form and function determine the accuracy of the approximate solution.

To implement the developed numerical scheme for solving the heat transfer problem considered above, was used the environment of multiphysical simulation COMSOL Multiphysics [6]. The use of the Heat Transfer Module in this software environment allows you to perform the necessary numerical calculations based on the developed model, taking into account accepted assumptions and mobile boundaries of the system.

The construction a grid of finite elements is one of the most important steps in the algorithm for calculating the heat transfer model. In the modern COSMOL Multiphysics simulation environment, the thermo exchange problem is solved using an adaptive grid. The different shape of the finite elements also directly affects the convergence rates. Therefore, to obtain a stable numerical solution to the problems of the model (1-5), a study when using three - and four-node elements was carried out. The density of the grid used and the view itself of the final element clearly affect the accuracy of the calculation and the convergence. The calculation accuracy is reduced if the sizes of the neighboring elements near the stress concentrator are differ significantly. Therefore, for the most important and small construction components of the model (areas of plasma, evaporation and sediment), a fine-mesh grid was used.

Contrastive analysis when using different types of grids of finite element for the numerical calculation of heat-exchange of the considered system (figure 1) is presented in table 1.

Table 1. Analysis of characteristics of calculate grids.

Criterion	Grid 1	Grid 2	Grid 3	Grid 4	Grid 5	Grid 6
View of finite element	Three-node	Three-node	Three-node	Three-node	Four-node	Four-node
Number of grid nodes	302	442	815	1183	1403	4704
Number of grid elements	580	860	1603	2320	1333	4562
Number of grid degrees of freedom	2581	3862	7573	9826	7209	24468
The number of elements grid of the plasma	269	368	543	1076	1152	3893
The number of elements grid of the anode	44	67	59	176	77	280
The number of elements of the grid of the cathode	197	303	739	788	69	289
The number of elements of the working chamber	56	66	224	224	26	86
Calculation time (min.)	6	7	8	14	2	8

3. Results and discussion

The analysis of the results of numerical calculations showed that the data obtained for the grids 2 - 6, characterizing the temperature values at the anode end and at the plasma-anode border, differ slightly. The relative error of discrepancy of the performed numerical calculations for these grids does not exceed 0.012 %. In the figure 2 studies of the convergence rate of solutions for the meshes of the finite elements considered above are presented.

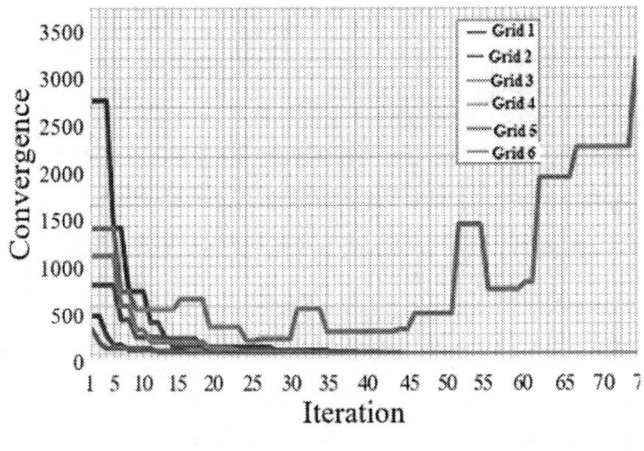

Figure 2. Convergence rate of problem-solving for different grids.

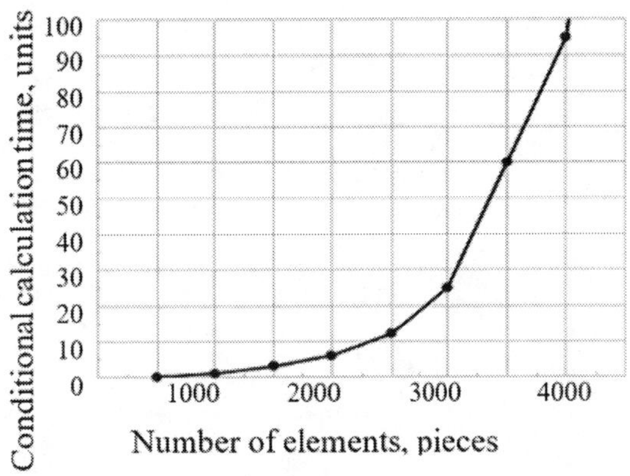

Figure 3. The influence of grid size on calculation time when using the finite-element method.

The best convergence rate of the solving the problem by the finite element method show grids 5 and 6. The analysis of influence of the elements quantity for four-node calculated grid on the total computation time when using the model is shown in figure 3. Based on the analysis of the convergence rate of the solution of the problem under consideration for different grids (figure 2), the influence of the grid size on the calculation time (figure 3), and the accuracy of the simulation results, grid 5 with the number of nodes –1403 is the best. For the numerical solution of the developed model, it is more rational to use four-node finite elements with a height of 1/5 R and with clotting in the region with the highest temperature gradient (of plasma) to 1/10 R (figure 1).

To verify the conformity of the constructed model to the physical process, a series of experiments was performed on the installation of an electric arc synthesis CNS. The installation is equipped with an automation control system [7], which allows maintaining stable synthesis parameters. The results of comparison of the calculated and experimental data of the time variation of the butt profile of anode and the temperature at a fixed point of the cathode surface are presented in figure 4 and 5.

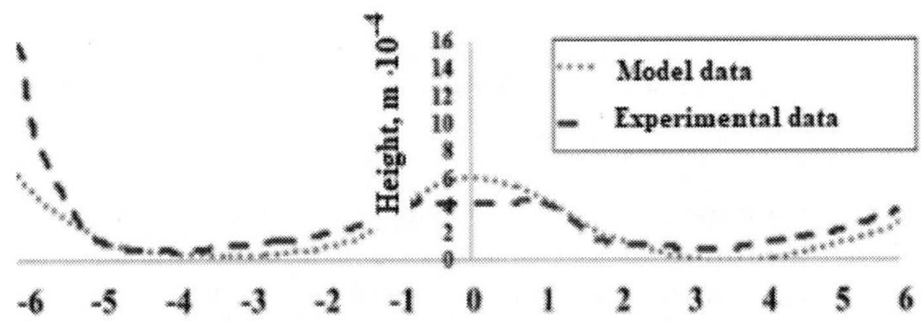

Figure 4. Changing of the profile of the burnup of the anode: the synthesis parameters: the time t = 120 s, the force of the arc current I = 150 A, voltage U = 25 V, the helium pressure in the camera P_{He} = 53.3 kPa.

The analysis showed that the average relative reduced error δ of the numerical calculation based on the heat-exchange model on the deviation of the anode evaporation profile is $\delta \approx 9$ % (figure 4), and for the temperature deviation on the cathode surface (figure 5) is $\delta an \approx 5{,}79$ %, which does not exceed the maximum permissible error value of 12 % accepted for this model, allowing to describe the influence of temperature fields on the evaporation of the anode and the formation of the final carbon fractions. This gives the right to talk about the adequacy of the heat-exchange model to the real process.

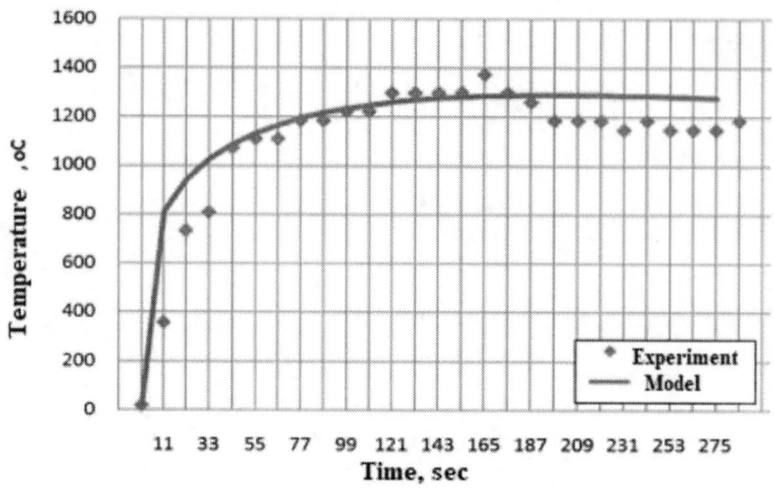

Figure 5. Temperature change in time on the cathode surface.

4. Conclusion

Thus, a method is proposed for mathematical modeling of the heat-exchange process during the electric arc synthesis of CNS, taking into account the total heat flux power in the heat equation of the buffer medium and the mobile boundaries of the system anode – plasma – cathode, which makes it possible to calculate the dynamics of the temperature field of the system taking into account the changing configuration of working zone. The rationally view and size of the grid for numerical calculation of the model under consideration by the finite element method is determined. This makes it possible to study the influence of the technological parameters of synthesis (of the arc current, voltage, etc.) on anode volatilization and the formation of final carbon fractions in the sediment deposition and on the walls of the operating chamber.

References

[1] Santiago E V, López S H, López M A C, Contreras D R, Farías-Mancilla R, Flores-Gallardo S G, Hernández-Escobar C A and Zaragoza-Contreras E A 2016 Optical properties of carbon nanostructures produced by laser irradiation on chemically modified multi-walled carbon nanotubes *Opt. Laser Technol.* **84** 53

[2] Nemchinsky V A and Raitses Y 2016 Anode sheath transition in an anodic arc for synthesis of nanomaterials *Plasma Sources Science and Technology* **25** (3) 035003

[3] Abramov G V, Gavrilov A N, Tolstova I S and Ivashin A L 2017 Formation of clusters of carbon structures in plasma under thermal destruction of graphite *Nanotechnologies in Russia* **12**(3-4) 139

[4] Morita T, Sakawa Y, Kuramitsu Y, Takabe H, Dono S, Tanji H, Aoki H, Waugh J N, Woolsey N C, Gregory C D and Koenig M 2017 Characterization of electrostatic shock in laser-produced optically-thin plasma flows using optical diagnostics *Physics of Plasmas* **24**(7) 072701

[5] Samogin Y N, Serkov S A and Chirkov VP 2017 *Finite Element Method in Dynamic Calculations of Turbomachinery* (Moscow: Fizmatlit) [in Russian]

[6] Tabatabaian M 2016 *COMSOL 5 for Engineers* (Herndon, VA: *Mercury Learning & Information*)

[7] Abramov G V, Gavrilov A N, Kravets O J and Varlamov N V 2019 The use of quantum-kinetic models in system design synthesis of carbon nanostructures *IOP Conf. Series: Materials Science and Engineering* **498** 012008

Electrodeposition of MWCNTs/silver composite coatings with enhanced mechanical characteristic

Y Litovka[1], I Dyakov[1], R Stolyarov[1,*], V Kulakov[1], A Zhigachev[2] and V Korenkov[2]

[1]Department "Technology and Methods of Nanoproducts Manufacturing", *Tambov State Technical University,* 106 Sovetskaya Street, 392000, Tambov, Russian Federation
[2]Scientific and Educational Center "Nanotechnologies and Nanomaterials", *Tambov State University*, 33 Internatsionalnaya Street, 392000, Tambov, Russian Federation

[*]E-mail: stolyarovra1985@gmail.com

Abstract. The paper describes the technology for producing silver electrochemical coatings containing multi-walled carbon nanotubes (MWCNTs). The MWCNT effect on the roughness, structure and tribological properties of silver coatings obtained with the addition of the nanotubes to an electrolyte in the low concentration range (0 to 80 mg/L). An increase in the microhardness and a decrease in the coefficient of friction were found for the modified coatings at the MWCNTs concentrations in the electrolyte of 40 and 60 mg/L, respectively.

1. Introduction

Silver has a low transition (contact) electrical resistance, high corrosion resistance, electrical and thermal conductivity, which ensures its wide application in coatings of movable contacts of electronic equipment. At the same time, silver is a soft metal and has great wear, which limits the resource of its functioning. Numerous methods for increasing wear resistance and reducing transition resistance based on the technologies of applying various kinds of coatings with a thickness of up to several micrometers, high microhardness, low friction coefficient and low surface roughness have been proposed. Among them, electroplating from electrolytes with the addition of carbon nanomaterials seems rather prospective [1-3]. The use of electrodeposition makes it possible to achieve homogeneous distribution and fix carbon nanomaterials in metal matrixes [4-6]. Nanocomposite-based electrochemical coatings possess high mechanical and technical characteristics, in particular, strength [7-12], hardness [7-9, 12], wear [13-16], and corrosion resistance [16, 17].To obtain high mechanical and functional properties of nanocomposite coatings and considerably increase the cost of final products in large-scale manufacturing, carbon nanoparticle concentrations in electrolytes significantly higher than 0.1 g/L are used.

Considering the aforementioned, the main goal of the present work was to synthesize silver electroplating coatings with improved mechanical and operating characteristics based on an electrolyte, enriched with low concentrations of multi-walled carbon nanotubes (MWCNTs).

2. Materials and methods

2.1. Multi-walled carbon nanotubes (MWCNTs)

A Taunit carbon nanomaterial, representing MWCNTs with a diameter of 20-50 nm and a length of up to 10 μm, was used herein. The nanomaterial was synthesized by NanoTechCenter Ltd. (Tambov, Russia) through the method of catalytic pyrolysis of natural gas on a Ni/Mg catalyst at atmospheric pressure and a temperature of 620°C.

2.2. Electrochemical deposition of the coatings

The coating was applied to a copper billet with a pre-polished surface size of 2x2 cm. The preparation of blanks included: degreasing in an organic solvent and drying; assembling parts on copper hook and isolation; electrochemical degreasing at a cathode current density of 5 A/dm^2 and a temperature of 60°C for 2 min in an aqueous solution of the following composition: NaOH – 2%, $NaHCO_3$ – 4%, $Na_3PO_4 \cdot 12H_2O$ – 4%; rinsing under warm running water; rinsing in cold running water; chemically activating a 10% aqueous solution of H_2SO_4 at 25°C for 1 min; and, again, rinsing in cold running water.

Electrochemical silvering was carried out using a ready-for-use 07-SG electrolyte acquired from Ecomet (Moscow, Russia), which represents a colorless transparent aqueous solution with a density of 1.11±0.02 kg/m^3 and includes a silver-containing compound. The pH value of this electrolyte was kept in the range of 8.6-10.0. The anode was made of platinum. The ratio of anode and cathode surfaces was 2:1. Electrochemical deposition of silver was carried out at a cathode current density of 0.2 A/dm^2 and a temperature of 55°C for 18 min. Under these conditions and deposition time, the estimated coating thickness was about 2-3 μm. The processing of the samples after applying the silver coating includes: capturing the electrolyte silvering in two baths; dismantling the hook; rinsing in cold running water; rinsing in hot running water; and drying.

The above described procedure additionally consisted in introducing the Taunit MWCNTs into the silver plating electrolyte at concentrations varying from 20 to 80 mg/L. The MWCNT-modified electrolyte was processed with an ultrasonic disperser at a frequency of 22 kHz, amplitude of 80 microns, and sound intensity of 786 W/cm^2 for 10 min. Ultrasonic treatment was performed immediately before the electrochemical deposition to minimize the aggregation of the nanotubes, increase their wettability, and achieve a uniform distribution in the electrolyte bulk. For each nanotubes concentration, a separate electrolyte was prepared, with which five coating samples (replicates) were obtained for all the studies presented in the work.

2.3. Characterization of the coatings

Morphology and microstructure of the silver coating surface, as well as single MWCNTs, was carried out using a Merlin scanning electron microscope (Carl Zeiss, Germany), and a JEM 2100F transmission electron microscope (JEOL, Japan).

To determine the crystalline size, internal strains and phase composition of the coatings, diffraction patterns were obtained using a D2 Phaser X-ray diffractometer (Bruker AXS, Germany). Diffraction patterns were recorded in the θ-2θ Bragg-Brentano reflection geometry within the angular range of 10-80 °2θ. The following crystallographic data were used for Ag when calculating peak positions and intensities: space group Fm-3m (225), and a = 4,085 Å. Values and errors for the data obtained for five samples at each MWCNT concentration were calculated using the Rietveld method.

An NT 9080 optical profilometer (Veeco, Город, USA) was used to study the surface topography of the samples. The roughness was measured in the central part of the samples in randomly selected three areas with a size of 460x615 μm. Each of them produced mapping of the surface and calculation of roughness parameter values – mean arithmetic deviation of profile Ra and the root mean square deviation of profile Rq.

Five coating samples were obtained for each MWCNT concentration in the electrolyte from the studied range to gather reliable information on the mechanical and tribological properties.

The tribological properties of the samples were studied by scratching (nanoindentation) of the surface layer using a G200 nanoindenter (MTS Nano Instruments, USA). Before mechanical testing, the coating surfaces were mechanically polished to reduce the influence of roughness on the accuracy of scratch width measurements. The scratch was applied to a Berkovich corner diamond indenter; its curvature radius is 20 nm, and a constant scratching speed was10 mm/s.

The normal load on the indenter, F_N = 15 mN, was selected experimentally considering that the scratch depth is no more than 40% of the silver coating thickness. The scratch length L was 500 µm. During the scratching, the tangential force F_T was measured.

Similar to the conventional friction coefficient [18], the value µ was calculated based on F_N and F_T:

$$M = F_T / F_N, \tag{1}$$

where, μ – friction coefficient, F_T – tangential force (friction force), F_N – normalload. The scratch cross-sectional area, multiplied by its length, is characteristic of volumetric and gravimetric wear (wear resistance).

If the scratch length was constant, then the scratch width was characteristic of wear resistance coatings, linearly depending on it. The scratch width was measured experimentally. The microhardness of the coating was calculated for pyramidal indenters according to the following equation:

$$H_s = 4\, F_N / b^2, \tag{2}$$

where, H_s – microhardness, F_N – normal load, b – scratch width [18].

To study the tribological characteristics of friction and wear (determined by the scratch width), the nanomodified coating samples were additionally polished, since the parameter Rq should to be less than 100 nm.

The porosity of the material affects its corrosion resistance and surface roughness and, accordingly, its wear resistance mainly during the second stage of wear (performance wear), which refers to the normal operation of the friction assembly and lasts for the longest time. The porosity was estimated based on the SEM images of the mechanically polished surfaces (Rq < 100 nm) of the silver coating samples. Only pores larger than 0.5 µm were considered.

3. Results and discussion

3.1. Morphology, roughness and structure of the nanomodified coatings

The SEM images of the coating samples obtained at different Taunit MWCNT concentrations (C_{MWCNTs}) are presented in figure 1. A qualitative analysis of figures 1a, 1b and 1c shows a decrease in the roughness of the coatings when increasing the nanotubes concentration in the electrolyte from 0 to 40 %, which is quantitatively confirmed by the results presented in figure 2. The samples obtained at C_{MWCNTs} = 60 mg/L (figure 1d) and C_{MWCNTs}=80 mg/L (figure 1e) have a more suitable morphology (close to the profile of the sliding surface) to ensure minimum wear during a long operational period.

The roughness effect on the wear resistance occurs due to the fact that the contact surfaces take place only at surface protrusions. The area of actual contact (especially, during the initial break-in period) is relatively small. On the contact pads, high local pressure may arise, under the influence of which elastic compression and plastic deformation take place. As a result, vertices of the irregularities of the contact areas are sheared off. Over the time, the height of the irregularities (roughness) decreases, the actual touch area increases, and the specific pressure on the contact pads decreases, thereby reducing the wear.

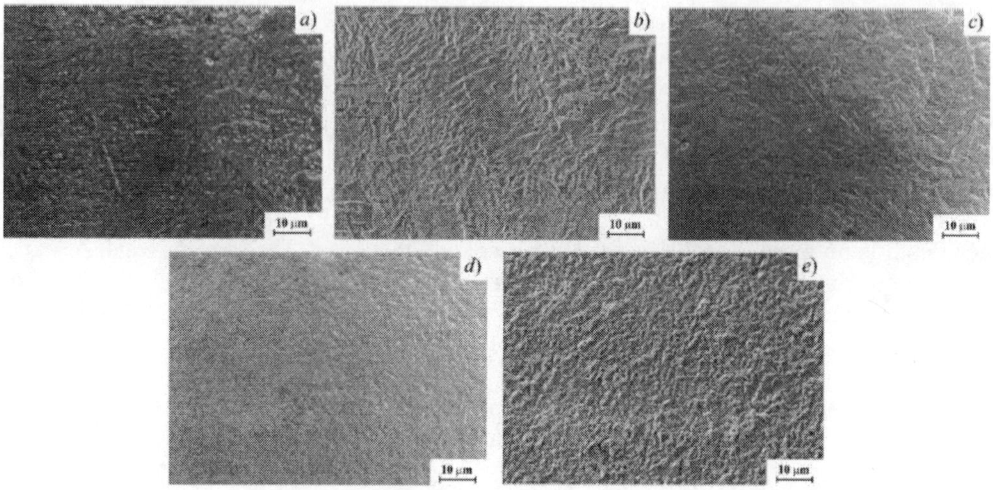

Figure 1. Surface morphology images of the coating samples obtained at different MWCNT concentrations: 0 (a), 20 (b), 40 (c), 60 (d), and 80 mg/L (e).

Figure 2 shows the dependence of the surface roughness on different MWCNT concentrations in the electrolyte. As can be seen, the error magnitude is large (10-20 %) and may be caused by the significant difference between the roughness values obtained for different samples with the same nanotube concentration. The *Ra* and *Rq* value scattering on different parts of the same sample was found to be considerably less and not to exceed 5 nm (figure 2).

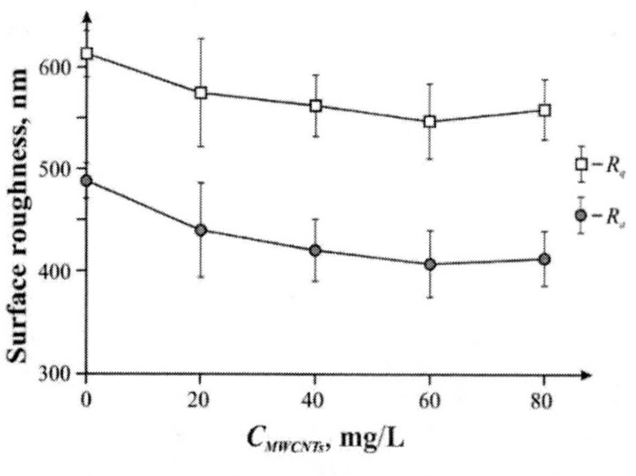

Figure 2. Influence of MWCNT concentration in electrolyte on the surface roughness.

The surface studies of the samples showed that the addition of the nanotubes to the coating solution leads to a 15-20% decrease in the surface roughness (at 60 mg/L MWCNTs). The reduction of the surface roughness of the nanomodified coatings at > 60 mg/L MWCNTs indirectly confirms the uniform distribution of the nanotubes in the metallic matrix and leads to a decrease on wear and tear. The obtained result lies in good agreement with the experimental data presented elsewhere [11].

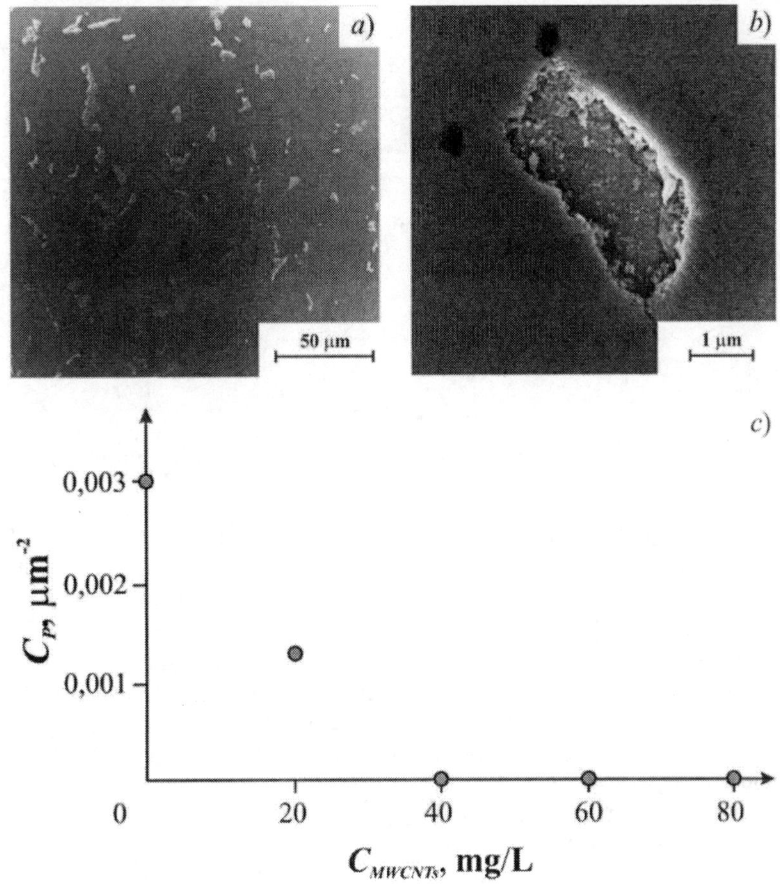

Figure 3. SEM images of pores in the reference coating at different magnifications (a, b). Influence of MWCNT concentration in electrolyte on pore surface concentration (c).

The dependence of the pore surface concentration C_P on the MWCNT concentration is shown in figure3. The presence of pores on the coating surface increases the surface roughness and reduces the corrosion resistance (figures 3a and 3b). The pores larger than 0.5 µm were not available in the coating samples obtained at C_{MWCNTs} of 40-80 mg/L. The addition of the MWCNTs at these concentrations to the electrolyte will contribute to an increase in the wear resistance and durability of the nanomodified coatings (figure 3c).

The results show that when introducing small nanotube into the electrolyte, there are no apparent changes in the crystalline size (figure 4). The relative deformation of the grains in all the coatings is small (hundredths of a percent) and virtually unchanged at varying MWCNT concentrations. Our experimental data differ significantly from the results of the works presented in [1,20], since in our case, the CNTs do not affect the crystallite size of the coating. This is probably due to the use of low CNTs concentrations in the electrolyte and the type of metal matrix.

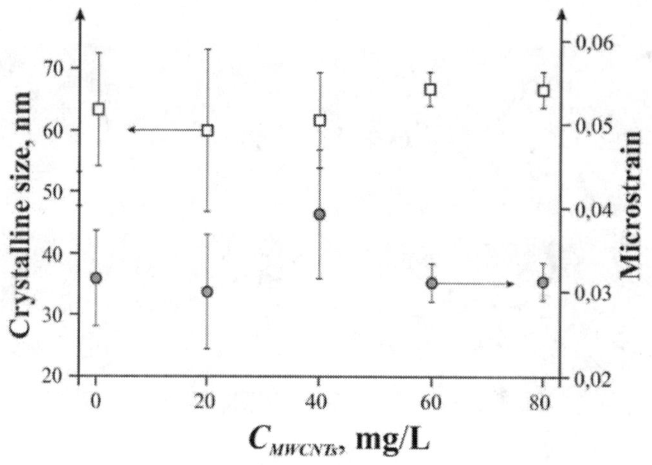

Figure 4. Influence of MWCNT concentration in electrolyte on crystalline size and microstrain.

In the paper [19], the authors describe the electrodeposition of nickel on the surface of MWCNTs. They demonstrate that with increasing current density the number of the nuclei of the crystal on the nanotube surface increases as a result of activation of larger number of surface defects, which could start the crystallization. Therefore, for large current density values (more than 0.5 A/dm^2) nanotubes acted as additional crystallization sites, resulting in decreased grain sizes of the coating. This effect was also noted by the authors of the works [9,11,20]. In our case, probably, there was no change in the crystalline size and internal stresses for the reference (non-modified) and nanomodified coatings, since the electrodeposition was carried out at a low current density (0.2 A/dm^2), which was not enough for activating the nanotubes.

3.2. Mechanical and tribological properties

The abrasion (wear) resistance (wear) and the transitional electric resistance of the moving contacts depend on many factors such as load, sliding speed, temperature, lubrication, media parameters, and properties of the rubbing materials. Under the conditions of our experiments, all the factors, except friction properties of the materials, remained unchanged, so their influence was neglected. The main characteristics and parameters affecting the wear resistance included surface roughness, microhardness, friction coefficient, and porosity.

It is known that the harder the friction material, the less the wear is [15]. Therefore, the microhardness is characteristic of coatings that indicates abrasion resistance. Due to constant shape of the indenter, and, consequently, the shape of the scratch profile, it was concluded that the scratches on various samples differ only by their width and depth, and the dependence of the scratch cross-sectional area is linear relative to its width. The scratch cross-sectional area, multiplied by its length, is characteristic of volumetric and mass wear (abrasion resistance).

Figure 5 demonstrates the dependences of the scratch width and microhardness/friction coefficient on the MWCNTs concentration in the electrolyte. The scratch width and amount of material removed from the friction surface decrease (figure 5a), which was confirmed experimentally and is consistent with the results obtained elsewhere [6,8,11-13,17]. This is the results of the fact that the lowest wear and the highest microhardness were reached in the samples obtained at C_{MWCNTS} = 40 mg/L, while the lowest friction coefficient value (0.46) was achieved at C_{MWCNTS} = 60 mg/L (figure 5b). Introducing the MWCNTs into the electrolyte hardens the coating and reduces its friction coefficient.

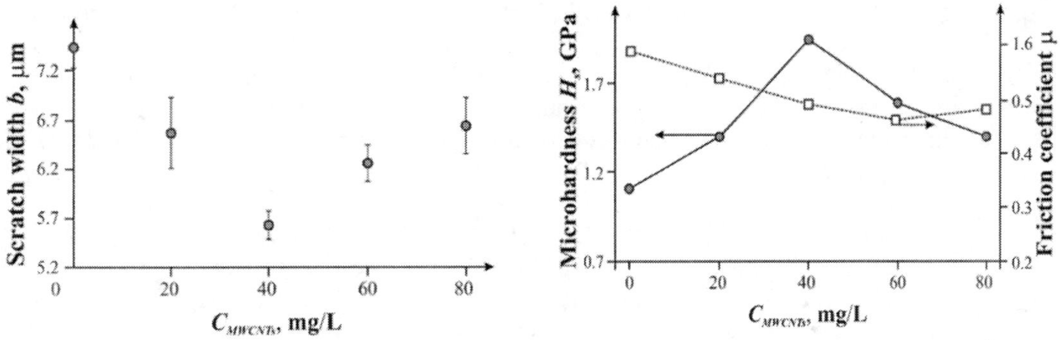

(a) (b)

Figure 5. Effect of MWCNTs concentration in electrolyte on the scratch width (a) and microhardness and friction coefficient (b).

The microhardness H_S of the modified coating is 40-50 % higher than that of the reference coating (figure 5b). The analysis of electron microscopic images shows that the porosity of the modified and reference coatings is very different (figure 3). The additive contribution of the MWCNTs to the microhardness of the obtained coatings may be 25 %, if the nanotube strength close to the theoretical value of ~100 GPa is assumed. However, due to their defects, the nanotubes used in the present work have strength well below 10 GPa. Accordingly, the additive contribution of the MWCNTs the the microhardness may not exceed 1...2%. The size of crystallites and internal stresses in the reference and nanomodified coating and of nanomodified coatings do not differ (figure 4). Consequently, the nanomodified coating is hardened due to the reduction in porosity.

4. Conclusion

The technology was developed to obtain silver electroplating coatings with improved quality indicators from the electrolyte with the addition of a nanomaterial. The differences from the conventional technology are as follows: 1) the MWCNTs were introduced into the electrolyte; 2) the electrolyte was treated using an ultrasonic disperser; 3) the MWCNT concentration in the electrolyte was kept within the specified limits. The technology made it possible to increase the wear resistance of the coatings. It was found that at the MWCNT concentrations of 40 and 60 mg/L, the best mechanical and tribological properties of the coatings and lower surface roughness were achieved. Thus, varying the concentration of MWCNT in electrolyte, one can selectively control the morphology, roughness and mechanical properties of nanomodified silver coatings.

Acknowledgments

The present study was funded by the Russian Foundation for Basic Research in the framework of Research Project No. 18-29-17047.

References

[1] Karimzadeh A, AliofkhazraeiM and Walsh F C 2019 A review of electrodeposited Ni-Co alloy and composite coatings: microstructure, properties and applications. *Surf. Coat. Tech.* **372** 463 doi.org/10.1016/j.surfcoat.2019.04.079

[2] He T, He Y, Li H, Fan Y, YangQ and He Z 2018 A comparative study of effect of mechanical and ultrasound agitation on the properties of pulse electrodeposited Ni-W/MWCNTs composite coatings. *J. Alloys Compd.* **743** 63 doi.org/10.1016/j.jallcom.2018.01.368

[3] Chakraborty R, Sengupta S, Das S, Saha P and Das K 2017 Synthesis and characterization of MWCNT reinforced nano-crystalline copper coating from a highly basic bath through pulsed electrodeposition. *Surf. Interface* **9** 28 doi.org/10.1016/j.surfin.2017.07.001

[4] Chu K and Jia C 2014 Enhanced strength in bulk graphene–copper composites. *Phys. Status Solidi*

(A) **211** 184 doi.org/10.1002/pssa.201330051
[5] Feng Y, Mc Guire G E, Shenderova O A, Ke H and Burkett S L 2016 Fabrication of copper/carbon nanotube composite thin films by periodic pulse reverse electroplating using nanodiamond as a dispersing agent. *Thin Solid Films* **615** 116 doi.org/10.1016/j.tsf.2016.07.015
[6] Feng Y and Burkett SL 2015 Fabrication and electrical performance of through silicon via interconnects filled with a copper/carbon nanotube composite. *Comp. Mater. Sci* .**97** 022004 doi.org/10.1016/j.commatsci.2014.10.014
[7] Jeon YS, Byun JY and Oh TS 2008 Electrondeposition and mechanical properties of Ni–carbon nanotube nanocomposite coatings.*J. Phys. Chem. Sol.* **69** 1391 doi.org/10.1016/j.commatsci.2014.10.014 10.1016/j.jpcs.2007.10.049
[8] Guo Ch, Zuo Y, Zhao X, Zhao J and Xiong J 2008 The effects of electrodeposition current density on properties of Ni–CNTs composite coatings. *Surf. Coat. Technol.* **202** 3246 doi.org/10.1016/j.surfcoat.2007.11.032
[9] Dai P, Xu W and Huang Q 2008 Mechanical properties and microstructure of nanocrystalline nickel-carbon nanotube composites produced by electrodeposition. *Mater. Sci. Eng. A* **483** 172 doi.org/10.1016/j.msea.2006.09.152
[10] Kim Y, et al 2013 Strengthening effect of single-atomic-layer graphene in metal–graphene nanolayered composites. *Nat. Commun.***4** 2114 doi.org/10.1038/ncomms3114
[11] Shi L, Sun C F, Gao P, Zhou F and Liu WM 2006 Electrodeposition and characterization of Ni–Co–carbon nanotubes composite coatings. *Surf. Coat. Tech.* **200** 4870 doi.org/10.1016/j.surfcoat.2005.04.037
[12] Khabazian S and Sanjabi S 2011 The effect of multi-walled carbon nanotube pretreatments on the electrodeposition of Ni–MWCNTs coatings. *Appl. Surf. Sci.* **257** 5850 doi.org/10.1016/j.apsusc.2011.01.118
[13] Susumu A and Taishi K 2014 Electroless deposition of cu/multiwalled carbon nanotube composite films with improved frictional properties. *J. Solid State Sci. Technol.* **3** 201 doi.org/10.1149/2.018406jss
[14] Susumu A, Akihiro F, Masami M and Morinobu E 2008 Excellent solid lubrication of electrodeposited nickel-multiwalled carbon nanotube composite films. *Mater. Lett.* **62** 3545 doi.org/10.1016/j.matlet.2008.03.047
[15] Carpenter C R, Shipway P H and Zhu Y 2011 Electrodeposition of nickel-carbon nanotube nanocomposite coatings for enhanced wear resistance. *Wear* **271** 2100 doi.org/10.1016/j.wear.2011.01.068
[16] Lee C K 2012 Wear and corrosion behavior of electrodeposited nickel–carbon nanotube composite coatings on Ti–6Al–4V alloy in Hanks' solution. *Tribol. Int.* **55** 7 doi.org/10.1016/j.triboint.2012.05.024
[17] Chen X H, Chen C S, Xiao H N, Cheng F Q, Zhang G and Yi G J 2005 Corrosion behavior of carbon nanotubes-Ni composite coating. *Surf. Coat. Tech.* **191** 351 doi.org/10.1016/j.surfcoat.2004.04.055
[18] Fischer-Cripps A C 2011 *Nanoindentation* (New York: Springer) p 279
[19] Golovin Y I, Stolyarov R AandShuklinov A V 2013 Morphology and growth kinetics of Ni nanoparticles on the surface of multiwalled carbon nanotubes at galvanostatic electrodeposition. *Tech. Phys.* **58** 1189 doi.org/10.1134/S1063784213080136
[20] Yang YL, Wang YD, Renb Y, He CS, Deng JN, Nan J, ChenJG and Zuo L 2008 Single-walled carbon nanotube-reinforced copper composite coatings prepared by electrodeposition under ultrasonic field. *Mater. Lett.* **62** 47 doi.org/10.1016/j.matlet.2007.04.086

IOP Conf. Series: Materials Science and Engineering **693** (2019) 012006 doi:10.1088/1757-899X/693/1/012006

Method of manufacturing of composite for 3D printing and the electrophysical properties of the obtained material

K Dorozhkin[1], A Tkachev[2], G Kuleshov[1], A Badin[1], E Galunin[2], T Shematilo[1] and V Suslyaev[1]

[1]Department of Radioelectronics, Faculty of Radiophysics, *National Research Tomsk State University*, 36 Lenina Ave., Tomsk 634050, Russia

[2]Department of Technique and Technology of Production of Nanoproducts of Institute of Technology, *Tambov State Technical University*, 106 Sovetskaya Street, Tambov 392000, Russia

[1]E-mail: yasbtk@yandex.ru

Abstract. The present paper presents the technology for the manufacture of filaments based on ABS plastic and Taunit-M nanomaterial for use in 3D printing. A test sample was printed from this filament on a 3D printer using the layered overlap method at 15% Taunit-M content. The transmittance spectra for the printed sample were obtained using a Mach-Zander interferometer in the frequency range 120-260 GHz. The complex dielectric constant of the studied composite was calculated.

1. Introduction

Currently, there is an increasing need for shielding materials for high-frequency electronics [1], for ecology and medicine, for antenna technology and for military purposes [2]. The problem of electromagnetic compatibility is relevant [3] and requires new technologies for designing small-sized matching devices of different geometric shapes. Additive technologies (3D printing), which are currently successfully used in various fields of science and technology, can be used in this case. A distinctive advantage of 3D printing is the production of individual products, which, if necessary, can be easily converted into mass customization [4]. Most often, 3D printing technology uses a single source material, such as plastic. However, composites based on polymers filled with active phase inclusions are promising for shielding. Polymers: polylactic acid (PLA) [5], Acrylonitrile butadiene styrene (ABS) [4], polycarbonate (PC) [6], polybutylene terephthalate (PBT) [7] provide strength characteristics, flexibility, adhesion,. resistance to external influences. The active component of the composites provides the required electromagnetic parameters. Carbon nanostructures: single-and multi-layer carbon nanotubes, graphenes and their combinations are successfully used in reducing the level of electromagnetic radiation in free space [8-9]. Magnetic materials: ferrites, carbonyl iron [9], effectively interact with electromagnetic radiation when composites based on them are located on a conductive surface. New materials are required in connection with the miniaturization of electronic devices, with the expansion of the operating frequency band.

The following technologies are used to implement 3D printing : inkjet printing, where a powder bound with a liquid binder is used [10], stereolithography (SL), laminated object fabrication (LOM), fused deposition modeling (FDM), selective laser sintering (SLS), and extrusion [4]. The FDM method is the most common 3D printing method in the world. Millions of 3D printers work on the basis of FDM, from the cheapest to industrial three-dimensional printing systems. FDM printers work with various types of plastics, the most popular and affordable of which is ABS. Plastic products are characterized by high strength, flexibility, perfect for product testing, prototyping, as well as for the manufacture of ready-to-use objects.

Additive technologies offer unlimited possibilities in the manufacture of objects of any geometry. The popular FDM printing technology is replenished with threads having unique mechanical and electrophysical properties. These radio-absorbing and electrically conductive filaments containing dielectric matrices with nanoscale carbon inclusions, such as carbon nanotubes [11,12], multi-wall carbon nanotubes (MWCNT) of the "Taunite" series [13], graphene [14], can serve for these purposes.

This paper presents a description of the methodology for manufacturing a polymer filament from ABS and the results of studying the electromagnetic response and electrophysical parameters of flat samples from carbon nanomaterials and ABS polymer.

2. Materials and methods

2.1. Starting material

Carbon nanomaterials of the company NanoTechCenter (Tambov, Russia) of the Taunit-M series (specification No. 2166-001-77074291-2012) [15,16] were selected for the manufacture of composite filaments. The material is a stable aqueous colloidal solution containing 20 wt.% MCNT “Taunit-M”. These are coaxial multilayer carbon nanotubes (MCNT) with an external diameter of 8–15 nm (inner diameter of 4–8 nm) and a length of more than 2 μm. The number of layers of one tube is about 6–10. The amount of non-carbon impurities is not more than 1 wt.%. The specific geometric surface is 300–320 m^2/g. The bulk density is 0.03–0.05 g/cm^3.

2.2. Method of manufacturing of radio-filament for 3D printing

The thermal technology of mixing and extruding was used in the overwhelming majority of studies for the production of radiofilaments. The chemical method of composite material manufacturing for the extruder filament was the most successful.

Block-scheme of the technological process of manufacture of the samples based on ABS with nanomaterial “Taunit-M” is shown in figure 1.

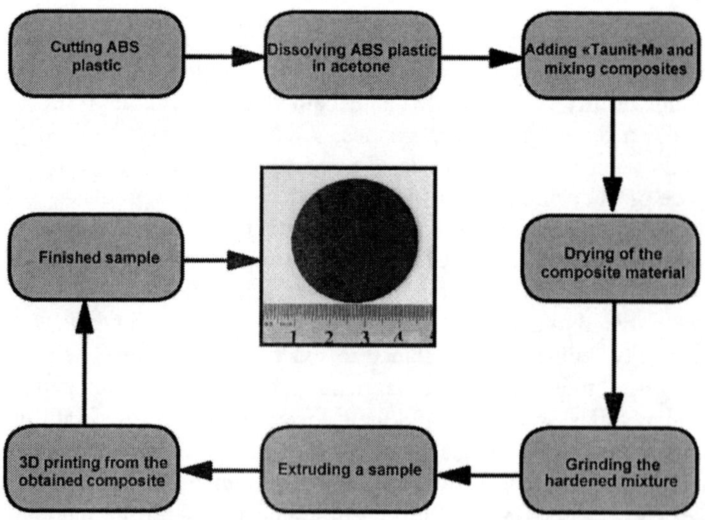

Figure 1. Block-scheme of the technological process of manufacture of the samples based on ABS with nanomaterial “Taunit-M”.

ABS plastic was used to conduct an experiment to create a filament from Taunit-M, since this polymer is the most affordable. Its mechanical strength in the scientific literature is noted as excellent. The authors of the article have experience using ABS plastic earlier to create 3D filaments from multi-walled carbon nanotubes.

The process of making of composite filament involves several steps.

1. At the first stage necessary quantity of ABS granules or small pieces of pure ABS were carefully weighed on an electronic scale Shimazu (accuracy ~ 1 mg).
2. The obtained material is dissolved by solvent acetone or V-646 and pours it into glass containers.
3. The next stage is adding carbon nanomaterial "Taunit-M" to the dissolved mixture and mix thoroughly by ultrasonic disperser during for 10 minutes at a power of 75 W.
4. The composite is removed from the mold after complete polymerization of the resulting mixture composition.
5. Then the polymerized mixture is grinded (figure 2). Further, this raw material was used for extruding, when manufacturing filament.

The extruder is a device for continuous processing of polymeric raw materials. Because squeezing, mixing and contact with a heated cylinder, the polymer raw material melts and turns into a homogeneous mass - the melt. The assembled extruder is shown in figure 3.

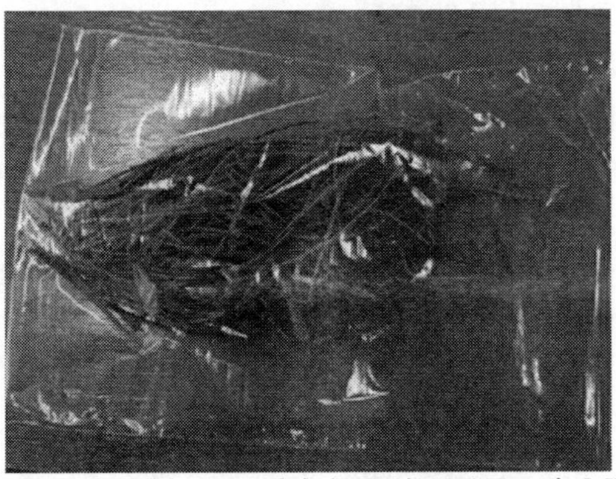

Figure 2. This material (ABS/15%"Taunit-M" composite) is prepared for extrusion.

Figure 3. Extrusion head of computerized system for the manufacturing of test filaments.

The production technology of the filament provides for at least 10 iterations of extrusion to form a homogeneous structure of the filament. Increase the number of iterations (*N*) leads to a decrease in the size of air pores inside the filament (figure 4). It should be noted that the porosity negatively effects on flexibility of the filament, increasing the probability of breaking during the 3D printing process.

The uniformity of filaments is a prerequisite for successful use in 3D-printers based on FDM technology. The filament is heated in the nozzle until it reaches a semi-liquid state, and then is extruded on a platform or over a previously printed layer.

2.3. Experimental equipment

The Mach-Zender interferometer for measuring transmission and phase shift was used [17,18]. Interferometer consists of two arms: working "Arm II" (with ABS/15%"Taunit-M" composite – "Sample") and reference "Arm I". Backward-wave oscillator (BWO) with water cooling was used as a source of monochromatic electromagnetic wave. Focusing of quasi-optical beam is carried out by Teflon lenses. As a detector of terahertz radiation optoacoustic converter (Golay cell) was used. "Amplitude modulator" was used for setting the amplitude modulation.

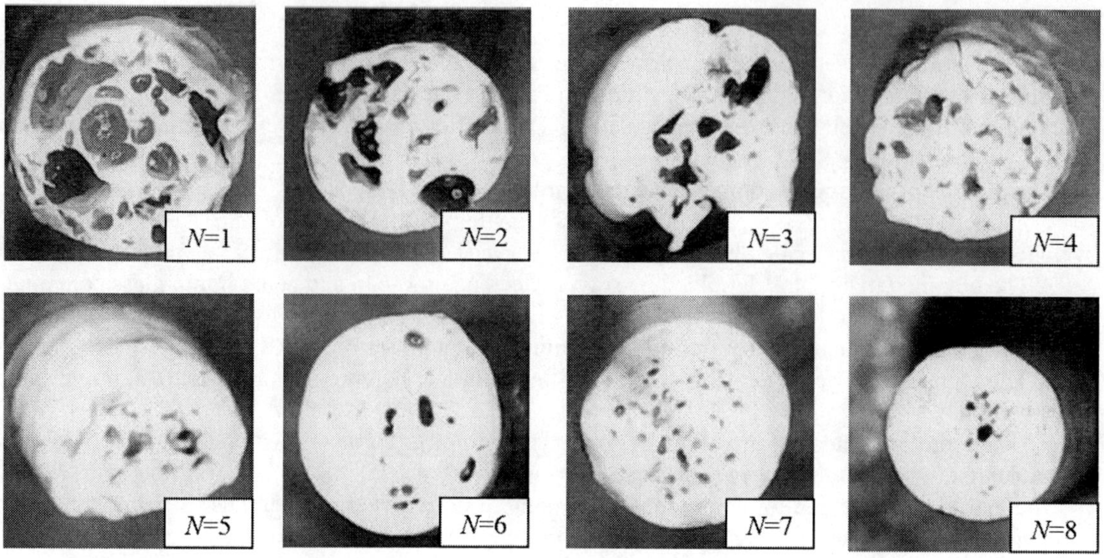

Figure 4. Micrographs of a filament section showing the effect of the number (*N*) of times extrusion on the porosity of the obtained material (filament diameter – 0.7 mm).

3. Results and discussion

The reflection coefficients *R* and transmission *T* of a flat sample 1.99 mm thick made from the obtained filament were measured in the frequency range 120-260 GHz (figure 5). It is seen that the transmitted signal decreased by 30-40%. when the reflection coefficient is relatively small (of the order of 10%) A composite material manufactured by the same technique based on MCNT was used to evaluate the effectiveness of the interaction of the new material with microwaves. Carbon nanotubes were obtained by catalytic gas-phase deposition of ethylene in the presence of a $FeCo/Al_2O_3$ catalyst at the Institute of Catalysis of the Siberian Branch of the Russian Academy of Sciences. The prepared nanotubes were purified by refluxing in HCl, filtered, washed to a neutral pH and dried for 24 hours at 40 °C. It is a light, downy black powder containing individual nanotubes, bundles and aggregates of nanotubes, impurity metal particles that were enclosed in nanotubes, particles of an oxide carrier. The average nanotube diameter is 9.4 nm.

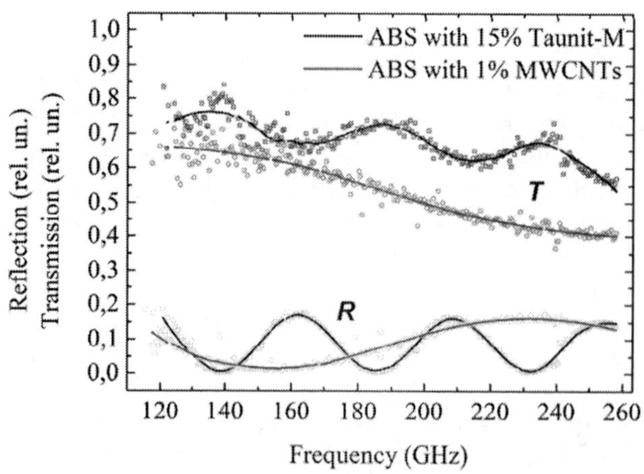

Figure 5. Frequency dependence of transmission and reflection coefficients for 3D-printed composites based on 15% carbon nanomaterial "Taunit" and 1% MWCNTs.

The concentration of MCNT was 1%, which approximately corresponds to the mass of carbon material in 15% liquid MCNT Taunit-M. This material is well known and well studied. The power of the transmitted signal through the test sample of the same thickness as the newly created one turned out to be lower (figure 5). Perhaps this is due to the inaccuracy in determining the mass of the active phase in the new material.

The dielectric permittivity (DP) spectra of materials (figure 6), calculated from the measured components of the electromagnetic response of the samples (figure 5), indicate another reason for the difference.

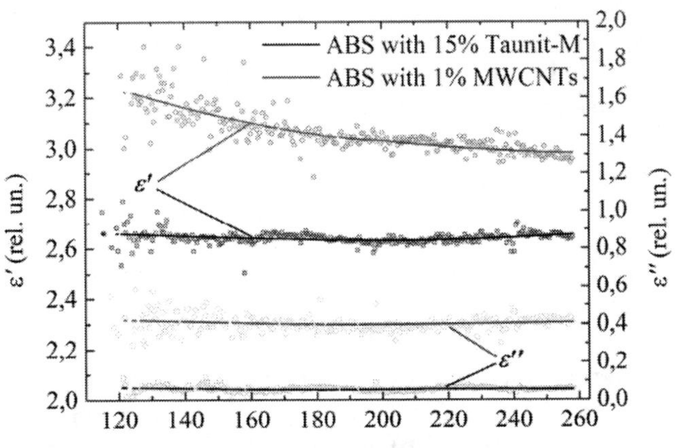

Figure 6. Frequency dependence of real and image parts of permittivity for 3D-printed composites based on 15% carbon nanomaterial "Taunit" and 1% MWCNTs.

The frequency dependence of the DP composite from MWCNTs shows the presence of conductivity, which occurs when the tubes contact. The conductivity of the MCNT Taunit-M composite is much lower, the frequency dispersion of the DP is not observed in this range. This means that the contacts between the particles of the conductive fraction are negligible.

It was not possible to increase the concentration of the liquid fraction to ensure electrical contacts in the new material. ABS plastic quickly solidified with increasing mass of liquid MCNT Taunit-M during mixing, which was caused by a conflict of acetone with the liquid fraction of the new material. It is likely that the effective use of this promising material requires a change in the method of mixing the components or replacement of the polymer. This issue will be resolved soon.

4. Conclusion

Filaments made according to the developed technology for 3D printing, based on "Taunit-M" nanomaterial in a polymer ABC matrix, can be used for the manufacture of low-reflectance screens in the terahertz range. For the manufacture of threads with a high content (more than 15%) of the active phase, a change in manufacturing technology or a change in the substance of the polymer matrix is required.

5. Acknowledgment

The reported study was funded by RFBR according to the research project No. 18-32-00810 and project No. 19-32-90150.

References

[1] Batrakov K, Kuzhir P, Maksimenko S, Volynets N, Voronovich S, Paddubskaya A, Valusis G, Kaplas T, Svirko Y and Lambin P 2016 Enhanced microwave-to-terahertz absorption in graphene. *Appl. Phys. Lett.* **108** 123101 DOI:10.1063/1.4944531

[2] He J, Luo H, He L, Yan S, Shan D, Huang S and Deng L 2018 Investigation on microwave dielectric behavior of flaky carbonyl iron composites. *J. Mater. Sci. Mater. Electron.* **29** 15112 https://doi.org/10.1007/s10854-019-00959-0
[3] Kuzhir P P, Letellier M, Bychanok D S, Paddubskaya O G, Suslyaev V I, Korovin E Y, Baturkin S A, Fierro V and Celzard A 2017 Electrical Properties of Carbon Foam in the Microwave Range *Russ. Phys. J.* **59** 1703 DOI:10.1007/s11182-017-0964-3
[4] Ngo T D, Kashani A, Imbalzano G, Nguyen K T Q and Hui D 2018 Additive manufacturing (3D printing): A review of materials, methods, applications and challenges *Compos. Part B Eng.* **143** 172 **DOI:** 10.1016/j.compositesb.2018.02.012
[5] Zhang D, Chi B, Li B, Gao Z, Du Y, Guo J and Wei J 2016 Fabrication of highly conductive graphene flexible circuits by 3D printing *Synth. Met.* **217** 79 https://doi.org/10.1016/j.carbon.2014.10.080
[6] Postiglione G, Natale G, Griffini G, Levi M and Turri S 2015 Conductive 3D microstructures by direct 3D printing of polymer/carbon nanotube nanocomposites via liquid deposition modeling *Compos. Part A Appl. Sci. Manuf.* **76** 110 http://dx.doi.org/10.1016/j.compositesa.2015.05.014
[7] Gnanasekaran K, Heijmans T, van Bennekom S, Woldhuis H, Wijnia S, de With G and Friedrich H 2017 3D printing of CNT- and graphene-based conductive polymer nanocomposites by fused deposition modeling *Appl. Mater. Today* **9** 21 https://doi.org/10.1016/j.apmt.2017.04.003
[8] Shuba M V, Yuko D I, Kuzhir P P, Maksimenko S A, Kanygin M A, Okotrub A V, Tenne R, Lambin Ph 2017 How effectively do carbon nanotube inclusions contribute to the electromagnetic performance of a composite material? Estimation criteria from microwave and terahertz measurements. *Carbon* **129** 688 DOI:10.1016/j.carbon.2017.12.067
[9] Shuba M V, Melnikov A V, Paddubskaya A G, Kuzhir P P and Maksimenko S A 2013 Role of finite-size effects in the microwave and subterahertz electromagnetic response of a multiwall carbon-nanotube-based composite: Theory and interpretation of experiments. *Physical Review* B **88** 045436 DOI: 10.1103
[10] Naiden E P, Zhuravlev V A, Suslyaev V I, Minin R V and Itin V I 2011 Structure parameters and magnetic properties of Me_2W^1 cobalt-containing hexaferrite systems synthesized by the SHS method. *Russian Physics J.* **53(**9) 87 https://doi.org/10.1007/s11182-011-9519-1
[11] Czyzewski J, Burzyński P, Gaweł K and Meisner J 2009 Rapid prototyping of electrically conductive components using 3D printing technology *J. Mater. Process. Technol.* **209** 5281 DOI 10.1007/s40964-017-0019-x
[12] Schmitz D P, Ecco, L G, Dul S, Pereira E C, Soares B G, Barra G M, Pegoretti A 2018 Electromagnetic interference shielding effectiveness of ABS carbon-based composites manufactured via fused deposition modelling. *Materials Today Communications* **15** 70 https://doi.org/10.1016/j.mtcomm.2018.02.034
[13] Kotsilkova R, Ivanov E, Todorov P, Petrova I, Volynets N, Paddubskaya A, Kuzhir P, Uglov V, Biró I, Kertész K, Márk G I and Biró L P 2017 Mechanical and electromagnetic properties of 3D printed hot pressed nanocarbon/poly (lactic) acid thin films. *J. Appl. Phys.* **121** 064105 http://dx.doi.org/10.1063/1.4975820
[14] Sweeney C B, Lackey B A, Pospisil M J, Achee T C, Hicks V K, Moran A G, Teipel B R, Saed M A, Green M J 2017 Welding of 3D-printed carbon nanotube–polymer composites by locally induced microwave heating. *Science Advances* **3** 1700262 DOI: 10.1126/sciadv.1700262
[15] Polyakov V, Mukhin V, Abramova I, Morozova S, Shubina N, Voropaeva N, Tkachev A, Burakov A, Romantsova I, Karpachev V, Figovsky O 2017 Sorption Activity of "Taunit"-Series Carbon Nanomaterials. *Int. Letters of Chemistry, Physics and Astronomy* **74** 1 doi: 10.1016/j.proche.2012.10.043
[16] Terukov E I, Babaev A A, Tkachev A G, Zhilina D V 2018 Radio-wave absorbing properties of

polymer composites on the basis of shungite and carbon nanomaterial Taunit-M. *Technical Physics* **63** 1044 https://doi.org/10.1134/S1063784218070289

[17] Badin A V, Dorozhkin K V, Suslyaev V I, Berdyugin A I, Vigovskiy V Y 2017 Quasi-optical 2D system for non-contact non-destructive testing of defects in natural and artificial crystals. *Proceedings of SPIE - The Int. Society for Optical Eng.* **10466** 1046625 https://doi.org/10.1117/12.2291960

[18] Dunaevskii G E, Suslyaev V I, Zhuravlev V A, Badin A V, Dorozhkin K V 2016 Ferromagnetic resonance in hexagonal ferrite $Ba_3Co_2Fe_{24}O_{41}$ at the THz frequency range. *Int. Conf. on Infrared, Millimeter, and Terahertz Waves, IRMMW-THz* 7758771 DOI: 10.1109/IRMMW-THz.2016.7758771

IOP Conf. Series: Materials Science and Engineering **693** (2019) 012007 doi:10.1088/1757-899X/693/1/012007

Calculation of parameters of the rotary apparatus for the production of graphene concentrate based on synthetic oils

V F Pershin[1], A G Tkachev[1], R A Al-Jarah[1], T V Pasko[1] and A A Osipov[2]

[1]Department of Technology and Methods of Nanoproducts Manufacturing, *Tambov State Technical University,* 106 Sovetskaya Street, Tambov 392000, Russian Federation
[2]LLC "NanoTechCenter", 51 Sovetskaya Street, Tambov 392000, Russian Federation

[1]E-mail: pershin_home@mail.ru

Abstract. This article presents the results of a study of the process of liquid-phase shear exfoliation of graphite in oil. The result is a suspension containing graphene structures. It was experimentally established that when designing industrial devices, the concentration of graphene structures in oil can be considered directly proportional to the following parameters: graphite concentration in the initial suspension; rotor radius; rotational speed; number of moving blades; suspension processing time. A relationship, allowing to calculate the performance or parameters of industrial rotary devices based on the results of experiments on a laboratory setup, was obtained.

1. Introduction

Friction is an important characteristic of the motion of contacting surfaces, because the friction are among the main causes of failures of engines, gears, bearings, etc. and wear of items, the uninterrupted performance of which depends on the quality of mechanical system operation, especially in the transport, industrial and energy sectors [1]. Analysis of the real situation showed that up to a third of the fuel energy is lost due to the friction of moving fragments, such as transmission, tires, brakes, etc. [2]. Such a large energy loss is associated with high values of friction and wear coefficients.

At the present level of development of science and technology, there are two options to reduce the friction coefficient and reduce energy loss: the creation of anti-friction materials for parts that are included in the friction units; the creation of new high-performance lubricants that reduce the coefficient of friction, friction and wear parts. Both of these problems are trying to solve thousands of scientists and engineers around the world. Since there are a huge number of mechanisms that work and will work for many years in the industry, the second solution to the problem of friction today seems to us to be more realistic and relatively quickly implemented in real life.

There are a huge number of lubricants of different consistency (liquid, plastic, etc.), composition and method of use. But all these lubricants contain a base (natural or synthetic) and antifriction additives. There are a huge number of bases and additives, but we focus on synthetic oils and graphene derivatives as additives.

Graphene is one of the newest nanoscale carbon allotropes, has unique properties associated with minimizing friction and wear, and also has good thermal and mechanical properties. So, this 2D material can serve as a solid or colloidal (liquid) lubricant [3,4]. It has now been established that the tribological characteristics of graphene and its derivatives depend on many factors, including morphology, a thickness of graphene layers, the method of manufacture and surface chemistry [5-7].

Research results both at the micro level [8] and at the macro level [9] showed that graphene and graphene oxide (GO) is really very much lower friction coefficient.

It should be noted that graphene and its derivatives are environmentally friendly products, and this fact increases the prospects of their use as an effective nano-additive for oils and plastic lubricants [10]. In this work, the tribological properties of GO in SAE20W-50 engine oil were investigated, and a decrease in wear rate of 60-80% was observed, and the minimum value of the friction coefficient (0.057) was determined by adding 0.5 wt.% GO. A review of recent advances in the production and study of the tribological characteristics of graphene is given in [11].

The results of studying the tribological characteristics of crumpled graphene balls seem to be very interesting [12]. It was found that these balls are very stable in the solid state and when dispersed in liquids, they do not open and do not collapse even after heating or granulation. Besides, they can be particularly dispersed in different solvents (for instance, in lubricating oils), without employing any additional procedure such as chemical functionalization, due to the fact that they are attracted to each other through the weak van der Waals forces [13]. It is noteworthy that base oils containing only 0.01-0.1 wt.% of these balls appear to be more efficient in decreasing friction and wear than conventional commercial lubricants. This fact was established after determining the tribological properties of graphene and the other carbon-based materials – graphite, reduced graphene oxide (r-GO, or chemically modified graphene), and carbon black – using a pin-on-disk tribometer. A base oil (*e.g.*, polyalphaolefin) was mixed with 0.01-0.1 wt.% of these materials and then sonicated until the solid particle residue disappears and complete dispersion occurs. It was found that the crumpled graphene balls, added to the oil at the above-mentioned concentrations, provide consistently good performance and the lowest friction and wear coefficients: their 20-and 85-% reduction is achieved, respectively, compared to the unmodified oil. Moreover, the graphene-modified base oil outstrips the 5W30 lubricant due to its reduced friction and wear.

Similar results on the reduction of friction and wear coefficients were obtained when modifying greases with graphene plates, which consist of several layers (up to 25 layers) [14,15].

Graphene plates were not balls, but their shape was very different from the plane, which was confirmed by their image, obtained by SEM.

Thus, an analysis of the literature on the use of graphene as an antifriction additive showed that graphene and its derivatives are very promising materials. In our opinion, the results of laboratory studies have so far not been widely applied on an industrial scale, mainly due to the high cost of graphene and its derivatives. The articles [16,17] published the results of the production of few-layers graphene using a high-speed stator-rotor mixer. The fact of obtaining double-, few- and multi-layered graphenes as one-atom-thick carbon platelets is confirmed by electron microscopy and Raman spectroscopy. In [18], we experimentally proved that using liquid-phase shear exfoliation it is possible to obtain graphene structures (a mixture of graphene, few-layer and multilayer graphene).

As a result of experimental studies, we found [15] that when 0.1-% multi-layered graphene (up to 25 layers) is added to a comprehensive calcium lubricant, the wear spot diameter decreases by 50%, the scoring index increases 2.9 times, and the bearing capacity increases 3.8 times.

Currently, there exist no methods for calculating the parameters of industrial rotary devices. Thus, the aim of the present research is to study the process of liquid-phase shear exfoliation of graphite and to obtain dependences for calculating the performance or main parameters of a rotary device intended for manufacturing graphene masterbatches.

2. Preliminary experiments

We used the improved rotary device, as described elsewhere [19]. Its main difference from the device used in [16] is that its rotor blades are able to move in the radial direction under the action of centrifugal forces during rotation of the rotor. This technical solution provides contact of the blades with the inner surface of the stator and increases the efficiency of shear exfoliation.

As parameters that can affect the intensity of the process of exfoliation of graphite were chosen: the radius of the inner surface of the stator; rotor speed; the number of moving blades. To determine the type of dependences of the effect of these parameters on the exfoliation intensity, several series of experiments were conducted with a discrete change in one of the parameters.

Experiments were performed as follows. Preparing 3-6 liters of oil suspension of crystalline graphite GSM-2 with a concentration from 2.5 to 10%. The suspension was poured into a cylindrical container, a rotary apparatus was installed, and the processing of the suspension began. Every 5 cycles of suspension processing, samples were taken to determine the concentration of graphene structures.

Figure 1 shows the dependence of the concentration of graphene structures on the number of suspension processing cycles. The rotor speed was 10,000 rpm, the concentration of graphite in the initial suspension was 50 mg/ml.

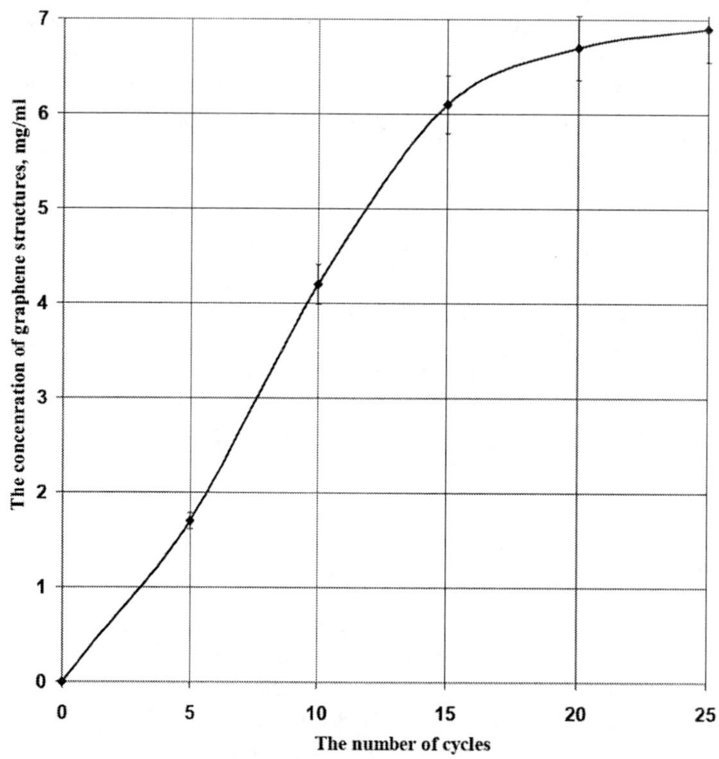

Figure 1. The dependence of the concentration of graphene structures on the number of suspension processing cycles.

As can be seen from the graph, at the beginning of the exfoliation process, during 15 processing cycles, the dependence is almost linear. In the case of industrial production, it is not economically feasible to process the suspension more than 15 times. Thus, to design industrial rotor devices, the dependence of the concentration of graphene structures on the number of processing cycles can be considered linear. It should be especially noted that since the time of one cycle is constant, the dependence of the concentration of graphene structures on time is also linear.

3. Determination of the main parameters of the exfoliation process

In our opinion, the intensity of liquid-phase shear exfoliation depends on the magnitude of shear stresses that arise between graphene layers when a graphite particle enters the zone between the stationary inner surface of the stator and the moving blade. Thus, in the industrial rotary device, it is required to ensure tangential stress values not less than those taking place in the laboratory device.

3.1 Determination of normal and tangential stresses

The graphite particle in the zone between the stator and the rotor is acted upon by centrifugal forces that move and ultimately press this particle to the inner surface of the stator. If the particle mass is equal to m, the angular velocity of rotation of the rotor is ω, the inner radius of the stator is *R*, then the centrifugal force F_{CP} acts on the particle, the numerical value of which is:

$$F_{CP} = m\omega^2 R. \tag{1}$$

Normal stresses σ_{CP}, which act between the graphene layers that make up a particle, generally depend on the shape of the particle and its lateral dimensions, more precisely, the area. In the first approximation, the magnitude of the tangential stresses can be estimated using the average value of the area of the graphene layers S_M. In this case, the normal stresses σ_{CP} will be equal to:

$$\sigma_{CP} = F_{CP} / S_M. \tag{2}$$

To determine the stresses arising between the graphene layers in the particle from the action of the moving rotor blade, we use the scheme of action of the forces from [19], which is given in figure 2.

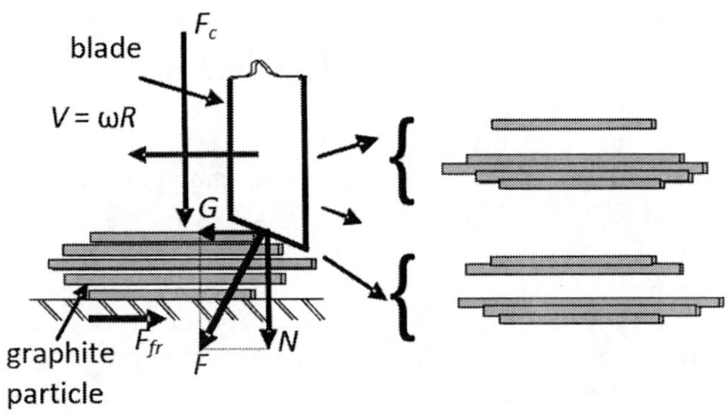

Figure 2. Diagram of forces acting on a particle located between the blade and the inner surface of the stator.

First of all, we define the centrifugal force F_C, which acts on the blade:

$$F_C = m_{PL}\omega^2 R_{CT}, \tag{3}$$

where m_{PL} is the mass of the blade, ω is the angular velocity of rotation of the rotor, R_{CT} is the distance from the center of gravity of the blade to the axis of rotation of the rotor

If we consider the forces acting on the particle, then $N = F_C$. Normal stresses in the contact zone, particles with a blade will be equal to:

$$\sigma_P = N / S_P, \tag{4}$$

where σ_P is the normal stresses in the contact zone, S_P is the contact area of the particle with the blade.

The average value of normal stresses that occur between graphene layers is equal to:

$$\sigma_G = N / S_G \tag{5}$$

where S_G is the contact area between adjacent graphene layers.

The shearing force *G* acting on the particle is equal to:

$$G = N \cdot \mathrm{tg}\alpha_P, \tag{6}$$

The tangential stresses that arise in the zone of contact of the blade with the particle will be equal to:

$$\tau_P = G / S_P. \tag{7}$$

The average value of the tangents stresses that occur between adjacent layers is:

$$\tau_G = G / S_G. \tag{8}$$

The obtained dependences (5) and (8) are necessary in order to ensure minimum values of normal and tangential stresses between graphene layers during exfoliation of graphite particles when designing new rotary apparatuses since we are successfully exfoliated under laboratory conditions at given values of stresses.

3.2 Calculation of the current concentration of graphene structures

Initially, it can be assumed that all the parameters of the process and the device for its implementation are interconnected. The main task is to find the functional dependence of the concentration of graphene nanoplates, which was formed in suspension as a result of the separation of particles of graphite. Taking into account that in practical calculations we will use a linear section of the dependence of the concentration of graphene structures on the number of cycles shown in Figure 2, we made the assumption that the current concentration linearly depends on: graphite concentration in the initial suspension c_O, mg/ml; radius of stator R, m; n is the number of blades; velocity of rotation of the rotor ω, rpm. Since, ultimately, the resulting dependence will need to be associated with similar parameters of the laboratory setup; we introduce the index to the above parameter designations – $_L$, and obtain the following designation parameters of the laboratory installation: t_L; c_{LO}; R_L; n_L; ω_L. We will look for the indicated dependence in the following form:

$$c = kc_L\left[(c_O / c_{LO})\times(R / R_L)\times(\omega / \omega_L)\times(n / n_L)\times(t / t_L)\ \right], \qquad (9)$$

where k is the proportionality coefficient.

The numerical value of the coefficient k, apparently depends on the properties of graphite and oil. If the designed apparatus will process a suspension of the same graphite and oil that was used in the laboratory setup, k is equal to 1.

4. Results and discussion

The parameters that we used when obtaining the dependency presented in figure 2 will be considered the parameters of the laboratory setup: $c_L = 6$ mg/ml; $c_{LO} = 50$ mg/ml; $R_L = 20$ mm; $\omega_L = 10000$ prm; $n_L = 4$; $t = 2700$ sec. It is these parameters that we will use when calculating the concentrations according to (9). The concentration of graphene structures was determined by the specific gravity of the suspension using the technique given in [20].

Figure 3 shows the dependence of the concentration of graphene structures on the rotor speed. In this graph, the points show the experimental values, and the line is a calculation by the formula (9). Line 1 is a calculation at a concentration of graphite in the initial mixture of 50 mg/ml, and a straight line 2, at a concentration of 100 mg/ml. As you can see, the calculation satisfactorily coincides with the results of the experiment, up to a rotor speed of 12,000 rpm, but with a further increase in speed, the real intensity of the exfoliation process of graphite decreases. Since from a practical point of view it is impractical to carry out the process with reduced intensity, dependence (9) can be used in the design of industrial rotary apparatuses.

The linear nature of changes in the concentration of graphene structures depending on the rotational speed of the rotor is experimentally confirmed by the results obtained for the graphite exfoliation in water using a rotor device of the similar design [21].

Figure 4 shows the dependence of the concentration of graphene structures on the concentration of graphite in the initial suspension. On this graph, the points stand for the experimental values, and the line represents the calculation according to equation (9). As can be seen, the calculated and experimental values satisfactorily coincide.

A comparison of the calculated and experimental values of the concentrations showed that the calculated values differ from the average experimental values by no more than 15%.

The results of experiments with two movable blades showed that the intensity of the exfoliation process decreased by approximately two times. Unfortunately, the design of our device did not allow us to change the diameter of the rotor, so we are going to decide on the influence of the number of blades and the diameter of the rotor in the near future.

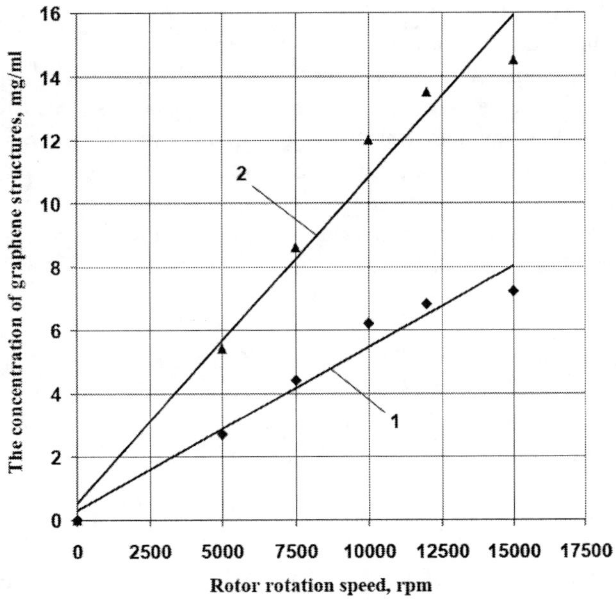

Figure 3. The dependence of the concentration of graphene structures on the rotor speed.

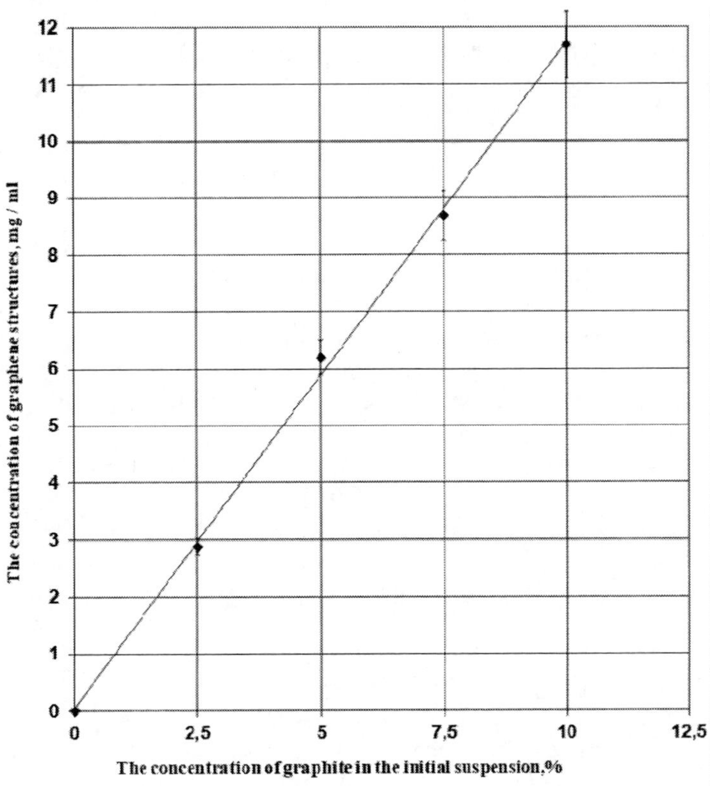

Figure 4. The dependence of the concentration of graphene structures on the concentration of graphite in the initial suspension.

5. Conclusion

The process of liquid-phase shear exfoliation of graphite in oil is investigated. Dependences are obtained that allow us to determine the normal and tangential stresses that occur in a graphite particle, when processed in a rotary apparatus with moving blades and which must be provided in the designed apparatus. It was established experimentally that the concentration of graphene structures in suspension is directly proportional to the speed of rotation of the rotor and the concentration of graphite in the initial suspension. A dependence is proposed for calculating the concentration of graphene structures from the results of a limited number of experiments on a laboratory setup. Comparison of calculated and experimental values of concentrations was carried out and it was found that the discrepancies between these values do not exceed 15%.

6. Acknowledgments

The work was funded by the Ministry of Science and Higher Education of the Russian Federation in the framework of the Federal Target Program “Research and Development in Priority Areas of Advancement of the Russian Scientific and Technological Complex for 2014–2020” (Agreement No. 14.577.21.0253, Unique Identification Number: RFMEFI 57717X0253).

References

[1] Demydov D, Adhvaryu A, McCluskey P and Malshe Ajay P 2010 Advanced lubricant additives of dialkyldithiophosphate (DDP)-functionalized molybdenum sulfide nanoparticles and their tribological performance for boundary lubrication. *Nanoscale Materials in Chemistry: Environmental Applications, ACS Symposium Series* **1045** 137 https://doi.10.1021/bk-2010-1045.ch008

[2] Holmberg K, Andersson P, Erdemir A 2012 Global energy consumption due to friction in passenger cars. *Tribol. Int.* **47** 221 https://doi.10.1016/j.triboint.2011.11.022

[3] Wahlisch F, Hoth J, Held C, Seyller T, Bennewitz R 2013 Friction and atomic-layerscale wear of graphitic lubricants on SiC (0001) in dry sliding. *Wear* **300** 78 https://doi.10.1016/j.wear.2013.01.108

[4] Kitt A, Qi Z, Remi S, Park H, Goldberg B and Swan A 2013 How graphene slides: measurement and theory of strain-dependent frictional forces between graphene and SiO_2. *Nano Lett.* **13** 2605 https://doi.10.1021/nl4007112

[5] Berman D, Erdemir A and Sumant A V 2014 Graphene: a new emerging lubricant *Mater. Today* **17** 31–42 https://doi.10.1016/j.mattod.2013.12.003

[6] Penkov O, Hae-Jin Kim, Hyun-Joon Kim, and Dae-Eun Kim 2014 Tribology of graphene: a review. *Int. J. Precis. Eng. Manuf.* **15** 577 https://doi.10.1007/s12541-014-0373-2

[7] Wanlin G, Jun Y, Hu Q, Yufeng G, Hongrong W, Minmin X 2014 Friction of low-dimensional nanomaterial systems. *Friction* **2** 209 https://doi.10.1007/s40544-014-0064-0

[8] Vasić B, Matkovic A, Gajic R, Stankovic I 2016 Wear properties of graphene edges probed by atomic force microscopy based lateral manipulation. *Carbon* **107** 723 https://doi.org/10.1016/j.carbon.2016.06.073

[9] Daoyi W, Yufu X, Lulu Y, Tao Y, Xianguo H 2018 Tribological behaviour of graphene oxide sheets as lubricating additives in bio-oil. *Industrial Lubrication and Tribology* **70(8)** 1396 https://doi.10.1108/ILT-11-2017-0356

[10] Kumar P and Wani M F 2017 Tribological characterisation of graphene oxide as lubricant additive on hypereutectic Al-25Si/steel tribopair. *Tribol. Trans.* **61** 335 https://doi.10.1080/10402004.2017.1322735

[11] Ali I, Basheer A A, Kucherova A, Memetov N, Pasko T, Ovchinnikov K, Pershin V, Kuznetsov D, Galunin E, Grachev V, Tkachev A 2019 Advances in carbon nanomaterials as lubricants modifiers. *J. of Molecular Liquids* **279** 251 https://doi.10.1016/j.molliq.2019.01.113

[12] Dou X, Koltonow A R, He X, Jang H D, Wang Q, Chung Y W, Huang J 2016 Selfdispersed crumpled graphene balls in oil for friction and wear reduction. *Proc.Natl. Acad. Sci. U. S. A.* **113** 1528 https://doi.10.1073/pnas.1520994113

[13] Narayan R and Kim S 2015 Surfactant mediated liquid phase exfoliation of graphene *Nano Convergence* **2(1)** 20 https://doi.10.1186/s40580-015-0050-x

[14] Al-Saadi D A Y, Pershin V F, Salimov B N and Montaev S A 2017 Modification of graphite greases graphene nanostructures. *J. Frict. Wear* **38(5)** 355 https://doi.10.3103/S1068366617050026

[15] Pershin V F, Ovchinnikov K A, Al-Hilo Z A A, Stolyarov R A and Memetov N R 2018 Development of environmentally safe lubricants modified by graphene. *Nanotechnol. Russ.* **13(5-6)** 340 https://doi.10.1134/S1995078018030138

[16] Keith R, Paton K R, Varrla E, Backes C, Smith R J, Khan U, O'Neill A, Boland C, Lotya M, Istrate O M, King K, Higgins T, Barwich S, May P, Puczkarski P, Ahmed I, Moebius M, Pettersson H, Long E, Coelho J, O'Brien S E, McGuire E K, Sanchez B M, Duesberg G S, McEvoy N, Pennycook T J, Downing C, Crossley A, Nicolosi N V and Coleman J N 2014 Scalable production of large quantities of defect-free few-layer graphene by shear exfoliation in liquids. *Nat. Mater.* **13(6)** 624 https://doi.10.1038/nmat3944

[17] Dimov D, Amit I, Gorrie O, Barnes M D, Townsend N J, Neves A I S, Withers F, Russo S and Craciun M F 2018 Ultrahigh performance nanoengineered graphene–concrete composites for multifunctional applications. *Adv. Funct. Mater.* 1705183 https://doi.10.1002/adfm.201705183

[18] Zhumagaliyeva G B, Pershin V F 2019 Production of Few-Layers Graphene in Synthetic Oil Using Rod Drum Mill. *Advanced Mater. & Technol.* **2** 59 https://doi.10.17277/amt.2019.02.pp.059-068

[19] Al-Shiblawi K A, Pasko A A, Pershin V F 2018 Modeling of the process of obtaining graphene structures using liquid phase graphite shift exfoliation. *Transactions TSTU* [in Russian – Vestnik TGTU] **24(4)** 717 https://doi.org/10.17277/vestnik.2018.04.pp.717-726

[20] Zhumagaliyeva G B, Osipov A A, Pershin V F 2019 The Effect of Mode and Geometric Parameters on Kinetics of Liquid Phase Shift Exfoliation Process Graphite at Rod Mill. *Transactions TSTU* [in Russian – Vestnik TGTU] **25(2)** 320 https://doi.org/10.17277/vestnik.2019.02.pp.320-328

[21] Pershin V F, Al-Shiblawi K A, Baranov A A, Pasko T V 2019 Production of low-layer graphene by liquid phase shear exfoliation. *St. Petersburg Polytechnic University J. of Eng. Sci. and Technol.* **25(01)** 143 https://doi.10.18721/JEST.25114

A nanostructuring approach for modification of the features of optical materials: lithium fluoride

N Kamanina[1,2], P Kuzhakov[1], A Kukharchik[1] and D Kvashnin[3,4,5]

[1]Laboratory for Photophysics of Media with Nanoobjects, *Vavilov State Optical Institute*, 5 Kadetskaya Liniya V O , bldg. 2, St.-Petersburg 199053, Russian Federation
[2]*St.-Petersburg Electrotechnical University* (LETI), 5 Prof. Popova Street, St.-Petersburg 197376, Russian Federation
[3]*Emanuel Institute of Biochemical Physics, RAS*, 4 Kosygina Street, Moscow 119334, Russian Federation
[4]*National University of Science and Technology MISiS*, 4 Leninsky Avenue, Moscow 119049, Russian Federation
[5]*Pirogov Russian National Research Medical University*, 1 Ostrovityanov Street, Moscow 117997, Russian Federation

[1,2]E-mail: nvkamanina@mail.ru

Abstract. It is well known that different technical methods and physical-chemical approaches can be used to improve the basic properties of optical materials operated in the UV-VIS-IR spectral range. Laser-based equipment and a nanotechnology approach can be jointly applied with good advantage. In this paper, the advantage of the laser-oriented deposition technique is discussed. Promising nanoobjects based on carbon nanotubes and LiF structures were considered, and their spectral, mechanical and wetting properties were evaluated. It was found the modification of the LiF with the carbon nanotubes improves the optical properties of the former. The quantum chemical simulations confirmed the experimental results obtained. Besides, the changes in the properties of the other classical materials (such as KCl, KBr, MgF_2, *etc*) were comparatively shown.

1. Introduction

Optical materials are promising and widely applied in the general optoelectronics, especially, in the laser schemes, in space and automobile industry as well as in the biomedicine. These materials are efficiently operated in the UV-VIS-IR wavelength ranges and at the extended range of the energy density [1-5]. At present, the laser structuring of the material surface allows to modify the physical and chemical properties of the matrix material with good advantage [6,7]. From this point of view, carbon nanotubes (CNTs) can be considered as the perspective nanoobjects due to their unique atomic structure, strong bonding between the carbon atoms and very small value of the refractive index as well [8-10].

Using CNTs for surface modification of optical materials may drastically tune its properties, causes with the fact that the atomic structure of such materials includes the substances that can provoke the different features against the radiation, such as: transmission, scattering, reflection, absorption, refraction, etc. These processes can be activated and modified via the propagation of the electromagnetic waves through the nanostructured optical media. Using structuration approach the spectral, refractive, mechanical properties, wetting phenomena, etc. can be essentially changed. For example, as the good evidences, the Fresnel losses can be changed, which can be easy shown based on the glass substrate. Indeed, using the classical method (PDV, CVD, etc.) one can deposit many layers of the thin coatings in order to reduce the losses at the reflection process and thus to improve the light transmission (aperture ratio) of the optics, the surface of the glass can be exposed to special treatment, which is called "enlightenment of optics". On the surface of the matrix materials (glass), the thin film should be deposited, the refractive index of which (n_{film}) should be less than the refractive index of glass (n_{glass}): $n_{\text{film}} = \sqrt{n_{\text{glass}}}$. In order to achieve the minimal reflection losses and maximum of the transmittance, the film should have the thickness determined by equation (1):

$$h = \frac{d}{n_{\text{film}}}, \tag{1}$$

where h is the geometric thickness of the film, d is an optical thickness of the film defined as $d = \frac{\lambda}{4}$ (λ is wavelength of the light in exact part of spectrum).

It is well known that the film with the thickness of $\frac{\lambda}{4}$ from the substance with the refractive index of $\sqrt{n_{\text{glass}}}$ decreases the reflective coefficient dramatically. If one considers the interface such as the air-glass with the reflective index of the glass material close to 1.5, the reflection from one surface of the substrate is approximately 4 %, and from two surfaces, it is approximately 8%.

Using the innovative laser structuration approach and forming the covalent bonding between the carbon nanotubes with the small refractive index (n ~1.1), the Fresnel loses via the reflection can be changed in one order of magnitude, at least.

Lithium fluoride [11] is the unique colourless cubic optical crystal with the unit cell parameter a close to 0.40279 nm. It has a very high transparency from the UV to the IR wavelength (from 0.12 to 6 microns), and already applied in the vacuum ultraviolet region and in the IR spectral range. Such advantages lead to using LiF for measuring the radiation doses by the method of the thermoluminescent dosimetry. Single crystals of lithium fluoride are used for X-ray monochromators and for the manufacture of the high-performance laser with the free colouring centres. Furthermore, in addition, the lithium fluoride crystals with colour centres are commonly used as the passive laser gates for the transfer of the total power generation in the monopulse one.

It should be mentioned that the refractive index n of the LiF materials can be changed by altering the wavelength as follows [12]: 1.77 (at 112.7 nm), 1.3978 (at 420 nm), 1.39 (at 620 nm), and 1.29 (at 6200 nm). The nanostructuration of the LiF surface makes it possible to change their optical and mechanical parameters, such as transparency, wetting angle and mechanical hardness.

In the present paper, the change in the Fresnel losses via transparency improving and the increase in the hardness and wetting angle are presented for the classical optical material based on the lithium fluoride (LiF). It should be noted that the use of the oriented laser-based deposition of carbon structures on the surface of the materials favorably differs from the CVD and PDV methods shown in the recently published works [13-15].

2. Methods and materials

Structural modification of the LiF surface was carried out with IR CO_2-laser with p-polarized irradiation at the wavelength of 10.6 μm and the power of 30 W. The laser system was connected to a vacuum hood containing the fixing unit samples and the device for placing substances deposited on the substrate. CNTs were placed at the materials interface under an additional electric field of 100-600 V×cm^{-1} in order to orient the nanotubes in the vertical position during the deposition process. Thus, the laser-oriented deposition (LOD) method was realised.

The spectra of the CNTs-treated materials were obtained using a Fourier FSM-1202 instrument as well as a VIS SP-26 spectrophotometer operated in the spectral range of 250-1200 nm. A POLAM-

P312 microscope was applied to make the image of the treated materials. Surface mechanical hardness (abrasive strength) was revealed using a CM-55 instrument, and the microhardness was measured via using the PMT-3M device produced by "LOMO" (Saint-Petersburg, Russia) with the ability to vary an indenter force as well. Moreover, the OCA 15EC device from LabTech Co. (Saint-Petersburg-Moscow, Russia) was applied to careful control the wetting angle change too. The modified surface analysis was carried out using a Solver Next atomic force microscope AFM (NT MDT Co., Zelenograd, Moscow Region, Russia).

To modify the LiF properties via their surface treatment, single-walled carbon nanotubes (SWCNTs), type #704121, with the diameter varied from 0.7-1.1 nm (Sigma-Aldrich Co., Steinheim, Germany) were used. The dimension of the nanotubes is important in order to combine their diameter directly with the unit cell of the model material. Moreover, Russian CNTs and nanofibers, type "Taunit-MD" (NanoTechCenter Ltd., Tambov, Russia) were applied as well.

All calculations were performed at the level of density functional theory as implemented in the VASP package [16,17] within the augmented plane-wave basis set [18]. Due to the large size of considered system the plane-wave cutoff energy was set to 250 eV, and the density of states was calculated only in the Γ-point. Several test calculations show that decreasing the cutoff energy from 400 to 250 eV leads to relatively small changes in the total energy of the system. Studied system consists of 200 lithium atoms, 200 fluorine atoms and 110 carbon atoms, which is the CNT together with 10 hydrogen atoms (passivation of the upper end).

3. Results, calculation and discussion

The spectral features of the modifications of the LiF surface are shown in figure 1 for the spectral range of 225-350 nm. It is important to note that the essential transmittance changes can be established in the UV range. It should be mentioned that the curve 1 (black line) indicates transmittance for the pure LiF (without modification), whereas curve 2 (red line) is related to the nanostructured one.

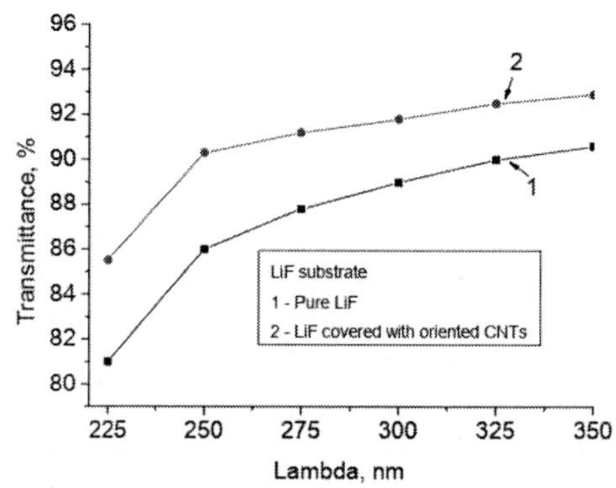

Figure 1. Transmittance of pure LiF surface (curve 1, black) and modified by CNTs (curve 2, red) in the UV-VIS spectral range.

It should be taken into account that the transparency and the reflective characteristics are coincided with each other. Thus, the decrease in the reflectance was found at the operated wavelength. One should note that the change in the reflectance is connected with the modified Fresnel losses. It was testified that the Fresnel losses (assigned to the relation between the refractive index of the CNTs and the matrix material) from the nanostructured surfaces can be decreased by using the laser oriented deposition technique as well as by the implantation of CNTs into the voids in the LiF surface layer. The fact of the possible strong binding of the CNTs with the LiF surface can be supported by the theoretical simulations.

In order to obtain more deep understanding of the influence of CNT deposited on LiF surface on its physical and chemical properties, the theoretical calculations within the density functional theory were performed. In figure 2 the studied atomic structure of the CNT/LiF interface is shown. The LiF (100) surface was considered, since it has termination containing two types of the atoms (Li, F) on the surface. Then, the CNT was deposited on the surface (figure 2, side and top views, respectively). Due to the large number of atoms in the considered unit cell of the substrate, the CNTs has the length of 1.2 nm, which is long enough to describe the effects related to the binding with LiF surface. The upper end of the tube was passivated by hydrogen atoms, since after the deposition of the CNT on the LiF surface, the main changes appear in the region close to the CNT/LiF interface.

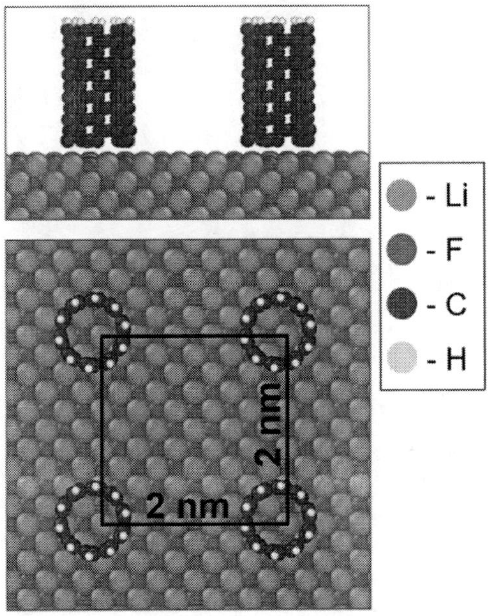

Figure 2. An atomic structure of the considered CNT/LiF interface.

The geometry relaxation of the interface structure shows that the deposition of the CNTs on the (100)-LiF surface does not lead to significant changes in the interface structure. The distance between the lower atoms of the CNTs and LiF surface was found to be 3 Å, which corresponds to the general van der Waals distance. Indeed, the calculations of the binding energy confirm it with the very small value of -0.66 meV/Å^2. Thus, to achieve the best binding between the CNTs and the substrate, it is necessary to deposit the former with some finite velocity to force them to penetrate into the latter.

Besides, the study of the electronic properties was performed, in particular, the electronic density of states (DOS) was calculated (figure 3). Only a tiny redistribution in the electronic density in the CNTs/LiF interface was found (see upper panel of figure 3). Moreover, the dependence of DOS on the energy also manifests about the weak interaction between the tube and substrate (see the bottom panel in figure 3). The deposition of the CNT leads to the shift of the main peak in the valence band towards the red region of visible spectra (~2 eV). The deposition of the CNTs also leads to the appearance of a certain number of energy levels close to the Fermi level, which can be seen from figure 3.

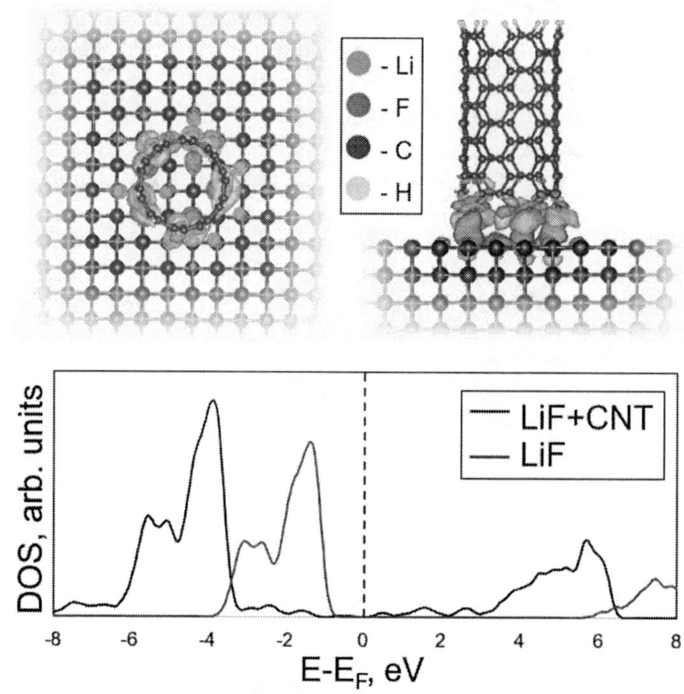

Figure 3. Distribution of the electronic density and a dependence of the electronic density of states on the energy for the CNT/LiF system.

This calculation (for example, the shift of the main peak in the valence band towards the red region of visible spectra) supports the transmittance and reflection spectral changes as well. According to the decreased Fresnel losses, one can expect for the change in the reflection parameters (figure 4).

Moreover, the wetting angle and the mechanical hardness can be successfully modified. The increase in the LiF material wetting angle after the LOD technique application was obtained. It was found that the wetting angle changes from 97.5-97.8° up to 107.3-107.5° after treating one side of the LiF substrate with the CNTs. It is worth noting that after processing on one side of the lithium fluoride sample, a Lotus effect, preventing water droplets from penetrating into the surface layers of the composite material, can be demonstrated. However, it remains possible to penetrate the water droplets from the other side of the sample, as well as from the ends. Therefore, the use of the laser oriented CNTs deposition in two-sided processing, as well as in view of the protection of the lateral surfaces of the sample, is more significant. It should be noticed that in order to study the structuration process influence on the physical and chemical characteristics of the LiF materials, the substrates with polished surfaces and the thickness of 5 mm are used.

The change in the mechanical characteristics is presented in table 1. Based on these data, one can testify that a 10-% increase in the microhardness can be found after treating the material surfaces with the CNTs through the laser oriented deposition technique. The data for the other optical materials are shown in table 1 for comparison. It should be remarked that increasing the mechanical strength of the LiF even by 10 % is very important from the point of view of using this crystal in passive mode generation when converting the total generation of laser radiation into a monopulse one.

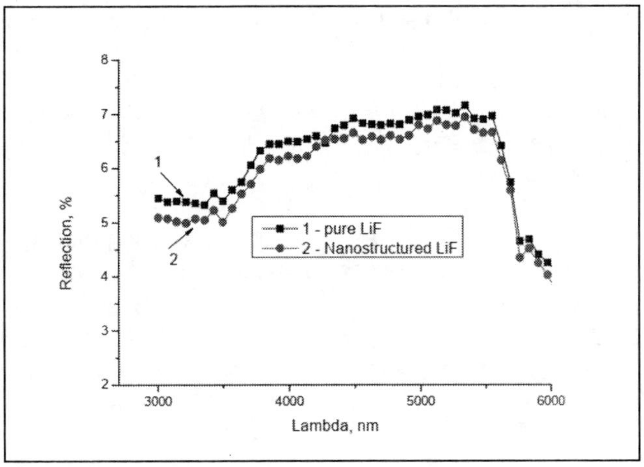

Figure 4. Reflection spectra spectra recorded for the LiF samples in the middle IR region.

Table 1. Comparative data on the microhardness change

Materials	Indentor, g	Microhardness increase, %
KCl	10	6
LiF	20	10
KBr	10	6-8
MgF_2	40	7

4. Conclusion

It was found that the laser oriented approach can be used with good advantage in order to deposit the CNTs in the vertical position on the surface of optical materials such as LiF. This approach allowis to arrange the covalent bonding between the carbon atoms and the near interface matrix material atoms. The experimental data were related to the theoretical explanation. The quantum chemical calculations showed the changes in the electronic properties (red shift of the valence band) of the LiF after the CNTs deposition, thereby supporting this bonding evidence. Moreover, the spectral and mechanical properties, as well as the wetting angle change, were found to be in good agreement with the modelling performed. Furthermore, there were no purpose to compare the LiF materials directly with the KBr, KCl and the other crystals because they have different applications. For example, potassium bromide is a key element of Fourier spectroscopy; potassium chloride can be used as a mirror when working with a CO_2-laser. Thus, the advantage and effective application of the laser-oriented processing for a large class of optical materials including lithium fluoride were demonstrated.

5. Acknowledgments

The obtained results were partially supported by the Project "Nanocoating-GOI" (2012-2015). The authors would like to thank their colleagues, especially, Dr. V. Studeonov and Dr. P. Vasilyev (Lab. for Photophysics of Media with Nanoobjects at the Vavilov State Optical Institute, St.-Petersburg, Russia) for the helpful discussion at the Lab seminars. Moreover, some part of the obtained results was discussed during the seminar at the Kurchatov Institute in Gatchina (Nuclear Research Institute, Gatchina, St.-Petersburg, Russia, October and November 2017) and at the Applied Optics Conference at the Vavilov State Optical Institute (December 2018). D.G. Kvashnin acknowledges the financial support of the RFBR under Research Project No.18-32-00682 mol_a.

References

[1] Born M and Wolf E 1999 *Principles of Optics: Electromagnetic Theory of Propagation, Interference and Diffraction of Light* (Cambridge: University Press) p 987

[2] Landsberg G S 2003 *Optics* [in Russian – Optika] (Moscow: Fizmatlit) p 848

[3] Feynman R, Leighton R, and Sands M 1977 *The Feynman Lectures on Physics. Mainly Mechanics, Radiation, and heat* (Boston: Addison-Wesley Publishing Company) p 515

[4] Akhmanov S A and Nikitin S Yu 1997 *Physical Optics* (Oxford: Clarendon Press) p 656

[5] Yariv A and Yeh P 2006 *Photonics: Optical Electronics in Modern Communications (Oxford Series in Electrical and Computer Engineering)* (New York: Oxford University Press) p 848

[6] Kamanina N V and Vasilyev P Ya, RU Patent No. 2 355 001 C2 (10 May 2009)

[7] Kamanina N V, Kuzhakov P V and Vasilyev P Ya, RU Patent No. 2013118962 A (27 October 2014)

[8] Robertson J 2004 Realistic applications of CNTs *Materialstoday* **7** 46 https://doi.org/10.1016/S1369-7021(04)00448-1

[9] Namilae S, Chandra N and Shet C 2004 Mechanical behaviour of functionalized nanotubes *Chem. Phys. Lett.* **387** 247 https://doi.org/10.1016/j.cplett.2004.01.104

[10] Taherpour A A, Aghagolnezhad-Gerdroudbari A and Rafiei S 2012 Theoretical and quantitative structural relationship studies of reorganization energies of [SWCNT(5,5)-Armchair-CnH_{20}] (n=20-310) nanostructures by neural network CFFBP method *Int. J. Electrochem. Sci.* **7** 2468 http://www.electrochemsci.org/papers/vol7/7032468.pdf

[11] Lisitsyna L A, Oleshko V I, Lisitsyn V M, Putintseva S N, Denisov G S 2006 The activator luminescence of LiF crystals *Higher Institutions News: Physics* [Izvestiya Vuzov: Fizika – in Russian] **49** 17-20

[12] Lide D R 2009 *CRC Handbook of Chemistry and Physics* (Boca Raton: CRC Press - Taylor & Francis) p 2828

[13] Yahya N 2011 *Carbon and Oxide Nanostructures: Synthesis, characteristics and applications* (Berlin Heidelberg: Springer) p 416

[14] Balasubramanyam N, Prasanthi Smt. G and Yugandhar M 2015 Study of coated TiN and TiC on cutting tools for the PVD and CVD coated tungsten carbide by sand blasting pretreatment of nickel and carbon *Int. J. Adv. Sci. Technol.* **75** 51 http://dx.doi.org/10.14257/ijast.2015.75.06

[15] Pahade V S, Chavan P S and Baisane V P 2016 A review paper on vapour deposition coating *Int. J. Eng. Appl. Sci.* **3** 75

[16] Kresse G, Furthmüller J. 1996 Efficient iterative schemes for ab initio total-energy calculations using a plane-wave basis set *Phys Rev B* **54** 11169 https://doi.org/10.1103/PhysRevB.54.11169

[17] Kresse G, Furthmüller J 1996 Efficiency of ab-initio total energy calculations for metals and semiconductors using a plane-wave basis set *Comput. Mater. Sci.* **6** 15 https://doi.org/10.1016/0927-0256(96)00008-0

[18] Blöchl P E 1994 Projector augmented-wave method *Phys. Rev. B* **50** 17953 https://doi.org/10.1103/PhysRevB.50.1795

IOP Conf. Series: Materials Science and Engineering **693** (2019) 012009 doi:10.1088/1757-899X/693/1/012009

Evaluation of the efficiency of lightweight concrete modified with additives based on nanostructures

R J Sldozian[1, 2], Z Mikhaleva[1] and A Tkachev[1]

[1]Department of Technology and Methods of Nanoproducts Manufacturing, *Tambov State Technical University,* 106 Sovetskaya Street, 392000 Tambov, Russian Federation
[2]Department of Applied Science, *University of Technology*, Iraq

[1]E-mail: rami_J_ag@yahoo.com

Abstract. The paper presents the results of studies on assessing the effectiveness of the modification of lightweight concrete with nanostructures. The results of these experimental studies show that the introduction of additives based on nanostructures, such as dispersions of carbon nanotubes and silica nanoparticles, improves the physical-mechanical characteristics of lightweight foam concrete (LWFC), even at low additive concentrations. As a result of chemical and physical interactions, nanosilica (NS) accelerates the pozzolanic reaction between cement and nanosilicate in the concrete mixture, filling the gaps. CNTs affect the hydration process and effectively increase the grain growth of calcium silicate hydrate (C-S-H). The presented experimental studies not only demonstrate the possibility of modifying construction materials with nanostructures in order to improve the physical and mechanical properties of the LWFC but also provide potential modification mechanisms that help in designing and manufacturing high-tech lightweight concrete of various compositions.

1. Introduction

Modern technologies for the production of construction materials are aimed at using nanosystems and providing for employing those materials, the basic properties of which are manifested at the nanoscale level. Firstly, this is due to the excellent physical-mechanical characteristics of nanomaterials. When developing composites for construction purposes, the following types of nanomaterials are actively studied and applied: nanosilica and carbon nanostructures such as graphene, fullerenes and carbon nanotubes (CNTs), *etc.* The use of CNTs leads to an increase in the physical-mechanical characteristics of construction composites, durability, moisture resistance and crack resistance, as well as a reduction in the consumption of expensive components, which makes this material attractive for the construction materials industry. At the present stage of nanotechnology development, the study of the effectiveness of CNTs-modified concretes is of particular scientific and practical importance. Modification with chemical additives, especially based on CNTs, improves the properties of lightweight foam concrete (LWFC), which usually consists of cement, sand, foaming agent and water [1]. The concrete body has a large number of pores with a diameter of 1-3 mm; therefore, its density usually ranges from 200 to 2000 kg/m^3.

A large number of research and scientific experience have shown that ultra-light foam concrete (ULFC) is not stable [2-3]. The instability inescapable of foam concrete often leads to a low mechanical characteristic such as compression and fracture strength, which limits the material applications. Various studies have estimated the compressive and fracture strength of foam concrete. Besides, more and more researchers in the field of materials science are trying to improve the physical-mechanical characteristics of construction materials, like compressive and tensile strength of ULFC using various fibers as modifiers. Moreover, the curing conditions not only significantly affect the mechanical properties of foam concrete, but also significantly affect its fracture energy. The properties of LWFC are significantly affected by the microstructural properties and are related to the distribution of bubbles and the development of hydrated products [4-7]. Foam concrete possesses excellent properties such as low density, thereby leading to a decrease in the structural dead load and operating costs. In addition, due to its textured surface and microstructure unit, it improves soundproofing and thermal conductivity, but the strength gradually decreases as the density decreases. However, in such a medium, at low density values (< 1000 kg/m^3), the above positive characteristics of the LWFC can be better manifested. Low-density LWFC elements are used in non structural applications and partitions in buildings in road construction and concrete floors, and for higher density it may be used in structural applications such as composite wall systems [8-10]. The flexural strength increases by the presence of the fibers of LWFC, but it also effectively reduces drying shrinkage [8,11].

Over the last years, investigators have shown that the addition of CNTs dispersions improves the compressive and fracture strength of concrete/foam concrete [12-15]. The nanoscale reinforcement potentially possesses the quality of the filler, contributing to the formation of a more dense material; besides, it makes it possible to slow down and prevent the development of cracks at the early stages of hardening, as well as to improve the quality of the phase boundary matrix aggregate as a whole. As a result, using CNTs can produce more durable and stiff concrete [16]. Concrete is a fragile material with a binder in the form of cement paste, which has a porous structure, including micro- and mesopores. The properties of the concrete mix and their changes mainly depend on the hydration of cement. The use of CNTs in cement composites, for the most part, is aimed at creating a reinforcing effect at the nanoscale level [17]. Some researchers have attempted to disperse CNTs using various methods: mechanical processing and surface modification. Surface modification methods using various acids can lead to destructive phenomena in concrete/foam concrete. The methods for dispersing CNTs employing surfactants can block the CNTs effect on the structure of foam concrete. Thus, the use of these conventional processes of dispersing CNTs in concrete/foam concrete is limited [18-20].

Considering the aforementioned, the aim of the present research was to study the efficiency of the modification of LWFC using various nanostructures.

2. Experimental part

In order to assess the effect of nanostructures on the physical-mechanical properties of LWFC, experimental studies were carried out with various nanomodifiers. The nano-SiO_2 (NS) was added to the cement composition in a dry format a ratio of 1, 1.5, 2, 2.5, and 3 % by weight of cement in order to determine the effect of the additive on the physical and mechanical properties. It was found that the optimal ratio is 2.5 %, while the pozzolan reaction between the cement and the NS is activated, and the nanosilica particles fill the gaps of C-S-H gel and contribute to a denser bond with C-S-H, which makes the matrix of the binder paste denser. Table 1 and figure 1 show the NS properties and XRD, respectively.

The following materials were used in the experiment with the nanomodifiers of lightweight concrete based on CNTs: Portland-type cement (M500, Eurocement, Belgorod, Russia), gradient sand as fine aggregate (collected in Tambov, Russia), as well as ordinary tap water (the water/cement ratio was 0.4); besides, a foaming agent acquired from MAXPEN (Voronezh, Russia) was used to obtain LWFC. Taunit-24 CNTs were manufactured at NanoTechCenter Ltd. (Tambov, Russia). Table 2 demonstrates their properties, and figure 2 presents their SEM images. Polyvinylpyrrolidone (PVP) was used as a surfactant to prepare the dispersion. Figure 3 shows the microsized agglomeration of the CNTs.

IOP Conf. Series: Materials Science and Engineering **693** (2019) 012009 doi:10.1088/1757-899X/693/1/012009

Table 1. Physical properties of the NS.

Property	Value
Particle Size	30-100 nm
Density	2.220 kg/m^3

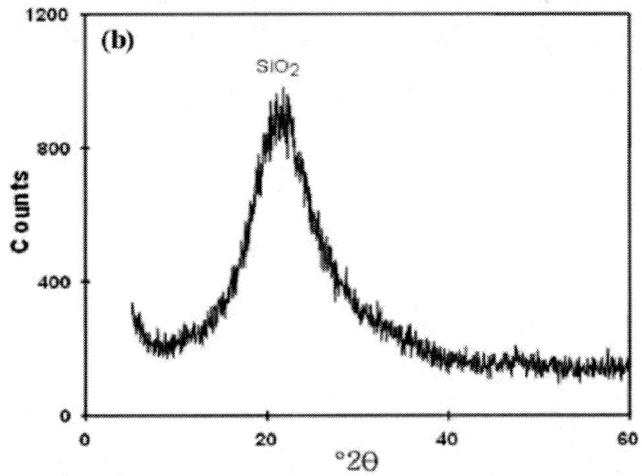

Figure 1. X-Ray diffraction pattern of the NS powder.

Table 2. Properties of the Taunit-24 CNTs.

Properties	values
External diameter (nm)	20-50
Inner diameter (nm)	10-20
Length (μm)	≥2
The total amount of impurities,% initial after cleaning	 ≤10 ≤1
Surface Area m^2/g	≥160
Density, g / cm^3	0.3-0.6

IOP Conf. Series: Materials Science and Engineering **693** (2019) 012009 doi:10.1088/1757-899X/693/1/012009

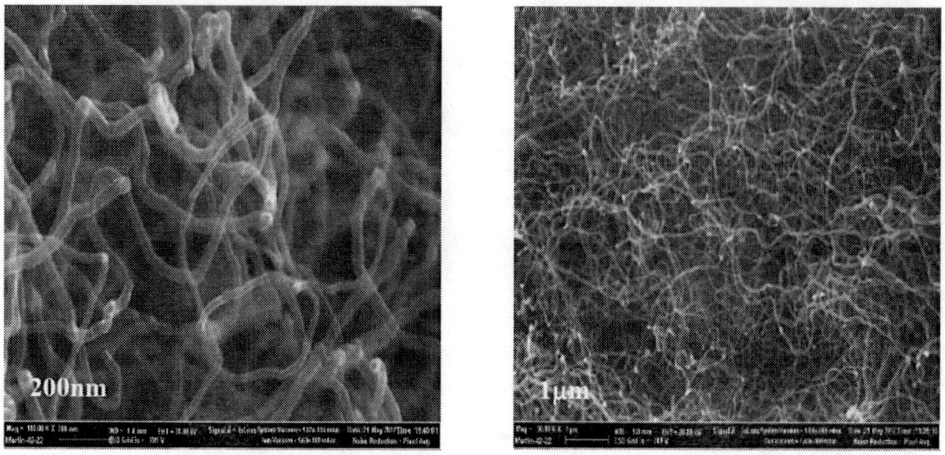

Figure 2. SEM images of the Taunit-24 CNTs [21].

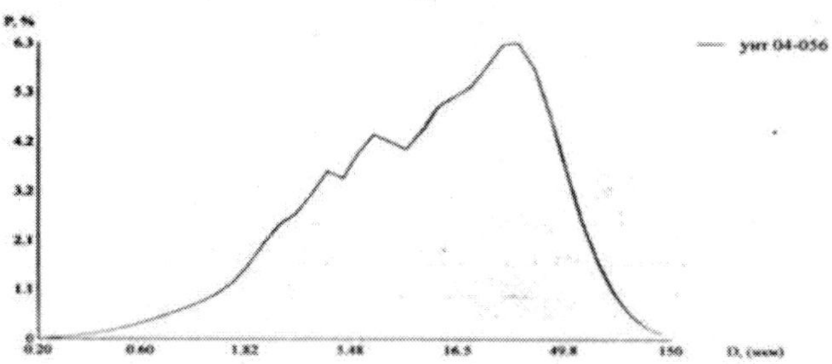

Table of Particle size correspondence (D.Micro) to specified weight fraction values.

CNTs	2.53	4.42	6.83	10.1	14.5	19.8	26.2	34.0	46.4	150
P%	10	20	30	40	50	60	70	80	90	100

Table of the weight fraction of particles (p.%) corresponding to the given values of particle sizes

CNTs	12.6	17.8	22.6	51.1	60.3	68.2	77.7	82.1	98.4	99.4
P%	3	4	5	15	20	25	32	36	80	100

The tables show the values of the weight fractions contained in the intervals of sizes smaller than the specified bottom.

Figure 3. Micro sized agglomerates of CNTs (Taunit-24).

The water-cement ratio was chosen depending on two parameters: sample density and compressive strength [22]. Polyvinylpyrrolidone was dissolved in (100 mL) of water and stirred with a magnetic stirrer for 10 min. Then the CNTs were added and mixed for 5 min by hand in a glass beaker. Ultrasonic treatment of the solution was carried out for 15 min at a frequency of 60 kHz [23]. Figure 4 shows the ultrasonic device and the CNTs dispersion after the sonication procedure.

IOP Conf. Series: Materials Science and Engineering **693** (2019) 012009 doi:10.1088/1757-899X/693/1/012009

Figure 4. Ultrasonic device and CNTs before mixing with water (a); CNTs after mix with water (b).

In accordance with the ratio of the components shown in table 3, the mixing process was carried out by mixing the cement with the sand, and then the mixing water was added to prepare the solution. The CNTs dispersion was added to the solution mixture. After that, the foaming agent was added to the mixture to obtain the LWFC.

Table 3. Mixture preparations.

Cement (g)	Sand (g)	Foam (g)	Mix No. With CNTs	**CNTs (%)	Mix No. With NS	**NS (%)
1200	1200	6	Mix 0	0	Mix 0	0
1200	1200	6	Mix 1_{CNTs}	0.0004	Mix 1_{NS}	1
1200	1200	6	Mix 2_{CNTs}	0.0006	Mix 2_{NS}	1.5
1200	1200	6	Mix 3_{CNTs}	0.0008	Mix 3_{NS}	2
1200	1200	6	Mix 4_{CNTs}	0.001	Mix 4_{NS}	2.5
1200	1200	6	Mix 5_{CNTs}	0.0012	Mix 5_{NS}	3

* W/C=0.4

** In reference to cement mass.

After the mixing, the concrete mixture was placed into a prism mold (40 x 40 x 160 mm). The samples were removed after 24 h of molding, and then were immersed into the water to cure at a temperature of 23°C±2°C according to State Standard of the Russian Federation No. 31108-2016. The average of three prisms was tested according to State Standard of the Russian Federation No.310.4-81) to determine the flexural strength and then the compressive strength at the age of 28 days. The testing machine was uniaxial (IP-M testing machine) with a capacity of 2000 kN, the applied loading rate was 0.4 MPa/s.

3. Results and discussion

The results of the experimental studies with the NS modifier showed an increase in the compressive and flexural strength of the LWFC. It was also found that the additives based on the NS and the CNTs slightly increase (by 3 %) the density of the concrete compared to the reference sample, and clearly increase the compressive and tensile strength. Figure 5 shows the distribution of the NS in the concrete matrix.

As a rule, resistance to the flexure under load, especially for the foam concrete which has many pores, leads to a low flexural strength value. The physical-mechanical characteristics of the concrete are enhanced by the addition of CNTs, since the strength of the C-S-H gel in the cement is enhanced by the presence of the CNTs. The reason for improving the mechanical properties of the concrete may be that the CNTs in the concrete form a structure that is resistant to cracking under load, especially in terms of flexural strength. Adding the nanomodifiers to concrete will reduce the consumption of cement, which will reduce the cost of lightweight concrete products. The increase in the compressive strength indicates that agglomeration of the nanotubes incement products is minimized due to the proper dispersion of the CNTs. The agglomeration is primarily opposed to the ultrasonic processing and magnetic stirring, if the nanotubes are not completely dispersed, the mechanical properties of concrete may not be improved. Figure 6 shows the distribution of the CNTs in the concrete body (matrix). The CNTs have a large specific surface area, which, in turn, provides interaction at the interface. The availability of the connecting cracks is one of the weakest links in the heterogeneous concrete mixture. The introduction of the CNTs fills this space and prevents the crack spreading. Figures 7 and 8 show the effect of NS and CNTs on mechanical properties.

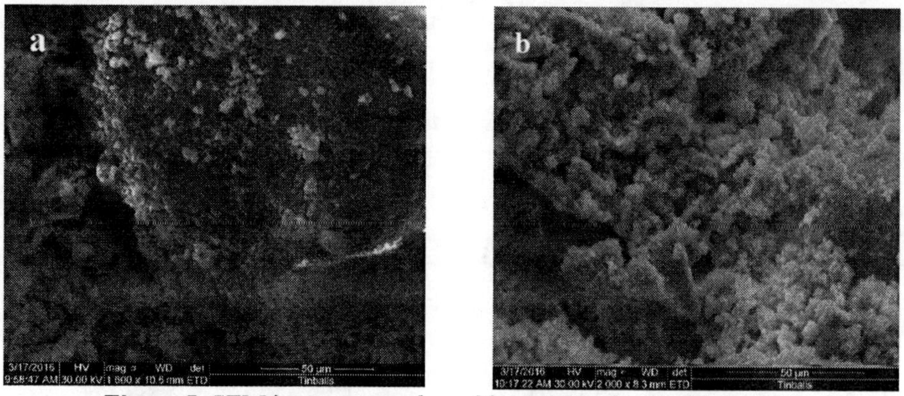

Figure 5. SEM images: samples without (a) and with (b) the NS.

Figure 6. SEM images: samples without (a) and with (b) the CNTs.

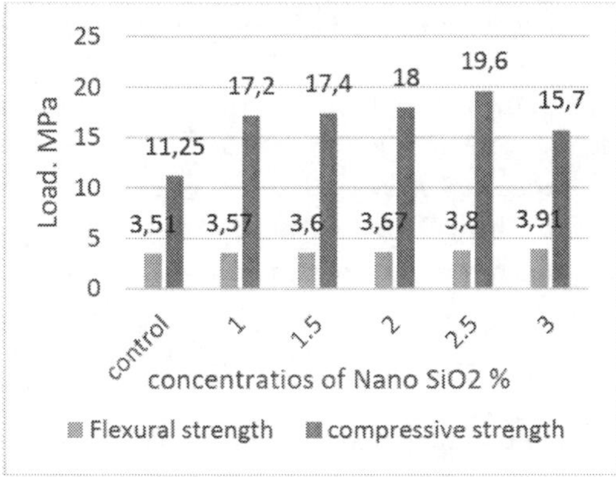

Figure 7. The NS concentration effecton theflexural and compressive strength.

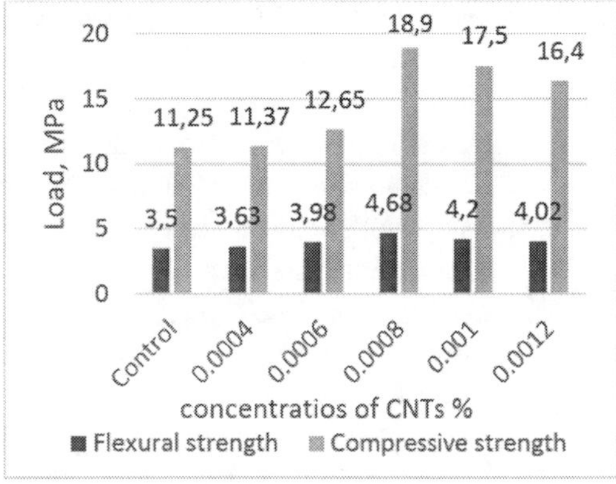

Figure 8. The CNTs concentration effecton the flexural and compressive strength.

The structure of concrete, as a rule, has pores, which are classified as internal pores in cement C-S-H, they are formed in the process of hydration and external pores. Testing of samples for water absorption was carried outaccording to State Standard of the Russian Federation No.31108-2016. Figure 9 presents the results of assessing the effect of the content of CNTs on water absorption. There is a decrease in water absorption of CNTs at (0.0008%) by weight of cement, due to the formation of a denser structure.

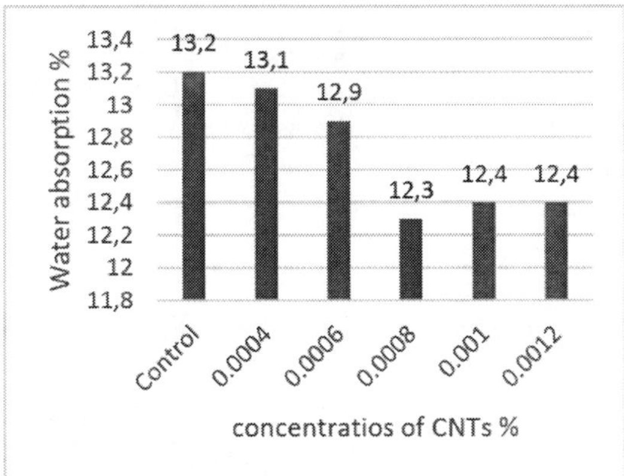

Figure 9. The CNTs effect on water absorption.

4. Conclusion

Nanomodifiers to the LWFC based on NS and CNTs can control and direct the process of hydration hardening, due to the formation of various structures affecting the packing density of the cement matrix. It was also found that the contents of CNTs (Taunit 24) and NS even in small doses in the ranges of (0.0004 - 0.0012 % and 1, 1.5, 2, 2.5 and 3%, respectively) can affect the microstructure, and, accordingly,the physical-mechanical properties of the LWFC, such as compressive strength (increase up to 68%) and flexural strength (increase up to 34 %), as well as water absorption (decrease by 8%). Increasing the strength characteristics of the lightweight concrete will reduce the consumption of cement.

References

[1] Zhang J and Liu X2018 Dispersion Performance of Carbon Nanotubes on Ultra-Light Foamed Concrete*Processes* **6** 194

[2] Falliano D, De Domenico D, Ricciardi G and Gugliandolo E 2018 Experimental investigation on the compressive strength of foamed concrete: effect of curing conditions, cement type, foaming agent and dry density *Constr. Build. Mater.* **165** 735

[3] She W, Du Y, Miao C, Liu J, Zhao G, Jiang J, and Zhang Y 2018 Application of organic-and nanoparticle-modified foams in foamed concrete: reinforcement and stabilization mechanisms *Cem. Concr. Res.* **106** 22

[4] Gunawan P and Setionob 2014 Foamed lightweight concrete tech using galvalumAz 150 fiber *Procedia Eng.* **95** 433

[5] Vesova L M 2016 Disperse Reinforcing Role in Producing Non-autoclaved Cellular Foam Concrete *Procedia Eng* **150** 1587

[6] Mahzabin M S, Hock L J, Hossain M Sand Kang L S 2018 The influence of addition of treated kenaf fibre in the production and properties of fibre reinforced foamed composite *Constr. Build. Mater.***178** 518

[7] Falliano D, De Domenico D, Ricciardi G, and Gugliandolo E 2019 Improving the flexural capacity of extrudable-foamed concrete with glass-fiber bi-directional grid reinforcement: An experimental study *Compos. Struct.* **209** 45

[8] Bing C, Zhen W, and Ning L 2011 Experimental research on properties of high-strength foamed concrete. *J. Mater. Civ. Eng.* **113** 118

[9] Falliano D, De Domenico D, Sciarrone A, Ricciardi G, Restuccia L, Tulliani J M C and Gugliandolo E 2019 Fracture behavior of lightweight-foamed concrete: the crucial role of curing conditions *Theor. Appl. Fract. Mech.* **103** 102297

[10] Sun H Y, Gong A M, Peng Y L, and Wang X 2011 The study of foamed concrete with polypropylene fiber and high volume fly ash *Appl. Mechan. Mater.* **90** 1039

[11] Falliano D, De Domenico D, Ricciardi G, and Gugliandolo E (2019) Compressive and flexural strength of fiber-reinforced foamed concrete: Effect of fiber content, curing conditions and dry density *Constr. Build. Mater.* **198** 479

[12] Mastali M, Kinnunen P, Isomoisio H, Karhu M and Illikainen M 2018 Mechanical and acoustic properties of fiber-reinforced alkali-activated slag foam concretes containing lightweight structural aggregates*Constr. Build. Mater.* **187** 371

[13] Eftekhari M, Ardakani S H, and Mohammadi S 2014 An XFEM multiscale approach for fracture analysis of carbon nanotube reinforced concrete *Theor. Appl. Fract. Mech.* **72** 64

[14] Zhang J and Liu X 2015 Dispersion performance of carbon nanotubes on ultra-light foamed concrete *Processes* **306** 78

[15] Luo J, Hou D, Li Q, Wu C, and Zhang C 2017 Comprehensive performances of carbon nanotube reinforced foam concrete with tetraethyl orthosilicate impregnation *Constr. Build. Mater.* **131** 512

[16] Pasynkov M V, Valimukhametova A R, Ariskina R A and Salakhov 2017 Research of the processes happening in ceramic masses when adding multilayered carbon nanotubes *Bulletin of Kazan Technological University* [in Russian] **20** 67

[17] Kwon Y H, Kang S H, Hong S G, and Moon J2017 Intensified pozzolanic reaction on kaolinite clay-based mortar *Appl. Sci.* 7 522

[18] Vaganov V, Popov M, Korjakins A and Šahmenko G 2017 Effect of CNT on microstructure and minearological composition of lightweight concrete with granulated foam glass *Procedia Eng.* **172** 1204

[19] Ferreira F V, Francisco W, Menezes B R, Brito F S, Coutinho A S, Cividanes L S, and Thim G P2016 Correlation of surface treatment, dispersion and mechanical properties of HDPE/CNT nanocomposites *Appl. Surf. Sci.* **389** 921
[20] Kim G M, Yang B J, Cho K J, Kim E M, Lee H K 2017 Influences of CNT dispersion and pore characteristics on the electrical performance of cementitious composites *Compos. Struct.* **164** 32
[21] Information available at http://www.nanotc.ru.
[22] Ali J Hamad 2014 Materials, production, properties and application of aerated lightweight concrete *Int. J. Mater.Sci. Eng.* **2** 152
[23] Tolchkov Y N, Mikhaleva Z A, Sldozian R D A, Memetov N R and Tkachev A G 2018 Effect of Surfactant Stabilizers on the Distribution of Carbon Nanotubes in Aqueous Media *J. Wuhan Univ. Tech.-Mater. Sci. Ed.* **33** 533

Fuzzy logical-linguistic model for assessing the qualitative composition of carbon nanomaterials

A Popov[1] **and D Polyakov**[2]

[1]Department of Technology and Methods of Nanoproducts Manufacturing, *Tambov State Technical University,* 106 Sovetskaya Street, 392000 Tambov Russian Federation
[2]Department of Information systems and Information Security Safety, *Tambov State Technical University,* 106 Sovetskaya Street, 392000 Tambov Russian Federation

[1]E-mail: olimp_popov@mail.ru

Abstract. The paper considers the development of methodological approaches to assessing the qualitative composition of carbon nanomaterials (nanotubes) and predicting their behavior in processes based on the theory of fuzzy sets. The state of the nanotechnology industry is analyzed, and the problematic moments of its development are highlighted. The necessity of proposing new approaches to evaluate nanomaterials due to the specificity of their properties is substantiated. The trends in the industrial use of carbon-based nanomaterials are identified, and the importance of their rational application based on an objective description of the qualitative composition of carbon nanomaterials for the development of the nanotechnology industry is shown. In this regard, it is proposed to use the mathematical apparatus of the theory of fuzzy sets, linguistic variables, and corresponding mathematical methods considering expert opinions and experimental results. A mathematical model created allows to predict the behavior of the nanotubes in various processes and solve the problem of the latters' optimization.

1. Introduction

The pace of formation of innovative economy and the active implementation of advanced scientific achievements in the manufacturing sector determine the level of the country's economic security and the quality of life of its citizens. The formation of the sixth technological structure necessitates structural changes in the national economy based on the use of research results in nanotechnology.

The intensive development of the scientific component in Russia's nanotechnology industry and the increasing volumes and ranges of nanostructured materials update the task of their large-scale use in the product manufacturing for final consumption. The solution for this task involves the acceleration of the transition from the laboratory searching of technologies for taking advantage of the unique properties of nanoscale objects to their universal applications in the key sectors of economy. In the development of nanoindustry, the following problematic points can be highlighted:

- insufficient theoretical basis for designing processes at the nanolevel, both in the production of materials and their use in industry; as a result, the predominantly empirical way of overcoming this issue is through getting new knowledge;
- it is difficult to obtain data on the properties of single nanoparticles and describing the cumulative properties of the entire system.

IOP Conf. Series: Materials Science and Engineering **693** (2019) 012010 doi:10.1088/1757-899X/693/1/012010

The formation of nanoindustry and the active use of nanomaterials to improve the quality of products require, among other things, developing methods and measurement tools, and searching for adequate options to assess the qualitative composition of such products.

Employing advanced diagnostic methods [1], standardization and certification of nanomaterials produced on an industrial scale will make it possible to control the quality of products thereof manufactured, and ensure national and international markets entry for nanoproducts [2,3]. The need for developing new approaches to the evaluation of nanomaterials is due to their special properties, considerable variation in the parameters of single components, and, accordingly, to the inability to completely apply the regulations of chemicals to these materials.

The significant gap between the quality of scientific research, scientific and technical groundwork and insufficient infrastructure of the country's nanoindustry and the incompliance with the time requirements on the availability and quality of work on standardization, metrology and assessment of the nanomaterials conformity pose a rather serious barrier to the innovative renewal of the country. It is necessary to bring the systemic classification of nanotechnology and products thereof based to a new level, and to develop mechanisms for assessing the quality of the nanomaterials obtained.

At present, the analysis of the state and development trends of nanotechnology objects makes it possible to conclude that synthesizing carbon nanomaterials (CNMs) – fullerene-like structures that represent a new allotrope of carbon in the form of closed, frame, macromolecular systems – is one of the most promising areas of nanotechnology [3-6]. Among these materials, carbon nanotubes (CNTs) occupy a special place, which, with a diameter of 1-50 nm and a length of several microns, form a novel class of quasi-one-dimensional nanoobjects. CNTs possess a number of unique properties due to the ordered structure of their nanofragments: good electrical conductivity and adsorption characteristics, the ability for cold electron emission and gas storage, diamagnetic characteristics, chemical and thermal stability, and high strength combined with high values of elastic deformation.

Materials developed on the basis of CNTs can be successfully used as structural modifiers of construction materials, hydrogen accumulators, radioelectronic elements, additives to lubricants, varnishes and paints, highly efficient adsorbents, gas distribution layers of fuel cells. The use of carbon nanostructures in fine chemical synthesis, biology and medicine is being widely discussed.

Being synthesized in reactors on a commercial scale, CNTs represent a multi-component mixture containing single-, two- and multi-walled formations of various lengths and diameters in a fairly wide range. For instance, the diameter of single-walled carbon CNTs ranges from 0.3 to 5 nm, whereas the diameter of two-walled CNTs varies from 1.8 to 7.1 nm. It is possible to make any CNTs type dominant by creating various technological synthesis modes in the reactors, but the synthesis procedure rarely yields highly homogeneous structures. They often represent mixtures of various nanocarbon formations possessing obviously different properties.

Developing multi-walled CNTs (Taunit series) with a predominantly conical shape of graphene layers, having an outer diameter of 10-60 nm, an inner diameter of 10-20 nm, and a length of more than 2 microns, is one of the promising areas of commercial CNTs production. These CNTs have shown a significant effect when used in various sectors of the economy (e.g., [3]). At the same time, the resulting product is quite heterogeneous, some of the parameters of its constituents – multi-walled CNTs – are inferior to those of single-walled ones; therefore, a certain problem lies in describing the nanoproduct properties by means of an integral characteristic.

Considering the aforementioned, the actual task is to assess the qualitative composition of CNTs based on the known characteristics: length, diameter, and number of layers. To construct the corresponding model, it is required to combine expert opinions presented in a natural language and experimental results with specific values of those characteristics.

Implementing the mathematical apparatus of the theory of fuzzy sets is one of the well-known and widely recommended approaches to the construction of the above-mentioned model [7,8]. For the expert opinion formalization within the framework of the chosen theory, linguistic variables are employed [7-9]. Taking into account the need for analyzing a significant amount of textual information, appropriate mathematical methods are actively used [10-15].

2. Materials and methods

Let us build a model for assessing the qualitative composition of carbon nanomaterials and predicting the result of their participation in further reactions using the theory of fuzzy sets. For this, several linguistic variables, described below, should be introduced.

Let $L = <l, Tl, R, Gl, Ml>$ be a linguistic variable that formalizes the nanotube length, where:

l = *"length of a tube"* is the name of the linguistic variable L;

Tl = *{"short", "medium", "long" (tube)}* is the term-set of L values;

Gl is the syntactic rule generating new L values;

Ml is the rule specifying the semantics for new, Gl-generated, statements.

Gl represents a procedure for creating new, semantically meaningful, statements using the syntactic connectives Svl = *{"and", "or"}* and the modifiers $Modl$ = *{"extremely", "very", "not"}*. The set of new statements generated by this syntactic rule will be denoted as $Gl(Tl)$.

Let us consider the use of the syntactic rule. If we take the arbitrary element $svl \in Svl$, and two arbitrary values t_1 and $t_2 \in Tl$, then the syntactic rule for constructing a new value will be as follows: $t_1\, svl\, t_2$. That is, for instance, let t_1 = *"short"*, t_2 = *"medium"* and svl = *"or"*, then $t_1\, svl\, t_2$ = *"short or medium (tube)"*. Now, we consider the arbitrary element $modl \in Modl$. The semantic rule for the arbitrary term $t \in Tl$ will be as follows: $modl\ t$. For instance, when t = *"short"* and $modl$ = *"not"*, the semantic rule builds = *"not short (tube)"*.

To construct statements, their consistent application to each other is admissible. For instance, the connective *"or"* with respect to t_2 is applied to the expression obtained in the previous example. Then, we will get the following statement: *"not short (n)or medium (tube)"*.

Ml is the semantic rule that assigns a membership function to each value of a fuzzy variable [10]. The membership function formalizes the meaning of this value; that is, for each $Y \in Gl(Tl)$, it defines: $\mu_Y: R \rightarrow [0;1]$.

Let us construct the membership functions μ_Y for Y = *"short"* and Y = *"long"*. For this, it is required to check 4 sizes with the expert: a_1, b_1, a_2, b_2, where a_1 (nm) is the size of a very short or even extremely short tube; b_1 (nm) is the minimum size of a tube, which under no circumstances can be called short; a_2 (nm) is the size of a very long or even extremely long tube; b_2 (nm) is the maximum size of a tube, which under no circumstances can be called long

Let $FL = \{f{:}R \rightarrow R \mid f(x) \downarrow \text{on } R, f(a_1) = 1, f(b_1) = 0\}$, and $FR = \{f{:}R \rightarrow R \mid f(x) \uparrow \text{on } R, f(a_2) = 1, f(b_2) = 0\}$. Then, the corresponding membership functions can be built with respect to the element FL or FR:

$$\mu_{short}(x) = \min\{1, \max\{0, f(x)\}\}\ \forall x \in R, \text{ where } f \in FL, \tag{1}$$

$$\mu_{long}(x) = \min\{1, \max\{0, f(x)\}\}\ \forall x \in R, \text{ where } f \in FR. \tag{2}$$

The choice of a specific function from a given wide class will be carried out at the final stage of model building. At the same time, it is easy to see that the membership function (1) will be equal to 1 for all values $x<a_1$, to 0 for all $x>b_1$, and continuously decrease between them. Similarly, the membership function (2) will be 0 for all values $x<b_2$, and 1 for all $x>a_2$, and continuously increase between them.

It is advisable to choose Archimedean T,S-norms and an operation of fuzzy negation as a semantic formalization of the syntactic connectives *"and"* and *"or"* and the modifier *"not"* [7]. On the one hand, such a choice is classical for the theory of fuzzy sets, and on the other, it provides a wide range of T,S-norms, thereby allowing to select a specific function type at the final stage, which will increase the adequacy of the model.

Thus, let t_1 and t_2 be some values of the linguistic variable with the membership functions μ_{t1} and μ_{t2}, respectively. Then, $\mu_{t_1 or t_2} = S(\mu_{t_1}, \mu_{t_2})$, $\mu_{t_1 and t_2} = T(\mu_{t_1}, \mu_{t_2})$, and $\mu_{not t_1} = n(\mu_{t_1})$, where $T(.,.)$ и $S(.,.)$ are the Archimedian T,S-norms, respectively, and $n(.)$ is the negation expressed as follows^

$$n(\mu(x)) = 1 - \mu(x), \forall x \in R. \tag{3}$$

It should be noted that number squaring is classically used to create the semantic rule for the modifier *"very"*. However, in our opinion, it is more expedient to take a more generalized form of this modifier. Thus, the functions *pow()* and *pow2()* will be chosen herein as a semantic rule formalizing the syntactic modifier *"very"* and its superlative form *"extremely"*:

$$pow(\mu(x)) = (\mu(x))^{\alpha}, \forall x \in R; \tag{4}$$

$$pow(\mu(x)) = (\mu(x))^{\alpha \cdot \beta}, \forall x \in R, \tag{5}$$

where α and $\beta \in \{r \in R \mid r > 1\}$.

Like in the case of the *FL* and *FR* function types, specific values of α and β will be determined at the final stage of model creation based on experimental data.

It should be noted that for complete definition of the linguistic variable *L*, setting the membership function to formalize the term *"medium"* is left. The tube length is referred to as medium if the tube is not short and not long, which, according to the accepted semantic rules, means:

$$\mu_{medium}(x) = T(n(\mu_{short}(x)), n(\mu_{long}(x))) \forall x \in R. \tag{6}$$

Let *D* =<*d, Td, R, Gd, Md*> be the linguistic variable for representing the tube diameter in the mathematical apparatus of fuzzy sets, where

d = "*diameter of a tube*" is the name of the linguistic variable *D*;

Td = {"*small*", "*medium*", "*large*"} is the term-set of *D* values;

Gd is the syntactic rule generating new *D* values and consisting of the syntactic connectives *Svd* = {*"and", "or"*} and the modifiers *Modd* = {"*extremely*", "*very*", "*not*"};

Md is the rule specifying the sense for *Gd*-generated statements. This rule is identical to *Ml* in the case of the linguistic variable *L*, and is given by formulas (1)-(6), with the only difference that formula (1) is related to the function describing the meaning "small", whereas formula (2) – to "large".

The main difference between the linguistic variable defining the number of tube layers (walls) is that the determination domain for the membership functions, specifying the semantic rules, lies in R, and due to the apparent discreteness of the number of layers, the determination domain for the corresponding functions of the considered variable will be *N*.

At the same time, it should be noted that for simplicity of treatment using processor devices, we will use continuous approximation of the membership functions formalizing the semantic rules. Let μ be a certain membership function that formalizes the arbitrary semantic rule of the linguistic variable describing the number of layers. Then, to set μ, $\mu: N \rightarrow [0; 1]$, we will use the approximation of μ_A, $\mu_A: N \rightarrow [0; 1]$ being such that $\forall i \in N \ \mu(i) = \mu_A(i)$. Due to the fact that the transformations made using formulas (1-6) for the continuous functions are identical to the corresponding transformations made for their values at each point of the determination domain, the approximation proposed herein will not affect the result of calculations. In other words, instead of saving a table of membership function values in integer arguments, we save a function μ_A with real arguments, which has the same values in integer numbers.

Let *S* =<*s, Ts, R, Gs, Ms*> be a linguistic variable to formalize the number of layers in the language of the mathematical apparatus of the theory of fuzzy sets, where:

s = "*number of tube layers*" is the name of the linguistic variable *S*;

Ts = {"*few*", "*medium*", "*many*"} is the term-set of *S* values;

R is the determination domain for the semantic rules.

Gs is the syntactic rule generating new *S* values and consisting of the syntactic connectives *Svs* = {"*and*", "*or*"} and the modifiers *Mods* = {"*extremely*", "*very*", "*not*"};

Ms is rule specifying the sense for *Gs*-generated statements. This rule is also given by formulas (1)-(6), with the only difference that formula (1) is related to the function describing the meaning "few", whereas formula (2) – to "many".

It is obvious that the constants, a_1, b_1, a_2, b_2 will be unique for each of the linguistic variables considered, since the very nature of the characteristics described by these variables is different. Then, for convenient sharing of these constants, we will denote them regarding each linguistic variable *L*, *D* and S as $a_1^L, b_1^L, a_2^L, b_2^L, a_1^D, b_1^D, a_2^D, b_2^D$ and $a_1^S, b_1^S, a_2^S, b_2^S$, respectively. On the other hand, the semantic rules that formalize the connectives and the modifiers, on the contrary, should be unified to create formulas, in which several different linguistic variables are present simultaneously.

Considering the expert opinions, we will make up a list of rules in a natural language using the linguistic variables values and the syntactic connectives between them from the set Sv = *{"and", "or"}*. As previously discussed, these connectives, are formalized through the *T,S*-norms.

We will develop a model simulating the behavior of a mixture of CNTs in subsequent chemical reactions based on the theory of fuzzy sets.

3. Results and discussion

Let us develop a fuzzy logical-linguistic model to assess the behavior of a CNTs mixture in subsequent chemical reactions based on the theory of fuzzy sets. The base of fuzzy production rules is one of the key components of this model [10].

The rules created by the experts will be of two types:

1) *"if V, then the reaction rate is acceptable"*, designated as $Y(V)$,
2) *"if V, then the reaction rate is not sufficient"*, designated as $X(V)$,

where V represents an expression composed of the values of the linguistic variables L, D и S connected with the modifier Sv.

For instance, let V = *"the tube length is short and the tube diameter is small or medium, and there are many tube layers (number of layers)"*. Then, $Y(V)$ = *"if the tube length is short and the tube diameter is small or medium, and there are many tube layers (number of layers), then the reaction rate is acceptable"*, and $X(V)$ = *" if the tube length is short and the tube diameter is small or medium, and there are many tube layers (number of layers), then the reaction rate is not sufficient"*. It should be noted that due to setting the semantic rules for each variable and formalizing the connectives "and" and "or" between the variables, we can put the expression V in accordance with the following membership function:

$$\mu_V(x,y,z) = T\Big(\mu^L_{short}(x), T\Big(S\Big(\mu^D_{small}(y), T\big(n(\mu^D_{short}(y)), n(\mu^D_{long}(y))\big)\Big), \mu^S_{many}(z)\Big)\Big),$$

where x, y, z are the length and the diameter of tubes, and the number of tube layers, respectively. The resulting membership function is multidimensional [5], since its determination domain is R^3.

In the same way, one can compose a membership function specifying the semantics of the arbitrary expression V created by the elements of term sets of linguistic variables, syntactic connectives and modifiers.

Let us suppose that according to the results of the expert survey, *n rules of the first type*, $Y(V_1)$, $Y(V_2)$, ..., $Y(V_n)$, and m rules of the second type, $X(V_{n+1})$, $X(V_{n+2})$, ..., $X(V_{n+m})$, were obtained, and the semantics of the expressions V_1, V_2, ..., V_{n+m} is formalized by the membership functions $\mu_1(x,y,z)$, $\mu_2(x,y,z)$, ..., $\mu_{n+m}(x,y,z)$, respectively.

Then, by virtue of the syntax and the semantics of the rules, the latter can be represented as two generalized rules:

1) *"if V_1 or V_2 or ...or V_n, then the reaction rate is acceptable"*;
2) *"if V_{n+1} or V_{n+2} or ...or V_{n+m}, then the reaction rate is not sufficient"*.

The membership function (μ_Y),evaluating the degree of truth of the expression *"V_1 or V_2 or ... or V_n"* is as follows:

$$\mu_Y(x,y,z) = S\big(\mu_1(x,y,z), \mu_2(x,y,z), \dots, \mu_n(x,y,z)\big), \tag{7}$$

By analogy, the degree of truth of the expression *"V_{n+1} or V_{n+2} or ... or V_{n+m}"* has the following form:

$$\mu_X(x,y,z) = S\big(\mu_{n+1}(x,y,z), \mu_{n+2}(x,y,z), \dots, \mu_{n+m}(x,y,z)\big). \tag{8}$$

It should be noted that in expressions (7) and (8), the form of the S-norm is rather exotic. Indeed, in these expressions, the S-norm is taken from a large number of arguments, although it usually represents a two-argument function. However, due to the fact that the Archimedean S-norm possesses the properties of associativity and commutativity [7], it is possible to generalize the S-norm as a function of several variables:

$$\underset{i=\overline{1,m}}{S}(x_i)=\begin{cases}S(x_1,x_2),m=2,\\ S(x_m,S(x_1,x_2,...,x_{m-1})),m>2.\end{cases} \quad (9)$$

It is worth mentioning that due to the semantics of generalized rules, the solution to the problem of predicting the reaction is carried out by calculating the values of expressions (7) and (8). Thus, in the ideal case, if $\mu_Y(x,y,z)=1$, and $\mu_X(x,y,z)=0$, then the reaction rate is acceptable. The reaction rate is not sufficient if $\mu_Y(x,y,z)=0$, and $\mu_X(x,y,z)=1$.

In practice, the values of $\mu_Y(x,y,z)$ and $\mu_X(x,y,z)$ will lie in the range from 0 to 1, and, according to one of the main classes of fuzzy implication [11], the decision on the acceptability of the reaction rate should be made on the basis of the following rule:

$$\underset{i=\overline{1,m}}{S}(x_i)=\begin{cases}\mu_Y(x,y,z)>\mu_X(x,y,z), reaction \quad rate \quad acceptable,\\ \mu_Y(x,y,z)\le\mu_X(x,y,z), reaction \quad rate \quad non-sufficient.\end{cases} \quad (10)$$

It remains to decide on some parameters, namely:

• select functions from *FL* and *FR* to build a semantic rule using formulas (1) and (2);

• determine the values of the real numbers α и β to formalize the modifiers *"very"* and *"extremely"*;

• choose the *T* and *S* norms.

The last choice merits a separate discussion. The fact is that there exist whole families of *T,S*-norms [13] given by a formula with the real parameter λ. Each λ value determined in the half-interval $[0,\infty)$ specifies its own pair of *T,S*-norms. For instance, let us consider the Hamacher *T,S*-norms [13] denoted as T_h and S_h, respectively:

$$T_h(x,y)=\begin{cases}\dfrac{xy}{\lambda+(1-\lambda)(x+y-xy)},\lambda\in[0,\infty),(\lambda,x,y)\ne(0,0,0),\\ 0,(\lambda,x,y)=(0,0,0),\\ T_D(x,y),\lambda=\infty;\end{cases} \quad (11)$$

$$S_h(x,y)=\begin{cases}\dfrac{x+y+(\lambda-2)xy}{1+(\lambda-1)xy},\lambda\in(0,\infty),(\lambda,x,y)\ne(0,1,1),\\ 1,(\lambda,x,y)=(0,1,1),\\ S_D(x,y),\lambda=\infty;\end{cases} \quad (12)$$

where

$$T_D(x,y)=\begin{cases}0,(x,y)\in[0,1)\times[0,1),\\ \min(x,y),x=1 \quad or \quad y=1\end{cases},$$

$$S_D(x,y)=\begin{cases}1,(x,y)\in[0,1)\times[0,1),\\ \max(x,y),x=1 \quad or \quad y=1\end{cases}.$$

The choice of the specific type of the parametric Archimedean T, S-norms (*e.g.*, expressions (11) and 12)) and the value of the parameter λ should be carried out to provide the best implementation of the prediction model (expressions (7)-(10)). To formalize such a choice, we will consider *A* — the set of the Archimedean *T,S*-norms.

Let the result of a particular reaction rate experiment be given by the tuple <*(x, y, z), k*>, where *x*, *y*, *z* is the length and the diameter of tubes, and the number of tube layers, respectively, and *k* is the result of the experiment: $k\in\{0,1\}$, if $k = 0$, then the reaction rate was insufficient for the given value of the characteristics (*x*, *y*, *z*); on the other hand, $k = 1$ means that the reaction proceeded at an acceptable rate. Then, the error (E) of the prediction model for the particular experiment will be calculated as follows:

$$E(\langle(x,y,z),k\rangle)=(|k-\mu_Y(x,y,z)|+|(1-k)-\mu_X(x,y,z)|)/2 \quad (13)$$

Let us suppose that q experiments were performed, then their results are represented by the set of the tuples $K\ \{<(x_1, y_1, z_1), k_1>, <(x_2, y_2, z_2), k_2>, ..., <(x_q, y_q, z_q), k_q>\}$. We want to find the parameters for the proposed model so that they minimize errors in all the experiments. In this case, it is convenient to minimize the maximum error. Let $R{:}K\to[0,1]$ be the model error function, then

$$R(K)=\max_{i=\overline{1,q}} E(\langle(x_i,y_i,z_i),k_i\rangle) \quad (14)$$

Since we have listed the model parameters that can be changed, and also have written down the total error, given by the model, in expression (14), it is advisable to set the following optimization problem for choosing the parameters:

$$R(K)\to\min,$$
$$f\in FL, g\in FR, \alpha,\beta\in R, \alpha,\beta>1, (T,S)\in A, \lambda\in[0,\infty) \quad (15)$$

The general form of the optimization problem (expression (15)) is unlikely to have an analytical or iterative solution. Indeed, the number of functions in *FL* and *FR* is infinite, as well as the values of α, β, λ. In practice, when solving the optimization problem (expression (15)), it is worth considering functions of a particular type only (*e.g.*, polynomial ones).

4. Conclusion

A fuzzy ligical-linguistic model was developed for assessing the quality of carbon materials, taking into account the variation of the key parameters (diameter, length, and number of layers), and conceptual approaches were proposed for predicting the behavior of CNTs when conducting a reaction based on the theory of fuzzy sets. The proposed model combines two approaches: it aggregates expert opinions through the formalization of rules created using linguistic variables and is configured by selecting a number of parameters considering the results of experiments.

References

[1] Burakova E A, Dyachkova T P, Rukhov A V, Tugolukov E N and Popov A I 2018 *Nanomaterials: Methods of Preparation and Diagnosis, and Application Areas* [in Russian] (Tambov: Izdatel'stvo "TGTU") p 80

[2] Romanenko A V and Popov A I 2012 Features of building a costly model of quality management in nanoindustry. *Bulletin of Michurinsk State Agrarian University* [Vestnik Michurinskogo Gosudarstvennogo Agrarnogo Universiteta – in Russian] 136

[3] Mishchenko S V and Tkachev A G 2008 *Carbon Nanomaterials. Production, Properties, Applications* [in Russian] (Moscow: Mashinostroenie) p 320

[4] Zhang B-T, Zheng X, Li H-F and Lin J-M 2013 Application of carbon-based nanomaterials in sample preparation: A review. *Anal. Chim. Acta* **784** 1 DOI: 10.1016/j.aca.2013.03.054

[5] Choudhary N, Hwang S and Choi W 2014 Carbon Nanomaterials: A Review. *Handbook of Nanomaterials Properties*, eds B Bhushan, D Luo, S Schricker, W Sigmund, S Zauscher (Berlin: Springer) pp 709–769 DOI:10.1007/978-3-642-31107-9_37

[6] Xu J, Cao Z, Zhang Y and Yuan Z 2018 A review of functionalized carbon nanotubes and grapheme for heavy metals adsorption from water: Preparation, application and mechanism. *Chemosphere* **195** 351 DOI: 10.1016/j.chemosphere.2017.12.061

[7] Yager R R 1986 *Fuzzy Set and Possibility Theory: Recent Developments* (New York: Pergamon) p 633 DOI: 10.1002/zamm.19840640817

[8] Yarushkina N G 2003 Fuzzy and hybrid systems: A review of results and trends. *Artificial Intelligence News* [Novosti Iskustvennogo Intellekta – in Russian] **6** 5

[9] Zadeh L A 1973 *The Concept of a Linguistic Variable and Its Approximation to Making Approximate Decisions* [in Russian] (Moscow: Mir) p 167

[10] Batyrshin I Z 2001 *Basic Operations of Fuzzy Logic* [in Russian] (Kazan': Otechestvo) p 100

[11] Zadeh L A 1975 The concept of a linguistic variable and its application to approximate reasoning – I. *Inform. Sci.* **8(3)** 199 DOI: 10.1016/0020-0255(75)90036-5
[12] Zadeh L A 1975 The concept of a linguistic variable and its application to approximate reasoning – II. *Inform. Sci.* **8(4)** 301 DOI: 10.1016/0020-0255(75)90046-8
[13] Zadeh L A 1975 The concept of a linguistic variable and its application to approximate reasoning – III. *Inform. Sci.* **9(1)** 43 DOI: 10.1016/0020-0255(75)90017-1
[14] Batyrshin I Z 1990 On fuzzinesstic measures of entropy on Kleene algebras. *Fuzzy Set. Syst.* **34(1)** 47 DOI: 10.1016/0165-0114(90)90126-Q
[15] Jan R-H, Hwang F-J and Chen S-T 1993 Topological optimization of a communication network subject to a reliability constraint. *IEEE T. Reliab.* **42** 63 DOI: 10.1109/24.210272

Polymer composites based on functionalized carbon nanotubes

T P Dyachkova[1,*], D V Tarov[1], E A Burakova[1], E N Tugolukov[1], A N Blokhin[1], E V Galunin[1], L V Rosenblum[2] and D E Kobzev[1]

[1]*Tambov State Technical University,* 106 Sovetskaya Street, 392000 Tambov, Russian Federation

[2]*Derzhavin Tambov State University,* 33 Internatsionalnaya Street, 392000 Tambov, Russian Federation

[*]E-mail: dyachkova_tp@mail.ru

Abstract. The present paper considers the efficiency of using carbon nanotubes (CNTs) functionalized by titanium stearate groups for modifying of polyethylene and polypropylene. For this purpose, the strength characteristics, Shore hardness, thermal diffusivity and conductivity of the obtained nanocomposites were determined. The weak effect of the CNTs (0.01-1.5 wt.%) on the values of thermal conductivity coefficients and thermal diffusivity was revealed. It was found that compared to the unmodified matrix, nanocomposites containing 0.01-0.1 wt.% of the titanium stearate-functionalized CNTs demonstrate, an increase in the Shore hardness by 12.5-14.5%, a decrease in the weight wear by 54.2%, and the daily water absorption - to the values below the threshold of sensitivity of the standardized method of determination. The functionalization by the titanium stearate groups made it possible to reduce the CNTs mass fraction in the composition of polyethylene and polypropylene composites by more than 10 times compared with the pristine nanotubes.

1. Introduction

Covalent and non-covalent functionalization of carbon nanotubes (CNTs) is one of the generally accepted methods for solving the problem of their distribution in polymers and solvents [1]. In particular, CNTs with oxygen- [2-3] and nitrogen-containing [4] groups are successfully used for modifying epoxy binders. Functionalized CNTs possessing the affinity for non-polar matrices are suitable for employing in composites based on polyethylene (PE) polypropylene (PP). In particular, alkylated [5], fluorinated [6] and silanized [7] CNTs belong to efficient PE and PP modifiers. However, the implementation of the CNTs functionalization methods presented is technologically complex due to a large number of stages and the use of expensive and/or toxic reagents. In this paper, it is proposed to use a more environmentally friendly and economical way of functionalizing CNTs with titanium stearate groups.

The purpose of this work was to study the effectiveness of using functionalized CNTs for the PE and PP modification. In this regard, the strength characteristics, Shore hardness, thermal stability, thermal conductivity and thermal diffusivity of composites with technical carbon, pristine and functionalized nanotubes were determined at various modifier concentrations.

2. Materials and methods

"Taunit-M" CNTs (d = 8 ÷ 15 nm, length ≥ 2 μm, and specific surface area 300-320 m^2/g) produced by "NanoTechCenter" Ltd. (Tambov, Russia) were used herein. For their functionalization, the nanotubes were first treated with concentrated nitric acid at 110 °C for 5 h, and then washed on the filter with distilled water to neutral pH. A 40-wt. % aqueous paste of CNTs, carboxylated with 1.0 mmol/g COOH-groups (according to titrimetry data), was obtained. 1.5 g of triethanolamine and 200 mL of distilled water were added to this paste, after which the mixture was mechanically stirred to form a homogeneous suspension, treated with ultrasound for 0.5 h and cooled. Then, with stirring, 100 mL of a solution containing 10 mmol of triethanolamine titanate (TEAT) and 150 mL of a solution containing 20 mmol of triethanolamine stearate (STEA) were added. The resulting mixture was stirred for 2 h. At the same time, carbon dioxide was passed into it at a rate of 1 L/min. The following chemical reactions took place:

$$Ti\{N(CH_2CH_2O-)_2CH_2CH_2OH\}_2 + 2(HOCH_2CH_2)_3NH(C_{17}H_{35}COO) + 3CO_2 + 3H_2O \rightarrow$$
$$\rightarrow \{N(CH_2CH_2O-)_2CH_2CH_2OH\}Ti(C_{17}H_{35}COO)_2 + 3(HOCH_2CH_2)_3NH(HCO_3); \quad (1)$$
$$2-COOH + \{N(CH_2CH_2O-)_2CH_2CH_2OH\}Ti(C_{17}H_{35}COO)_2 + H_2O + CO_2 \rightarrow$$
$$\rightarrow (-COO)_2Ti(C_{17}H_{35}COO)_2 + (HOCH_2CH_2)_3NH(HCO_3). \quad (2)$$

The second reaction represents the interaction between the carboxyl groups on the CNTs surface and diethanolamine titanate distearate formed as a result of the TEAT-STEA interaction.

At the end of the process, the precipitate was washed on the filter with distilled water and dried in a vacuum oven at 80 °C until the weight loss stopped. As a result, the functionalized CNTs were obtained, with titanium distearate mass fraction of about 26 %. The transmission electron microscopy (TEM) images of the pristine and functionalized CNTs are shown in figures 1a and 1b, respectively. A modifying layer up to 2 nm thick can be identified on the CNTs surface.

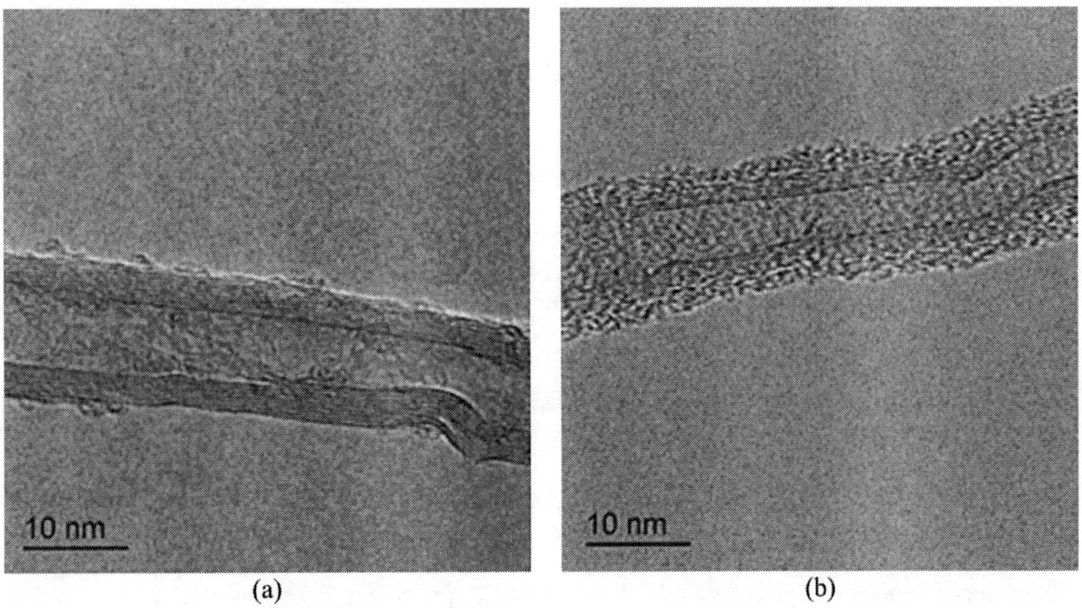

(a) (b)

Figure 1. TEM images of the pristine (a) and titanium stearate-functionalized (b) CNTs.

A typical IR spectrum (figure 2) recorded for the functionalized CNTs contains peaks inherent in the vibrations of the C–H bonds in alkyl (at 2952; 2914; 2848; 1463, and 1429 cm^{-1}). A series of narrow bands at 1100–1300 cm^{-1} is due to the transposition of the methylene groups of stearic acid. The intense peak at 1697 $cm^{-1}$is assigned to the vibrations of the C=O bonds in carboxyl. According to [8], the peaks at 1415; 939 and 803 cm^{-1} characterize the vibrations of the Ti–O bonds.

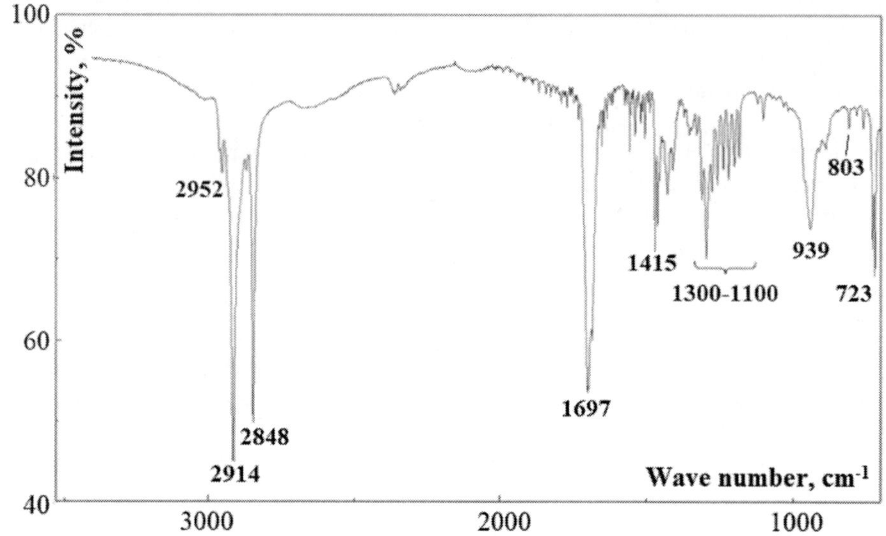

Figure 2. IR-spectra of the titanium stearate-functionalized CNTs.

Heat-resistant PE (PE-RT SP 988, LG Chem, South Korea) and PP (Polykom LLC, Russia) were subjected to modification. The concentration of the filler (pristine CNTs, functionalized CNTs, and carbon black - CB) was 0.01-1.5 wt. %. The modifying additives were mixed with the granular polymer (PE, PP) in melt in a Brabender rotary mixer. Samples with the required geometric parameters were obtained by injection molding.

The studies on the hardness of composites according to Shore (scale D) were carried out on a 5069 UHL 4.2 IT model device (self-made). Strength characteristics were evaluated on a DBBMTCL-500 kg polymer testing machine (Testometric, UK). The measurement error on this instrument was ±5 %.

The study on the tribological characteristics was performed on a friction machine (Korvet-41, Russia) by determining the weight wear. Cylindrical samples with a diameter of 35 mm and a height of 6.5 mm were fixed with a screw in the center of the friction machine table. The abrasive mesh P150 (75-106 μm) was used as a counterbody, the diameter of its working area was 120 mm. The weight wear was determined by changing the sample mass before and after testing. The time of the experiment was 20 min, the load on the sample was set with a force of 0.5 kg with a shoulder of 200 mm from the axis of the mechanism for lowering the cartridge, and was controlled by spring weights throughout the experiment. The counter body rotation frequency was 12.5 rpm. The measurement error on this instrument was ± 5 %.

The daily water absorption of composites was evaluated by increasing the mass of the standard composite sample after exposure to distilled water for 1 day. The measurement error on this instrument was ±2 %

The studies on the thermal conductivity and thermal diffusivity were carried out using the information-measuring system of thermophysical properties of solid materials (self-made), which was previously calibrated on the reference materials (polymethylmethacrylate). The cylindrical sample with a diameter of 35 mm was placed on a laboratory table, a measuring probe was placed on top. After 10 min of temperature control, the thermal conductivity was measured. The time between experiments was 15 min. The measurement error on this instrument was ±5 %

The thermal stability of the composites was estimated on the basis of the analysis of the TG/DSC curves obtained on the STA 449 F3 Jupiter synchronous thermal analysis device (Netzsch, Germany). During the measurements, the furnace was filled with air at a flow rate of 30 mL/min. The temperature

IOP Conf. Series: Materials Science and Engineering **693** (2019) 012011 doi:10.1088/1757-899X/693/1/012011

program included holding at 30 °C for 10 min, heating from 30 to 900 °C at a rate of 10 °C/min, and cooling the samples to room temperature at a rate of 15 °C/min.

3. Results and discussion

The results of mechanical tests of the PE composites modified with various types of carbon nanomaterials (CB, pristine and functionalized CNTs) are presented in table 1.

The introduction of both carbon black and CNTs, as a rule, contributes to the improvement of the studied physical and mechanical properties of polyethylene. An increase in Shore hardness can reach 3.5 and 7 units in the presence of CB and CNTs, respectively. However, this indicator is observed at 1.5% concentration of the initial CNTs in the composite. To achieve the same effect, only 0.1 % of the functionalized CNTs is required, i.e. 15 times less. The maximum hardness jump is observed at the lowest content of the modifier, and then it stabilizes. The weight wear resistance also increases markedly in the presence of even 0.01-% carbon materials. And the most effective ones are the functionalized CNTs with a minimum concentration.

Table 1. Physical and mechanical properties of the PE composites.

Modifier	Mass fraction (%)	Shore hardness	Weight wear (g)	Maximum load (N / mm^2)	
				under cutting	under bending
-	-	56	0.0409	17.1	37.9
CB	0.01	58.0	0.0356	17.2	38.9
	0.10	57.5	0.0321	16.6	41.5
	0.50	57.5	0.0289	17.1	37.1
	1.00	59.5	0.0336	18.2	42.5
	1.50	59.5	0.0347	17.1	42.2
Pristine CNTs	0.01	60.0	0.0293	17.2	38.4
	0.10	61.0	0.0261	17.2	37.7
	0.50	61.0	0.0241	17.5	38.0
	1.00	62.0	0.0238	17.5	37.5
	1.50	63.0	0.0248	17.5	38.0
Functionalized CNTs	0.01	62.5	0.0187	18.5	38.9
	0.10	63.0	0.0191	17.3	41.5
	0.50	62.5	0.0219	17.5	37.1
	1.00	62.0	0.0227	16.5	42.5
	1.50	62.0	0.0223	17.5	42.2

The parameters of the maximum PE load for bending and cutting under the action of carbon modifiers do not change so unequivocally. At some concentrations of CB, the initial and functionalized CNT, these properties may deteriorate. At best, it is possible to achieve an increase in the maximum stress in a cutting condition of 8.2% in the presence of 0.01 wt. % functionalized CNTs, in a bending condition of 12.1 % in the presence of 1 % CB or functionalized CNTs.

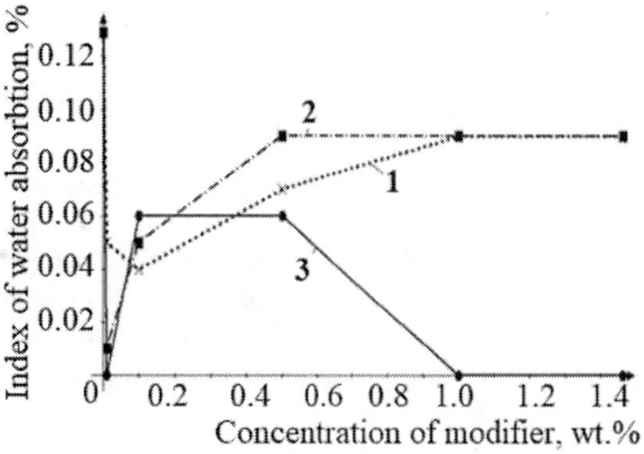

Figure 3. The effect of the concentration of the CB (1), pristine (2), and functionalized (3) CNTs on the daily water absorption of the PE composites.

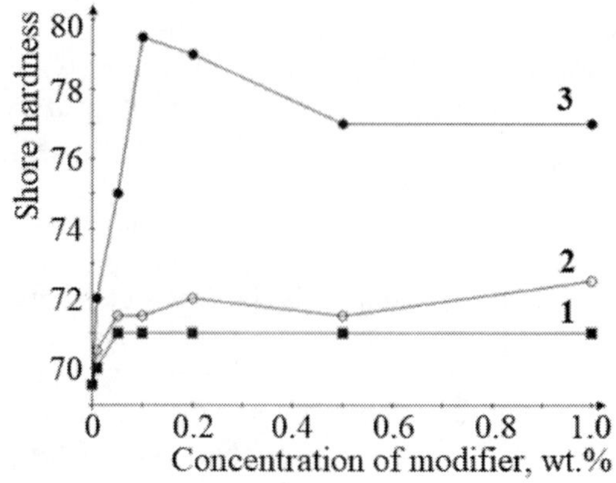

Figure 4. The effect of adding the CB (1), pristine (2), and functionalized (3) CNTs on the Shore hardness of the PP.

The dependence of daily water absorption of the polyethylene composite on the concentration of the carbon modifier is shown in figure 3. The introduction of CB and CNTs contributes to the hydrophobization of the material. Moreover, the dependence of the index of water absorption of the composite on the concentration of the filler may be different. In the presence of the CB, it passes through a minimum at 0.1 wt.%. When the non-functionalized CNTs are added, the minimum water absorption of the composite is observed at 0.1 wt. %. The dependence of this parameter on the concentration of functionalized CNTs is variable. The first minimum of water absorption (0%) is observed at 0.01% modifier content, then, there is a slight increase to 0.06%, and at 1 and 1.5 % content of functionalized CNTs, a zero indicator is again observed.

A number of positive effects of using CNTs in the composition of PP-based composites are noted. At the functionalized CNTs content of 0.1 %, the Shore hardness increases up to 10 units (from 69.5 to 79.5). The use of the initial CNTs and CB at the same and higher concentrations does not allow achieving the same indicators (figure 4). This indicates a high affinity of the functionalized CNT to the PP matrix and good dispersibility within it.

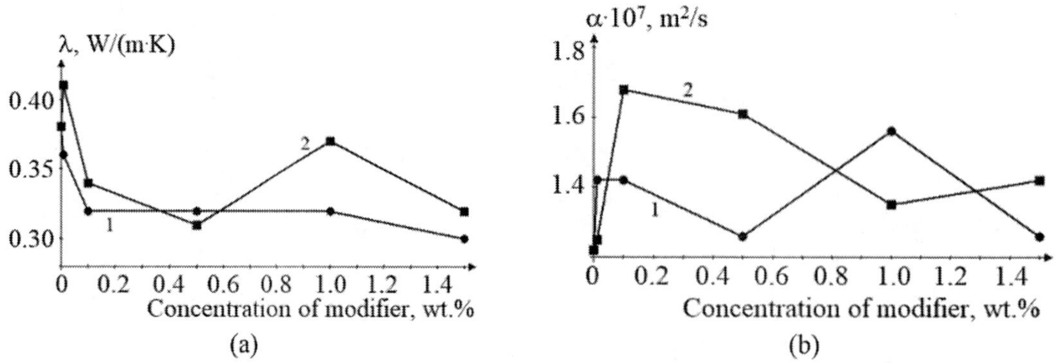

Figure 5. Changes in the coefficients of thermal conductivity λ (a) and thermal diffusivity α (b) of the PE in the presence of the CB (1) and functionalized CNTs (2).

According to the data presented in figure 5, the thermophysical characteristics of the PE when modifying with both the technical carbon and functionalized nanotubes change slightly. Instead of the expected increase in the coefficient of thermal conductivity (λ) in most cases, with the introduction of

CB or CNTs, there is a slight decrease (figure 5a). The thermal diffusivity (α) of the material increases by about 34% in the presence of 1 % functionalized nanotubes (figure 5b).

Analyzing the TG / DSC curves of the composites "PE-functionalized CNTs", it can be observed that their thermal stability is high. The material is melted at 140-150°C. This significantly exceeds the temperature range of PE-RT SP 988 PE products.

Thus, the carbon materials contribute to the improvement of a number of physical-mechanical parameters of the PE and PP. However, the performance parameters depend on the type of the carbon modifier being used, its concentration and chemical pretreatment (functionalization).

As compared to the CB from the standpoint of improved mechanical properties, the higher efficiency of the CNTs is due to their structure and unique properties. The functionalized CNTs have a higher chemical affinity to the modifiable matrix, so that to achieve the required performance of the composites, consumption is required, an order of magnitude and more reduced compared to the initial CNTs [9]. It should be noted that in earlier published works [10], in order to increase hardness and resistance to wear, up to 5-10 wt. % of carbon nanotubes were introduced into polyethylene-based composites.

Another positive point in the application of the functionalized CNTs in the composition of polyethylene and polypropylene composites is the possibility of increasing resistance to moisture. Already with a minimum content of carbon nanotubes with titanium stearate groups (0.01 wt.%). The daily moisture absorption can be reduced to zero. This may be due to both more uniform dispersion of the functionalized CNTs in the polymer matrix, and their higher hydrophobic properties. For comparison, in [11], in order to make the polyethylene resistant to moisture, 1.5 wt.% of the CNTs is introduced into it.

The weak influence of the carbon modifiers used in the work on the thermophysical characteristics of the composites can be explained by two reasons. First, this is a high agglomeration of CB and the original CNTs. In the case of the use of functionalized CNTs, in addition to this reason, an enhanced interaction with the matrix can have an effect, as a result of which the CNTs no longer come into contact with each other, being separated by layers of a less conductive polymer. However, given that too high thermal conductivity of the material may contribute to heat loss to the environment when using polyethylene, for example, for the manufacture of hot water pipes, low thermal conductivity is rather an advantage than a disadvantage.

Summary

The functionalization of the CNTs with titanium stearate groups contributes to their adaptation to a PE-based polymer matrix and effective use in composites with increased hardness, wear resistance and moisture. These effects are quite pronounced at the functionalized CNTs concentrations equal to 0.01-0.1 wt.%. Similar results with the initial (non-functionalized) CNTs can be achieved at concentrations that are an order of magnitude higher than those indicated herein.

Thus, the use of the functionalized CNTs to modify PE and PP is appropriate and contributes to the improvement of the performance characteristics of products on their basis.

Acknowledgments

The research was supported by the Russian Foundation for Basic Research (Grant No. 18-29-19150 mk).

References

[1] Punetha V D, Rana S, Yoo H J, Chaurasia A, McLeskey J T, Ramasamy M S, Sahoo N G and Cho J W 2017 Functionalization of carbon nanomaterials for advanced polymer nanocomposites: a comparison study between CNT and graphene *Progr. Polym.Sci.* **67** 1 DOI: 10.1016/j.progpolymsci.2016.12.010.

[2] Gantayat S, Rout D and Swain S K 2017 Mechanical properties of functionalized multiwalled carbon nanotube/epoxy nanocomposites *Mater. Today Proc.***4** **(2)** 4061 DOI: 10.1016/j.matpr.2017.02.308.

[3] Cha J, Jin S, Shim J H, Park C S, Ryu H J and Hong S H 2016 Functionalization of carbon nanotubes for fabrication of CNT/epoxy nanocomposites *Mater. Des.* **95** 1 DOI: 10.1016/j.matdes.2016.01.077.

[4] Neves J C, Castro V G, Assis A L, Veiga A G, Rocco M L and Silva G G 2018 In situ determination of amine/epoxy and carboxylic/epoxy exotermic heat reaction on surface of modified carbon nanotubes and stuctural verification of covalent bond formation *Appl. Surf. Sci.* **436** 495 DOI: 10.1016/j.apsusc.2017.12.031.

[5] Ferreira F V, Franceschi W, Meneces B R, Brito F S, Lozano K, Countinho A R, Cividanes L S and Thim G P 2017 Dodecylamine functionalization of carbon nanotube to improve dispersion, thermal and mechanical properties of polyetylene based nanocomposites *Appl. Surf. Sci.* **410** 266. DOI: 10.1016/j.apsusc.2017.05.098.

[6] Kharitonov A P, Maksimkin A V, Mostovaya K S, Kaloshkin S D, Gorshenkov M V, D'yachkova T P, Tkachev A G and Alekseiko L N 2015 Reinforcement of bulk ultrahigh molecular weight polyethylene by fluorinated carbon nanotubes insertion followed by hot pressing and orientation stretching *Compos. Sci. Technol.* **12** 26 DOI: 10.1016/j.comscitech.2015.10.009.

[7] Kuan C F, Kuan H C, Ma C C, Chen C H and Wu H L 2007 The preparation of carbon nanotube/linear low density polyethylene composite by a water-crosslinking reaction *Mater. Lett.* **61** 2744. DOI: 10.1016/j.matlet.2006.01.112.

[8] Egerton T A, Everall N J, Mattinson J A, Kessell L M and Tooley I R 2008 Interaction of TiO_2 nano-particles with organic UV absorbers *J.Photochem. Photobiol.A Chemistry.* **193(1)** 10 DOI: 10.1016/j.jphotochem.2007.06.001.

[9] Khan M W, Asif S U, Rehman K M U, Uddin W, Mubasher, Ahmed S, Khan E U, Tagliaferro A, Jagdali P and Fakhar-e-Alam M 2019 The electrical behavior of functionalized multiwall carbon nanotubes decorated with polymer nanocomposites *Phys. B Cond. Matter* **556** 17 DOI: 10.1016/j.physb.2018.12.029.

[10] McNally T, Pötschke P, Halley P, Murthy M, Martin D, Bell S, Brennan G P, Bein D, Lemoine P and Quinn J P 2005 Polyethylene multiwalled carbon nanotube composites *Polymer* **46** 8222 DOI: 10.1016/j.polymer.2005.06.094.

[11] Azam M U and Samad M A 2018 UHMWPE hybrid nanocomposite coating reinforced with nanoclay and carbon nanotubes for tribological applications under water with/without abrasives *Tribol. Inter* **124** 145 DOI: 10.1016/j.triboint.2018.04.003.

Improved microhardness of chrome galvanic coatings with combining nanodiamonds and nanotubes

Yu Litovka[1], M Nasraoui[1] and V Dolmatov[2]

[1]Department CAD, *Tambov State Technical University,* 106 Sovetskaya Street, Tambov 392000, Russian Federation
[2]Scientific-research laboratory, *FSUE "SKTB "Technolog",* 33-a Sovetsky Avenue, Saint-Petersburg 192076, Russian Federation

[1]E-mail: polychem@list.ru

Abstract. The method is developed for obtaining nanomodified galvanic coatings by adding a mixture of nanomaterials to the electrolyte. An experimental study of the microhardness of the precipitate of the technological process of obtaining nanomodified by a combination of nanodiamonds and multilayer carbon nanotubes; by a combination of single-layer carbon nanotubes and multilayer carbon nanotubes of chromium galvanic coating from a standard electrolyte was carried out. When nanodiamonds 12 g/l and multilayer carbon nanotubes "Taunit" 80 mg/l are added to the electrolyte, the microhardness of the chromium coating increases by 27% compared to the chromium coating obtained from the standard chromium plating electrolyte without additives. When single-layer carbon nanotubes 50 mg/l and multilayer carbon nanotubes "Taunit" 80 mg/l are added to the electrolyte, the microhardness of the chromium coating increases by 22% compared to the chromium coating obtained from the standard chromium plating electrolyte without additives. The service life of the parts, the chrome coating on which is obtained from the electrolyte with a mixture of nanodiamonds and multilayer carbon nanotubes "Taunit", single-layer carbon nanotubes and multilayer carbon nanotubes is significantly higher than when using a traditional chrome coating, as well as using multilayer carbon nanotubes, single-layer carbon nanotubes or nanodiamonds separately.

1. Introduction

Chrome galvanic coatings are used to give products special properties. Microhardness and, as a consequence, the wear resistance of the chrome coating are important characteristics in the case working of coated parts in friction mode [1]. Modern industry places ever higher demands on the quality of galvanic coatings. In this regard, the task of increasing the microhardness (an important qualitative indicator of galvanic coating) is relevant.

To increase the microhardness, various methods are used: changing the current modes; using additives in electrolytes. In recent years, due to the development of nano industry and a significant reduction in the cost of nanomaterials, have developed the methods of increasing microhardness is the use of nano-additives in chromium plating electrolytes. Good results were obtained using nanodiamonds [2-4] as an additive. Depending on the brand of nanodiamonds, the value of microhardness reached 1100 – 1300 kg/mm^2. Good results were obtained with multilayer carbon nanotubes (MCNT) [5] (microhardness by a concentration of 0.06 g/l exceeded 1100 kg/mm^2) and single-layer carbon nanotubes (SCNT) [6] (microhardness by a concentration of 0,05 g/l exceeded 850 kg/mm^2).

The effect of additives in electrolytes on the microhardness of galvanic coatings of nanodiamonds, SCNT and MCNT has different mechanisms. Due to the introduction of nanodiamonds in the crystal lattice of metals, a composite electrochemical coating is obtained, which has improved characteristics compared to the original metals. The presence in the coating of extremely developed in area and strong chemical bonds boundary layers "metal-diamond" provides not only wear resistance, but also increased microhardness [7]. The mechanism of influence of MCNT and SCNT on microhardness consists in emergence of the additional centers of crystallization on defects of carbon nanotubes. As a result, the size of crystals decreases [5]. Because of that, the microhardness increases.

In this regard, we suggest that the combined use of nanodiamonds and MCNT, SCNT and MCNT can give an effect that exceeds the sum of the effects of each additive.

The aim of the work is to study the effect of additives in the electrolyte in the form of a combination of nanodiamonds and MCNT, combination of SCNT and MCNT on the microhardness of the chromium coating.

2. Experimental part

The plan of experimental studies included the following stages.

a) application of chrome plating from a standard electrolyte without additives;

b) coating with additive nanodiamonds of different additive concentrations;

c) identification of the nanodiamonds concentration at which the highest microhardness of the chromium coating is observed;

d) coating with the addition of MCNT with different concentrations of the additive;

e) identification of the MCNT concentration at which the highest microhardness of the chromium coating is observed;

f) coating with nanodiamonds and MCNT with concentrations providing the maximum microhardness detected in stages (c) and (e).

g) coating with the addition of SCNT with different concentrations of the additive;

h) identification of the SCNT concentration at which the highest microhardness of the chromium coating is observed;

i) coating with SCNT and MCNT with concentrations providing the maximum microhardness detected in stages (e) and (h).

At each stage, at each concentration of the additive, a coating must be applied to at least 5 details.

2.1 Methods and materials

Getting a galvanic chrome coating was performed using the most common in industry of standard sulfate electrolyte chromium of the following composition: chromic anhydride CrO_3 250 g/l; sulfuric acid H_2SO_4 2.5 g/l.

The first nano-additive is an aqueous suspension of diamond charge containing 62 wt. %, and detonation nanodiamonds obtained by detonating charges TG 50/50 (alloy of TNT and hexogen). The contents of detonation nanodiamond in the diamond material is 71.0 wt. % The suspensions were obtained by treatment of diamond charge in an aqueous medium using a cavitation disintegrator. Before the introduction of the suspension into the electrolyte was treated with ultrasound on the installation of IL 100-6 / 4, frequency 22 kHz, sound intensity 786 W/cm^2; processing time - 30 min.

The concentration of nanodiamonds in the electrolyte varied from 4 g/l to 16 g/l.

The second nano-additive is fullerene-like multilayer carbon nanotubes (MCNT) - a nanocarbon material registered under the brand name "Taunit" [8]. It is a long hollow fiber consisting of graphene layers (no more than 30), 10-60 nm in diameter, up to 1 μm in length. For the distribution of MCNT "Taunit" in the volume of the electrolyte and to obtain a stable colloidal solution, the technology of soluble effervescent tablets was used [9]. MCNT "Taunit" was mixed with the following components: surface – active agent, soda and citric acid, and then pressed into tablets. Pressing force of 10 tons. Pressure 32 kg/mm^2.

Tablets with MCNT "Taunit" were added to the electrolyte. Division of agglomerates of nanotubes was carried out by the action of carbon dioxide released during dissolution of soda. The state of a highly dispersed metastable colloid was achieved by using a surfactant.

The concentration of MCNT "Taunit" in the electrolyte varied from 10 to 125 mg/l.

Square plates made of St3 steel with an area of 0.1 dm^2 (30x30 mm) were used as the cathode. Only the side facing the anode was covered; the reverse side was insulated. As an anode, a plate with a size of 30×30 mm of the following composition was used: 10% tin and 90% lead. Anode-cathode area ratio 1:1.

When applying the chrome coating, the electrolyte temperature was automatically maintained at 55°C.

Change in time of cathode electric current density i_k is shown in figure 1.

The microhardness of the Hμ coating was measured using the PMT-3M device. Microhardometer PMT-3M is designed to measure the microhardness of materials by pressing into the test material of the diamond tip of Vickers with a square base of a four-sided pyramid, which provides geometric and mechanical similarity of the prints as the indenter deepens under load. The measurement of the diagonals of the prints produced by a photoelectric ocular micrometer FOM-1-16 with the automatic processing of the measurement results. Measurement error 2%.

On each sample, the microhardness was measured at 5 points, in which the prints were symmetrical, after which the result was averaged. Then averaging of all samples of each experiment was carried out.

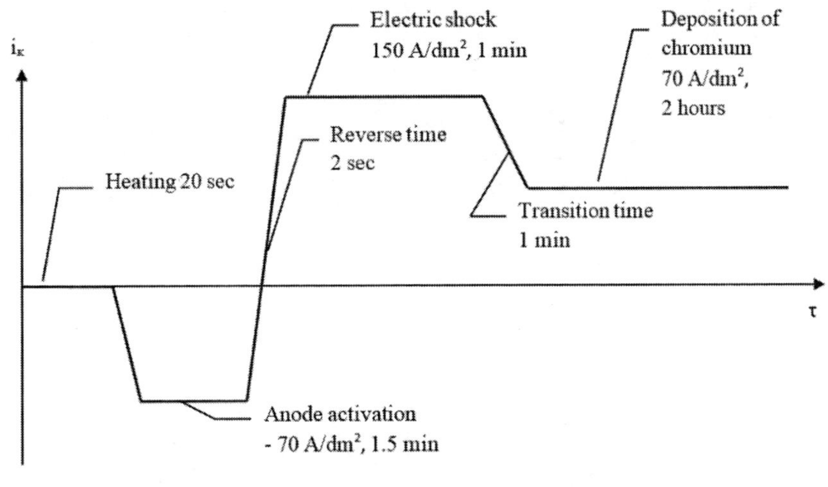

Figure 1. The change in cathode current density in the chrome plating process.

3. Results and discussion

The results of the studies dependence determine the microhardness of chromium coating with the addition of nanodiamonds are presented in figure 2.

The highest value of the microhardness of the chromium coating was obtained at a concentration of nanodiamonds 12 g/l. The result is in good agreement with the previously published experimental data [4,10]. In these studies, the maximum microhardness is also achieved at a concentration of 12 g/l nanodiamonds.

The results of the studies dependence determine the microhardness of chromium coating with the addition of MCNT are presented figure 3.

The highest value of the microhardness of the chromium coating was obtained at a concentration of MCNT 80 mg/l. The result is in good agreement with the previously published experimental data [5]. In this work also the maximum microhardness is achieved at a concentration of MCNT 60-80 mg/l.

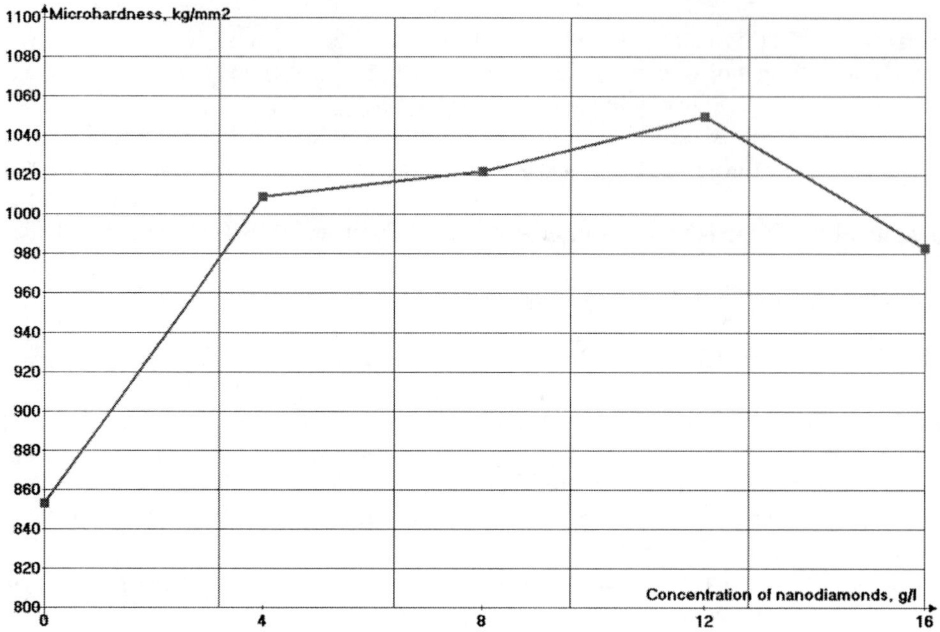

Figure 2. Dependence determine the microhardness of chromium coating with the addition of nanodiamonds.

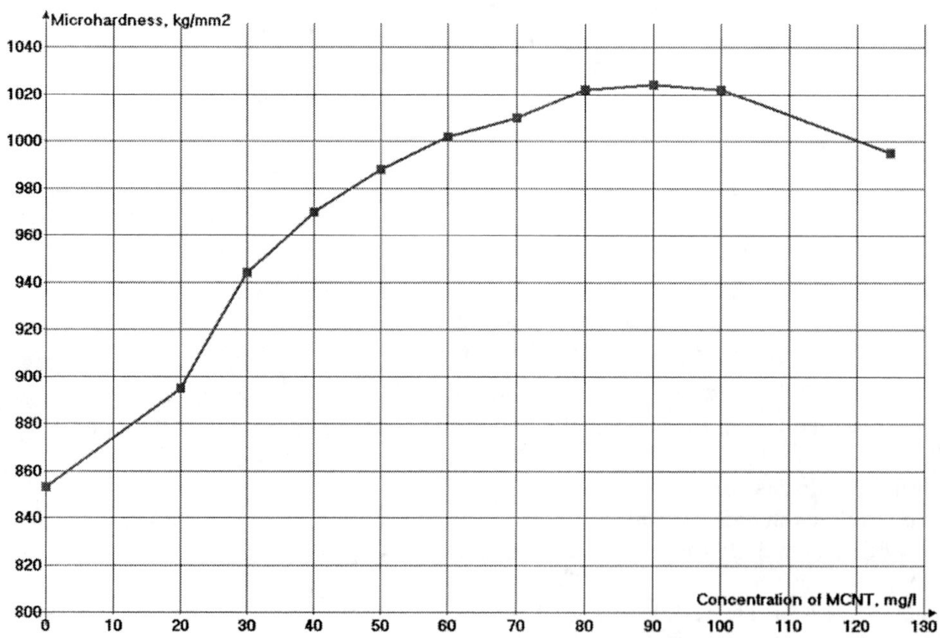

Figure 3. Dependence determine the microhardness of chromium coating with the addition of MCNT.

The analysis of the experimental results shows that the highest value of the microhardness of the chromium coating was obtained at a concentration of nanodiamonds 12 g/l and at a concentration of MCNT 80 mg/l. The next series of experiments was carried out with the addition of a mixture of nanodiamonds (with a concentration of 12 g/l) and MCNT (with a concentration of 80 mg/l). The results are summarized in Table 1.

Table 1. Results of experiments to determine the microhardness of chromium coating with various additives.

No	Additives	Microhardness, kg/mm^2	Microhardness (relative significance), %
1	0	853	100
2	Nanodiamonds, 12 g/ l	1050	123
3	MCNT "Taunit" 80 mg/l	1024	120
4	Mixture nanodiamonds (12 g/ l) + MCNT "Taunit" (80 mg/l)	1084	127

When adding a mixture of nanodiamonds 12 g/l and MCNT "Taunit" 80 mg/l, the microhardness of the chrome coating increases to 1084 kg/mm^2 (compared with the chrome coating obtained from the standard electrolyte chromium plating without additives, the increase in microhardness is 27 %). Thus, the addition of a mixture of nanodiamonds and MCNT allow to obtain a microhardness value exceeding on 4-7% the values of this qualitative indicator when using nanodiamonds and MCNT separately [1,5].

The results of the studies dependence determine the microhardness of chromium coating with the addition of SCNT are presented figure 4.

The highest value of the microhardness of the chromium coating was obtained at a concentration of SCNT 50 mg/l. The result is in good agreement with the previously published experimental data [6]. In this work, the maximum microhardness is also achieved at a concentration of SCNT 50 mg/l.

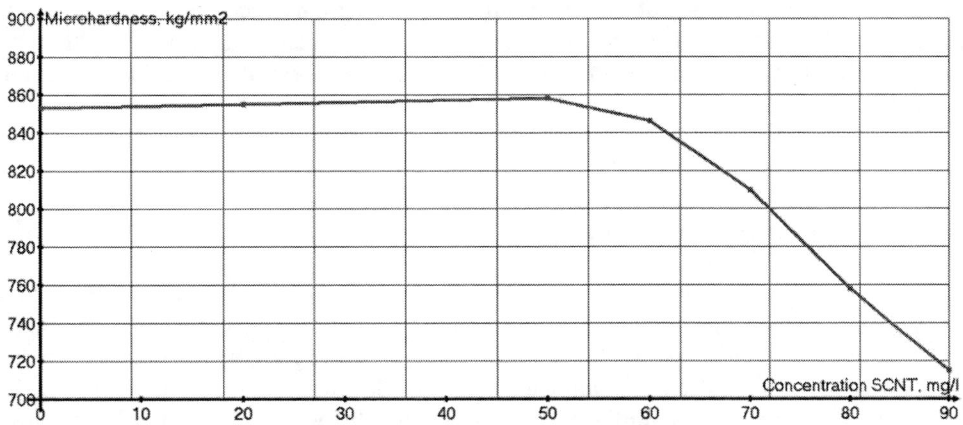

Figure 4. Dependence determine the microhardness of chromium coating with the addition of SCNT.

The analysis of the experimental results shows that the highest value of the microhardness of the chromium coating was obtained at a concentration of SCNT 50 mg/l and at a concentration of MCNT 80 mg/l. The next series of experiments was carried out with the addition of a mixture of SCNT (with a concentration of 50 mg/l) and MCNT (with a concentration of 80 mg/l). The results are summarized in Table 2.

When adding a mixture of SCNT 50 mg/l and MCNT "Taunit" 80 mg/l, the microhardness of the chrome coating increases to 1043 kg/mm^2 (compared with the chrome coating obtained from the standard electrolyte chromium plating without additives, the increase in microhardness is 22%). Thus, the addition of a mixture of SCNT and MCNT allow to obtain a microhardness value exceeding on 2 – 20% the values of this qualitative indicator when using SCNT and MCNT separately [5,6].

Table 2. Results of experiments to determine the microhardness of chromium coating with various additives.

No	Additives	Microhardness, kg/mm^2	Microhardness (relative significance), %
1	0	853	100
2	SCNT, 50 mg/ l	858	100.6
3	MCNT "Taunit" 80 mg/l	1024	120
4	Mixture SCNT (50 mg/ l) + MCNT "Taunit" (80 mg/l)	1043	122

4. Conclusion

As a result of the study. the tendency is revealed to increase the microhardness of the chrome coating by adding a mixture of carbon nanomaterials (nanodiamonds, single - and multi-layer nanotubes) to the electrolyte. The applied significance of the developed technologies for machine-building enterprises is as follows. If products with nanomodified chrome coating are the final products of the enterprise, they have a competitive advantage over similar products due to increased wear resistance. If products with nanomodified chrome coating are used in technological processes within the enterprise, production costs are reduced. So, at enterprise «TPZ-Tool», Tula, due to increase in wear resistance of the tool covered with the nanomodified chrome covering, the number of the tool decreased by 20% a year. The developed technologies are becoming more relevant as the cost of carbon nanomaterials produced in the Russian Federation decreases.

The increase in the microhardness of the chromium coating is caused by a combination of two mechanisms: the introduction of nanodiamonds into the crystal lattice of the coating metal and the appearance of additional crystallization centers on defects in carbon nanotubes.

The operation life of the obtained parts is significantly higher than when using a traditional chrome coating, as well as when using multilayer carbon nanotubes, simple-layer carbon nanotubes or nanodiamonds separately.

The obtained value of the microhardness of the chromium galvanic coating exceeds this figure obtained by known technologies.

The developed method opens up prospects for the joint use of various nano-additives in galvanic electrolytes to improve the quality of coatings.

References

[1] Tseluikin V N 2014 Nanoparticle modified composite coatings: structure and properties. Review. *Russian Nanotechnology* **9** (1-2) 25 https://doi.org/10.1134/S1995078014010145

[2] Alexandrova G S, Burkat G K, Dolmatov V Y and Gmysin E V 2016 On chromium deposition in the presence of boron-modified detonation nanodiamonds in a standard chromium plating electrolyte. *Chemical industry* **XCIII** 1 15

[3] Vinokurov E G 2014 Synthesis and properties of inorganic composite coatings containing detonation nanodiamonds. *Protection of Metals and Physical Chemistry of Surfaces.* **50** 4 480 https://doi.org/10.1134/S2070205114040194

[4] Tseluikin V N 2009 Composite electrochemical coatings: production, structure, properties. *Surface Physicochemistry and Material Protection.* **45** 3 287

[5] Litovka Yu V, Dyakov I A, Kuznetsova O A, Tkachev A G, Popov D Yu, Stolyarov R A 2011 Nanomodified galvanic chrome plating. *Electroplating and Surface Treatment.* **XIX** 4 29

[6] Litovka Yu V, Nasraui M, Kuznetsova O A, Maistrenko N V 2018 Modified by single-layer carbon nanotubes chrome galvanic coatings. *Reinforcing Technologies and Coatings* **14** (10) 463

[7] Ryjov E V, Koscheev Yu L, Marusina T M, Kuznetsov S A, Belonogov A S 2009 Nanodiamonds in galvanic chrome coatings. *Galvanic World* **3** 5

[8] Mishchenko S V, Tkachev A G 2008 *Carbon nanomaterials. Production, properties, application.* (Moscow: Engineering) p 320

[9] Patent 2477341 Russian Federation 2013, MPK C25D 15/00, B82B 1/00

[10] Vinokurov E G, Arsenin A M, Grigorovich K V, Bondar V V 2006 Electrodeposition of the modified dispersed particles of chromium coatings and their physicomechanical properties. *Protection of Metals.* **42** 3 312

IOP Conf. Series: Materials Science and Engineering **693** (2019) 012013 doi:10.1088/1757-899X/693/1/012013

Conductivity of a carbon nanotubes-epoxy resin nanocomposite

A Blokhin [1,*], I Zaytsev[1], A Sukhorukov[1], R Stolyarov[1], A Popov[1], I Burmistrov[2], D Kobzev[1] and V Yagubov[1]

[1]Laboratory of Materials Science, *Tambov State Technical University,* 106 Sovetskaya Street, 392000, Tambov, Russian Federation
[2]*Yuri Gagarin State Technical University of Saratov*, 77 Politechnicheskaya Street, 410054, Saratov, Russia Russian Federation

[*]E-mail: cha-cha@rambler.ru

Abstract. Multi-walled carbon nanotubes (CNTs) were used to reduce the electrical resistivity (*i.e.*, increase the conductivity) of industrial epoxy. Samples loaded with 6 wt.% CNTs showed a 9-fold increase in the conductivity. The starting epoxy resin is dielectric. The electrical resistivity data showed a percolation threshold between 0.5 and 2 wt.% CNTs loading. The temperature coefficient of electrical resistivity of the samples containing 2 wt.% in the temperature range from 20 to 60°C was $11.5 \cdot 10^{-3}$ K^{-1}. These results suggested that the conductivity of the CNTs-epoxy composites is improved, without any need for chemical functionalization of the nanotubes.

1. Introduction

The interest in developing novel hybrid composite materials based on carbon nanotubes (CNTs) is caused by the unique properties of the latter [1-4], which leads to a wide range of applications, from electronic elements to antistatic coatings. These materials include conductive composites [5-8], high-strength lightweight construction materials [9-12], and composites used as sensors [10,13-15] and even as sorbents [12,16-18]. When developing new materials, one should focus on components manufactured on an industrial scale. CNTs are mainly produced in laboratory amounts, but in the present work, we address the manufacturer of CNTs in large amounts.

The methods used for a long time to impart electrical conductivity to polymeric materials are based on introducing large quantities (60-80 %) of electrically conductive materials (graphite, metals, and various titanium compounds), which results in a softening of the matrix and a decrease in strength and performance characteristics. Considering this fact, the guideline of this work is the development of a physico-mechanical method for introducing and dispersing CNTs (trademark Taunit and Taunit-M) in an epoxy matrix (trademark ED-20) in order to achieve electrical conductivity of about 2-3 Ohm·m, with small amounts of the nanocarbon additive (6-9 %). The proposed method should combine the filling of CNTs with conventional fillers to positively affect its industrial application due to the conductivity indicators comparable with that of competitors' and reduced cost.

In the present paper, we measured electrical resistivity, but we also talk about electrical conductivity, implying that the former is the inverse of the latter.

Therefore, the aim of the work is to determine the effect of CNTs (their type and concentration) on the conductive properties of a polymer matrix.

2. Materials and methods

Taunit and Taunit-M milti-walled CNTs (hereinafter referred to as CNTs-t and CNTs-m, respectively) commercially produced by NanoTechCenter Ltd. (Tambov, Russia) were used as conductive additives. Their characteristics are presented in Table 1, whereas their structure is shown in figures 1 and 2.

Table 1. The main characteristics of the used carbon nanotubes.

Parameters	**CNTs-t**	**CNTs-m**
Outer diameter, nm	20-50	10-30
Internal diameter, nm	10-20	5-15
Length, μм	≥ 2	≥ 2more
Bulk density, g/cm^3	0.3-0.6	0.025-0.06
Specific surface area (nitrogen sorption), m^2/g	160	270

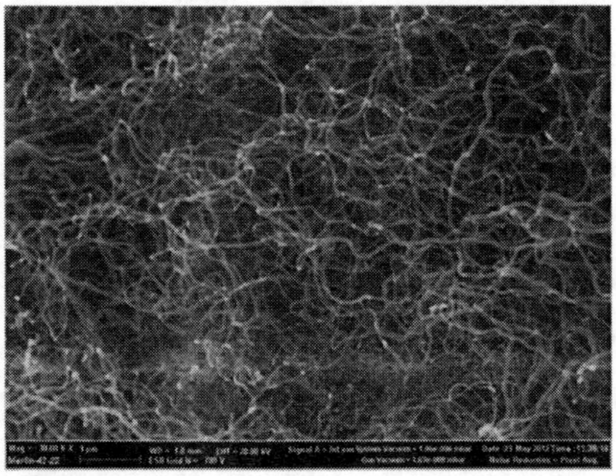

Figure 1. SEM image of the CNTs-t microstructure (scale bar is 1 μm).

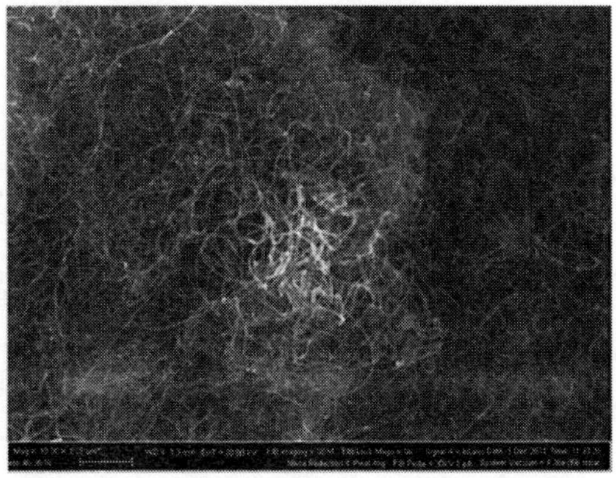

Figure 2. SEM image of the CNTs-m microstructure (scale bar is 1 μm).

The difference in the CNTs diameters means various quantities of layers, and as a result, larger CNTs amounts contained in 1 g of substance and, respectively. Consequently, when introducing the same weight percent amounts of the CNTs, the CNTs effect on the composite conductivity will be different.

The percent of the CNTs introduction into the epoxy matrix was 0.5-6 wt.%. Since the CNTs represent agglomerates, which can be seen in the scanning electron microscopy (SEM) images of the pristine CNTs (figures 1 and 2), their introduction was accompanied by dispersion in a viscous matrix that impeded their agglomeration. The dispersion and distribution in the monomer bulk occurred in a gap of 5 μm, with a shear flow at a speed of 120 rpm for 30 min, and with subsequent sonication at a power of 2 kW for 30 min.The effect of the method proposed on the particle dispersion was assessed through dynamic light dispersion (*i.e.*, dispersion of light over particles in the course of Brownian motion) on a "Nicomp 380 ZLS" analyzer (Particle Sizing Systems inc., USA). The dynamic light dispersion technique makes it possible to estimate the effective size of disperse phase which would be the true size if these particles were spherical. Since the CNTs and the agglomerates formed by them are not spherical, the results obtained allow assessing only changes in the disperse composition of the analyzed suspensions of the CNTs in the epoxy monomer. The analysis time was 5 min. The initial dispersion of the CNTs-t and CNTs-m was within the region of 5-35 μm, and after their introduction in the proposed way, it was 500 nm.

To determine specific electrical resistivity, a "Teraommetr E6-13A" measuring instrument (Punane RET, Estonia) was used. Samples represented cylinders with a diameter of 4 mm and a length of 40 mm.

The contacts were connected by applying the same downforce from the sample ends. Knowing the sample resistivity and geometry, the specific volume resistivity was calculated using the standard formula. The temperature coefficient of resistivity was evaluated according to the standard approach. The specific resistivity of the sample was determined at 20 °C and 60 °C and the temperature coefficient of resistance was estimated according to equation (1):

$$\alpha = \frac{1 \cdot \Delta\rho}{\rho \cdot \Delta T} \tag{1}$$

where, ρ – the electrical resistivity; $\Delta\rho$ – the difference in electrical resistivities at ΔT; and ΔT – the temperature difference, at which the resistance was measured.

The error of all the studies carried out was < 10 % (six replicates per one experimental point).

3. Results and discussion

The SEM images of the CNTs-t and CNTs-m incorporated into the epoxy matrix structure are represented in figures 3 and 4, respectively. From these images, it can be concluded that the CNTs distribution in the epoxy matrix is fairly uniform and has no air cavities.

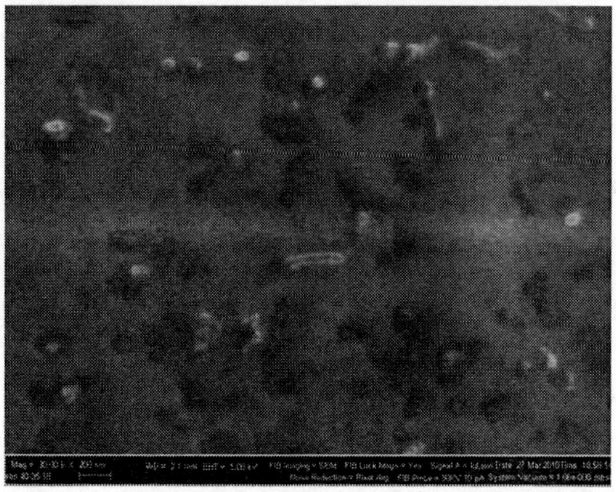

Figure 3. An SEM image of the CNTs-t (5 wt.%) microstructure in the epoxy resin (sample fracture, scale bar is 200 nm).

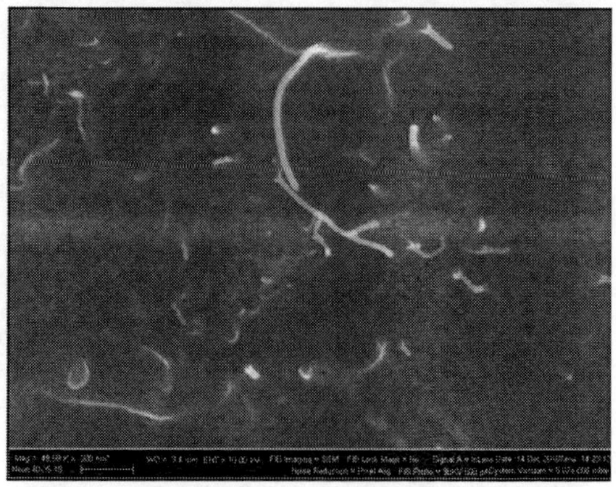

Figure 4. SEM image of the CNTs-m (5 wt.%) microstructure of in the epoxy resin (sample fracture, scale bar is 200 nm).

The research conducted showed that the CNTs-m possess the best conductivity, in comparison with the CNTs-t (figure 5), presumably they had various chirality. Besides, it was confirmed that using the CNTs-m made it possible to develop novel conductive polymer nanocomposites with a resistivity of about 2.5 Ohm·m and filling degree of 6 wt.%. The introduction of more than 6 wt.% of the CNTs makes the monomer considerable thicker (i.e., they increase viscosity). It can be assumed that being an electrical conductor, the CNTs form a percolation circuit which lines up on the boundary of macromolecules during their polymerization, thereby imparting the conducting properties to the material (this occurs when 2 wt.% of CNTs are introduced).

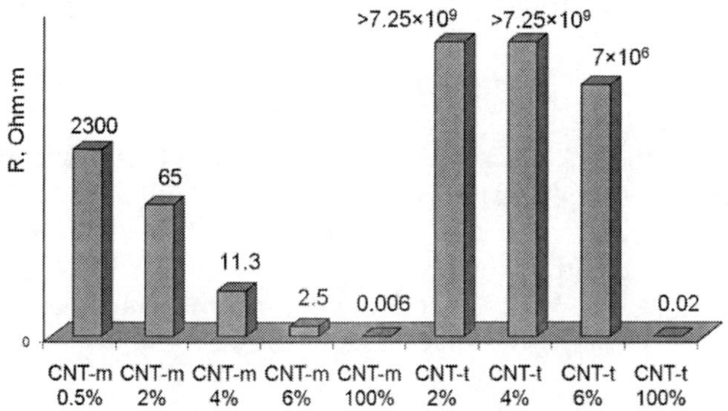

Figure 5. Specific resistivity changes depending on the additive type and content.

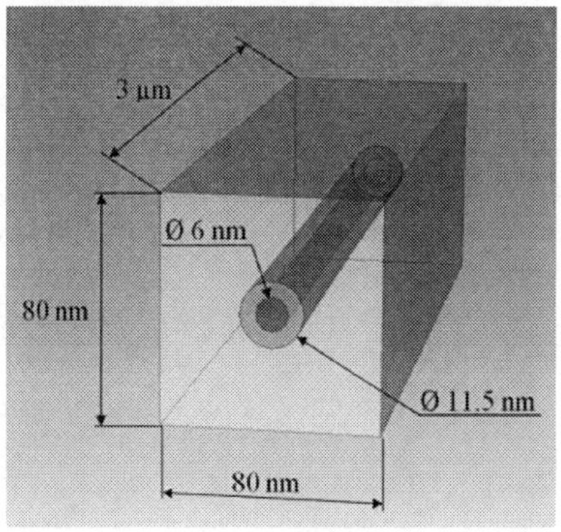

Figure 6. The average volume of the epoxy resin corresponding to 1 CNTs-m one in the 2-wt.% nanomodified epoxy binder.

According to the estimates, in the nanocomposite reinforced with 2 wt.% of the CNTs, the epoxy polymer volume of 0.5 10^{-20} corresponds to 1 CNTs-m (figure 6). In the same way, the temperature coefficient of resistivity of the samples containing 0.5 wt. % of the CNTs was measured. The coefficient values were found to be as follows: $\alpha_{0.5\%}=8.8\cdot10^{-3}$ K^{-1}, and $\alpha_{2\%} = -11.5\cdot10^{-3}$ K^{-1}.

The negative coefficient value refers to semiconductors and hybrid materials. In the nanocomposite developed, the CNTs were separated with contact barriers, and the temperature increase caused the electrical resistivity decrease, which lies in accordance with the fluctuation induced tunneling model described elsewhere. In case of the sample containing 0.5 wt.% of the CNTs, the increase in the resistivity with raising the temperature can be explained with a negative effect of the rather large coefficient of bulk thermal expansion found for the epoxy matrix on the unstable percolation circuit.

4. Conclusion

The effect of CNTs-t and CNTs-m effect on the electrical conductivity of the epoxy matrix as a function of concentration was determined. Only CNTs-m can be used effectively to provide conductivity for the epoxy-based composites. The electrical resistivity data showed a percolation threshold between 0.5- and 2-wt.% CNTs loading. The methods proposed for the introduction and dispersion of CNTs showed their effectiveness with the aim of obtaining an electrically conductive polymer nanocomposite that is not inferior in the properties to analogs, but significantly differs in price, due to the use of CNTs-m (price 1.6 USD per gram) [20].

Acknowledgement

The reported study was funded by RFBR (Russian Foundation for Basic Research) according to the research project No. 18-29-19121/18.

References

[1] Salvetat J P, Bonard J M, Thomson N B, Kulik A J, Forró L, Benoit W and Zuppiroli L 1999 Mechanical properties of carbon nanotubes *Applied Physics A: Materials Science and Processing* **69** 255 doi.org/10.1007/s003390050999

[2] Kaushik B K and Majumder M K 2015 Carbon Nanotube Based VLSI Interconnects (New Delhi: Springer) p 37 doi.org/10.1007/978-81-322-2047-3_2

[3] Dai H 2002 Carbon nanotubes: synthesis, integration, and properties *Acc. Chem. Res.* **35** 1035 doi.org/10.1021/ar0101640

[4] Reich, S, Thomsen C and Maultzsch J 2007 *Carbon Nanotubes: Basic Concepts and Physical Properties* (Wiley Blackwell) p 215

[5] Kwon Y and Tománek D 1998 Electronic and structural properties of multiwall carbon nanotubes *Physical Review B - Condensed Matter and Materials Physic* **58** 16001 doi.org/10.1103/PhysRevB.58.R16001
[6] Collins P G 2000 Nanotubes for electronics *Scientific American* **283** 67 doi.org/10.1038/scientificamerican1200-62
[7] Vasylenko A, Wynn J, Medeiros P, Morris A J, Sloan J and Quigley D 2017 Boosting electronic transport in carbon nanotubes *Phys Rev Lett* **95** 121408 doi.org/10.1103/PhysRevB.95.121408
[8] Charlier J C, Blase X and Roche S 2007 Electronic and transport properties of nanotubes *Reviews of Modern Physics* **79** 677 doi.org/10.1103/RevModPhys.79.677
[9] Cao Q, Kim H-, Pimparkar N, Kulkarni J P, Wang C and Shim M 2008 Medium-scale carbon nanotube thin-film integrated circuits on flexible plastic substrates *Nature* **454** 495 doi.org/10.1038/nature07110
[10] Andrews R, Jacques D, Qian D and Rantell T 2002 Multiwall carbon nanotubes: synthesis and application *Acc. Chem. Res.* **35** 1008 doi.org/10.1021/ar010151m
[11] Chen W X, Tu J P, Wang L Y, Gan H Y, Xu Z D and Zhang X B 2003. Tribological application of carbon nanotubes in a metal-based composite coating and composites *Carbon* **41** 215 doi.org/10.1016/S0008-6223(02)00265-8
[12] Burakova I V, Babkin A V, Neskoromnaya E A, Burakov A E, Kurnosov D A and Tkachev A G 2018 Adsorption of the Methylene Blue Dye on Carbon Nanocomposites Under Dynamic Conditions: A Kinetic Study *Journal of Physics: Conference Series* **1124** 1 doi.org/10.1088/1742-6596/1124/8/081029
[13] Wong H-S P and Akinwande D 2012 *Carbon Nanotube and Graphene Device Physics* (Cambridge: Cambridge University Press) p 86
[14] Baughman R H, Zakhidov A A, and De Heer W A 2002 Carbon nanotubes - the route toward applications *Science* **297** 787 doi:10.1126/science.1060928
[15] Thostenson E T, Ren Z and Chou T W 2001 Advances in the science and technology of carbon nanotubes and their composites: A review *Composites Science and Technology* **61** 1899 doi.org/10.1016/S0266-3538(01)00094-X
[16] Burakova I V, Burakov A E, Tkachev A G, Troshkina I D, Veselova O A, Babkin A V, Wei M A and Imran A 2018 Kinetics of the adsorption of scandium and cerium ions in sulfuric acid solutions on a nanomodified activated carbon *J. Mol. Liq.* **253** 277 doi.org/10.1016/j.molliq.2018.01.063
[17] Burakov A, Romantsova I, Babkin A, Neskoromnaya E, Kucherova A and Kashevich Z 2015 Ecotoxicology of heavy metals: liquid-phase extraction by nanosorbents *IOP Conference Series: Materials Science and Engineering* **98** 1 doi.org/10.1088/1757-899X/98/1/012023
[18] Babkin A, Neskoromnaya E and Burakova I 2017 Implementation of carbon nanomodification for sorption materials *MATEC Web of Conferences* **129** 1 doi.org/10.1051/matecconf/201712906021
[19] Ksenevich V K, Gorbachuk N I, Viet H, Shuba M V, Paddubskaya A G and Wieck A D 2017 Non-linear electrical conductivity of carbon nanotubes/WS2 nanotubes (nanoparticles) hybrid films *Nonlinear Phenomena in Complex Systems* **20** 360
[20] Bauhofer W and Kovacs JZ 2009 A review and analysis of electrical percolation in carbon nanotube polymer composites. *Compos. Sci. Technol.* **69** 1486

Physical-chemical properties and catalytic activity of the Ni-Co-CB material

O Andrienko[1,2], N Kobotaeva[1,3], T Skorokhodova[1,3], E Marakina[1] and V Sachkov[1]

[1]Laboratory of chemical technologies, *National Research Tomsk State University*, 36 Lenina Avenue, 634050 Tomsk Russian Federation
[2]Quantum Electronics Laboratory, *Zuev Institute of Atmospheric Optics of the Siberian Branch of the RAS,* 1 Zuev Square, 634021 Tomsk Russian Federation
[3]Laboratory of catalytic processing of light hydrocarbons, *Institute of Petroleum Chemistry of the Siberian Branch of the RAS,* 4 Akademichesky Avenue, 634055 Tomsk Russian Federation

[1]E-mail: vioes@mail.ru

Abstract. Carbon materials are widely used in catalysis as a support for active catalyst components due to their chemical inertness and developed surface. The most common carbon carriers are represented by activated carbons, carbon black and various carbon-carbon composites. In petrochemistry, carbon materials are also employed as catalysts to oxidize sulfur compounds contained in heavy fractions of oil being refined. In this regard, the present paper discusses the issues of metal-carbon composite material manufacturing. Technical carbon, or carbon black (CB), was used as a carbon support, the surface of which was modified with nickel and cobalt. The resulting composite material, Ni-Co-CB, has a high specific surface area comparable to that of the pristine carbon material, thereby favoring its catalytic activity. The high catalytic activity of the composite material developed was observed during oxidative desulfurization of heavy oil feedstock. Hence, the degree of sulfur removal for fuel oil was found to be 82.7 %, whereas for oil it reached 70.7 %.

1. Introduction

Currently, carbon materials (activated carbon, carbon black, graphite, carbon nanotubes, fullerenes, etc.) and their products are increasingly used in catalytic, adsorption, electrochemical processes, medicine, hydrometallurgy, *etc*. The advantages of carbon materials are their resistance and stability in aggressive environments at elevated temperatures, large specific surface area, the ability to adjust the parameters of the porous structure and physical and mechanical properties in a wide range [1].

In the works [2-3] it was found that the structural characteristics of carbon materials (specific surface area, pore size, porosity) affect the catalytic activity of carbon catalysts. In the study of the catalytic wet air oxidation of acetic acid, the graphite-supported Ru catalyst showed much more activity than those based on activated carbon [2].

Catalytic properties of catalysts on carbon carriers are often superior to those on oxide carriers. Thus, in the catalytic wet air oxidation of phenol, Ru and Pt catalysts deposited on the carbon black are more selective than catalysts based on the Si-Ti ones in this process [4].

In addition, carbon materials are used as catalysts in the absence of any metals. Carbon materials have shown good catalytic properties in the processes of methane decomposition, p-toluidine oxidation, aniline conversion, and catalytic ozonation of organic compounds [5-6].

The production of catalysts is of great importance for the development of the oil refining, petrochemical and chemical industries. In petrochemistry, catalysts based on rare and noble metals (platinum, palladium, rhodium, rhenium, etc.) and metal compounds (cobalt, nickel, molybdenum, etc.) are used. At the same time, for more efficient use of the active component (metal, metal compound), it is applied to carriers – materials with high specific surface area and mechanical strength. The most widespread are oxide (aluminum oxide, silicon oxide and zeolites, etc.) and carbon carriers.

In particular, carbon materials and their composites are used as catalysts for oxidation of sulfur compounds contained in various oil fractions [7-8].

Considering the aforementioned, the aim of the present research was to study the physical-chemical properties and catalytic activity of a material composed of carbon black and metal (Ni, Co) nanoparticles.

2. Experimental part

Carbon black (CB) of grade P 354, obtained at the Institute for Hydrocarbon Processing of the Siberian Branch of the Russian Academy of Sciences (Omsk, Russia) and containing C (96.4 %), O (2.5 %), S (0.5 %), N (0.3 %), and H (0.3 %) [1], was used herein as a carbon support for preparing a composite material.

In this starting material, the functional groups are formed already at the stage of its production, *i.e.* during thermal oxidative pyrolysis of hydrocarbons followed by oxidation of the mixture using a 20-% hydrogen peroxide solution at 100 °C. Being the CB component, oxygen is found in hydroxyl, carbonyl, ether and carboxyl groups [9]. The size and shape of CB particles were examined on a JEM-2100 transmission electron microscope (JEOL, Tokyo, Japan). The CB units dispersed are represented by globule aggregates, as shown in figure 1. The texture characteristics of the CB are given in table 1.

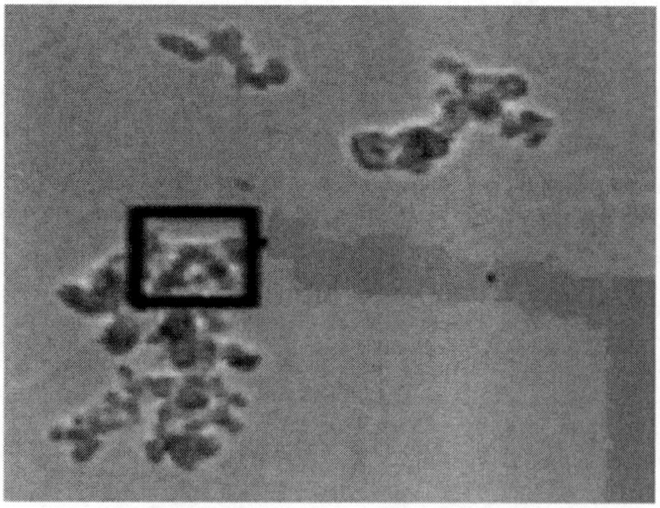

Figure 1. An image of CB particles (P354) as globule aggregates obtained using transmission electron microscopy.

Table 1. Texture characteristics of the CB.

Sample	Preparation method	Average aggregate size, nm	Aggregate surface zeta potential, mV
CB (P354)	Furnace	154±70	-6.37

To prepare a metal-carbon composite material (Ni-Co-CB), transition metal (nickel combined with cobalt) nanoparticles were deposited over the CB surface. For this purpose, the method of chemical reduction of metals from solutions of their salts with sodium hypophosphite was used. This technique is widely employed in electroplating to deposit nickel (chemical nickel plating) on compact items (metals and alloys) [9]. Its advantages include obtaining a dispersion coating and the uniformity of metal deposition on a substrate.

The physical-chemical properties of the composite material and CB were studied through powder X-ray diffraction and infrared spectroscopy. The powder X-ray diffraction was measured using a XRD 7000 diffractometer (Shimadzu, Kyoto, Japan). Diffraction patterns were obtained using CuKα radiation in the Bragg–Brentano configuration with steps of 0.03, exposures of 6 s, and angles in the range of 10-80°. Calculations of the amorphous phase fraction were performed using the standard software supplied with the corresponding instrument. IR-spectra were recorded using a Nicolet 5700 Fourier transform infrared spectrometer (Thermo Scientific, Waltham, MA, USA), in a KBr matrix in the frequency range of 400–4000 cm^{-1} with a resolution of 4 cm^{-1} and 32 scans. Infrared spectra of attenuated total reflectance (ATR) were registered using the same spectrometer equipped with a Smart Performer ATR device (Thermo Scientific). The spectra were captured on a ZnSe crystal in the range of 4000–650 cm^{-1} with a resolution of 4 cm^{-1}; 640 scans were collected for each spectrum.

The specific surface area of the CB and the Ni-Co-CB was determined using the method of thermal nitrogen desorption using a Sorbtometer M unit (Katakon, Novosibirsk, Russia), and calculation were performed through the multipoint BET method in an automatic mode in accordance with Russian National Standard (GOST) No. 23401-90. The method allows to estimate the volume of gas (nitrogen) first pre-adsorbed on the surface of the analyzed material at a temperature of liquid nitrogen, and then desorbed from the material at increasing temperature. An isotherm of the adsorption of molecular nitrogen at 77.4 K was obtained using anASAP 2020 surface and porosity analyzer (Micromeritics, Norcross, GA, USA).

The catalytic activity of the Ni-Co-CB composite material was studied in the course of oxidative desulfurization of heavy oil feedstock (heavy oil residue or fuel oil and heavy crude oil). This process included the following stages: oxidation of sulfur-containing compounds of the oil feedstock, and extraction of oxidized sulfur-containing compounds from the oil feedstock. To carry out the oxidation, a 500-mL reaction container equipped with a stirrer and a thermometer was loaded with a portion of the oil stock and heated with stirring to the reaction temperature (50 °C). When the desired temperature was reached, the catalyst was thoroughly admixed, and then a 15-wt.% hydrogen peroxide solution was added. The molar ratio of the sulfur contained in the oil feedstock and the S/H_2O_2 oxidizing agent was equal to 1:4. The reaction time was 90 min. Then, the oxidized sulfur-containing compounds were extracted with a polar solvent and water (content 10 v.%) at 35-40 °C. Dimethylformamide was used as an extractant. The solvent:oil feedstock ratio was 2:1. The reaction mass was kept under a layer of the extractant for 60 min at a given temperature. Finally, the extract-oxidized sulfur-containing compounds and the raffinate-desufurized oil stock phases were separated.

For the extraction, high-grade dimethylformamide (EKROS-1, Moscow, Russia), manufactured according to Russian National Standard (GOST) No. 20289-74, was used.

The feedstock used represented the residue of West Siberian oil from CJSC 'Chernigov Refinery' (Novaya Balakhonka, Kemerovo Region, Russia) at 370-560 °C, with a total sulfur content of 1.04 wt.%, and oil from the Usinsk Deposit of CJSC "Chernigov Refinery", with a total sulfur content 1.64 wt.%.

The sulfur content in the feedstock and desulfurized petroleum stock was determined by burning in a lamp in the presence of oxygen. The principle of the lamp method consists in burning a feedstock sample in a platinum holder (basket) of the cork of a Sheniger flask with an oxygenated absorption solution. Upon burning, the absorption solution was titrated with a barium nitrate solution in the presence of an indicator until a stable color appeared. By analogy, a 'blank' experiment was conducted (without a weighed feedstock sample) under the same conditions, and the total sulfur content in the analytical sample of feedstock was calculated.

The feedstock degree of desulfurization (sulfur removal) X (%) was calculated according to equation (1):

$$X = \frac{(\omega_0 - \omega_k)}{\omega_0} \cdot 100\%, \qquad (1)$$

where ω_0 is the initial sulfur content, wt.%; ω_k is the sulfur content in the feedstock after the desulfurization (residual content), wt.%.

3. Results and discussion

The availability of the Ni and Co nanoparticles in the composite material samples was confirmed by IR-spectroscopy, XRD, and HRTEM. Figure 2 demonstrates the IR spectra of the CB activated by the above mentioned nanoparticles.

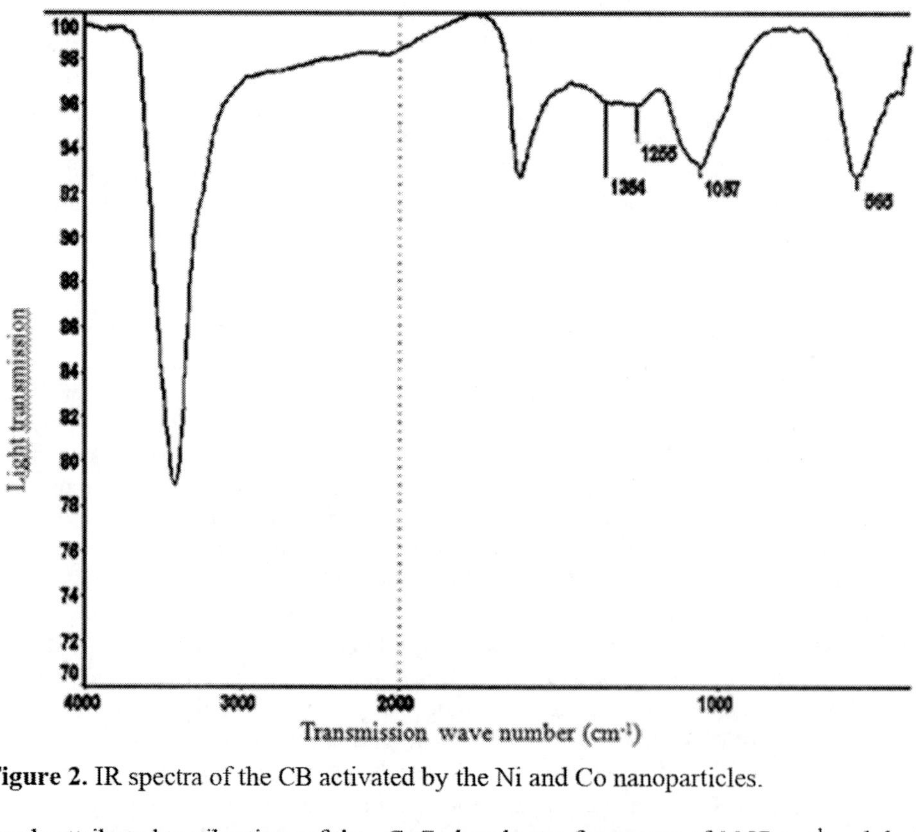

Figure 2. IR spectra of the CB activated by the Ni and Co nanoparticles.

The bands attributed to vibrations of the –C–C– bonds at a frequency of 1057 cm^{-1} and the bands at frequencies 563 and 580 cm^{-1} (vibrations of the –M–C bond) were observed.

The Raman spectra of a Ni-CNTs composite material were considered in [10], where it was shown that in addition to the structured carbon bands at frequencies 1595 and 1282 cm^{-1} (G and C bands, respectively), an additional broad band was observed within the range of 600-500 cm^{-1}. The quantum-chemical calculation of the NiC_2 model compound using the B3LYP/DGDZVP method yielded Ni-C stretching vibrations in the range of 480–500 cm^{-1} and bending vibrations at 580 cm^{-1}.

Thus, on the basis of the available IR spectrum (figure 2), as well as the calculated and published data, it can be stated that there is a metal-carbon bond (M-C) in the resulting Ni-Co-CB composite material.

The X-ray pattern of the Ni-Co-CB composite material (figure 3) shows broadened and fuzzy peaks of the Ni (at 2θ = 38.42; 44.555; 58.462; 71.074) and the cobalt (at 2θ = 43.838, 51.731).

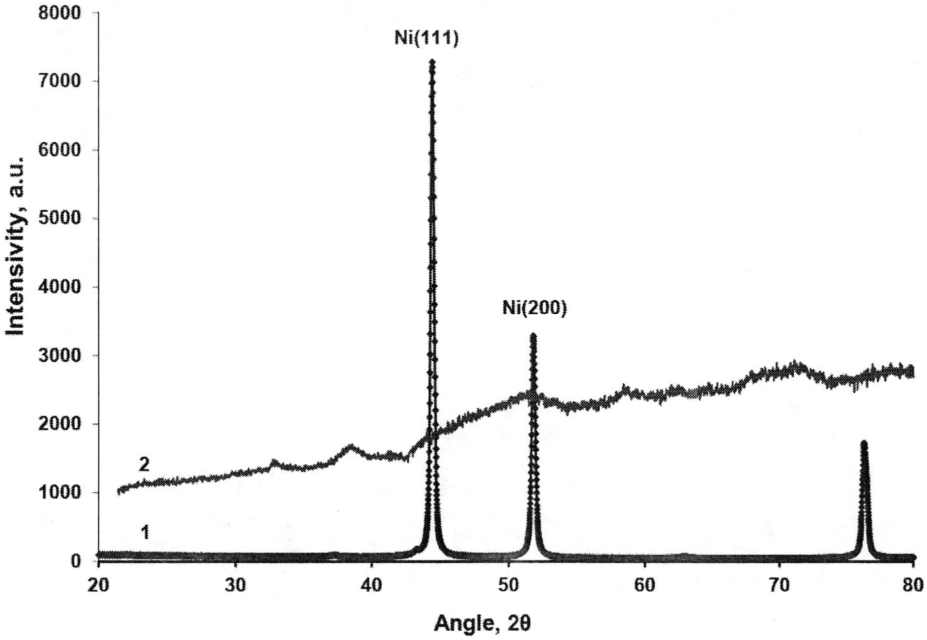

Figure 3. X-ray diffraction patterns of: **1** – the crystalline Ni; and **2** – the Ni-Co-CB composite material.

These obviously broadened peaks suggest the amorphous nature of the coating obtained. Such broadening was observed earlier when Ni and Co were deposited on CNTs via chemical reduction from a solution using sodium hypophosphite [10]. This is due to the fact that when using sodium hypophosphite [11], during reduction, the phosphorus precipitates simultaneously with the nickel according to the following reaction: $2H_2PO_2 \rightarrow H_2PO_3^- + P + H_2O$. The phosphorus penetrates into the nickel crystal lattice, and a so-called nickel-phosphorus coating (Ni-P), with a phosphorus content of about 7–10 %, is formed.

The resulting composite material possesses a developed surface, the parameters of which are comparable with those of the surface of the pristine carbon material (table 2).

Table 2. Specific surface area and pore size of the CB-based composite materials.

Sample	Physical-chemical parameters		
	Specific surface area, m^2/g	Pore diameter, nm	Pore volume, cm^3/g
CB	110.0	1.9	0.034
Ni-Co-CB	107.4	1.8	0.031

Table 3 presents the results on the oxidation of the heavy oil residue (fuel oil) and heavy oil using the hydrogen peroxide in the presence of the Ni-Co- CB material as catalyst.

Table 3. Oxidation of the fuel oil using 15-wt.% hydrogen peroxide in the presence of the Ni-Co-CB catalyst.

Oil feedstock	Ni-Co-CB concentration wt.%	V_{H2O2}, cm^3	Reaction temperature, °C	Reaction time, min	Initial sulfur content in oil feedstock wt.%	Residual content of sulfur in the oil stock, wt%	Degree of sulfur removal (X), %
Fuel oil	1.0	6	35	90	1.04	0.20	80.8
	1.0	6	35	120	1.04	0.22	78.8
	1.0	6	50	60	1.04	0.28	73.1
	1.0	6	50	90	1.04	0.18	82.7
	1.0	6	50	120	1.04	0.20	80.8
	1.0	6	95	90	1.04	0.30	71.7
	1.0	6	95	120	1.04	0.32	69.2
Oil	1.3	6	35	90	1.64	0.52	68.2
	1.3	6	35	120	1.64	0.5	69.5
	1.3	6	50	60	1.64	0.6	63.4
	1.3	6	50	90	1.64	0.48	70.7
	1.3	6	50	120	1.64	0.55	66.5
	1.3	6	95	90	1.64	0.72	56.1
	1.3	6	95	120	1.64	0.7	57.3

As can be seen from this table, under the conditions considered, the high degree of sulfur removal was achieved for the fuel oil and oil – 82.7 and 70.7 % – when using the Ni-Co-CB catalyst at 1.0 and 1.3 wt.%, respectively. This is probably due to the fact that this metal-modified material has a developed surface, as mentioned above.

4. Conclusion

In the present work, a metal-carbon composite material was developed by depositing metal nanoparticles (Ni and Co) over the CB surface via chemical reduction of metals from solutions of their salts.

In the IR spectra of the metal-CB composite material, not only bands related to vibrations of the −C−C− bonds at the frequency of 1057 cm^{-1} but also bands at the frequencies of 563 and 580 cm^{-1} (vibrations of the −M−C bond) were observed. This can be adequately explained by intermolecular interactions in the system of conjugated graphene bonds.

The XRD method revealed that the nickel-phosphorus coating formed on the CB surface has an amorphous character, thereby suggesting that the particles deposited on the CB surface are nanosized.

The resulting composite material has a high specific surface area comparable to that of the pristine CB, which favors its catalytic activity.

The Ni-Co-CB composite material exhibited a high catalytic activity in the course of oxidative desulfurization. The degree of sulfur removal was found to be 82.7 and 70.7 % for the fuel oil and oil, respectively.

5. Acknowledgments

The work was performed under the financial support of the Ministry of Science and Higher Education (formerly, the Ministry of Education and Science) of the Russian Federation in the framework of Grant Agreement No. RFMEFI57817X0225.

References

[1] Likholobov V A, Surovikin V F, Plaksin G V, Tsekhanovich M S, Surovikin Yu V and Baklanova O N 2008 Nanostructured carbon materials in catalysis and adsorption. *Catalysis in Industry* [Kataliz v promyshlennosti – in Russian] **S** 63

[2] Cybulski A and Trawczyński J 2004 Catalytic wet air oxidation of phenol over platinum and ruthenium catalysts. *Appl. Catal. B: Environ.* **47** 1 https://doi.org/10.1016/S0926-3373(03)00327-8

[3] Quintanilla A,Casas J A and Rodríguez J J 2007 Catalytic wet air oxidation of phenol with modified activated carbons and Fe/activated carbon catalysts. *Appl. Catal. B: Environ.* **76** 135 https://doi.org/10.1016/j.apcatb.2007.05.019

[4] Yang S, Cui Y, Sun Y and Yang H 2014 Graphene oxide as an effective catalyst for wet air oxidation of phenol. *J. Hazard. Mater.* **280** 55 https://doi.org/10.1016/j.jhazmat.2014.07.051

[5] Pereira M F R, Figueiredo J L,Órfão J J M, Serp P, Kalck P and Kihn Y 2004 Catalytic activity of carbon nanotubes in the oxidative dehydrogenation of ethylbenzene. *Carbon* **42** 2807 https://doi.org/10.1016/j.carbon.2004.06.025

[6] Yang S X, Li X, Zhu W P, Wang J B and Descorme C 2008 Catalytic activity, stability and structure of multi-walled carbon nanotubes in the wet air oxidation of phenol. *Carbon* **46** 445 https://doi.org/10.1016/j.carbon.2007.12.006

[7] Yu G, Lu Sh, Chen H and Zhu Zh 2005 Oxidative desulfurization of diesel fuels with hydrogen peroxide in the presence of activated carbon and formic acid. *Energ. Fuel.* **19** 447 https://doi.org/10.1021/ef049760b

[8] Xiao J, Wu L, Wu Y, Liu B, Dai L, Li Zh, Xia Q and Xi H 2014 Effect of gasoline composition on oxidative desulfurization using a phosphotungstic acid/activated carbon catalyst with hydrogen peroxide. *Appl. Energy* **113** 78 https://doi.org/10.1016/j.apenergy.2013.06.047

[9] Petrova T P 2000 Chemical Coatings. *Soros Educational Journal* [Sorosovskiy Obrazovatel'nyi Zhurnal – in Russian] **6** 57

[10] Kobotaeva N S, Skorokhodova T S and Ryabova N V 2015 Catalytic systems of cumene oxidation based on multiwalled carbon nanotubes. *Russ. J. Phys. Chem. A* **89** 462 https://doi.org/10.1134/S0036024415030164

[11] Kobotaeva N S and Skorokhodova T S 2015 A study of the possibility of the application of nickel-activated multilayer carbon nanotubes at the development of catalysts. *Kauchuk and Rubber* [Kauchuk i Rezina – in Russian] **1** 16

IOP Conf. Series: Materials Science and Engineering **693** (2019) 012015 doi:10.1088/1757-899X/693/1/012015

Carbon nanofillers used in epoxy polymeric composites: a brief review

A Blokhin, A Sukhorukov, R Stolyarov, I Zaytsev, N Yashchishin and V Yagubov

Laboratory of Materials Science, *Tambov State Technical University,* 106 Sovetskaya Street, Tambov 392000, Russian Federation

E-mail:dr.reningan@yandex.ru

Abstract. The paper discusses scientific publications regarding polymer composites based on epoxy resins, which are the main competitor of many traditional structural materials. The review on carbon nanofillers for composite materials, such as graphene, graphene oxide, carbon nanotubes, etc., is given. Methods for introducing these nanomaterials into different binders are considered, and comparative results of improving the strength characteristics of nanocomposites are presented.

1. Introduction

Polymer composite materials such as fiberglass, carbon fiber, organoplastics, powder-filled polymers, textolites possessing unique characteristics [1,2] are the main component in manufacturing products used for solving diverse problems, ranging from the production e of sports equipment and jewelry up to their use in the aviation aerospace industries [3-5].

Due to the widespread use of polymers, the requirements for their strength characteristics (tensile strength, Young's modulus, crack resistance, resistance to impact loads) [6] are becoming increasingly high. Carbon nanomaterials (CNMs) used as a modifier of polymer matrices can help in meeting the increased requirements for the strength characteristics.

Most often, when developing polymer composite materials (PCMs), an epoxy resin is used; it is one of the most common binders in the industry due to the variety of available grades [7] and curing agents [8], thereby making it possible to obtain materials with different property combinations [9]. With the advent of the CNM industrial production [10], there is a tendency for employing them as a PCM nanofillers which can increase tensile strength from +2.5 up to +47% [11,12].

Carbon fillers are of great interest in scientific publications. According to the database of peer-reviewed scientific literature, there are more than 10,000 papers, of which more than 1,000 publications are devoted to carbon nanofillers which possess high strength, an ordered structure, and unique geometry. Therefore, over the years, a large number of experiments and studies have accumulated in the scientific and technical literature that describe the modification of epoxy matrices with various nanomaterials. This makes it possible to identify an important trend, namely, that in recent years a large number of publications have been devoted to solving specific technological problems, and due attention has not been paid to the review work, which would enable the comparison of all the process parameters and properties of the nanocomposites obtained.

Using carbon nanofillers, different in structure, properties, percent of their application and methods of their introduction, allows obtaining materials with the desired combination of properties. In scientific research, much attention is paid to carbon nanotubes (CNTs), both single- [13-16] and multi-walled [16-20], which can be used as a filler to improve the strength and conductive properties. The papers [21-25] discuss the introduction of graphene into the polymer matrix, whereas articles [26-28] deal with nanoparticles such as graphene oxide and reduced graphene.

Considering the aforementioned, the aim of the work was to study the current state of the issue of choosing CNMs and methods for their introduction to develop a nanocomposite with improved strength properties compared to the analogs.

2. Methods for introducing nanofillers into the epoxy matrix

Since carbon nanomaterials are subject to agglomeration, dispersion and uniform distribution of nanofillers in polymer binders appear to be one of the main problems that many researchers are facing. This is reflected in many scientific papers [18-20,27-44].

Based on the simplicity of the hardware design, introducing CNMs into the epoxy binder using mechanical action (dispersants, mixers) is the most common technique [12,19,25-31]. However, it is not always possible to destroy small agglomerates and distribute the CNMs using only the mechanical method. In the papers [11,20,22,29,32-34], the distribution of the nanofiller in the epoxy binder was carried out using an ultrasonic device, but without preliminary mechanical treatment, complete deagglomeration of the nanomaterial occurs over a long period of time. In the studies [22,35-40], a combination of several methods is described. At the first stage, the CNM is introduced into the binder using a mixing device, and then, the material is dispersed in an ultrasonic unit. In the works [20,27,28,41,42], CNMs were dispersed in solvents using ultrasound. After that, the resulting mixture was added to the epoxy binder, and the solvent was removed. However, the use of the solvent in this method adversely affects the strength properties of the final product. In [43,44], preliminary grinding of the CNMs in a ball mill and an ultrasonic device was used, then, the CNMs were introduced into the epoxy binder. The disadvantage of this method is grinding itself, meaning the ingress of grinding material particles.

The choice of the technique for introducing the nanofillers depends on the quality and structure of the nanomaterial or the viscosity of the polymer matrix. For instance, in [45-49], the CNMs are introduced gradually, using a three-roll mill, which functions on the principle of applying a high shear force to the material between two rolls rotating in opposite directions and at different speeds. Further, the material is dispersed in an ultrasonic setup.

As a whole, the CNM effect on the PCM depends on the uniform distribution of the nanofiller in the binder; therefore, the technique for introducing the CNMs into the polymer material through preliminary mechanical action with subsequent or simultaneous ultrasonic treatment is recommended.

3. Studying the carbon nanomaterial effect on the physical-mechanical properties of the PCMs

The results of using various nanofillers to increase the strength characteristics of epoxy matrices are shown in Table 1. From the data presented, it can be concluded that the use of the carbon nanofillers may significantly improve the strength characteristics of the composite material. Among the whole variety of nanoadditives, CNTs are considered as fillers with great potential for increasing the physical-mechanical properties. Moreover, they showed a better effect on the strength than the other fillers presented in the review. Besides, a great potential of modification and functionalization of the carbon nanofillers [50-52] was revealed after reviewing the scientific literature on their full potential as composite fillers.

Table 1. The effect of the carbon nanofillers on the physical-mechanical properties of the PCMs.

Nanomaterial	Tensile strength increase/decrease, +/-%	Ref.
CNTs (1.5%)	+7	[45]
CNTs (0.3%)	+5	[46]
CNTs (0.3%)	+37	[47]
Multi-walled CNTs (0.51%)	+9	[11]
Multi-walled CNTs (0.3%)	+35	[16]
Multi-walled CNTs (0.17%)	+47	[11]
Single-walled CNTs (0.3%)	+44	[16]
Double-walled CNTs (0.3%)	+37	[16]
ABS/CNTs (6%)	+12	[17]
Graphene nanoplatelets (0.25%)	+20	[22]
Graphene nanoplatelets (6%)	+1	[17]
Graphene oxide (1%)	-18	[12]
Graphene oxide (0.75%)	-9	[12]
Graphene oxide (0.5%)	+71	[48]
Graphene oxide (0.5%)	+2.9	[12]
Graphene oxide (0.25%)	+2.5	[12]
Graphene oxide (0.1%)	+38	[48]
Carbon nanofiber (1.5%)	+11	[49]
Carbon nanofiber (0.5%)	+18	[50]
Carbon fiber+CNTs	+19	[51]
Glass fiber+carbon nanofiber	+26	[52]

4. Conclusion

Through the use of nanofillers, the PCMs can compete in the physical and mechanical properties with the other materials such as ceramics, glass, or even metal. It is worth noting that the conducted literature review on scientific publications does not cover the whole variety of nanoparticles used as PCM fillers. This is due to a large number of nanomaterials and ongoing studies to improve the physical-mechanical properties of different polymer matrices for various purposes. The recommended technique herein is represents a complex that includes the combination of several the methods considered with the aim of maximizing the uniformity of localization of the nanofiller in the binder, and this, in turn, has a positive effect on the nanoPCM operational characteristics such as elastic modulus and tensile strength. This method is easily scalable for the commercial production of nanocomposites.

5. Acknowledgment

The presentresearch was funded by RFBR (Russian Foundation for Basic Research) in the framework of Project No.18-29-19121\18.

References

[1] Kolosova A S, Sokol'skaya M K, Vitkalova I A, Torlova A S and Pikalov E S 2018 Modern polymeric composite materials and their application. *Int. J. of Applied and Basic Res.* **5** 245 [in Russian]

[2] Bondaletova L I, Bondaletov V G 2013 *Polymer Composite Materials* (Tomsk: TPU) p 118

[3] Ravishankar B, Nayak S K and Kader M A 2019 Hybrid composites for automotive applications – Areview. *Journal of Reinforced Plastics and Compos.* **38** 835 https://doi.org/10.1177/0731684419849708

[4] Barile C, Casavola C and De Cillis F 2019 Mechanical comparison of new composite materials for aerospace applications. *Compos. part B: Eng.* **162** 122 https://doi.org/10.1016/j.compositesb.2018.10.101
[5] Matei A, Tucureanu V, Tîncu B C, Marculescu C V, Burinaru T A and Avram M 2019 Polymer nanocomposites materials for aerospace applications. *AIP Conf. Proc.* **2071** 1 https://doi.org/10.1063/1.5090064
[6] Polilov A N 2016 *Experimental Mechanics of Composites* (Moscow: MGTU) p 375
[7] *Epoxy resins. Active diluents*, Chimex LTD, available at: http://www.chimexltd.com/catalog/category/814
[8] *Hardeners. Curing accelerators*, Chimex LTD, available at: http://www.chimexltd.com/catalog/category/806
[9] Ovchinnikov I I, Ovchinnikov I G, Mandrik-Kotov B B and Mihaldykin E S 2016 Application problems polymer composite materials in transport construction. *Science* **8** 2223[in Russian]
[10] *Production,* LLC "NanoTechCenter", available at: http://www.nanotc.ru/content-category1
[11] Mehdi R and Saeed F 2018 Mechanical and low-velocity impact properties of epoxy-composite beams reinforced by mwcnts. *Journal of Compos. Mater.* 1 https://DOI: 10.1177/0021998318790049journals.sagepub.com/home/jcm
[12] Huskić M, Bolka S, Vesel A, Mozetič M, Anžlovar A, Vizintine A and Žagar E 2018 One-step surface modification of graphene oxide and influence of its particle size on the properties of graphene oxide / epoxy resin nanocomposites. *Europ. Polym. Journal.* **101** 211 https://doi.org/10.1016/j.eurpolymj.2018.02.036
[13] Zabihi O, Ahmadi M, Nikafshar S, Chandrakumar P K and Naebe M 2018 A technical review on epoxy-clay nanocomposites: Structure, properties, and their applications in fiber reinforced composites. *Compos. B* **135** 1 https://doi:10.1016/j.compositesb.2017.09.066
[14] Sook Y M and Woo S K 2018 High mechanical properties of super aligned carbon nanocomposite by polyurethane based crosslinking molecules. *Compos. Sci. Technol.* **161** 100 https://doi.org/10.1016/j.compscitech.2018.04.011
[15] Hsieh T H, HuangY S and Shen M Y2018 Carbon nanotube size effect on the mechanical properties and toughness of nanocomposites. *Polym. Compos.* **39** 1072 https://doi.org/10.1002/pc.24505
[16] Aldousari S M, Hedia H S, Al Thobiani and Fouda N 2018 Influence of different nanomaterials on the mechanical properties of epoxy matrix composites. *Mater. Tesning.* **60** 639 https://doi.org/10.3139/120.111199
[17] Dul S, Pegoretti A and Fambri L 2018 Effects of the nanofillers on physical properties of acrylonitrile-butadiene-styrene nanocomposites: comparison of graphene nanoplatelets and multiwall carbon nanotubes. *Nanomater.* **8** 674 https://doi: 10.3390/nano8090674
[18] Faridirad F, Barmar M and Ahmadi S 2018 The effect of MWCNT on dynamic mechanical properties and crystallinity of in situ polymerized polyamide 12 nanocomposite. *Polym. for Advan. Technol* .**29** 2134 https://doi.org/10.1002/pat.4322
[19] Noor N A, Razak J A, Ismail S, Mohamad N, Yaakob M Y and Theng T H 2017 Influence of mwcnts addition on mechanical and thermal behaviour of epoxy/kenaf multi-scale nanocomposite. *Mater. Sci. Eng.* **210** 126 https://DOI: 10.1088/1757-899X/210/1/012058
[20] Minoo N, Jing W, Abbas A, Hamid K, Nishar H, Lu H L, Ying C and Bronwyn F 2014 Mechanical property and structure of covalent functionalised graphene/epoxy nanocomposites. *Front. Mater.* **4** 1 https://DOI: 10.1038/srep04375
[21] Hadden C M, Klimek-McDonald D R, Pineda E J, King J A, Reichanadter A M, Miskioglu I, Gowtham S and Odegard G M 2015 Mechanical properties of graphene nanoplatelet/carbon fiber/epoxyhybrid composites: multiscale modeling and experiments. *Journal of Carbon.* **95** 100 https://doi.org/10.1016/j.carbon.2015.08.026
[22] Shen M Y, Chang T Y, Hsieh T H, Li Y L, Chiang C L, Hsiharng Y and Yip M C 2013 Mechanical properties and tensile fatigue of graphene nanoplate reinforced polymer nanocomposites.

Journal of Nanomater. **2013** 9 http://dx.doi.org/10.1155/2013/565401

[23] Mostapha T, Khalid L, Imane B, Debora D and Ali M 2018 Effect of graphene nano-additives on the local mechanical behavior of derived polymer nanocomposites. *Journal of Polym.* **10** 667 https://DOI: 10.3390/polym10060667

[24] Zabihi O, Ahmadi M, Nikafshar S, Chandrakumar P K and Naebe M 2018 A technical review on epoxy-clay nanocomposites: structure, properties, and their applications in fiber reinforced composites. *Compos. part B: Eng.*, **135** 1 https://doi:10.1016/j.compositesb.2017.09.066

[25] Lai W, Xuelong C, Kailiang H, Zhijuan Y, Lun W, Hongqing W, Zhen Q, Xue W, Zewen L and Zhe W 2018 Enhancement in mechanical properties of epoxy nanocomposites by styrene-ethylene-butadienestyrene grafted graphene oxide. *Compos. Interfaces.* **26** 141 https://doi.org/10.1080/09276440.2018.1481303

[26] Aradhana R, Mohanty S and Sanjay K N 2018 Comparison of mechanical, electrical and thermal properties in graphene oxide and reduced graphene oxide filled epoxy nanocomposite adhesives. *Journal of Polym.* **141** 109 https://doi.org/10.1016/j.polymer.2018.03.005

[27] Menezes de B R, Ferreira F V, Silva B C, Simonetti E A, Bastos T M, Cividanes L S and Thim G P 2018 Effects of octadecylamine functionalization of carbon nanotubes on dispersion, polarity, and mechanical properties of CNT/HDPE nanocomposites. *Journal of Mater.* **53** 14327 https://DOI: 10.1007/s10853-018-2627-3

[28] Domun N, Paton K R, Hadavinia H, Sainsbury T, Zhang T and Mohamud H 2017 Enhancement of fracture toughness of epoxy nanocomposites by combining nanotubes andnanosheets as fillers. *Journal of Mater.* **10** 1179 https://doi: 10.3390/ma10101179.

[29] Dimitrios G P, Kinloch I A and Young R J 2017 Mechanical properties of graphene and graphene-based nanocomposites. *Mater. Sci.* **90** 75 https://doi.org/10.1016/j.pmatsci.2017.07.004

[30] Kim J A, Seong D G, Kang T J and Youn J R 2006 Effects of surface modification on rheological and mechanical properties of cnt/epoxy composites. *Journal of carbon.* **44** 1898 https://doi.org/10.1016/j.carbon.2006.02.026

[31] de Villoria R G, Miravete A, Cuartero J, Chiminelli A and Tolosana N 2006 Mechanical properties of swnt/epoxy composites using two different curing cycles. *Compos. part B: Eng.* **37** 273 https://doi.org/10.1016/j.compositesb.2006.01.002

[32] Buketov A V, Dolgov N A and Babich N V 2017 Mechanical characteristics of epoxy nanocomposite coatings with ultradisperse diamond particles. *Strength of Mater.* **49** 473 https://DOI 10.1007/s11223-017-9888-y

[33] Shekar K C, Prasad B A and Prasad N E 2014 Effect of amino multi walled carbon nanotubes reinforcement on the flexural properties of neat epoxy. *Appl. Mech. Mater.* **592** 912 https://doi.org/10.4028/www.scientific.net/AMM.592-594.912

[34] Rostamiyan Y, Fereidoon A and Mashhadzadeh A H 2015 Experimental study on the mechanical properties of an epoxy-based nanocomposite using polymeric alloying and different nanoreinforcements: nanofiber, nanolayered and nanoparticulate materials. *Sci. Eng. Compos. Mater.* **22** 591 https://DOI:10.1515/secm-2013-0305

[35] Muhammad A, Muhammad A and Abraiz K 2017 Fabrication, mechanical, thermal, and electrical characterization of epoxy/silica composites for high-voltage insulation. *Sci. Eng. Compos. Mater.* **25** 4 https://doi.org/10.1515/secm-2015-0445

[36] Afrouzian A, Movahhedi A H and Liaghat G H 2017 Effect of nano-particles on the tensile, flexural and perforation properties of the glass/epoxy composites. *Journal of reinforced Plast. and Compos.* **36** 900 https://doi.org/10.1177/0731684417694753

[37] Sung H S 2018 The effect of clay/multiwall carbon nanotube hybrid fillers on the properties of elastomer nanocomposites int. *J. of Polym. Sci.* **2018** 8 https://doi.org/10.1155/2018/5295973

[38] Rana S et al 2012 Effect of carbon nanofiber functionalization on the inplane mechanical properties of carbon/epoxy multiscale composites. *Journal of Appl. Pol. Sci.* **125** 1951 https://doi.org/10.1002/app.36302

[39] Rana S, Alagirusamy R and Joshi M 2011 Effect of carbon nanofiber dispersion on the tensile

properties of epoxy nanocomposites. *Journal of Compos. Mater.* **45** 2247 https://doi.org/10.1177/0021998311401076

[40] Zhou Y et al 2008 Improvement in mechanical properties of carbon fabric-epoxy composite using carbon nanofibers. *Journal Mater. Process. Technol.* **198** 445 https://doi.org/10.1016/j.jmatprotec.2007.07.028

[41] Rana S, Alagirusamy R and Joshi M 2010 Mechanical behavior of carbon nanofibre reinforced epoxy composites. *Journal Appl. Polym. Sci.* **118** 2276 https://doi.org/10.1002/app.30861

[42] Ahn S N, Lee H J, Kim B J, Tan L S and Baek J B 2008 Epoxy/amine-functionalized short-length vapor-grown carbon nanofiber composites. *Journal of Polym. Sci.* **46** 7473 https://doi.org/10.1002/pola.23052

[43] Wangab Q, Wenb G, Chenab J and Dang S S 2018 Reinforcing epoxy resin with nitrogen doped carbon nanotube: apotential lightweight structure material. *Journal of Mater. Sci. Technol.* **34** 2205 https://DOI: 10.1016/j.jmst.2018.02.021

[44] Jaemin C, Gwang H J, Jong K P, Jung C K, Ho J R and Soon H H 2017 Improvement of modulus, strength and fracture toughness of cnt/epoxy nanocomposites through the functionalization of carbon nanotubes. *Compos. part B: Eng.* **129** 169 https://doi.org/10.1016/j.compositesb.2017.07.070

[45] Chen W, Shen H, Auad M L, Huang C and Nutt S 2009 Basalt fiber–epoxy laminates with functionalized multi-walled carbon nanotubes. *Compos. part A: Appl. Sci. Manufact.* **40** 1082 https://doi.org/10.1016/j.compositesa.2009.04.027

[46] Gojny F H, Wichmann M H, Fiedler B, Bauhofer W and Schulte K 2005 Influence of nano-modification on the mechanical and electrical properties of conventional fibre-reinforced composites. *Compos. part A: Appl. Sci. Manufact.* **36** 1525 https://doi.org/10.1016/j.compositesa.2005.02.007

[47] Rahmanian S, Suraya A R, Shazed M A, Zahari R and Zainudin E S 2014 Mechanical characterization of epoxy composite with multiscale reinforcements: carbon nanotubes and short carbon fibers. *Mater. & Design.* **60** 34 https://doi.org/10.1016/j.matdes.2014.03.039

[48] Aradhana R, Mohanty S and Kumar S N 2018 Comparison of mechanical, electrical and thermal properties in graphene oxide and reduced graphene oxide filled epoxy nanocomposite adhesives. *Journal in Polym.* **141** 109 https://DOI: 10.1016/j.polymer.2018.03.005

[49] Bhattacharyya A, Rana S, Parveen S, Fangueiro R, Alagirusamy R and Joshi M 2013 Mechanical and thermal transmission properties of carbon nanofiber-dispersed carbon/phenolic multiscale composites. *Journal of Appl. Polym. Sci.* **129** 2383 https://doi.org/10.1002/app.38947

[50] Rana S, Alagirusamy R and Joshi M 2011 Development of carbon nanofibre incorporated three phase carbon/epoxy composites with enhanced mechanical, electrical and thermal properties. *Compos. part A: Appl. Sci. Manufact.* **42** 439 https://doi.org/10.1016/j.compositesa.2010.12.018

[51] Bekyarova E et all 2007 Multiscale carbon nanotube–carbon fiber reinforcement for advanced epoxy composites. *Langmuir* **23** 3970 https://DOI: 10.1021/la062743p

[52] Ali S, Boming Z and Changchun W 2014 Mechanical characterization of glass/epoxy polymer composites sprayed with vapor grown carbon nano fibers. *Conf. on Appl. Sci. & Technol* ***1*** 52 https://DOI: 10.1109/IBCAST.2014.6778120

The heat treatment duration effect on the efficiency of a catalyst for carbon nanotubes synthesis

E Burakova*, G Besperstova, A Tkachev, T Dyachkova, E Tugolukov, I Gutnik and N Orlova

Department "Technology and Methods of Nanoproducts Manufacturing", *Tambov State Technical University,* 106 Sovetskaya Street, 392000, Tambov, Russian Federation

*E-mail: elenburakova@yandex.ru

Abstract. The present paper shows that the duration of the pre-catalyst heat treatment stage implementation for obtaining a metal oxide Co-Mo/Al_2O_3 catalyst through thermal decomposition has a significant impact on its effectiveness in the synthesis of carbon nanotubes (CNTs) via chemical vapor deposition (CVD). The size of active sites formed and the specific surface area of the Co-Mo/Al_2O_3 catalyst were found to depend on the pre-catalyst heat treatment conditions. It was experimentally proved that the characteristics of both the catalyst (specific surface area and efficiency) and the CNTs (diameter and degree of defectiveness) can be controlled by the mode parameters of the thermal decomposition stage. The research results confirm the possibility of obtaining an effective catalyst system without implementing the secondary heat treatment stage - calcination. Moreover, the use of the Co-Mo/Al_2O_3 catalyst, obtained according to this procedure, in the CVD process allows to reduce the cost of CNTs synthesized.

1. Introduction

Carbon nanotubes (CNTs) are widely used as modifiers of many construction and functional materials. Various composites possessing improved performance characteristics [1, 2], as well as materials absorbing electromagnetic waves [3, 4], accumulating electrochemical energy [5] and hydrogen, and carrying out targeted drug delivery, have been developed on the basis of these nanostructures. There exist research works, in which CNTs are involved in recent neurocomputer developments to create connections between biological neurons and electronic devices. CNTs are employed in many areas due to their unique properties. They are being produced on a commercial scale; however, for each application area, nanostructures with a specific morphology are required, so it is very important to be able to manage their characteristics (diameter, degree of defectiveness, length, *etc*) at the synthesis stage.

There are various methods for synthesizing CNTs, but chemical vapour deposition (CVD) is the most preferable option for their industrial manufacturing [6-8]. To implement this method a catalyst is needed since the formation and growth of nanotubes occurs at its active sites. Therefore, the structure and morphology of CNTs are affected not only by the mode parameters of the synthesis procedure (temperature and pressure), nature [7] and consumption of carbon-containing raw materials [9, 10], but also by the characteristics of the catalyst [11-13]. According to the paper [14], the CNTs diameter depends on the size of the active catalyst particles; therefore, properly selected catalysts and synthesis conditions make it possible to obtain nanoproducts with specified characteristics. In recent years, many catalyst compositions have been developed for CNTs synthesis, but obtaining an effective catalytic system to directly create nanostructures (by length, diameter, structure, degree of defectiveness, *etc*) still remains an urgent issue.

There exist several methods for producing catalysts, but thermal decomposition is the most popular one. Its widespread use in obtaining catalysts for synthesizing CNTs can be explained by the simplicity of its implementation, as well as by the ability to control the formation of active sites of the catalyst system at the stage of its manufacturing [15].

The main stages of obtaining the catalyst through the thermal decomposition are as follows: 1) selection of reagents and preparation of a solution from them, 2) heat treatment of a pre-catalyst (decomposition and/or calcination), and 3) mechanical activation. In the work [16], it is shown that the quality of the initial reagents and the heat treatment conditions for the solution on the basis thereof significantly affect the stability and efficiency of the catalyst synthesized. The methods of initial components validation proposed by the authors have made it possible to exclude the use of low-quality reagents at the initial stage of catalyst preparation, thereby contributing to the improvement of the quality of the nanoproduct obtained.

Since the structure of the catalyst and its active grains are formed as a result of the thermal decomposition, special attention should be paid to studying the heat treatment stage. Therefore, in the present work, the effect of the mode parameters of the heat treatment (thermal decomposition) of an initial components solution on the characteristics of Co-Mo/Al_2O_3 catalyst systems and the morphology of CNTs synthesized was experimentally investigated.

2. Materials and methods

A cobalt-containing catalyst system (Co-Mo/Al_2O_3) was chosen for studies. The catalyst was obtained by thermal decomposition. In order to increase its efficiency, it was chemically activated (promotion); molybdenum was used as a promoter, since its introduction to the catalyst in small amounts contributed to an increase in the efficiency of the metal oxide system considered. $Co(NO_3)_2{\cdot}6H_2O$ (chemically pure), $Al(NO_3)_3{\cdot}9H_2O$ (chemically pure), $(NH_4)_6Mo_7O_{24}{\cdot}4H_2O$ (pure), and tribasic carboxylic acid served as initial reagents to prepare a solution of catalyst initial components (a so-called pre-catalyst).

The catalyst, obtained from the pre-catalyst solution after heat treatment (thermal decomposition), was employed to synthesize CNTs through the CVD method, and ethylene was used as a carbon-containing raw material. The synthesis of nanostructures was carried out in a laboratory batch reactor at 650 °C.

The catalyst, obtained from the pre-catalyst solution after heat treatment (thermal decomposition), was employed to synthesize CNTs through the CVD method, and ethylene was used as a carbon-containing raw material. The synthesis of nanostructures was carried out in a laboratory batch reactor at 650 °C.

The X-ray phase analysis of the obtained Co-Mo/Al_2O_3 catalyst samples was carried out using a Difrey-401 X-ray diffractometer (Scientific Instruments, St. Petersburg, Russia) with a Bragg-Bretano focus.

The specific CNTs yield (g_C/g_{cat}, where C means "carbon", and cat – "catalyst") was used as a parameter characterizing the efficiency of the Co-Mo/Al_2O_3 catalyst. The specific surface area of the catalyst system was determined using a "Sorbtometr-M" analyzer (KATAKON, Novosibirsk, Russia), the principle of which is based on the thermal desorption of an adsorbate gas from the surface of the sample under dynamic conditions.

A scanning electron microscopy (SEM) instrument (Merlin, Carl Zeiss, Germany) was used to assess the morphology of the catalyst system and the CNTs synthesized on it. The degree of the CNTs defectiveness CNTs was determined with a DXR Raman microscope (Thermo Scientific, USA) based on the intensity ratio of the D and G peaks.

The initial reagents intended for preparing the Co-Mo/Al_2O_3 catalyst were dissolved in a small amount of distilled water by vigorous stirring (100-150 rpm) for 15 min at 55 °C. At this stage, a problem of incomplete dissolution of the tribasic carboxylic acid occurred due to the presence of white crystals in the solution. The availability of such crystals in the pre-catalyst at the thermal decomposition stage is highly undesirable, since this leads to a decrease in the efficiency of the catalyst system synthesized. In this case, it is possible to increase the mixing intensity to~200 rpm and the dissolution time to maximum 5 min, since partial removal of the solvent takes place with a significant increase in the duration of this process, thereby leading to undesirable crystallization of the solution of the initial components even before their dissolution. This was enough to completely dissolve the components of the catalyst system under study; after that, the 50-mL solution was thermally decomposed at 350 °C. The duration of the pre-catalyst decomposition stage was varied from 15 to 50 min, after which the resulting catalyst systems were ground and used for the CNTs synthesis. During the experiment, it was found that the solution decomposition for 5-10 min at 350 °C is impractical, since this time is not enough to completely decompose the organic matter and form an active catalyst system.

3. Results and discussion

The results obtained during the experiment are presented in figure 1. The effectiveness of the obtained Co-Mo/Al_2O_3 catalyst in the CVD process was evaluated by the specific CNTs yield.

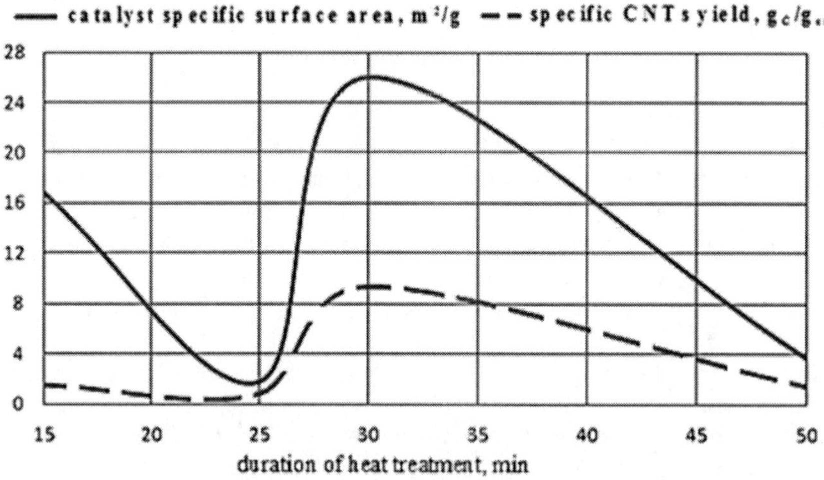

Figure 1. The effect of the heat treatment (thermal decomposition) of the initial components solution of the Co-Mo/Al_2O_3 catalyst on its specific surface area and efficiency during the CNTs synthesis.

Analyzing these data, one can conclude that the most efficient metal oxide Co-Mo/Al_2O_3 catalyst systems with higher specific surface areas (16.8 and 26.0 m^2/g) are formed as a result of the thermal decomposition of the pre-catalyst solution during 15 and 30 min at 350 °C. With further increase in the duration of this process, the formation of Co_3O_4 particles on the surface of the catalytic system can be observed. After the reduction of these particles, metal cobalt particles, responsible for nanotubes formation and growth, are formed. The diffractogram obtained for these catalyst conditions is shown in figure 2.

IOP Conf. Series: Materials Science and Engineering **693** (2019) 012016 doi:10.1088/1757-899X/693/1/012016

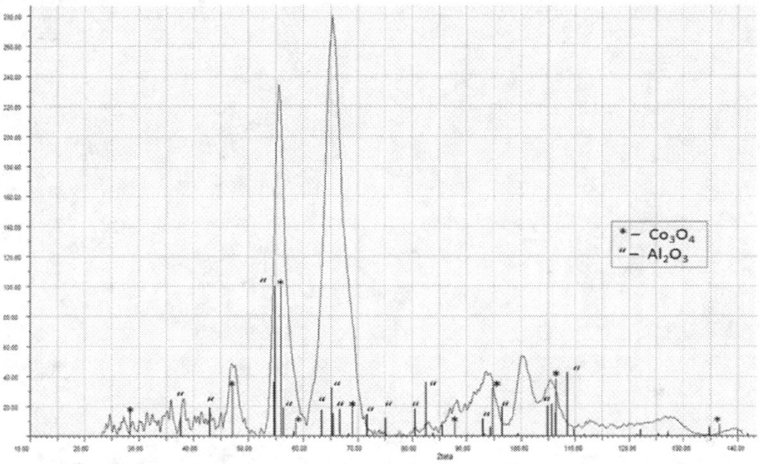

Figure 2. A diffractogram of the Co-Mo/Al_2O_3 catalyst system.

The analysis of figure 2 confirms the availability of Co_3O_4 in the Co-Mo/Al_2O_3 catalyst samples obtained.

The morphology of the resulting catalyst systems obtained under these conditions is shown in figures 3(a) and 3(b), respectively.

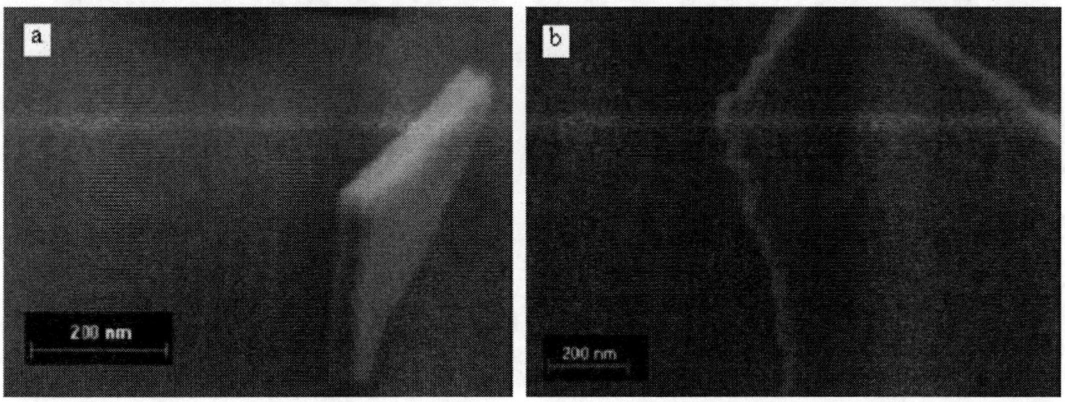

Figure 3. The morphology of the Co-Mo/Al_2O_3 catalyst systems obtained as a result of thermal the decomposition at 350 °C for 15 min (a) and 30 min (b).

A further increase in the solution decomposition duration leads only to a significant decrease in the specific surface area and the efficiency of the catalyst synthesized. The use of the Co-Mo/Al_2O_3 catalyst system, formed as a result of the thermal decomposition for 15 min at 350 °C, in the CNTs synthesis via CVD allows to get CNTs having diameters of 20-80 nm (figure 4(a)). The catalyst obtained after the thermal decomposition for 30 min contributes to the synthesis of 5-20-nm CNTs (figure 4(b)).

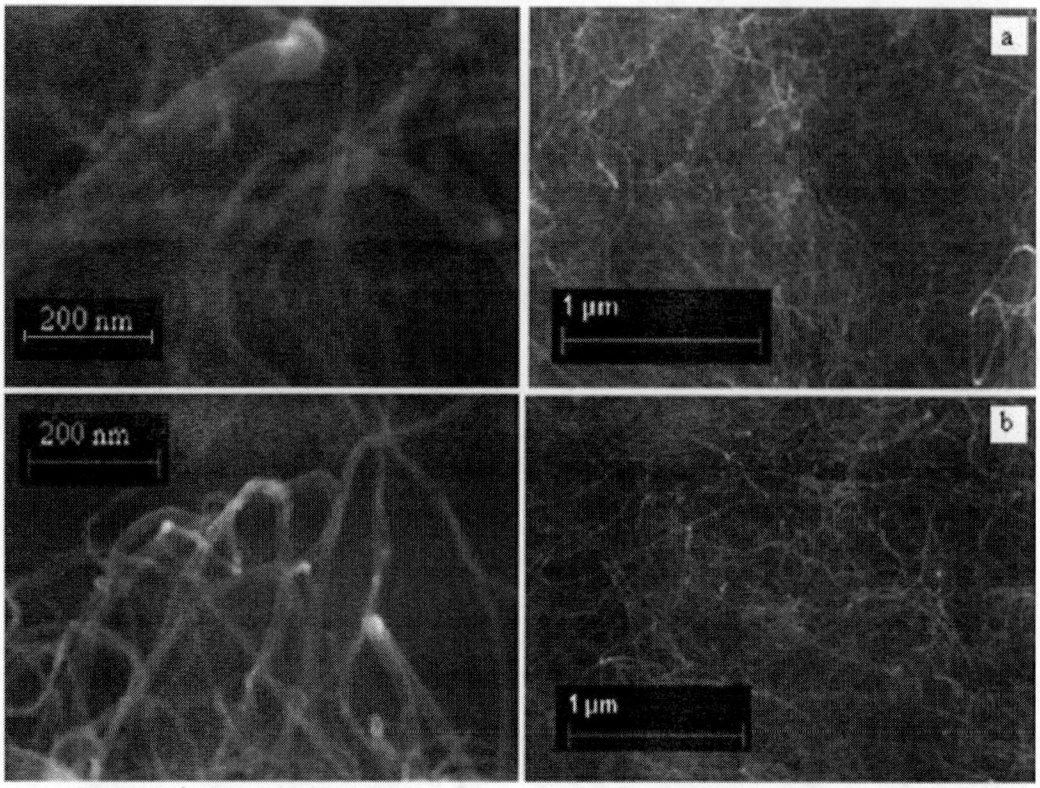

Figure 4. SEM images of the CNTs synthesized on the Co-Mo/Al_2O_3 catalyst obtained after the thermal decomposition of the pre-catalyst solution at 350 °C for 15 min (a) and 30 min (b).

The analysis of the SEM images recorded shows that the catalyst sample synthesized after the pre-catalyst thermal decomposition at 350°C for 30 min is more uniform, and in both cases, the CNTs formed have a length of more than 2 microns.

The degree of defectiveness (presented by the D/G ratio) of the CNTs synthesized on the catalyst systems obtained was assessed using Raman spectroscopy, the Raman spectra recorded are given in figure 5.

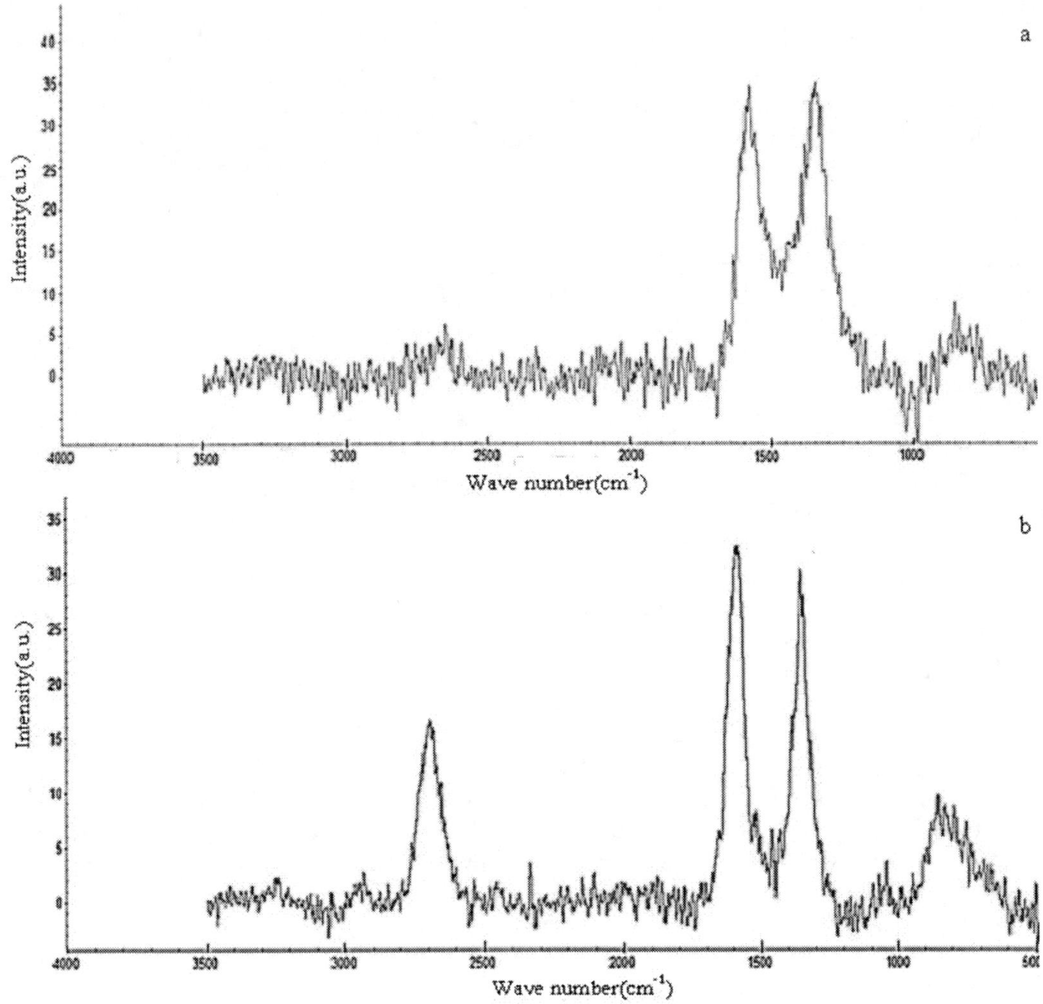

Figure 5. Raman spectra recorded for the CNTs synthesized on the Co-Mo/Al_2O_3 catalyst obtained after the thermal decomposition of the pre-catalyst solution at 350 °C for 15 min (a) and 30 min (b).

The analysis of the Raman spectra makes it possible to estimate the D/R ratios values – 1.05 (figure 5(a), and 0.784 (figure 5(b)). Thus, it was experimentally shown that, by eliminating the second heat treatment stage – calcination – in obtaining the metal oxide catalyst systems through the thermal decomposition, an efficient catalyst for CNTs synthesis can be obtained.

The increase in the pre-catalyst heat treatment duration over 30 min leads to getting a catalyst which provides the synthesis of CNTs with a greater degree of defectiveness. Thus, it seems more rational to carry out the pre-catalyst heat treatment at 350 °C for 30 min.

4. Conclusion

Usually, when a catalyst system is obtained via nitrates thermal decomposition, the pre-catalyst heat treatment is implemented in two stages: thermal decomposition, and calcination. It can be assumed that at the first stage, the formation of active sites takes place, whereas at the second stage, they are activated. After having studied in details the method for producing Co-Mo/Al_2O_3 catalyst system, it became possible to synthesize an effective catalyst avoiding the second heat treatment stage – calcination. Performing the pre-catalyst thermal decomposition at 350 °C makes it possible to develop catalysts with a small specific surface area (~ 2.0-26.0 m^2/g). By varying the pre-catalyst heat treatment duration (10-50 min), one can control not only the characteristics of the catalytic system, but also the parameters of the CNTs synthesized on it. As a result of the study, rational modes of the thermal decomposition stage implementation were selected. The specific yield of the CNTs synthesized on the catalytic system obtained after the pre-catalyst thermal decomposition for 30 min was found to be ~9.4 g_C/g_{cat}. Using the Co-Mo/Al_2O_3 catalyst in the CVD process makes it possible to synthesize CNTs with a diameter of 5-30 nm and a degree of defectiveness of 0.784. Besides, reduces the cost of these CNTs.

Acknowledgments
The work was supported by the Russian Foundation for Basic Research (RFBR) within the framework of the research project No. 18-43-680005.

References
[1] Vanyorek L, Sikora E, Balogh T, Román K, Marossy K, Pekker P, Szabó T J, Viskolcz B and Fiser B 2019 Nanotubes as polymer composite reinforcing additive materials – A comparative study *Arab. J. Chem.* https://doi.org/10.1016/j.arabjc.2019.01.001
[2] Rukhov A V, Dyachkova T P and Anosova I V 2017 Application of the solution of the inverse problem of heat exchange mathematical modeling to study the kinetics of the processes of modifying the carbon nanotubes surface with polyaniline *Fundamental Research* [in Russian] **10** 278
[3] Liu L, Zhang S, Yan F, Li C, Zhu C, Zhang X and Chen Y 2018 Three-dimensional hierarchical MoS_2 nanosheets/ultralong N-doped carbon nanotubes as high-performance electromagnetic wave absorbing material *ACS Appl Mater Interfaces* **10**(16) 14108
[4] Zhao T, Ji X, Jin W, Wang C, Ma W, Gao J, Dang A, Li T,Shang S and Zhou Z 2017 Direct in situ synthesis of a 3D interlinked amorphous carbon nanotube/graphene/$BaFe_{12}O_{19}$ composite and its electromagnetic wave absorbing properties *RSC Adv* **7** 15903
[5] Yang Y, Qiao S, Jin R, Zhou J and Quan X 2018 A novel aerobic electrochemical membrane bioreactor with CNTs hollow fiber membrane by electrochemical oxidation to improve water quality and mitigate membrane fouling *Water Res* **151** 54
[6] Sengupta J 2018 *Handbook of Nanomaterials for Industrial Applications* (Amsterdam: Elsevier) p 172
[7] Ghaemi F, Ali M, Yunus R and Othman R N 2019 Synthesis of carbon nanomaterials using catalytic chemical vapor deposition technique *Synthesis, Technology and Applications of Carbon Nanomaterials* 1–27 doi:10.1016/b978-0-12-815757-2.00001-2
[8] Malekimoghadam R and Rafiee R 2018 Carbon nanotubes processing *Carbon Nanotube-Reinforced Polymers: From Nanoscale to Macroscale* 41–59 doi:10.1016/b978-0-323-48221-9.00003-0
[9] Mohammed I A, Bankole M T, Abdulkareem A S, Ochigbo S S, Afolabi A S, Abubakre O K 2017 Full factorial design approach to carbon nanotubes synthesis by CVD method in argon environment *S. Afr. J. Chem. Eng.* **24** 17
[10] Oyewemi A, Abdulkareem A S, Tijani J O, Bankole M T, Abubakre O K, Afolabi A S and Roos WD 2019 Controlled syntheses of multi-walled carbon nanotubes from bimetallic Fe–Co catalyst supported on kaolin by chemical vapour deposition method *Arab. J. Sci. Eng.* doi: 10.1007/s13369-018-03696-4
[11] Bernardo G P O, Abella L, Auresenia J and Hinode 2017 Ab initio molecular dynamics thermal decomposition of methane on Ni-Cu/γ-Al_2O_3 catalysts for CNT production: Effect of Cu promoter *Comput. Cond. Mat.* **11** 20
[12] He L, Hua S, Jianga L, Liao G, Chen X, Han H, Xiao L, Ren Q, Wanga Yi, Sua Sh and Xiang J 2018 Carbon nanotubes formation and its influence on steam reforming of toluene over Ni/Al_2O_3 catalysts: Roles of catalyst supports *Fuel Process. Technol.* **176** 7-14

[13] Guo Yo, Zhai G, Ru Yu, Wu Ch, Jia X, Sun Ya, Yu J, Kang Zh and Sun B 2018 Effect of different catalyst preparation methods on the synthesis of carbon nanotubes with the flame pyrolysis method *AIP Adv.* **8** 035111

[14] Shunin Yu, Bellucci S, Gruodis A and Lobanova-Shunina T 2017 CNT and graphene growth: growing, quality control, thermal expansion and chiral dispersion *Nonregular Nanosystems: Theory and Applications* **26** 207

[15] Rukhov A V, Burakova E A, Bakunin E S, Besperstova G **S**, Tugolukov E N and Han Yu A 2017 Features of technology of preparation of catalytic systems by thermal decomposition for synthesis of carbon nanotubes *Inorg. Mater. Appl. Res.* **8** 802

[16] Burakova E, Dyachkova T, Besperstova G, Rukhov A and Tugolukov E 2017 Peculiarities of obtaining a catalyst for the synthesis of nanostructured carbon materials via thermal decomposition *AIP Conf. Proc.* **1899** 020008

IOP Conf. Series: Materials Science and Engineering **693** (2019) 012017 doi:10.1088/1757-899X/693/1/012017

TEM studies of conical scroll carbon nanotubes formed by aerosol synthesis

B Kulnitskiy[1,2], A Karaeva[1], V Mordkovich[1,2], S Urvanov[1] and A Bredikhina[1,2]

[1]*Technological Institute for Superhard and Novel Carbon Materials*, 7a Centralnaya Street, Troitsk 108840, Moscow, Russia
[2]*Moscow Institute of Physics and Technology State University*, 9 Institutskiy Lane, Dolgoprudny141700, Moscow Region, Russia

E-mail: boris@tisnum.ru

Abstract. Conical carbon nanotubes were obtained by the method of gas-phase chemical decomposition of a carbon precursor on the surface of a catalytic particle suspended in a carrier gas stream (also known as "aerosol method"). TEM study at the JEM-2010 microscope showed that the observed angles at the vertex of the cones (apex angles) were predominantly 5°, 10°, 15° and 20° regardless of the presence of a catalyst particle inside the cone. The analysis of the apex angles witnesses that the nanotubes have a scroll type of structure. Two alternative explanations are proposed for the formation of an unconventional apex angle 5°. The peculiarities of destruction of the catalyst particles inside the nanotubes are revealed.

1. Introduction

Carbon nanostructures have been in the focus of researchers' attention for several decades. This interest is determined in particular by their interesting electronic and mechanical properties. A detailed structural investigation of carbon nanotubes was given by Iijima [1]. The formation of carbon nanotubes follows different organizational schemes. For example, where they either nest within each other as coaxial graphene cylinders or cones, or where a graphene sheet rolls up into a scroll. Numerous types of helical nanostructures were described such as helix-shaped graphite nanotubes [2], graphitic nanocones [3], nanohorns [4], conical crystals and hollow conical structure [5,6], conical graphitic nanofibres [7] and the helically coiled nanofiber with conical cross-section [8]. There are suggestions that the conical structures can be used in biomedical applications as containers for precision delivery of medicine (nanodroppers) or, as another example, in scanning probe technique applications as cantilevers.

Various conical structures were earlier described in literature [3, 9-11]. The work [3] stipulates that only five angles can exist at the vortex of a single isolated cone, namely: 112.9°, 83.6°, 60°, 38.9° and 19.2°. This phenomenon is related with the limitation in the number of possible pentagon insertions into a graphene sheet, which forms a single cone by scrolling itself. It was shown in [10], however, that a scroll implies larger variety of apex angles. All that variety includes the pentagon-related angles mentioned above. The authors of [10] suggested that a scroll is based on the same graphene packing sequence as in graphite, i.e. ABABAB. This sequence means that neighboring sheets are rotated 30° relative to each other, while every third sheet is rotated 60° relative to the first one. This work reports an attempt of electron microscopy investigation of structural peculiarities of conical scroll carbon nanotubes formed by the aerosol synthesis.

2. Experimental part

The synthesis of carbon nanotubes was performed in a quartz reaction tube (with the inner diameter of 35 mm, 1350 mm long) placed in a vertical tube furnace. Carbon nanotubes were obtained by the method of gas-phase chemical decomposition of a carbon precursor on the surface of a catalytic particle suspended in a carrier gas stream (also known as "aerosol method") at a temperature of 1150–1200 °C and at a pressure of 1 atm. A schematic representation of the setup is shown in figure 1.

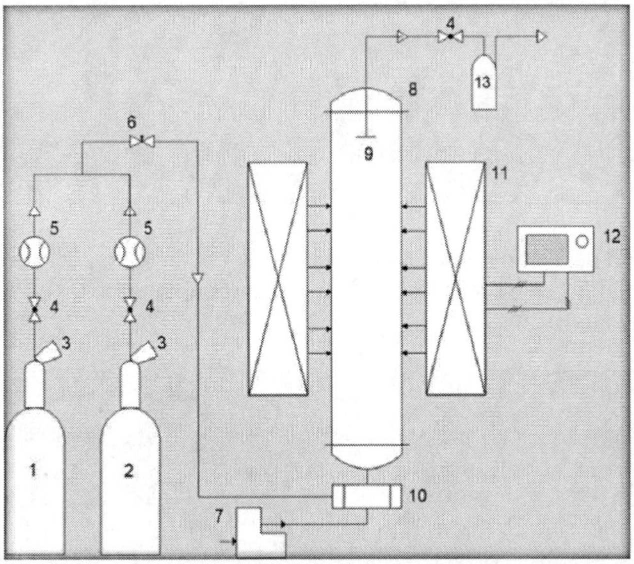

Figure 1. Schematic representation of the laboratory setup: 1 – He cylinder, 2 – H_2 cylinder, 3 – gas pressure reducer, 4 – valve, 5 – mass flow controller, 6 – fine regulation valve; 7 – input device, 8 – quartz tube reactor; 9 – "catcher"; 10 – nebulizer, 11 – furnace; 12 – temperature regulator; 13 – filter system.

Ethanol was used as the carbon precursor, and the nanotube growth activator was thiophene, while Fe nanoparticles served as a catalyst. Fe nanoparticles were generated from a ferrocene $(C_2H_5)_2Fe$ precursor in the same way as in our earlier works [12,13]. The main difference with earlier works was that the liquid component was introduced in the amount of 1% by a nebulizer device, i.e. it was sprayed into a fine aerosol using an ultrasonic nebulizer. Fine aerosol was directed to the synthesis zone by the carrier gas flow. The use of the nebulizer provided different reaction zone conditions, which allowed us to synthesize a specific nanotube product with structural differences with the results of [12,13].

The obtained samples were studied by high-resolution electron microscopy. The research results showed that the most interesting was the carbon deposit obtained using ferrocene (1 wt.%) as a catalyst precursor. The carbon deposit was analyzed by means of transmission electron microscopy.

3. Results and discussion

TEM showed the presence of conical nanotubes mainly. There were also onion-like structures and fragments of amorphous carbon. Conical nanotubes can either consist of multiple cones stacked inside each other or represent conical scrolls also known as cone-helix structures. The nanotubes of cone-helix structure are visually indistinguishable from stacked cone nanotubes in TEM images. Apex angle is the only reliable distinction between the two types. The measurement of apex angles showed that 5°±1° and 10°±1° are the most common. The former was detected in 17 cases, while the latter was detected in 18 cases. Apex angles of 20°±1° and 30°±1° occurred 11 times each. Apex angles of 25°±1° and 49°±1° were detected 3-4 times each. Some of the nanotubes contained catalyst inside, while the others were hollow. It is interesting that identical apex angles were observed in both filled and hollow nanotubes. Figure 2 demonstrates a fragment of a hollow conical nanotube with apex angle of 10° accompanied with corresponding FFT images. Figure 3 shows a filled nanotube with apex angle of 20°.

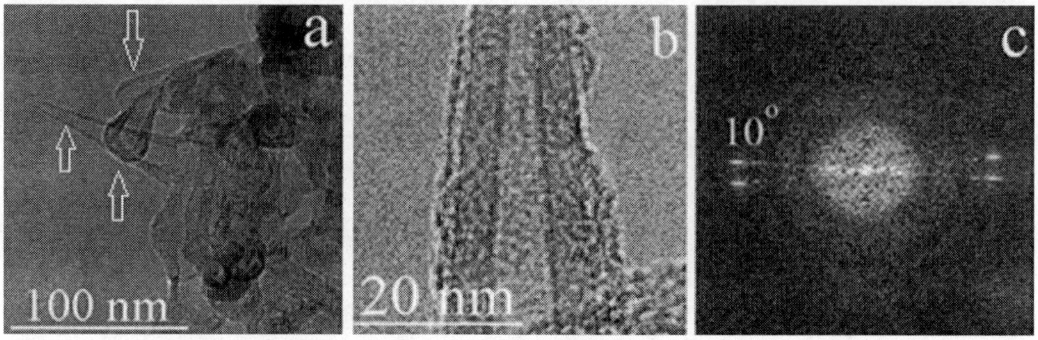

(a) (b) (c)

Figure 2. Carbon deposit obtained using ferrocene: (a) three conical nanotubes are shown by arrows; (b) the fragment of a hollow conical carbon nanotube with apex angle of 10°; (c) FFT image from (b).

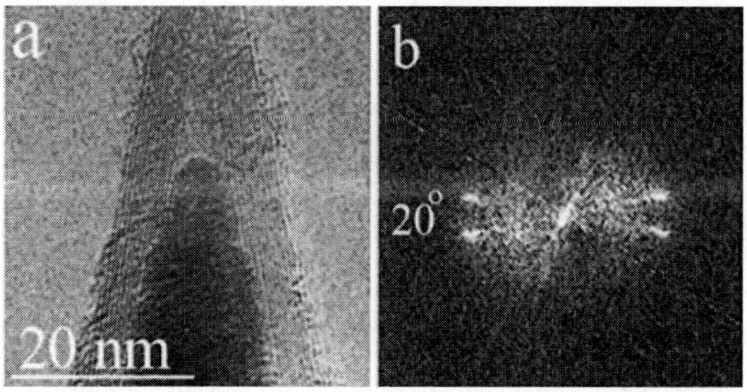

(a) (b)

Figure 3. (a) The fragment of a conical carbon nanotube, filled by α- Fe with the apex angle of 20°; (b) FFT image from (a).

The formation of multiwalled conical structures instead of double-walled cylindrical nanotubes as in our earlier works [12,13] can be explained taking into account different conditions of formation of the reaction mixture. The nebulizer device, being a very precise mass-flow regulation device, gives, nevertheless, an aerosol of small droplets instead of a homogeneous vaporized reaction mixture as in the earlier used saturator device. The presence of droplets may lead to formation of the catalyst precursor particles as a result of evaporation of ethanol. The formation of Fe nanoparticles from these ferrocene particles would occur not like the formation from vaporized molecules of that same ferrocene and result in larger, differently crystallized catalyst nanocrystals and hence to different structure of the carbon nanotubes.

One can suppose several ways of formation of scroll-type conical structures. First of all, it can be done by insertion of pentagons or heptagons into a graphene sheet. It allows for the formation of structures with five different apex angles. The work [14] considered a conical scroll, where neighboring layers are rotated by constant value θ relative to each other. The θ angle is directly related to the semi-apex angle φ_o of the cone by the relation:

$$\Theta=2\pi sin\varphi_o.$$

One observes pseudo-hexagonal symmetry with ABABAB stacking. This stacking is different from graphite stacking by sheet-to-sheet rotation. It is known, however, that pseudo-orthorombic symmetry may occur in carbon nanotubes with cone stacking ABCABC [15]. In such case, the sheet rotation angle should be multiple of 20°. This packing can be considered as a kind of a turbostratic structure, where the packages of parallel graphenes are inclined by chaotic angles relative to each other.

Braga [6] has shown that a rolled graphene sheet becomes more stable than the undistorted one, and that the curling process becomes self-sustaining, when a minimum overlap of successive layers has occurred. Therefore, in order to start the sheet-to-scroll transition input of a certain threshold energy is required in a first step (excitation by high temperature, sonication, irradiation etc.) [16].

Semi-apex angles for cone-shaped graphene were investigated in [14,17,18]. Table 1 shows the values of semi-apex angles for scrolled nanotubes as a function of the hexagon rotation angle (φ) and as a function of the number of pentagons.

Table 1. Semi-apex angles in conical scroll structure, formed by two methods

Rolled graphene sheet θ=2π sinφ			Introduction of pentagons
θ(°) - the rotation angle of the hexagons		Semi-apex angles φ(°)	φ(°) Semi-apex angles; the number of pentagons (*P*) is shown in brackets *sin*φ =1-P/6
ABCABC packing	ABAB packing		
20		3,18	
	30	4.78	
40		6,36	
	60	9.59	9.59 (5)
80		12.83	
	90	14.47	
100		16.12	
	120	19.47	19.47 (4)
	150	24.62	
	180	30	30 (3)
	210	35.68	
	240	41.81	41.81 (2)
	270	48.59	
	300	56.44	56.44 (1)
	330	66.44	
	360	90	

The apex angle of 10°±1° (semi-apex angle appr. 5°) indicates exactly a simple scrolling of a graphene sheet. The apex angle of 30° (semi-apex angle 15°) manifests the scrolling without pentagon insertion as well. As far as the 20° apex angle (semi-apex angle 10°) is concerned, it can be formed by either insertion of five pentagons or by simple scrolling of a graphene sheet. The value of 5°±1° (semi-apex angle 2.5°-3°) can be found in Table 1 as manifestation of pseudo-hexagonal symmetry in a conical nanotube. However, another explanation can be proposed for the 5°±1° apex angle, namely, the insertion of one heptagon as suggested by [10]. Conical nanotubes can also be formed by graphene cones nested inside each other, which can also require the pentagon insertion. The semi-apex angle data shown in Table 1 cannot help distinguishing nested structure from scrolls for all the cases listed in

IOP Conf. Series: Materials Science and Engineering **693** (2019) 012017 doi:10.1088/1757-899X/693/1/012017

the third column of the table. However, the data in the first and second columns witness a scrolled structure.

The presence of conical nanotubes with identical apex angle regardless of the presence of catalyst particles inside witnesses that the formation of cones is not related with a catalyst (figure 4). TEM images in figure 5 show particles inside a nanotube (two particles in figure 5a and three particles in figure 5b). The particle morphology in figure 5a suggests that the observed pattern is the result of destruction of an initial big particle into smaller splinters as cup-cone. Figure 5b manifests a more complicated pattern, i.e. an α-Fe particle in the middle with Fe_3C particles surrounding it. It is necessary to note that a cup-cone pattern is typical for tensile destruction of some cylindrical samples. Investigations show that the destruction proceeds through formation of a neck and consequent fracture forming a crack, which is symmetrical relative to the axis of a sample. Hardness of iron is 180 GPa. Cementite is very brittle but hard (HB more than 800) and is characterized by zero plasticity. Due to three-dimensional stress the neck material is capable of any plastic deformation, so the stress leads to cracking in the neck area, while the other parts of that same particle keep elongating. It explains why the central fracture area of cup-cone type is different from the peripheral areas, where plastic deformation keeps going [19].

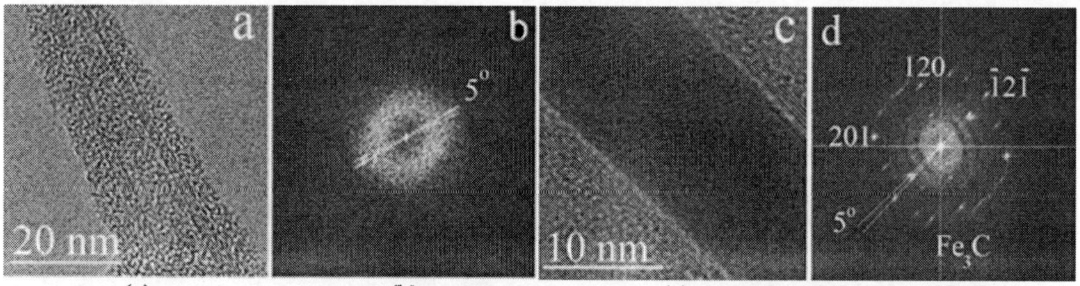

(a) (b) (c) (d)

Figure 4. (a) the fragment of a hollow conical carbon nanotube with the apex angle of 5°; (b) FFT image from (a); (c) the fragment of a conical carbon nanotube filled with Fe_3C with the apex angle of 5°; (d) FFT image from (c).

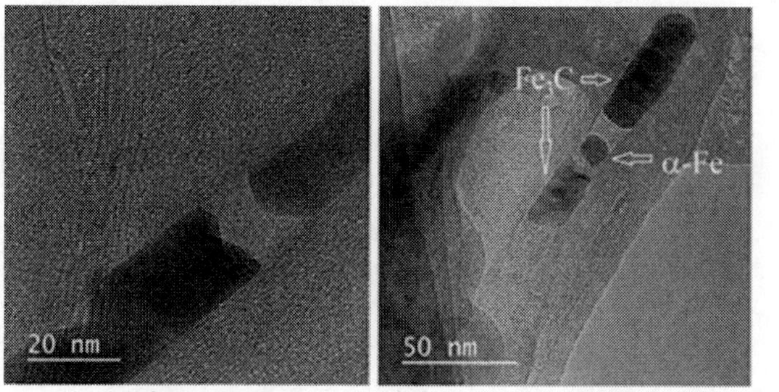

(a) (b)

Figure 5. (a) Ductile fracture. Cup and cone; (b) Three particles inside a carbon nanotube. An α-Fe particle is in the center and two Fe_3C particles are on the edges.

The metal (α-Fe) takes the shape as determined by the surrounding nanotube. The metal is pressurized by carbon sheets, thus subjecting it to tensile stress up to destruction of the particle. The fracture of a cup-cone type is a sign of plastic deformation, which is followed by carbide formation.

4. Conclusion

Conical carbon nanotubes were formed by the aerosol synthesis. A wide range of apex angles of nanotubes was revealed by TEM methods. Analysis of the values of apex angles shows that at least a part of the nanotubes have a scroll type of structure. The formation of a scroll with ABCABC stacking sequence can explain the unusual apex angle of 5°. It is shown that the metal (α-Fe) inside the nanotube is deformed by elongation along the nanotube axis and shrinking across. This deformation leads to a fracture caused by tensile stress.

5. Acknowledgments

This work was supported by Ministry of Science and Higher Education of the Russian Federation (project ID RFMEFI57417X0162; the agreement No. 14.574.21.0162).

References

[1] Iijima S 1991 Helicalmicrotubules of graphite carbon. *Nature* **354** 56 DOI: 10.1038/354056a0
[2] Amelinckx S, Zhang X B, Bernaerts D, Zhang X F, Ivanov V and Nagy J 1994 A Formation Mechanism for Catalytically Grown Helix-Shaped Graphite Nanotubes. *Science* **265** 5172 635 DOI:10.1126/science.265.5172.635
[3] Krishnan A, Dujardin E, Treacy M, Hugdahl J, Lynum S and Ebbesen T 1997 Graphitic cones and the nucleation of curved carbon surfaces. *Nature* **388** 451 https://doi.org/10.1038/41284
[4] Karousis N, Suarez-Martinez I, Ewels C and Tagmatarchis N 2016 Structure, Properties, Functionalization, and Applications of Carbon Nanohorns. *Chem. Rev.* **116** 4850 DOI: 10.1021/acs.chemrev.5b00611
[5] Mani R, Li X, Mahendra M and Rajan K 2003 Carbon Nanopipettes. *Nanoletters* **3** 5 671 DOI: 10.1021/nl034125o
[6] Braga S, Coluci V, Legoas S, Giro R, Galvao D and Baughman R 2004 Structure and Dynamics of Carbon Nanoscrolls. *Nanoletters* **4** 5 881 DOI: 10.1021/nl0497272
[7] Terrones H, Hayashi T, Munoz-Navia M, Terrones M, Kim Y, Grobert N, R.Kamalakaran R, Dorantes-Davila J, Escudero R, Dresselhays M and Endo M 2001 Craphitic cones in palladium catalysed carbon nanofibers. *Chem. Phys. Lett.* **343** 241 https://doi.org/10.1016/S0009-2614(01)00718-7
[8] Blank V and Kulnitskiy B 2004 Proposed formation mechanism for helically coiled carbon nanotubes formation. *Carbon* **42** 3009 DOI :10.1016/1.carbon.2004.06.004
[9] Dumpala S, Bhimarasetti G, Gubbala S, Meduri, Kona S and Sunkara M 2008 Carbon Microtubes and Conical Carbon Nanotubes. *Smart Materials* **22** 15 DOI: 10.1201/9781420043730.ch22
[10] Blank V, Polyakov E, Kulnitskiy B, Nuzhdin A, Alshevskiy Yu, Bangert U, Harvey A and Davock H 1999 Nanocarbons, formed in a hot isostatic pressure apparatus. *Thin Solid Films* **346** 86 https://doi.org/10.1016/S0040-6090(98)01453-9
[11] Eksioglu B and Nadarajah A 2006 Structural analysis of conical carbon nanofibers. *Carbon* **44** 360 DOI:10.1016/j.carbon.2005.07.007
[12] Karaeva A, Khaskov M, Mitberg E, Kulnitskiy B, Perezhogin I, Ivanov L, Denisov V, Kirichenko A and Mordkovich V 2012 Longer Carbon Nanotubes by Controlled Catalytic Growth in the Presence of Water Vapor. *J. Fullerenes, nanotubes and Carbon nanostructures* **20** 411 http://dx.doi.org/10.1080/1536383X.2012.655229
[13] Mordkovich V, Kazennov N, Ermolaev V, Zhukova E and Karaeva 2018 A Scaled-up process for producing longer carbon nanotubes and carbon cotton by macro-spools. *Diamond and Related Mater.* **83** 15 https://doi.org/10.1016/j.diamond.2018.01.017
[14] Amelinckx S, Devouard A and Baronnet A 1996 Geometrical aspects of the diffraction space of serpentine rolled microstructures: their study by means of electron diffraction and microscopy. *Acta Crystallogr.* **A52** 850 DOI:10.1107/s0108767396006605

[15] Blank V, Kulnitskiy B, Kirichenko A, Memetov N, Dyachkova T and Tkachev A 2017 Features of structures obtained by grapheme nanoplatelets treatment in a diamond anvil high pressure cell. *Fullerenes, Nanotubes and Carbon Nanostructures.* **25** 8 488 https://doi.org/10.1080/1536383X.2017.1339694

[16] Schaper A, 2008 Carbon Nanoscrolls in Carbon nanotubes and related structures, eds. V D Blank, B A Kulnitskiy, Research Signpost, Trivandrum (Kerala, India) pp 1–48

[17] Iijima S, Ichihashi T and Ando Y 1992, Pentagons, heptagons and negative curvature in graphite microtubule growth. *Nature* **356** (6372) 776 DOI:10.1038/356776a0

[18] Ajayan P, Ichihashi T and Iijima S 2002 Distribution of pentagons and shapes in carbon nano-tubes and nano-particles. *Chem Phys Lett.* **5** (29) 284 https://doi.org/10.1016/0009-2614(93)90057-8

[19] Timoshenko S P 1947 *Strength of Materials Parts II: Advanced Theory and Problems* (D. Van Nostrand Company, Inc., Lancaster, P A) p 510

IOP Conf. Series: Materials Science and Engineering **693** (2019) 012018 doi:10.1088/1757-899X/693/1/012018

Flat electric heaters with the effect of self-regulation based on nanomodified polymer composite

V Yagubov[*], Alexander Shchegolkov, Alexey Shchegolkov, A Tkachev and A Sukhorukov

Technology and Methods of Nanoproducts Manufacturing, *Tambov State Technical University,* 106 Sovetskaya Street, 392000 Tambov Russian Federation

[*]E-mail: vitya-y@mail.ru

Abstract. The present paper considers the effect of multi-walled carbon nanotubes (acting as a part of a silicon-organic composite) on the properties of a self-regulating electric heater. Flat electric heaters were manufactured with a heat exchange area of 1600 m^2. The heater operating parameters were studied in the temperature range from -60°C to +70°C. The dynamics of the electric current consumed was investigated under self-regulation in the same temperature range It was revealed that when cutting the self-regulating heater into separate parts, the properties of self-regulation and power are maintained in proportion to the size.

1. Introduction

The improvement of electric heating devices is aimed at developing "smart" materials that cause the reduction of energy costs with the possibility of the formation of new functional properties. Currently, there has been a rapid transition from the known technologies based on the principles of electric heating associated with the use of metal conductors [1,2]. The latter are widely used in anti-slip devices [3], flexible heaters for space technology [4], local heaters, and pre-start heaters of motor vehicle components [5]. The transition to new materials allows the development of electric heaters which have a controlled heating rate at different values of the supply voltage of both direct and alternating current. Carbon nanotubes (CNTs) can be attributed to this type of materials [6,7]. They are used to form coatings on flexible substrates [8] or as a component which, in turn, can be made on the basis of various polymer matrices [9-11].

The use of CNTs as a modifier in composites makes it possible to develop novel electrically conductive polymers, the characteristics of which can be altered depending on external influencing factors (mechanical, climatic, *etc*) [12,13]. Many nanomodified polymer compositions have a positive temperature coefficient (PTC) [14]. Materials with the PTC properties can change the value of electrical resistance depending on its temperature. Such materials can be applicable for heating elements with the effect of temperature self-regulation [15].

Today on the world market there are a huge number of companies (*e.g.*, Eltrace, Ceilhit, Lavita) in the world market engaged in the production of heating cables with the effect of self-regulation. However, heating large surfaces using such cables leads to installation difficulties, which is economically disadvantageous. In this regard, the aim of the present research was to develop samples of flat heaters (SFH) with the effect of self-regulation, having a heating area larger than that of heating cables.

IOP Conf. Series: Materials Science and Engineering **693** (2019) 012018 doi:10.1088/1757-899X/693/1/012018

2. Materials and methods

2.1. Materials
As the basis for SFHs, a Silagerm 8030 silicone compound (Tekhnologiya-Plast LLC, Lyubertsy, Russia) was used. It consists of two components – a base and a hardener. This compound is used for manufacturing elastic forms. For the elastomer production, the base and the hardener must be mixed in a ratio of 1/1. Full cure occurs within 12 h. Taunit MWCNTs (NanoTechCenter LLC, Tambov, Russia) were employed as an electrically conductive filler for the base. This nanomaterial have an average outer diameter of 35±15 nm, an average inner diameter of 15±5 nm, and a length of about 2 μm [16]. It was synthesized through the CVD method.

2.2. Manufacturing a nanomodified elastomer
To prepare 960 g of an elastomeric composite containing 3 wt.% MWCNTs, 28.8 g of the MWCNTs were added to a 2-L beaker. 465.6 g of the silicone compound base were added to the MWCNTs sample and mixed using an HT-120 DX vertical rotary agitator (DAIHAN Scientific, Seoul, South Korea) at a rotation speed of 1000 rpm 1 h. Then, 465.6 g of the hardener was added to the MWCNTs-base mixture and the stirring procedure was repeated for 10 min.

The resulting composite was applied to a plate made of aluminium foil (GRIFF, Moscow, Russia) with a thickness of 100 μm, which was previously mechanically pre-treated using emery paper with a grain size of 100 μm. This made it possible to impart artificial surface roughness to the aluminum foil, thereby improving the adhesion of the composite to it. After uniform application of the composite, an identical aluminium plate was installed on it. The resulting assembly was molded between two flat surfaces. After polymerization at a temperature of + 80 ° C for 4 h, the electric heater sample was removed from the assembly, cut to the required size, and isolated using a fluoroplastic-based polymer shell based. From the front part, the electrical leads were left to connect supply voltage. Using this technology, the sample shown in figure 1 was made.

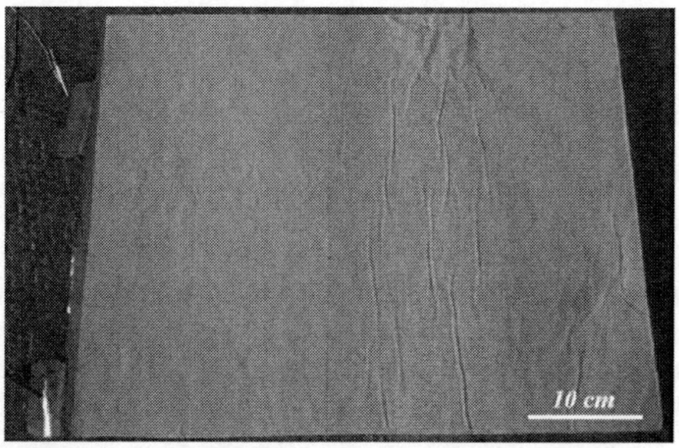

Figure 1. Appearance of SFHs based on a nanomodified elastomer.

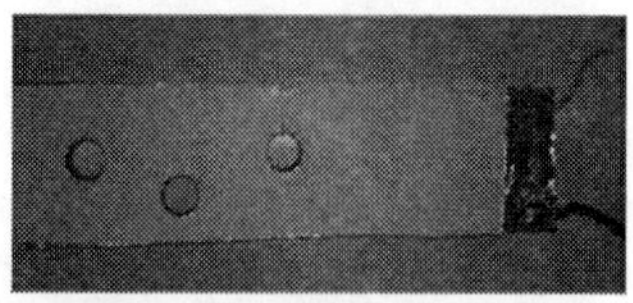

(a)

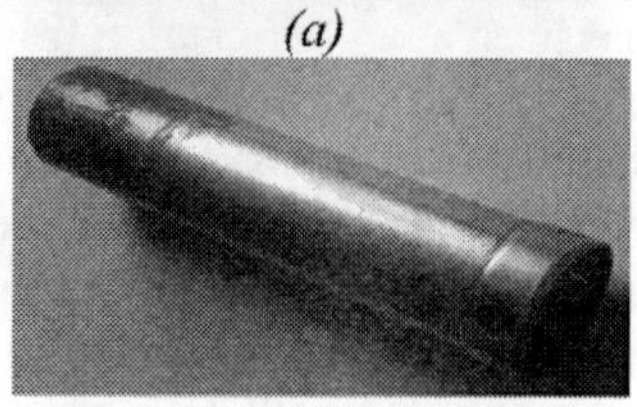

(b)

Figure 2. SFH with holes: a) sample of a nano-modified heater, b) punch.

2.3. Imparting artificially localized damages to the SFHs
To carry out this work, in accordance with the previously described technology, an SFH was made with dimensions of 200 × 300 mm (figure 2a). Holes were made in the center of the obtained sample using a piercer made of a cylindrical tube with a wall thickness of 0.5 mm and an outer diameter of 9 mm. The lower part of the piercer was ground in order to obtain a sharp edge (figure 2b). Wires were soldered to the electrodes.

2.4. Studying the electrophysical parameters of the SFHs based on the nanomodified elastomers.
The studies were performed on a measuring bench that consisted of a "LATR TDGC2-3" laboratory autotransformer (LATR) (Resanta, Moscow, Russia), with which the alternating electric current voltage from 0 to 220V was applied to the sample. For temperature measurements, electric current and power, two multimeters UNI-T UT61E and UNI-T UT71E (Uni Trend Group, Dongguan, China) were used; they were synchronized via USB interfaces with a personal computer (PC) based on "Intel Xeon E5450" processor (Intel, Santa Clara, USA).

Climatic tests of SFHs were carried out using a "KTH-1000" certified heat and cold chamber (NPF Tekhnologia, St. Petersburg, Russia) in the temperature range from - 60 to +70 ° C. Temperature field distribution measurements on the SFHs surface were carried out using a Testo-875 thermal imager (Testo, Lenzkirch, Germany).

3. Results and discussion

The studies showed that the SFH possess the effect of temperature self-regulation, since when the ambient temperature changes, the power consumption also changes. This power change occurs in proportion to the energy required to stabilize the temperature on the SFH surface (figure 3). For the SFH with dimensions of 40 × 40 mm, the specific power was found to be 800±10% W/m^2 at a chamber temperature of +10 °C. When the temperature in the chamber was reduced to -40 °C in order to maintain the required thermal conditions on the surface, the specific power of the SFH was 1500 ± 20 % W/m^2.

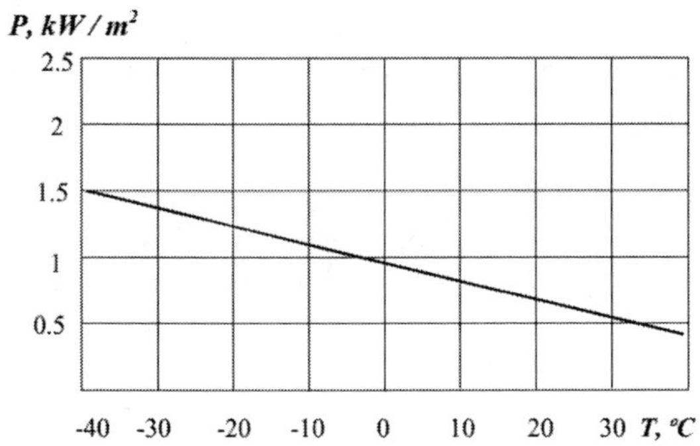

Figure 3. Changes in the SFH power at different temperatures.

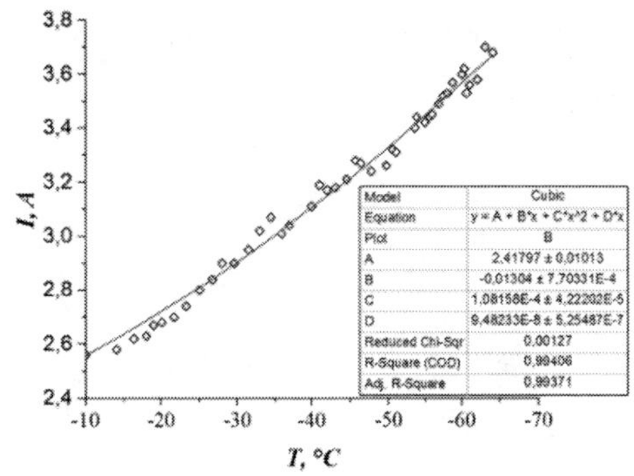

Figure 4. A dependence of the change in the current of the SFHs connected to an AC source with a voltage of 220 V when it is cooled.

Based on the data obtained, it was established that with a decrease in the temperature from -10 down to -70 ° C, the value of the current flow in the SFHs with dimensions of 40 × 40 cm increases from 2.6 up to 3.65 A (figure 4).

The technology for manufacturing functional nanomodified materials for the SFH, as well as their electro physical properties, made it possible to change the properties of self-regulation - namely, the possibility of changing the degree of increase in power with decreasing ambient temperature.

It has been shown that the operational and technical characteristics of the SFH are preserved after cutting it into separate parts of arbitrary shape and cutting holes in it (3 holes with a diameter of 9 mm), which is confirmed by photographs taken using a thermal imager (figure 5).

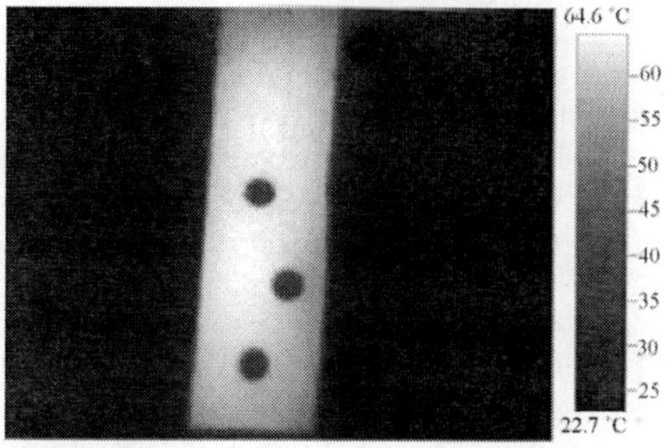

Figure 5. Performance of the artificially damaged SFH.

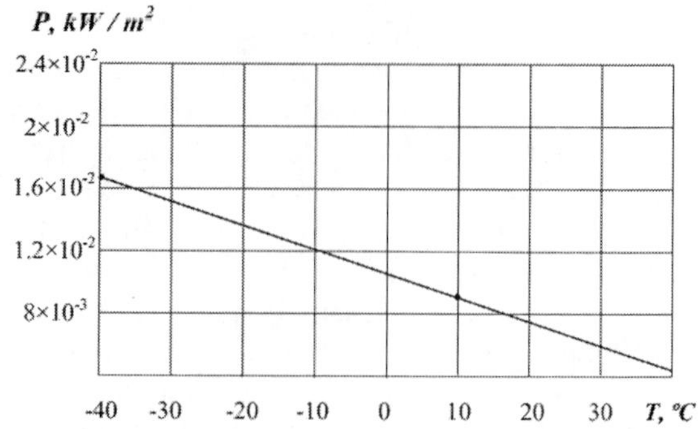

Figure 6. Changes in the power of the part cut off from the SFHs at different temperatures.

The heat release of the SFH with holes of arbitrary shape made in it does not change its performance, and the material does not deteriorate, which follows from the uniformity of the temperature field and the absence of any obvious temperature deviations in the boundary zone of the holes (figure 5).

The SFH cutting leads to the fact that the single parts obtained also have the property of self-regulation, when the power is proportional to the cutting area (figure 6).

The temperature field is uniform over the entire SFH surface. It was found that the temperature difference on the SFH surface between any two points is not more than 10 °C, which can be confirmed using a thermal imaging photograph (figure 7a) and a histogram of temperature values at randomly selected points (figure 7b). The experimental studies showed that the best mode of heat generation on the SFH surface was observed at an AC supply voltage of 220 V.

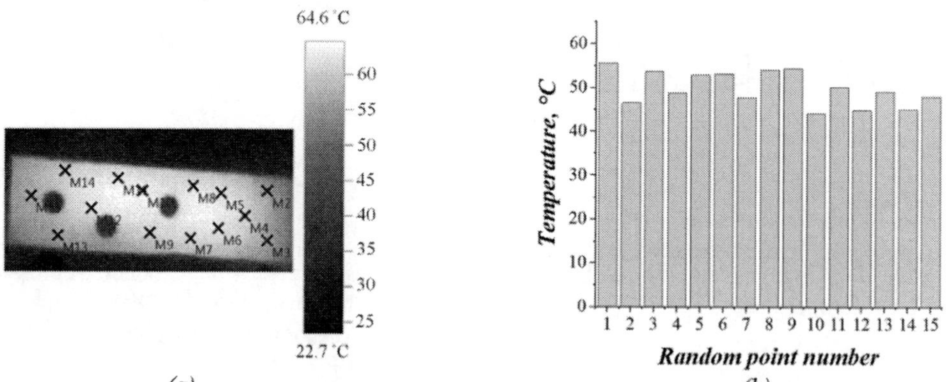

(a) *(b)*

Figure 7. Analysis of the temperature field on the surface of the damaged SFH: a) thermal image of the SFH with randomly selected points, b) histogram of the temperature distribution at randomly selected points)

Compared to ceramic heaters (figure 8) presented in [19], the ones developed herein are distinguished by their flexibility, since they are made on the basis of elastomers. The dimensions of ceramic electric heaters are strictly limited, so the production of a heater with a large heat transfer area is difficult. Self-regulating cables [20] differ from ceramic heaters in flexibility, which is similar to the developed heaters. However, the heating of large areas leads to installation difficulties and a large length of self-regulating cable, which is not economically viable.

Figure 8. Heaters made using robotics [19].

4. Conclusion

A technique was developed to manufacture a nanomodified elastomer used as a basis for the SFH production. A large SFH, possessing the effect of power self-regulation, was produced using this technique. The SFH allows heating surfaces with a large area in contrast to the existing self-regulating cables. So, when doing this using self-regulating cables, it is required to carry out complex installation, establish heat-conducting contact with the heated surface, *etc*.

In the present work, it is proved that the SFH remains operational when damaged. This makes it possible to manufacture technological holes in the SFH, which are intended for mounting it on the heated surface.

Acknowledgement

The reported study was funded by RFBR within the framework of Research Project no. 18-53-00032 BEL−A.

References

[1] Kozlov A V, Shapovalov A S and Yavchunovsky V Y 2008 Comparative analysis of the characteristics of open spirals and heating elements as part of electric heaters *Issues of Applied Physics*. [in Russian] **15** 53

[2] Li P, Li H, Ma J, Zhou Y, Zhang W, Cong L, Xu H and Liu Y 2018 Structural optimization of oxide/metal/oxide transparent conductors for high-performance low-emissivity heaters. *Adv. Mater. Inter.* **5** 24 doi.org/10.1002/admi.201801287

[3] Heymsfield E, Daniels J W, Saunders R F and Kuss M L 2020 Developing anti-icing airfield runways using surface embedded heat wires and renewable energy. *Sustain. Cities. Soc.* **52** 101712 https://doi.org/10.1016/j.scs.2019.101712

[4] Alemour B, Badran O and Hassan M R 2019 A Review of using conductive composite materials in solving lightening strike and ice accumulation problems in aviation. *J. Aerosp. Technol. Manag.* **11** 1919 doi.org/10.5028/jatm.v11.1022

[5] Kim Y, Lee H R, Saito T and Nishi Y 2017 Ultra-thin and high-response transparent and flexible heater based on carbon nanotube film. *Appl. Phys. Lett.* **110** 153301 https://doi.org/10.1063/1.4978596

[6] Kinloch I A, Suhr J, Lou J, Young R J and Ajayan P M 2018 Composites with carbon nanotubes and graphene: an outlook. *Science* **362** 547 doi:10.1126/science.aat7439

[7] Guadagno L, Raimondo M, Vertuccio L, Naddeo C, Barr G, Longo P and Nobile M R 2018 Morphological, rheological and electrical properties of composites filled with carbon nanotubes functionalized with 1-pyrenebutyric acid. *Compos. Part B-Eng.* **147** 12 doi.org/10.1016/j.compositesb.2018.04.036

[8] Janas D and Koziol K K 2013 Rapid electrothermal response of high-temperature carbon nanotube film heaters. *Carbon.* **59** 457 doi.org/10.1016/j.carbon.2013.03.039

[9] Chen W, Liu P, Liu Y, Wang Q and Duan W 2018 A temperature-induced conductive coating via layer-by-layer assembly of functionalized graphene oxide and carbon nanotubes for a flexible, adjustable response time flame sensor. *Chem. Eng. J.* **353** 115

doi.org/10.1016/j.cej.2018.07.110

[10] Shchegolkov A, Zavolokina L and Diesperova N 2019 Prospects for the use of electric heaters with self-installing thermal contact based on nano-modified materials for aerospace engineering. *Journal of Physics: Conference Series* **1145** 012026 doi.org/10.1088/1742-6596/1145/1/012026

[11] Wang T, Zhang Y, Liu Q, Cheng W, Wang X, Pan L and Xu H 2018 A self-healable, highly stretchable, and solution processable conductive polymer composite for ultrasensitive strain and pressure sensing. *Adv. Funct. Mater.* **28** 1705551 doi.org/10.1002/adfm.201705551

[12] Duan H, Wang X, Xie G, Hu R and Ma H 2018 Effect of thermosetting resin matrix on resistance-temperature stability of carbon fiber conductive composites. *Mater. Res. Express.* **6** 015312 https://doi.org/10.1088/2053-1591/aae909

[13] Liu Y, et al. 2018 Tailored pyroresistive performance and flexibility by introducing a secondary thermoplastic elastomeric phase into graphene nanoplatelet (GNP) filled polymer composites for self-regulating heating devices. *J. Mater. Chem. C.* **6** 2760 doi:10.1039/c7tc05621d

[14] *CNTs "Taunit" and Synthesis ThereofPotentiometric Methods*, "NanoTechCenter" Ltd, available at: http://nanotc.ru/producrions/87-cnm-taunit

[15] Rowlands W and Vaidhyanathan B 2019 Additive manufacturing of barium titanate based ceramic heaters with positive temperature coefficient of resistance (PTCR). *J. Eur. Ceram. Soc.* **39** 3475 doi.org/10.1016/j.jeurceramsoc.2019.03.024

[16] Jenkins S B and Jones J, U.S. Patent No. 15,885,270 (2 August 2018)

IOP Conf. Series: Materials Science and Engineering **693** (2019) 012019 doi:10.1088/1757-899X/693/1/012019

Structure and electrical conductivity of heat treated iodine-doped multi-walled carbon nanotubes

R Stolyarov[1], A Blohin[1], N Gorshkov[2], A Tkachev[1], B Kulnitskiy[3], T Pasko[1], A Sukhorukov[1] and I Burmistrov[2]

[1]Department of Technology and methods of nanoproducts manufacturing, *Tambov State Technical University,* 106 Sovetskaya Street, Tambov 392000, Russia
[2]Department of Chemistry and chemical technology of materials, *Yuri Gagarin State Technical University of Saratov*, 77 Politechnicheskaya Street, Saratov 410054, Russia
[3]Department of structural research, *Technological Institute for Superhard and Novel Carbon Materials*, 7a Tsentralnaya Street, Troitsk 108840, Russia

E-mail: stolyarovra1985@gmail.com

Abstract. The present paper discusses the possibility of simultaneously using two methods for the modification of multi-walled carbon nanotubes - high-temperature annealing and halogen-assisted modification. The nanotube modification was carried out during the high-temperature annealing in the presence of iodine. A change in the structure and an increase in the electrical conductivity of the modified multi-walled carbon nanotubes was observed.

1. Introduction

The electrical properties of carbon nanotubes (CNTs) are largely related to the structural features of their structure [1]. It is generally accepted to distinguish nanotubes by the number of concentric graphene cylinders forming them: single- (SWCNTs), double- (DWCNTs), and multi-walled (MWCNTs). The main structural feature of single-walled nanotubes is chirality. MWCNTs differ from SWCNTs considerably by a wider variety of shapes and configurations. A variety of structures is manifested both in the longitudinal and in the transverse directions. The electric conductivity of cheaper MWCNTs is much lower compared to the SWCNTs, and in some cases, its values may differ by several orders of magnitude. This is primarily due to the presence of a much larger number of structural defects on the surface of the MWCNTs compared with the SWCNTs. In this regard, several researchers have tried to increase the electrical conductivity of the MWCNTs through various methods, in particular, through doping with silver [2,3], nitrogen [4,5], polyaniline [6,7], iodine [8,9], *etc*. The best result is achieved with iodine-modified MWCNTs and SWCNTs; their electrical conductivity increases by 60 % and 2000 % compared to the initial one [8,9]. On the other hand, it is possible to reduce the number of MWCNTs defects and increase their electrical conductivity using high-temperature annealing [10,11].

Thus, using both the high-temperature treatment and halogen-assisted modification, it is possible to significantly increase the electrical conductivity of the MWCNTs. Therefore, the goal of the present work was to increase their electrical conductivity by changing their structure and modifying with iodine.

2. Materials and methods

2.1. Multi-walled carbon nanotubes (MWCNTs)
Taunit-M carbon nanomaterial, representing MWCNTs with a diameter of 10-30 nm and a length of up to 2 μm (figure 1), was used in the present research. The material was synthesized via catalytic pyrolysis of natural gas at NanoTechCenter LLC (Tambov, Russia).

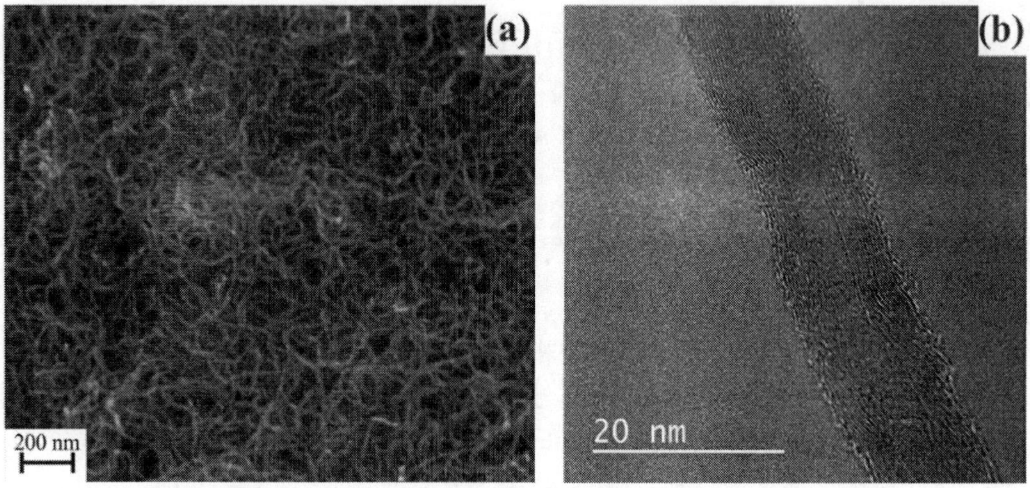

Figure 1. SEM (a) and TEM (b) images of the Taunit-M MWCNTs.

2.2. High temperature modification
The MWCNTs structure was modified by high-temperature treatment in the presence of iodine. The MWCNTs were placed in an isopropanol solution of chemically pure iodine. Then, the components were mechanically mixed and dried in a fume hood under normal conditions until complete evaporation of the isopropanol. After the dry mixture was distributed over the substrate, it was loaded into a sealed capacitive reactor equipped with a gas supply system. High-temperature modification of the MWCNTs with adsorbed iodine was carried out at a temperature of 620°C in argon for 2 h.

2.3. Characterization of MCNTs
To characterize the MWCNTs structure, the following techniques were employed: high-resolution transmission electron microscopy (HRTEM), scanning electron microscopy (SEM), and characteristic electron energy loss spectroscopy (EELS). Measurements of the electrical conductivity of the MWCNTs were performed using the method of impedance spectroscopy (IS).

The TEM studies of the material obtained were performed using a JEM-2010 high-resolution transmission electron microscope (JEOL, Japan) with energy dispersive spectroscopy (EDS) analyzer and a GIF Quantum attachment for EELS. The morphology and microstructure of the MWCNTs surface was studied using a Merlin scanning electron microscope (Carl Zeiss, Germany).

The electrical conductivity of the pristine and modified nanomaterials was measured using the IS on an Alpha AN impedance meter (NovoControl, Germany). Measurements were carried out for samples placed and pressed at a pressure of about 100 MPa between titanium electrodes with a diameter of 8 mm.

3. Results and discussion

3.1. Morphology and structure of the modified MWCNTs
The SEM and TEM images of the modified MWCNTs are presented in figure 2.

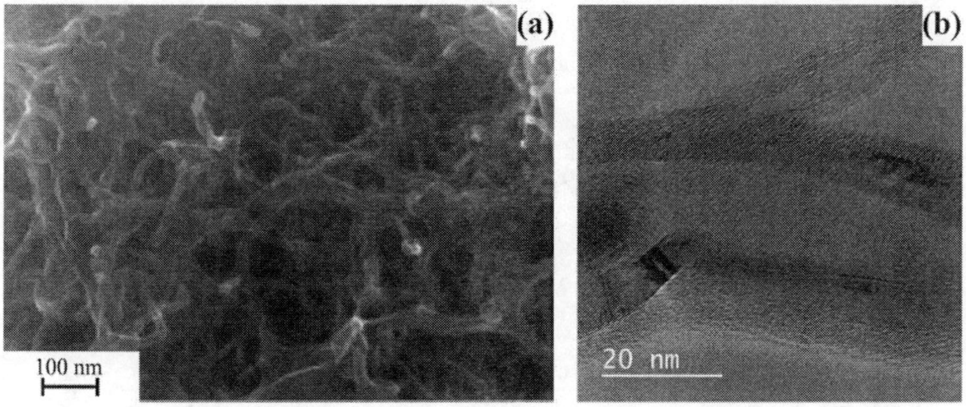

Figure 2. SEM (a) and TEM (b) images of the modified Taunit-M MWCNTs.

As a result of the iodine-assisted modification, the MWCNTs structure becomes deformed, and the outer layers are destroyed until a destructured carbon film is formed instead of them (figure 2b). Besides, during the modification, the compounds of single, closely spaced nanotubes are formed by means of the destructured outer layers (figure 2a). The mechanism of the nanotubes deformation during the thermal modification is schematically shown in figure 3.

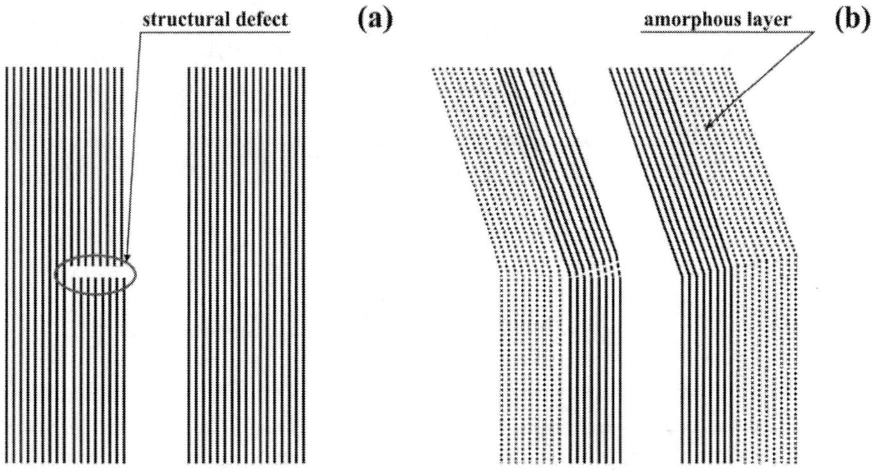

Figure 3. Mechanism of the nanotube deformation during thermal modification.

Figures 2a and 2b clearly show the connections of individual nanotubes into a single monolith. Agglomerates of carbon nanotubes usually represent objects from a few to hundreds of micrometers. The MWCNTs are tangled in them and are linked to each other through simple contact. In thermally modified agglomerates, the individual nanotubes are firmly interconnected by the destructed outer layers and form a single electrically conductive structure. This morphology provides improved electrical conductivity within individual clusters and reduces the electrical resistance of bulk materials, as will be shown below.

The key role in increasing the electrical conductivity is played by changes in the MWCNTs morphology, which was indirectly shown during the elemental analysis (Table 1). The research data obtained showed almost complete absence of iodine in the modified MWCNTs (Table 1). As a result

IOP Conf. Series: Materials Science and Engineering **693** (2019) 012019 doi:10.1088/1757-899X/693/1/012019

of heat treatment, iodine is completely evaporated and removed by a stream of inert gas from the MWCNTs sample, and thus, cannot exert a direct influence on the charge transfer process.

Table 1. Elemental composition of the modified MWCNTs.

Element	Content	
	wt.%	at.%
C	97.44	99.50
O	0.10	0.03
Co	1.24	0.26
Fe	0.66	0.14
I	0.56	0.05
Sum	100.00	100.00

3.2. Electrical conductivity of the nanomodified MWCNTs

Figure 4 shows the dependence of the electrical conductivity on the frequency of the modified and the original (pristine) nanotubes.

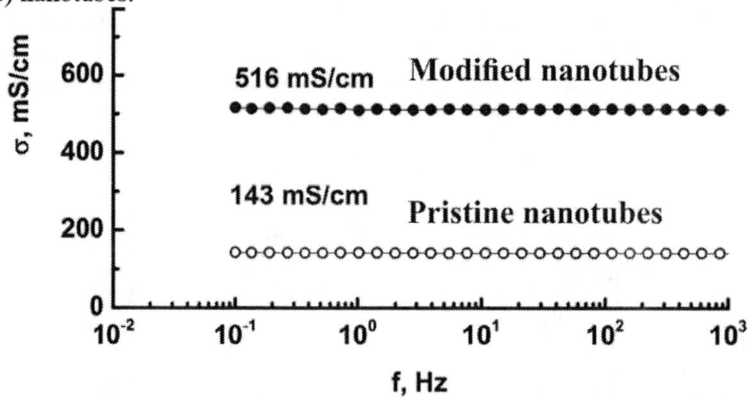

Figure 4. Electrical conductivity of the pristine and modified MWCNTs.

The increase in the electrical conductivity of the modified nanotubes compared to the pristine ones can be explained by several known mechanisms: intercalation or adsorption of iodine [8] and oxygen [1,12], or changes in the MWCNTs structure [10]. The oxygen or iodine adsorption should cause an increase in the conductivity and strong p-doping of the MWCNTs. The increase in the conductivity due to the oxygen and iodine modification of the MWCNTs cannot exceed tens of percent [8,12]. Moreover, their high content in the samples was not recorded (Table 1). The obtained EELS spectra showed a small peak of oxygen in the region of 532 eV (figure 5), which indirectly confirms its low concentration and, consequently, the low contribution to the electrical conductivity changes.

Moreover, the linear form of the frequency dependences of the electrical conductivity is characterized by the absence of p-doping of the modified MWCNTs (figure 4), and they still possess electronic conductivity.

Thus, the TEM and SEM studies demonstrated that the nanotubes were linked to each other as a result of high-temperature modification in the presence of iodine. In this structure, mechanical conductive contacts are possible only between the individual agglomerates. As a result of the nanotubes connection, the contact resistance is significantly reduced and, consequently, the conductivity of the modified material increases by more than 3.5 times (figure 4).

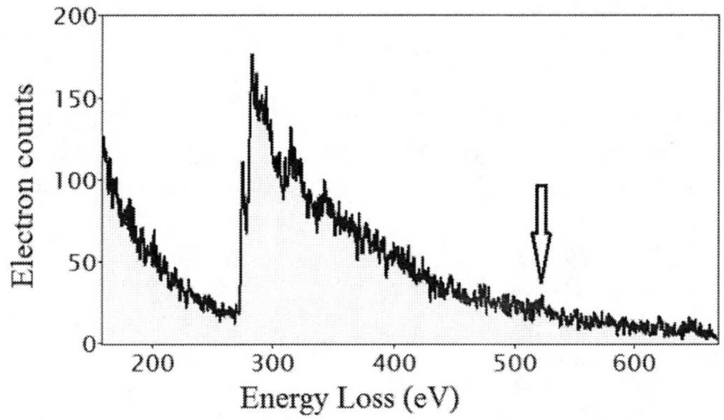

Figure 5. EELS spectra of the modified MWCNTs.

4. Conclusion

An easily scalable technique for the thermal modification of MWCNTs in the presence of iodine was developed. The use of halogen-assisted high-temperature annealing made it possible to obtain a carbon nanomaterial superior in the electrical conductivity by 360 %. The structure of the modified MWCNTs was studied. As a result of the iodine-assisted modification, the MWCNTs structure becomes deformed, and the outer layers are destroyed till the formation of a degraded carbon film instead. The MWCNTs cross-linked in the entire volume by means of this amorphous film could find wide practical application, for instance, as the main component of electrically conductive products based on polymers obtained through vacuum infusion.

5. Acknowledgments

The present study was funded by the Russian Foundation for Basic Research (RFBR) in the framework of Project No. 18-29-19121\1.

References

[1] Pérez-Rodríguez S, Torres D, Lázaro M J 2018 Effect of oxygen and structural properties on the electrical conductivity of powders of nanostructured carbon materials. *Powder Technol* **340** 380 https://doi.org/10.1016/j.powtec.2018.09.038

[2] Oluwalowo A, Nguyen N, Zhang S, Park J G and Liang R 2019 Electrical and thermal conductivity improvement of carbon nanotube and silver composites. *Carbon* **146** 224 https://doi.org/10.1016/j.carbon.2019.01.073

[3] Kharlamova M V, Yashina L V and Lukashin A V 2013 Comparison of modification of electronic properties of single-walled carbon nanotubes filled with metal halogenide, chalcogenide, and pure metal. *Appl. Phys. A.* **112** 297 https://doi.org/10.1007/s00339-013-7808-y

[4] Behzad S and Chegel R 2019 Engineering thermal and electrical properties of B/N doped carbon nanotubes: Tight binding approximation. *J. Alloys and Comp.* In Press https://doi.org/10.1016/j.jallcom.2019.03.381

[5] Talla J A and Ghozlan A A 2018 Effect of boron and nitrogen co-doping on CNT's electrical properties: Density functional theory. *Ch. J. Phys.* **56** 740 https://doi.org/10.1016/j.cjph.2018.01.009

[6] Zhang L, Zujovic Z D, Peng H, Bowmaker G A, Kilmartin P A and Travas-Sejdic J 2008 Structural Characteristics of Polyaniline Nanotubes Synthesized from Different Buffer Solutions. *Macromolecules. (A)* **41** 8877 https://doi.org/ 10.1021/ma801728j

[7] Dong J-Q and Shen Q 2009 Enhancement in Solubility and Conductivity of Polyaniline with Lignosulfonate Modified Carbon Nanotube. *J. Pol. Sci. Part B: Pol Phys.* **47** 2036

https://doi.org/10.1002/polb.21802

[8] Qiu L, Zou H, Zhu N, Feng Y, Zhang X and Zhang X 2018 Iodine nanoparticle-enhancing electrical and thermal transport for carbon nanotube fibers. *Appl. Therm. Eng.* **141** 913 https://doi.org/ 10.1016/j.applthermaleng.2018.06.049

[9] Tonkikh A A, Sanjabi S 2015 Metallization of single-wall carbon nanotube thin films induced by gas phase iodination. *Carbon.* **94** 768 https://doi.org/ 10.1016/j.carbon.2015.07.062

[10] Usanov D A, Skripal A V and Romanov A V 2014 Effect of annealing on the microwave characteristics of carbon nanotubes and the nanocomposite materials based on them. *Tech. Phys.* **59** 837 https://doi.org/10.1134/S1063784214060231

[11] Castillejos E, Bachiller-Baeza B, Perez-Cadenas M, Gallegos-Suarez E, Rodriguez-Ramos I, Guerrero-Ruiz A, Tamargo-Martinez K, Martinez-Alonso A, Tascon J M D 2012 Structural and surface modifications of carbon nanotubes when submitted to high temperature annealing treatments. *J. Alloy. Compd.* **62** 460 https://doi.org/10.1016/j.jallcom.2011.11.007

[12] Lekawa-Raus A, Patmore J, Kurzepa L, Bulmer J, Koziol K 2011 Electrical Properties of Carbon Nanotube Based Fibers and Their Future Use in Electrical Wiring. *Adv. Func. Mater.* **24** 3661 https://doi.org/10.1002/adfm.201303716

The method and the device for measuring thermophysical properties of liquids

A Divin, S Ponomarev, M Petrasheva, D Lyubimova, G Mozgova, P Belyaev and G Shishkina

Department of Mechatronics and technological measurements, *Tambov State Technical University*, 106 Sovetskaya Street, Tambov 392000, Russian Federation

E-mail: kafedra@uks.tstu.ru

Abstract. This paper describes the new method and the device for measuring the effective thermophysical characteristics of the investigated fluid under shear flow conditions at a fixed shear rate in the gap between coaxial cylinders. The shear flow is provided by the rotation of the outer cylinder while the inner cylinder remains stationary. The measuring unit can measure the complex rheological and thermophysical characteristics of liquids in less than 30 minutes. This article describes a two-step measurement method. The first step involves the measurement of the excess temperature established because of the heat source action due to the viscous friction in shear flow of the test liquid in a layer between the cylinders; the function of the dissipative heat source in the fluid at this stage is based on the power-law. The second step includes the thermal effect on the liquid layer from the heat source (electric heater) in the inner cylinder and the registration of the temperature response in the inner cylinder layer near the heater. The experimental determination results of polyoxyethylene solution thermophysical characteristics with the addition of nanocarbon material "Taunit" are presented here.

1. Introduction

A lot of attention has been paid by scientists around the world to studying the heat transfer processes in suspensions containing nanoparticles.

Scientists of Russia and CIS countries are actively working in this direction, in particular, at the Institute of computational modeling of the Siberian branch (SB) of the Russian Academy of Sciences (RAS), Siberian Federal University (Krasnoyarsk), at the S S Kutateladze Institute of Thermophysics (Siberian Division of the Russian Academy of Sciences, Novosibirsk, Russia). The analytical and numerical study of forced convection in nanofluid flows based on various liquid media has been of most importance for the research. Professor M M Safarov (Dushanbe, Tajikistan) and his followers investigate the influence of nanoparticles on the thermophysical properties of hydrazine and its derivatives depending on the concentration of nanoparticles and pressure.

The results of experimental determination of the suspensions' thermal conductivity dependence on the shear rate was described in [1]. With an increase in the shear rate, the thermal conductivity of the liquid increases asymptotically, and the higher the volume of concentration and the larger the volume of particles, the more intense the increase in thermal conductivity. This effect is explained by the appearance of microconvection due to the increase in the angular velocity of the particle and by the appearance of "micro-swirls" around it. Thus, the measuring device used in the conditions of shear flow allows to measure the effective thermal conductivity of the liquid with the presence of microparticles in it, which determines the heat transfer due to both – the thermal conductivity and the convection.

The intensity of heat exchange in liquids with nanoparticles at their boiling state was also confirmed in [2,3]. The effective thermal conductivity of suspensions based on oil and carbon nanotubes was studied in [4]. The authors of the article [4] found an abnormal increase in the effective thermal conductivity in comparison with the calculated values as compared with other nanosuspensions.

This paper discusses the method and the device, which differ from the known analogues in the ability of evaluating not only the thermal conductivity, as well as the thermal diffusivity, heat capacity, and rheological characteristics of the fluids, including the non-Newtonian ones.

2. Materials and methods

Figure 1 shows a diagram of the measuring device, which has two coaxial cylinders between which the test liquid is located. To ensure the measurements under conditions of high temperatures, the inner cylinder made of heat-resistant material – Polyether ether ketone (PEEK). It can withstand prolonged thermal exposure, while maintaining their mechanical properties, has resistance to thermal oxidative degradation, chemically inert, easily shaped and processed [5-7].

Cylinder 1 has primary and protective heaters, and its design provides the ability to measure the temperatures of these heaters [6]. The placement of the heaters in the inner fixed cylinder allows to stabilize the flow regime of the investigated liquid in the gap between the cylinders, while they are protected from possible impact from the investigated liquid by a sleeve made of heat-conducting material (aluminum) [8]. It should be noted that the thermal resistance converters capable of registering the temperature over time are located in the layers of the primary 5 and the protective 4 heaters.

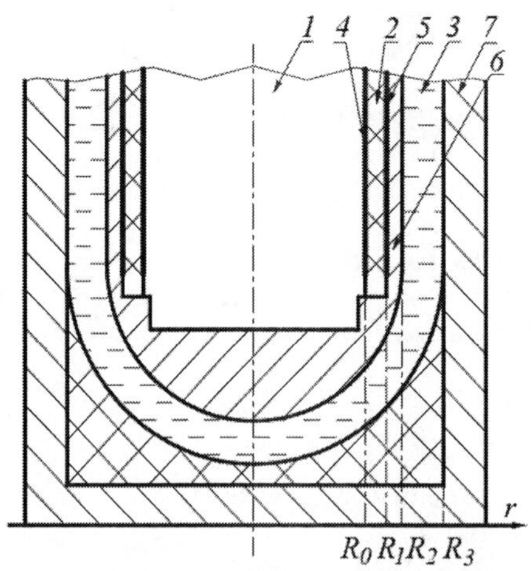

Figure 1. Diagram of the measuring device: *1* – the inner (stationary) cylinder; *2, 6* – protection sleeves; *3* – the investigated viscous liquidcontaining nanoparticles; *4* – protective heater; *5* – primary heater; *7* – the outer (rotating) cylinder.

The measurement of thermophysical characteristics of “a viscous liquid containing nanoparticles” (further nanofluid) were measured in several stages.

After establishing the temperature field that occurs due to the dissipative heating of the nanofluid layer during the shear flow (the first stage), the main heater is switched on, creating a temperature field in the measuring device, which is superimposed on the existing flow (the second stage) (figure 2) [7]. It is advisable to separate in time the onset moments action of these heat sources, applying the principle of superposition temperature fields [9-11].

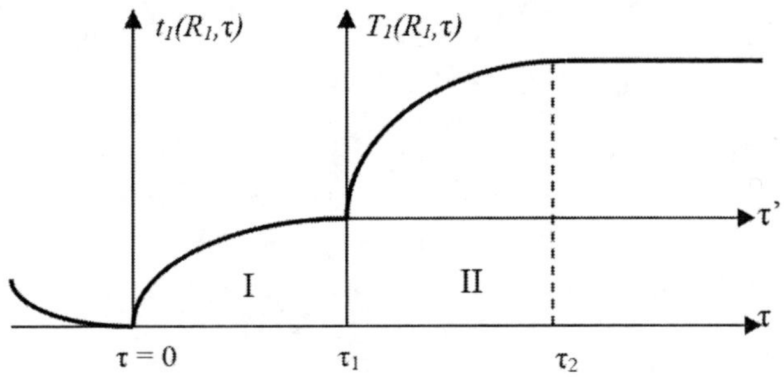

Figure 2. Temperature field in the measuring device during the experiment.

At the preparatory step of measurement considering the outer cylinder rotating at a known angular velocity Ω, the shear rate γ can be estimated as follows:

$$\gamma \approx \frac{\Omega \cdot R_3}{R_3 - R_2}.$$

After a certain interval of time τ_1, a stationary temperature field is established in the layers of the measuring device, conditioned by the dissipation of the energy of the viscous friction forces and microconvection due to the availability of nanoparticles in the liquid layer during the shear flow [6,7]. We will consider this temperature field as the base one.

Differential equations describing the stationary temperature field in layers 6 (protective sleeve) and 3 (investigated nanofluid) of the measuring device (figure 1) in the steady-state stationary mode of the preparatory stage of measurements will take the following form [6]:

$$\begin{cases} \dfrac{d}{dr}\left(r\dfrac{dt_1(r)}{dr}\right) = 0, \quad R_1 < r < R_2 \\ \dfrac{d}{dr}\left(r\dfrac{dt_2(r)}{dr}\right) = W(r), \quad R_2 < r < R_3. \end{cases} \tag{1}$$

The boundary condition is as follows:

$$-\lambda_1 \frac{dt_1(R_1)}{dr} = q_1(R_1) = 0, \tag{2}$$

$$t_1(R_2) = t_2(R_2), \quad \lambda_1 \frac{dt_1(R_2)}{\partial r} = \lambda_x \frac{dt_2(R_2)}{\partial r}, \tag{3}$$

$$t_2(R_3) = t_0, \tag{4}$$

where r is the radial coordinate;

$W(r) = mr^{n+1}\left(2\omega \dfrac{r^{-2/n}}{nr(R_2^{-2/n} - R_3^{-2/n})}\right)^{n+1}$ – the function of the heat source due to the dissipation of viscous friction energy and microconvectionon account to the presence of nanoparticles in the layer of the studied liquid, subject to the power rheological law [7,8]; m – the coefficient of consistency of a viscous liquid, Pa·s^n; n – the index of the fluid flow [13].

The additional condition can be written like this:

$$t_1(R_1) = t_e. \tag{5}$$

Where t_e is the experimentally determined temperature of the main heater winding.

The general solutions to the differential equations (1) obtained after integration can be expressed in the following form:

$$\begin{cases} t_1(r) = C_1 \ln(r) + C_2, \; R_1 < r < R_2, \\ t_2(r) = \dfrac{n^2 r^{-2/n-1}}{(n+2)^2} m \left[\dfrac{2\omega}{n(R_2^{-2/n} - R_3^{-2/n})} \right]^{n+1} + C_3 \ln(r) + C_4, \; R_2 < r < R_3. \end{cases} \quad (6)$$

The following function is introduced: $\Phi(\omega) = \dfrac{n^2}{(n+2)^2} m \left[\dfrac{2\omega}{n(R_2^{-2/n} - R_3^{-2/n})} \right]^{n+1}$.

Applying the general solutions (6) and the boundary conditions (2)-(4) we obtain the calculated dependence for the thermal conductivity coefficient of a viscous non-Newtonian fluid:

$$\lambda_x = \frac{C_1 \lambda_1}{C_3 - R_2^{2/n} \cdot \Phi(\omega)(2/n+1)}. \quad (7)$$

The results of thermal conductivity measurements calculated by this formula have significant uncertainty (error) in the case of small values of the consistency coefficient of the liquid material.

The non-stationary temperature field in the measuring device after switching on the heater will be described by the following mathematical model:

$$\begin{cases} \dfrac{\partial T_1(r,\tau)}{\partial \tau} = a_1 \dfrac{1}{r} \dfrac{\partial}{\partial r} [r \dfrac{\partial T_1}{\partial r}], \; R_1 < r < R_2, \\ \dfrac{\partial T_2(r,\tau)}{\partial \tau} = a_x \dfrac{1}{r} \dfrac{\partial}{\partial r} [r \dfrac{\partial T_2}{\partial r}], \; R_2 < r < R_3. \end{cases} \quad (8)$$

The initial condition can be written like this:

$$T_1(r, \tau_1) = 0, \quad (9)$$

$$T_2(r, \tau_1) = 0, \quad (10)$$

where τ_1 – conventionally taken as the starting point ($\tau_1 = 0$).

The boundary condition is as follows:

$$-\lambda_1 \frac{dT_1(R_1, \tau)}{dr} = q_1(R_1, \tau) = \frac{P}{S}, \quad (11)$$

Where P is the heater power, W; S – surface area of the main heater, м^2.

$$T_1(R_2, \tau) = T_2(R_2, \tau), \quad \lambda_1 \frac{dT_1(R_2, \tau)}{dr} = \lambda_x \frac{dT_2(R_2, \tau)}{dr}, \quad (12)$$

$$T_2(R_3, \tau) = 0. \quad (13)$$

The additional condition is given in the following form:

$$T_1(R_1, \tau) = T_э(\tau). \quad (14)$$

Using the integral Laplace transform $T^*(r,p) = \int_0^\infty T(r,\tau) \cdot e^{-p\tau} d\tau$, this model can be reduced to:

$$\begin{cases} pT_1^*(r,p) = a_1 \dfrac{1}{r} \dfrac{d}{dr} [r \dfrac{dT_1^*(r,p)}{dr}], \quad R_1 < r < R_2, \\ pT_2^*(r,p) = a_x \dfrac{1}{r} \dfrac{d}{dr} [r \dfrac{dT_2^*(r,p)}{dr}], \quad R_2 < r < R_3. \end{cases}$$

The boundary condition is as follows:

$$-\lambda_1 \frac{dT_1^*(R_1, p)}{dr} = \frac{q_1}{p},$$

$$T_1^*(R_2, p) = T_2^*(R_2, p),$$

$$\lambda_1 \frac{dT_1^*(R_2, p)}{dr} = \lambda_x \frac{dT_2^*(R_2, p)}{dr},$$

$$T_2^*(R_3, p) = 0,$$

where T^* – time integral temperature characteristic, °C·s; p – Laplace transform parameter, s^{-1}.

The additional condition represents the result of experimental temperature measurement in a layer of the internal cylinder with radius R_1:

$$T_1^*(R_1, p) = T_e^*.$$

The General solution of differential equations (15) can be written according to [12]:

$$T_i(r,p) = C_{2i-1} I_0\left(r\sqrt{\frac{p}{a_i}}\right) + C_{2i} K_0\left(r\sqrt{\frac{p}{a_i}}\right), \tag{15}$$

where $C_1, C_2 \ldots C_4$ – constant coefficients; $I_0(x)$, $K_0(x)$ – modified zero-order Bessel and Hankel functions, and $I'_0(x) = I_1(x)$, $K'_0(x) = -K_1(x)$.

After substituting into the boundary and additional conditions (9–15), the following equations can be obtained:

$$\begin{cases} \lambda_1 \cdot \left(C_2 K_1\left(R_1 \cdot \sqrt{\frac{p}{a_1}}\right) - C_1 I_1\left(R_1 \cdot \sqrt{\frac{p}{a_1}}\right)\right) = \frac{q_1}{p}; \\ C_1 \cdot I_0 \cdot \left(R_2\sqrt{\frac{p}{a_1}}\right) + C_2 K_0\left(R_2 \cdot \sqrt{\frac{p}{a_1}}\right) = C_3 I_0\left(R_2 \cdot \sqrt{\frac{p}{a_x}}\right) + C_4 K_0\left(R_2 \cdot \sqrt{\frac{p}{a_x}}\right); \\ \lambda_1 \cdot \left[C_1 I_1\left(R_2 \cdot \sqrt{\frac{p}{a_1}}\right) - C_2 K_1\left(R_2 \cdot \sqrt{\frac{p}{a_1}}\right)\right] = \lambda_x \cdot \left[C_3 I_1\left(R_2 \cdot \sqrt{\frac{p}{a_x}}\right) - C_4 K_1\left(R_2 \cdot \sqrt{\frac{p}{a_x}}\right)\right]; \\ C_3 \cdot I_0 \cdot \left(R_3\sqrt{\frac{p}{a_x}}\right) + C_4 K_0\left(R_3 \cdot \sqrt{\frac{p}{a_x}}\right) = 0; \\ C_1 \cdot I_0 \cdot \left(R_1\sqrt{\frac{p}{a_1}}\right) + C_2 K_0\left(R_1 \cdot \sqrt{\frac{p}{a_1}}\right) = T_e^*. \end{cases}$$

This system of five equations has six unknowns: C_1, C_2, C_3, C_4, λ_x, a_x. Given that the characteristics λ_x, a_x should not depend on the parameter p, it's possible to set two values $p_1 = p$ and $p_2 = kp$, $k > 0$, to obtain ten equations with ten unknowns: $C_1(p_i),\ldots C_4(p_i), \lambda_x, a_x$, where $i = 1, 2$ and solve this system with respect to λ_x and a_x.

At the stationary step of the measurement, the desired thermal conductivity of the liquid can be estimated by solving the following inverse boundary value problem:

$$\begin{cases} \frac{\partial}{\partial r}\left(r\frac{\partial T_1(r)}{\partial r}\right) = 0, \quad R_1 < r < R_2, \\ \frac{\partial}{\partial r}\left(r\frac{\partial T_2(r)}{\partial r}\right) = 0, \quad R_2 < r < R_3. \end{cases}$$

The boundary condition is as follows:

$$-\lambda_1 \frac{\partial T_1(R_1)}{dr} = q_1(R_1) = \frac{P}{S},$$

$$T_1(R_2) = T_2(R_2),\ \lambda_1 \frac{dT_1(R_2)}{dr} = \lambda_x \frac{dT_2(R_2)}{dr},$$

$$T_2(R_3) = T_0 = 0.$$

The additional condition can be expressed like this:

$$T_1(R_1)=T_e.$$

The solution of this problem gives the following calculated relationships for thermal conductivity:

$$\lambda_x=\frac{R_1\lambda_1\cdot\ln\left(\frac{R_2}{R_3}\right)}{R_1\ln\left(\frac{R_1}{R_2}\right)-\frac{ST_e\lambda_1}{P}}. \tag{16}$$

The method for evaluating the thermal conductivity and thermal diffusivity values of the test material at a given shear rate according to the experimental data of the non-stationary measurements phase can be described as follows:

1. Setting the input variables: R_1, λ_1, a_1, R_2, P, l, R_3, p, k.
2. Calculating the parameters (functions):

$$g_{11}(p)=R_1\cdot\sqrt{\frac{p}{a_1}};\ g_{21}(p)=R_2\cdot\sqrt{\frac{p}{a_1}};$$

$$h=R_3/R_2;\ q=P/\pi 2R_1 l.$$

3. Reading temperature values from an array file $T_j(\tau_i)$, where j=1, 2; i = 1,2,…, n; n – number of measurements.

4. Calculating the time integral characteristics of the temperature $T_1^*(R_1,p)=\int_0^\infty T(R_1,\tau_i)\cdot e^{-p\tau}d\tau$,

$T_2^*(R_1,kp)=\int_0^\infty T(R_1,\tau_i)\cdot e^{-kp\tau}d\tau$ (i=1, 2,…, n) using the Simpson method:

$$T_j^*(R_1,p_j)=\frac{\Delta\tau_2}{3}\left(T(R_1,0)e^{-p_j\tau_0}+T(R_1,\tau_n)e^{-p_j\tau_n}+\sum_{i=2}^{n/2}(4T(R_1,\tau_{2i-1})e^{-p_j\tau_{2i-1}}+2T(R_1,\tau_{2i})e^{-p_j\tau_{2i}})\right)+$$
$$+\frac{T(R_1,\tau_n)e^{-p_j\tau_n}}{p_j},\ j=1,2,\ p_1=p,\ p_2=kp.$$

5. Calculating the function values: in order to simplify the calculation, we set

$$f(x)=I_0(x)\cdot K_1(x)+I_1(x)\cdot K_0(x),$$

where I_0, I_1 – modified zero-order and first-order Bessel functions; K_0, K_1 – modified Hankel functions of zero and first order; x – is a variable, $x=p$, kp.

$$C_1(p)=-\frac{R_1\cdot q_1\cdot K_0(g_{11}(p))-T_1^*(R_1,p)\cdot p\cdot\lambda_1\cdot g_{11}(p)\cdot K_1(g_{11}(p))}{p\cdot\lambda_1\cdot g_{11}(p)\cdot f(g_{11}(p))};$$

$$C_2(p)=\frac{R_1\cdot q_1\cdot I_0(g_{11}(p))+T_1^*(R_1,p)\cdot p\cdot\lambda_1\cdot g_{11}(p)\cdot I_1(g_{11}(p))}{p\cdot\lambda_1\cdot g_{11}(p)\cdot f(g_{11}(p))};$$

$$C_1(kp)=-\frac{R_1\cdot q_1\cdot K_0(g_{11}(kp))-T_2^*(R_1,kp)\cdot p\cdot\lambda_1\cdot g_{11}(kp)\cdot K_1(g_{11}(kp))}{k\cdot p\cdot\lambda_1\cdot g_{11}(kp)\cdot f(g_{11}(kp))};$$

$$C_2(kp)=\frac{R_1\cdot q_1\cdot I_0(g_{11}(kp))+T_2^*(R_1,kp)\cdot k\cdot p\cdot\lambda_1\cdot g_{11}(kp)\cdot I_1(g_{11}(kp))}{k\cdot p\cdot\lambda_1\cdot g_{11}(kp)\cdot f(g_{11}(kp))}.$$

The g_x parameter is calculated as the root of the equation:

$$\frac{C_2(p)\cdot K_1(g_{21}(p))-C_1(p)\cdot I_1(g_{21}(p))}{C_4(p)\cdot K_1(g_x)-C_3(p)\cdot I_1(g_x)}\times\frac{C_4(kp)\cdot K_1(h\cdot g_x\cdot\sqrt{k})-C_3(kp)\cdot I_1(h\cdot g_x\cdot\sqrt{k})}{C_2(kp)\cdot K_1(g_{21}(kp))-C_1(kp)\cdot I_1(g_{21}(kp))}=1,$$

where $C_3(p) = -\dfrac{C_1(p)\cdot K_0(h\cdot g_x)\cdot I_0(g_{21}(p)) + C_2(p)\cdot K_0(h\cdot g_x)\cdot K_0(g_{21}(p))}{I_0(h\cdot g_x)\cdot K_0(g_x) - I_0(g_x)\cdot K_0(h\cdot g_x)}$,

$$C_4(p) = \frac{C_1(p)\cdot I_0(h\cdot g_x)\cdot I_0(g_{21}(p)) + C_2(p)\cdot I_0(h\cdot g_x)\cdot K_0(g_{21}(p))}{I_0(h\cdot g_x)\cdot K_0(g_x) - I_0(g_x)\cdot K_0(h\cdot g_x)}.$$

6. The desired values of the thermal diffusivity and the thermal conductivity are defined as follows:

$$a_x = \frac{p\cdot R_2^2}{g_x}, \tag{17}$$

$$\lambda_x = \frac{C_2(p)\cdot K_1(g_{21}(p)) - C_1(p)\cdot I_1(g_{21}(p))}{C_4(p)\cdot K_1(g_x) - C_3(p)\cdot I_1(g_x)}. \tag{18}$$

The effective thermophysical characteristics of the fluid under study are determined under shear flow conditions and are related to a fixed shear rate.

3. Results and discussion

Using the described measurement setup and measurement method were investigated the thermophysical and rheological characteristics of a 10 % solution of polyoxyethylene in water containing 0.5 % of carbon nanomaterial "Taunit" which is a quasi-one-dimensional, nano-scale, filamentary formations polycrystalline graphite predominantly cylindrical shape with an inner channel (carbon nanotube). The inner diameter of the nanotubes is 10-20 nm, the outer diameter is 40-50 nm, and the length is more than 2 μm.

Polyoxyethylene (POE) at room temperature – is a flexible thermoplastic white color with a melting point ≈70 °C. It is produced by catalytic polymerisation of ethylene oxide C_2H_4O. Used as binder and modifier for different materials. At high shear rates, the destruction of molecular chains is observed.

When measuring the test material with a volume of 25±1 ml was placed in the gap between the inner and outer cylinders of the measuring device, after which the measurements were performed based on the technique described elsewhere [6]. For each shear rate, a series of three measurements were carried out, the results of which were calculated arithmetic mean values of thermal conductivity and thermal diffusivity.

Joint measurement of the shear rate γ and the torque transmitted to the inner cylinder due to forces of viscous friction of the investigated liquid material allow us to find the dependence of the shear stress σ versus γ, which makes possible the determination of the rheological characteristics (parameters of power-law): the consistency coefficient m and the flow index n, using method of approximation.The shear rate of the investigated fluid in the gap between the cylinders varied from 0 to 150 s^{-1}. The dependence of the shear stress σ versus shear rate in rheology is called the "flow curves". Their analysis for the test material leads to the conclusion that the fluid has significant non-Newtonian properties and flow curves can be approximated by the exponential equation $\sigma = m\cdot\dot{\gamma}^n$.Here are the values of m and n for POE solution at a temperature of 30° with concentrations of nanomaterials "Taunit" 0 % and 1.5 % (Table 1).

Table 1. Rheological characteristics of the investigated liquid

Concentration of "Taunit, %	The consistency coefficient m, Pa·s^n	The flow index n
0	20	0.92
1.5	31	0.85

Figure 3 shows the dependences versus the shear rate of the thermal conductivity calculated by the formula (15). The values calculated by the formula (18) had a much larger spread in the sample of three dimensions in the series.

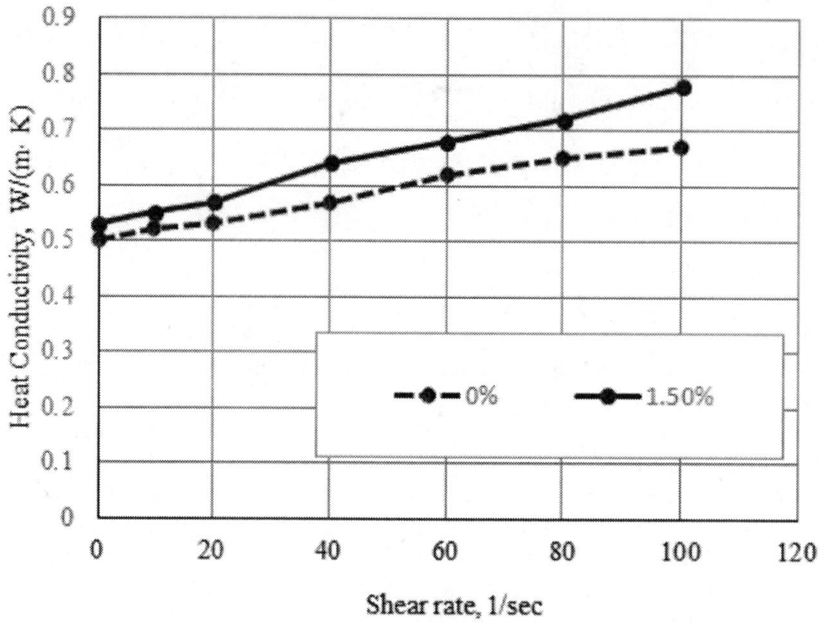

Figure 3. The dependence of the modified polyoxyethylene solution' thermal conductivity versus the shear rate.

The thermal diffusivity values of the studied material at zero shear rate, calculated by the formula (16), for the "Taunit" solutions with concentration 0 % and 1.5 % were equal to $(1.2\pm0.1)\cdot10^{-7}$ m^2/s and $(1.3\pm0.1)\cdot10^{-7}$ m^2/s, respectively. The dependence type of the thermal conductivity versus the shear rate (at various shear rates) qualitatively coincided with the dependencies shown in figure 3. Such dependence pattern confirms the measurement results of the suspensions' thermal characteristics at shear flow [1].

The effective values increase of the thermophysical characteristics with the growing shear rate can be explained by the heat transfer intensification in the liquid layer due to violation of the laminarity flow and the convective component flow.The presence of suspended particles (carbon nanotubes) in the liquid layer increases the value of the apparent thermal conductivity, which can be used to increase the efficiency of heat transfer from heated surfaces in the heat exchangers.

The analysis of arising measurement errors was carried out taking into account the results of experimental determination of thermophysical and rheological liquid characteristics with well-known thermophysical and rheological characteristics (glycerin, distilled water, 96 % ethyl alcohol), as well as the results of extended uncertainty (error) measurement evaluation. The results of this analysis showed that the developed measuring device has the following metrological characteristics, as presented in Table 2.

The thermal conductivity calculation using the formula (7) is possible only for fluids, as their flow curves follow the power law, and the heating of the liquid due to forces of viscous friction is greater than 2 °C. This is possible for liquids with high consistency and at significant shear rates. The measurement error in this case will be so significant that makes this method applicable only for preliminary evaluation of the thermal conductivity coefficient.

The thermophysical characteristics calculation using the formulas (15-18) is possible for liquids with any flow law. A significant measurement time remains as one of the disadvantages (at least half an hour).

Table 2. Metrological characteristics of the measuring unit.

Measured physical quantity	Duration of measurement	Measuring range	δ, %
Thermal conductivity (non-stationary stage) – λ, W/(m·K)	up to 20 min	0.1-0.8	±9
Thermal conductivity (stationary stage) – λ, W/(m·K)	up to 60 min	0.1-0.8	±6
Coefficient of thermal diffusivity, – $a \cdot 10^{-7}$, m^2/sec	up to 20 min	0.8-5	±7
Volumetric heat capacity– $c\rho \cdot 10^6$, $J/(m^3 \cdot K)$	up to 20 min	1-5	±10
Shear stress– σ, Pa	up to 1 sec	$10\text{-}10^4$	±0.5
Temperature – T, °C	-	20-250	±0.5
Shear rate – γ, s^{-1}	-	0-150	±5

4. Conclusion

Thus, hus, a method and apparatus have been developed which allow measuring with an error of not more than 10% the effective thermophysical and rheological characteristics of liquid phase materials and suspensions under shear flow conditions. Experimental data are presented for 10% polyoxyethylene solution containing additives of carbon-containing nanomateril "Taunit", as well as without them.

The results of experimental studies show an increase in effective values of thermal conductivity and coefficient of thermal conductivity at shear flow and at presence of nanoparticles in the liquid layer.

The presented method and measuring device are relevant for enterprises, technological processes of which include non-isothermal flows of liquid materials with an unstable structure through channels of different shapes.

Knowledge of thermophysical characteristics of processed materials will help to optimize these processes.

5. Acknowledgment

The work was conducted with the financial support by the Ministry of science and higher education of the Russian Federation (state task. The project 8.13393.2019/13.1)

References

[1] Li S X Q 2003 Shear rate dependent thermal conductivity measurement of two fruit juice concentrates. *J. of Food Eng.* **57** 217 https://doi.org/10.1016/S0260-8774(02)00300-X

[2] Minakov A V, Lobasov A S, Rudyak V Y, Guzei D V, Pryazhnikov M I 2014 Measuring of critical density of heat flow during boiling of nanoliquids on a cylindrical heater. *Technical Physics Letters* **40**(7) 562 https://doi.org/10.1134/S1063785014070062

[3] Shi M H, Shuai M Q, Chen Z Q, Li Q Xuan Y 2007 Study on pool boiling heat transfer of nano-particle suspensions on platesurface. *J. of Enhanced Heat Transfer* **14**(3) 223 https://doi.org/10.1615/JEnhHeatTransf.v14.i3.40

[4] Choi S U S, Zhang Z G, Yu W, Lockwood F E, Grulke E A 2001 Anomalous thermal conductivity enhancement in nanotube suspensions. *Applied Physics Letters* **79**(14) 2252 https://doi.org/10.1063/1.1408272

[5] Zhang J, Tian W, Chen J, Yu J, Zhang J, Chen J 2019 The application of polyetheretherketone (PEEK) implants in cranioplasty. *Brain Research Bulletin.* **153** 14149 https://doi.org/10.1016/j.brainresbull.2019.08.010

[6] Divin A G, Ponomarev S V, Belyaev P S, Petrasheva M A, Lyubimova D A 2014 Measurement setup for investigation of thermophysical and rheological characteristics of non-Newtonian fluids. *Bulletin of Kazan Technological University* [Vestnik Kazanskogo Tekhnologicheskogo Universiteta – in Russian] **17**(19) 293

[7] Vallejo J P, Żyła G, Fernández-Seara J, Lugo L 2018 Rheological behaviour of functionalized

graphene nanoplatelet nanofluids based on water and propylene glycol:Water mixtures. *Int. Comm. Heat Mass Transf.* **99** 43 https://doi.org/10.1016/j.icheatmasstransfer.2018.10.001

[8] Divin A G, Mishchenko S V, Ponomarev S V, Mozgova G V, Tkachev A G 2008 An automated measurement setup for studying the dependences of the thermal conductivity and rheological characteristics of non-newtonian fluids on the shear rate. *Instruments and Experimental Techniques* **51(3)** 480

[9] Yi J, Jin T, Deng Z 2019 The temperature field study on the three-dimensional surface moving heat source model in involute gear form grinding. *Int. J. of Adv. Manufact. Technol.* **103**(5-8) 3097 https://doi.org/10.1007/s00170-019-03752-9

[10] Vlasov V V 1975 Thermophysical measurements. *Reference manual on methods of calculation of fields, characteristics of heat and mass transfer and measurement automation* (Tambov) p 256

[11] Belyaev V P, Mishchenko S V, Belyaev P S 2015 Selecting Optimal Operating Parameters for Measuring the Solvent Diffusion Coefficient in Non-Destructive Testing of Products Made from Capillary-Porous Materials. *Measurement Techniques* **58(5)** 574

[12] Yanke E, Emde F, Lesh F 1964 *Special functions* (formulas, graphs, tables) p. 344

[13] Garlotta D 2001 A literature review of poly(lactic acid). *J. of Polymers and the Envir.* **9(2)** 63 https://doi.org/10.1023/A:1020200822435

IOP Conf. Series: Materials Science and Engineering **693** (2019) 012021 doi:10.1088/1757-899X/693/1/012021

Impedance spectroscopy of electrochromic films of nanocrystalline tungsten (VI) oxide

A V Shchegolkov[1], E N Tugolukov, A V Shchegolkov and V S Yagubov

Department of Technology and Methods of Nanoproducts Manufacturing, *Tambov State Technical University,*106 Sovetskaya Street, 392000 Tambov Russian Federation

[1]E-mail: alexxx5000@mail.ru

Abstract.The present paper study electrochromic films of nanocrystalline tungsten oxide (VI) by electrochemical impedance spectroscopy. Electrochromic films were obtained using the simplified technique of deposition of tungsten (VI) oxide from an aqueous dispersion on surface of transparent electrode In_2O_3. To obtain a uniform composition, the aqueous dispersion was treated with ultrasonic radiation, while the average particle size of tungsten oxide in the dispersion volume was from 30 to 70 nm. The prepared films had an amorphous structure and a crystalline structure was obtained by pyrolysis at a temperature of 250, 300, and 400 °C corresponding to different samples. The obtained nanocrystalline films were studied by electrochemical impedance spectroscopy in the frequency range from 1 to 106 Hz with an amplitude of an alternating voltage of 10 mV to determine the effect of temperature on the impedance of the films. The analysis data allowed us to determine the dependence of electrical conductivity on the volume of grains and the boundaries between crystals on the influence of temperatures on the structure properties of electrochromic films of nanocrystalline tungsten (VI) oxide.

1.Introduction

Transition metal oxides are one of the most common materials of micro- and optoelectronics [1-3]. The main characteristic of the material determining its physicochemical properties (optical, electronic, electrochemical, etc.) is the structure. Amorphous and crystalline films of metal oxides are known. Most transition metal oxides are dielectrics, but under certain conditions associated with structural changes, they become semiconductors, and with the addition of a conductive additive, they acquire electronic conductivity [4,5].

Currently, researchers are focused on the study of the electrochromic properties of transition metal oxides of materials and their application in various optoelectronic devices [6-8]. Among these materials, tungsten (VI) oxide (WO_3) is a well-known electrochromic oxide of n-type semiconductor, which attracts great scientific interest due to its physicochemical properties and numerous potential applications: smart windows, displays, sensors, thermoelectric devices and ferroelectric application [9,10].

By its nature of WO_3 is an insulator with a band gap of 3 eV [11,12], but when a stoichiometric state changes it becomes a semiconductor. The amorphous or crystalline structure of WO_3-based thin-film devices for micro- and optoelectronics as a resultis determined by the production technology, cost, and operating conditions of the manufactured device. Amorphous WO_3 films are more studied and have the greatest distribution due to low-temperature production methods, which makes their manufacture cheaper. Crystalline films of WO_3 can be used for light filters [13]. Further development of WO_3 electrochromic film technology involves the use of nanosized oxide powder to form nanocrystalline electrochromic films.

In this regard, the main aim of the present research was to study WO_3 nanocrystalline electrochromic films using cheap sputtering and pyrolysis technology (at temperature of 250, 300, and 400 °C) on conductive ITO glass and studying the impedance by electrochemical impedance spectroscopy (EIS).

2. Materials and methods

Electrochromic films of nanopowder WO_3 (Sigma-Aldrich, Germany) were produced by the spray pyrolysis technique to temperatures 250, 300 and 400 °C. Particle size of nanopowder WO_3 <100 nm (TEM) [13]. Schematic crystal lattice of WO_3 nanopowder below shows (figure 1). The physical basis of the method is written in detail [14]. In the temperature range from +890 to –190 °C, tungsten (VI) oxide has five polymorphic modifications (figure 1), in which the crystal lattice symmetry varies from tetragonal to monoclinic [11]: $D_{4h}^7 \xrightarrow{737\,^\circ C} D_{2h}^{16} \xrightarrow{327\,^\circ C} C_{2h}^5 \xrightarrow{27\,^\circ C} C_i^1 \xrightarrow{-43\,^\circ C} C_s^2$.

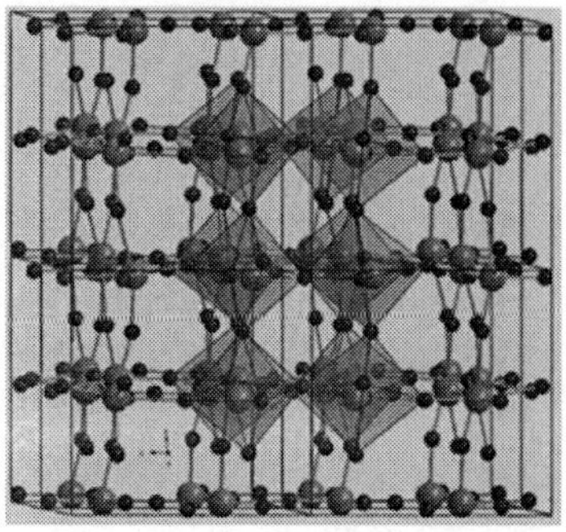

Figure 1. View of crystalline structure of WO_3.

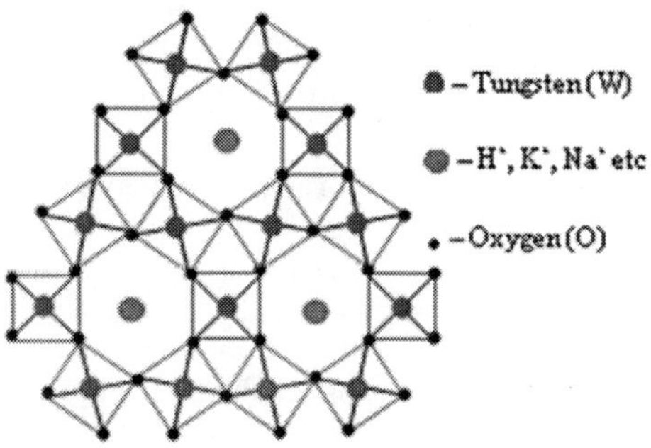

Figure 2. Hexagonal structureof WO_3formed at temperature 250 and 300 °C.

At temperatures from 250 to 400 °C, the hexagonal structure of WO_3 is formed (figure 2), the best electrochromic properties of which are characteristic of this structure [2]. When the temperature rises above 400 °C, the ITO conductive electrode degradation.

2.1. *The method spray pyrolysis for preparation electrochromic thin films based of WO_3*

The metal nanopowder of the electrochromic compound WO_3 was added to water (0.05 g/ml) and then dispersed in an ultrasonic unit (UP200S, Germany) for 5 min. until the complete distribution of nanoparticles in the water volume. The electrochromic composition was applied to a transparent In_2O_3 (ITO) electrode (Sigma-Aldrich, Germany) using a spraying machine. The surface of the ITO electrode was cleaned with isopropyl alcohol and washed with distilled water before applying the electrochromic composition. In the next step, the electrochromic coating was annealed in a vacuum oven at a temperature of 250, 300, and 400 ° C for 24 h before the formation of films. As a result, films with an average WO_3 particle size of 50 nm were obtained due to the insolubility of the metal nanopowder in water.

2.2. *Method EIS for the prepared electrochromic films*

The EIS was measured in a classical three-electrode electrochemical cell (figure 3-4) in the frequency range (ω) from 1 to 106 Hz with an amplitude of an alternating voltage of 10 mV using a potentiostat (PalmSens 4, Holland) of an electrochemical measuring complex, as a reference electrode (RE) we used the classic electrode Ag/AgCl in 3 M KCl (BVT Technologies, Czech).The films obtained on a glass substrate were used as a working electrode (WE) in a three-electrode system in figure 4.

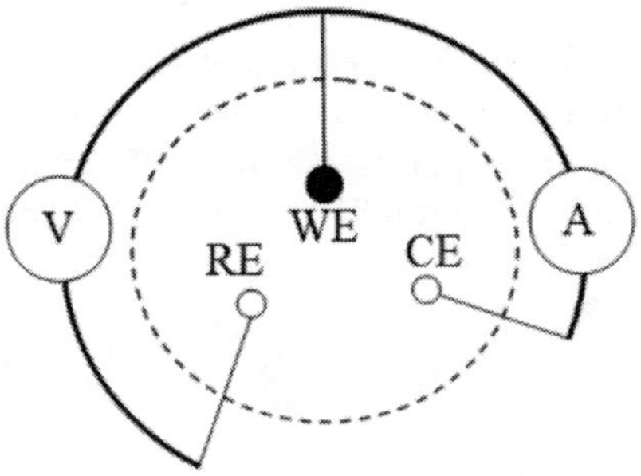

Figure 3. Schematic three-electrode cell: WE-working electrode; RE-reference electrode; CE-Counter electrode.

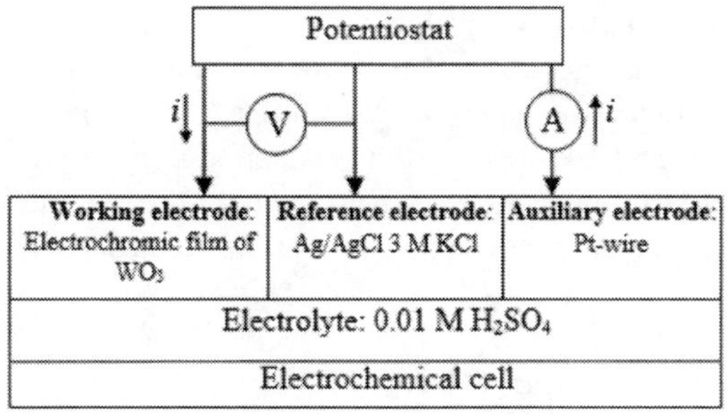

Figure 4. Three-electrode electrochemical system for researching electrocromic films of WO_3 in 0.01 M H_2SO_4 and reference electrode by Ag/AgCl 3 M KCl.

The size of the substrate base on ITO-glass was 0.2×0.8 cm and the Pt-wire (d~1 mm) was the auxiliary electrode (CE). All impedance measurements for the WO_3 electrochromic film were made in 0.01 M H_2SO_4 electrolyte relative to the Ag/AgCl reference electrode in 3 M KCl. The measurement data were recorded after the electrode was kept in solution to establish a quasistationary potential (2 min). Before each measurement, the reference electrode was calibrated by method written in [15].

2.3. Technique of EIS

The strength of impedance spectroscopy as an analytic technique rests on circuit models that are able to adequately represent the experimental data [16,17]. The impedance measurements did in potenciostat mode by PalmSens4 instrument. A sine wave is superimprosed on the specified DC-potential, so

$$E = E_{dc} + E_{ac}$$

This results in a dc-current with a superimposed ac-current,

$$I = I_{dc} + I_{ac}$$

The measurements yield the impedance defined by

$$Z(\omega) = \frac{E_{ac}(\omega)}{I_{ac}(\omega)}$$

The impedance is a complex number $Z(\omega) = Z'(\omega) - jZ''(\omega)$, where Z' – is the real part; Z'' – imaginary part of the impedance and $\omega = 2\pi f$, where f is the applied frequency. The phase shift φ – is defined as $\tan(\varphi) = \frac{-Z''}{Z'}$.

The impedance data are often presented in the Nyquis plot, with Z'' vs Z' or in a Bode plot, with the decimal logarithm of the magnitude of Z and Phase shift versus the logarifm of the frequency. The magnitude Z is defined as

$$|Z|^2 = (Z')^2 + (-Z'')^2$$

The inverse of the impedance is the admittance, which is also comlex numb $Y = Y' + jY''$,

$$Y = Y' + jY'' \text{, where:}$$

$$Y = \frac{1}{Z},\ Y' = \frac{Z'}{\left(Z'^2 + Z''^2\right)} \text{ and } Y' = \frac{Z''}{\left(Z'^2 + Z''^2\right)}$$

Processing the results of impedance measurements was performed using the Originlab2018 software.

3. Results and discussion

As a result, electrochromic films of nanosized powder (<100 nm) of WO_3 powder were obtained under low-temperature sputtering. In order to evaluate the effect of the contribution of the granular structure shown in [5,13,18] on the total conductivity, we compared three polycrystalline WO_3 samples for differential annealing of nanopowder at temperatures of 250, 300, and 400 °C.Figures 5-7 shows the difference nature of the change in the electrochromic impedance WO_3 films.

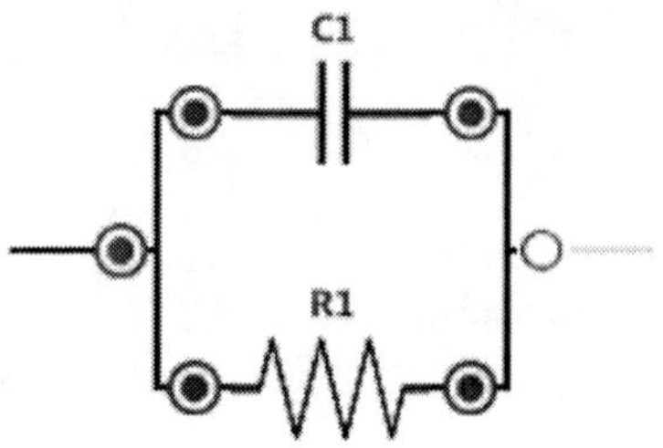

Figure 5. Equivalent circuit (R1 – resistance, C1 – capacitance).

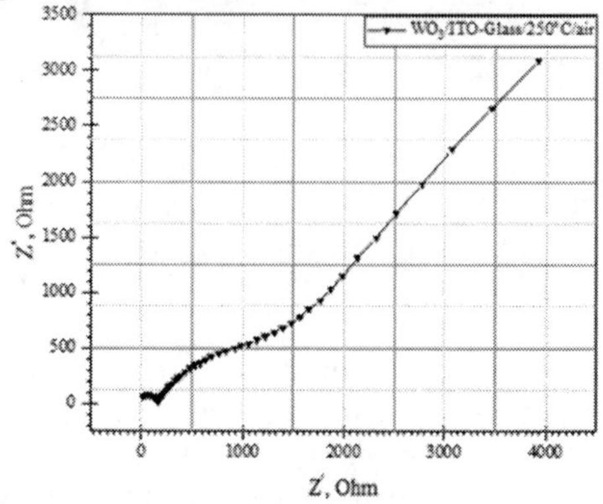

Figure 6. Hodographelectrochromic film WO_3 obtained by spray pyrolysis method at temperature of 250°C.

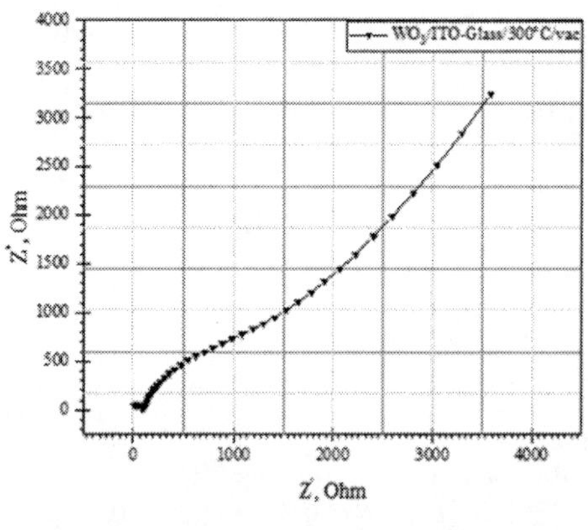

Figure 7. Hodographelectrochromic film WO_3 obtained by spray pyrolysis method at temperature of 300 °C.

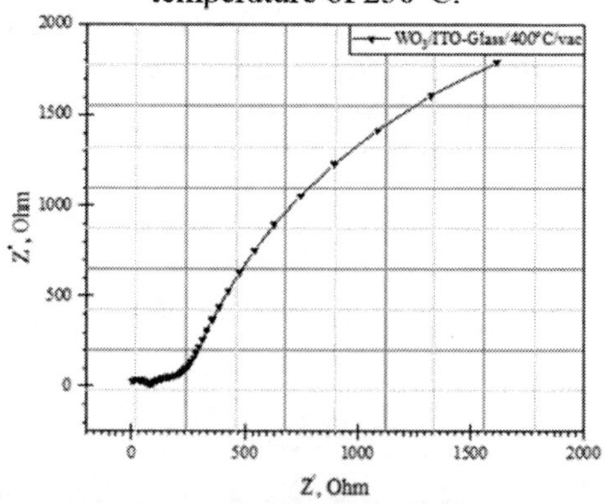

Figure 8. Hodographelectrochromic film WO_3 obtained by spray pyrolysis method at temperature of 400 °C.

The studies showed that during the temperature crystallization of the amorphous structure of the prepared WO_3 electrochromic films, the annealing temperature affects the change in impedance hodographs [19] as shown in figures 5-8. The type of circuit used is shown in figure 5. The graphs show that the impedance of the samples obtained at different temperatures has the same character, but different in resistance value.

So, with an increase in the annealing temperature of WO_3 films from 250 to 300 °C, a slight increase in conductivity is observed, and with annealing at 400 °C this increase is more pronounced. This difference is explained by different mobility of charge carriers with different microstructure of the samples. It should also be noted that this effect is associated with the formation of a denser surface packing of W atoms on the ITO substrate, and as a result, better carrier transport. Based on what, it can be assumed that the formation of an ordered structure improves ion transport on the film surface, and therefore the rate of the electrochemical reaction increases. According to [20], the size of WO_3 nanoparticles ranged from 30 to 70 nm.

As a result of EIS studies, the influence of annealing temperature on the electrical conductivity of WO_3 electrochromic films was proved. For the maximum annealing temperature, 400 °C was the lowest resistance value of 135 Ohm/m^2 and the maximum resistance of 1000 Ohm/m^2 for samples obtained at 250 °C and 300 °C. Thus in the temperature range of 250–400 °C better conductivities have been for smaller grain size, then higher grain size the samples showed. Therefore, an electrochromic film with a maximum conductivity of 400 °C was obtained optimally due to the formation of a denser packing of W, O atoms, as well as a fine-grained structure on the surface of the samples.

4. Conclusion

Electrochromic films of the WO_3 nanocrystalline structure were fabricated due to the simplified pyrolysis spray technology at different temperatures, which does not require complicated hardware design. As a result of the EIS tests, it was revealed that the samples obtained at different temperatures have different impedances, which is possible due to different levels of conductivity. Different temperatures make it possible to form different symmetries of WO_3 crystal lattices at a temperature of 250, 300 °C close to hexoganal, and at 400 °C to tetragonal. Using the impedance spectroscopy method in the study of the resulting structures allows us to judge the nature of their occurrence. Which makes the process of studying WO_3 thin films cheaper, in contrast to more expensive techniques and methods.What is very important in mass production.

References

[1] Granqvist C G 1992 Electrocromism and Smart Window Design. *Solid State Ionics* **53** 479 https://doi.org/10.1016/0167-2738(92)90418-O

[2] Granqvist A 1995 *Handbook of Inorganic Electrochromic Materials* (Amsterdam: Elsevier) p 617

[3] Monk P M S, Mortimer R J and Rosseinsky D R 2007 *Electrochromism and Electrochromic Devices* (Cambridge University Press, New York) p 483

[4] Panagopoulou M, Verhardou D, Koudoumas E, Katsarakis N, Tsoukalas D. and Raptis Y 2012 Tunable Proporties of Mg-Doped V_2O_5 Thin Films for energy Applications: Li-Ion Batteries and Electrocromics. *J. Phys. Chem. C.* **121** 70 doi: 10.1021/acs.jpcc.6b09018

[5] Vernardou D Psifis K Louloudakis D Paradimitropoulos G Davazoglou D Katsarakis N and Koudomas E 2015 Low Pressure CVD of Electrochromic WO_3 at 400 °C. *J. Electrochem. Soc.* **162** 579 https://doi.org/10.3390/coatings7020024

[6] Vijayakumar E Yun Y Quy V Lee Y Kang S Ahn K and Lee S W 2019 Development of Tungsten Trioxide Using Pulse and Continuous Electrodeposition and Its Properties in Electrochromic Devices. *J. Electrochem. Soc.* **166** 86

[7] Kwong W L Savvides N and Sorrell C C 2012 Electrodeposited nanostructured WO_3 thin films for photoelectrochemical applications. *Electrochim. Acta.* **75** 371

[8] Mukherjee R and Sahay P P 2015 Effect of precursors on the microstructural, optical, electical and electrochromic properties of WO_3 nanocrystallyne thin films. *J Mater Sci: Mater Electron.* 6293 doi: 10.1007/s10854-015-3216-8

[9] Petterson H Gruszecki T Johanson L-H Edwards M Hagfeldt A and Matuszczyk T 2004 Direct-driven electrochromic displays based on nanocrystalline electrodes. *Displays* **25** 223 https://doi.org/10.1016/j.displa.2004.09.015

[10] Sonami P R and Radhakrishman S 2002 Electrochromic materials and devices: present and future. *Mater. Chem. Phys.* **77** 117 10.1016/s0254-0584(01)00575-2

[11] Hubolov B M Podlinov V P and Musukov R A 2015 Electronic structure of single crystals and films of composite Tungsten Oxides. *Phase transitions, interphase boundaries and nanotechnology* **4** 22

[12] Khyzhun O Y and Solonin Y M 2007 Electronic stucture of nanoparticles of substoichometric

hexagonal tungsten oxides. *J. of Physics: Confernce Series*. **61** 534 doi:10.1088/1742-6596/61/1/108

[13] Bange K 1999 Colouration of tungsten oxide films: A model for opticaly active coatings. *Sol. Energy Mater. Sol. Cells*. **58** 131 10.1155/2013/279398

[14] Arakaki J Reyes R Horn M and Estrada W 1995 Electrochromism in NiO_x and WO_x obtained by spray pyrolysis. *Sol. Energy Mater. Sol. Cells*. **37** 33 https://doi.org/10.1016/0927-0248(94)00195-2

[15] Neghmouche N Khelef A and Lanez T 2010 Investigation of diffusion of ferrocene and ferricenium in aqueous and organic medium using voltammetry techniques. *Res. J. of Pharmaceutical, Biological and Chem. Sci.* **76**

[16] Macdonald J R 2005 *Impedance Spectroscopy. Theory, Experiment and Applications* (New Jersey: A John Wiley & Sons, Inc.) ed E Barsoukov – 2nd ed p 595

[17] *The Impedance Measurement Handbook. A Guide to Measurement Technology and Techniques. Agilent Technologies Co. Ltd.* 2006 p 140

[18] Vernardou D Drosos H Spanakis E Koudoumas E Savvakis C and Katsarakis N 2011 Electrochemical and photocatalytic properties of WO_3 coatings at low temperatures. *J. Mater. Chem*. **513** doi:10.1039/c0jm02413a

[19] Boukamp B A and Bouwmeester H J M 2003 Interpretaation of the Gerischer impedance in solid state ionics. *Solid State Ionics*. **157** 29 https://doi.org/10.1016/S0167-2738(02)00185-6

[20] *Properties tungsten (VI) oxide* Sigma Aldrich at https://www.sigmaaldrich.com

Twinning formation in nanodiamonds after treatment in a planetary mill: HRTEM studies

T Gordeeva[1,2], B Kulnitskiy[1,2], D Ovsyannikov[1], M Popov[1,2,3] and V Blank[1,2,3]

[1]Laboratory of Electron Microscopy, *Technological Institute for Superhard and Novel Carbon Materials,* 7a Centralnaya Street, 108840 Troitsk New Moscow Russian Federation
[2]Department of Physics and Chemistry of Nanostructures, *Moscow Institute of Physics and Technology (National Research University)*, 9 Institutskiy Per., 141700 Dolgoprudny Moscow Region Russian Federation
[3]*National University of Science and Technology MISiS*, 4 Leninskiy Avenue, 119049 Moscow Russian Federation

[1]E-mail: gordeeva.ta@phystech.edu

Abstract. The structure of a powder obtained as a result of processing a mixture of silicon and nano-diamond (1:4 weight ratio) in a planetary mill was studied by transmission electron microscopy. It was found that the obtained material includes diamond nanoparticles with an average grain size of 10 nm, having twins and lonsdaleite interlayers. In the present work, the plastic deformation of the diamond was detected at a temperature below 420 K by the mechanism of mechanical twinning, and the maximum stress in the diamond under the experimental conditions was found to not exceed 6 GPa, which is below the critical shear stresses for diamonds (55 GPa).

1. Introduction

Plastic deformation of a diamond at temperatures below ~1200 K is of particular scientific interest, since diamond is a brittle material, and practically any mechanical effect leads to the formation of cracks [1]. Diamond is unique in that its strength is equal to theoretical yield limit (55 GPa) [2,3], which coincides with the experimental value of yield limit [3]. Yield limit is a shear stress, at which a diamond lattice loses its stability and begins a phase transition [4] to the intermediate carbon phase. In addition, the theoretical yield limit is equal to the stress, at which the dislocation movement begins [2].

Plastic deformations in diamond are associated with dislocation plasticity [2]; therefore, it is usually observed at temperatures above 1200-1400 K [1]. At room temperature, the plastic deformations of the diamond were observed earlier in a high-pressure diamond anvil cell with a high nitrogen content at a pressure of 170 GPa [5], as well as during indentation [6,7] (the average contact pressure under the indenter for diamond was ~ 150 GPa [6]). Moreover, in contact with the material, the pressure of which is lower than the hardness of the diamond, in the general case, only the formation of cracks in the diamond is possible [8]. However, questions remain about the mechanism of plastic deformation.

One of the possible mechanisms of the diamond plastic deformations at low (below the Debye temperature) temperatures is twinning [9,10]. Recently, researchers have suggested that it is the twinning process that causes enhanced mechanical properties of the so called "nanotwinned diamond"

[11-14]. Plastic deformations through mechanical twinning are described in the works [15,16]. However, this mechanism was earlier observed only at high (1300 K) temperatures [17].

Despite the fact that the diamond plastic deformations at "low" temperatures are observed, this mechanism has not been studied experimentally, except in the case of a phase transition [14]. In this regard, the issue of the mechanism of the diamond plastic deformations at temperatures below 1200K needs to be clarified.

The objective of this research was to study the structural features of diamond nanoparticles deformation after processing in a planetary mill. In the present work, the unconventional method of obtaining nanopowder – mechanical grinding in a planetary mill – was used. In the grinding process, the powder particles experience a high mechanical load, which leads to phase transformations.

2. Experimental part

A powder-like mixture containing plastically deformed diamond was obtained by processing in a Micro Mill PULVERISETTE 7 Premium Line planetary mill (FRITSCH, Idar-Oberstein, Germany) with a steel drum and steel milling balls. The processing time in the experiments was 2 h (the processing cycle was 1 min of grinding and 3 min of cooling). After that, the powder obtained after processing in the mill, was subjected to processing in a mixture of HNO_3 and HF acids (EKOS-1, Moscow, Russia) to partially dissolve silicon. The remaining diamond powder was examined by high resolution transition electron microscopy (HRTEM) methods on a JEM-2010 instrument equipped with EDS attachment (JEOL, Tokyo, Japan).

In [18], it was shown that the maximum temperature of the sample during the processing cycle used by us in a planetary mill does not exceed 420 K, therefore, we can assume that under the experimental conditions, the sample is certainly not heated above the Debye temperature $T_\vartheta = 2100$ K. In [19], it was shown that the mechanical stresses arising from the collision of steel balls during processing in a planetary mill do not exceed 6GPa. It can also be considered that the contact pressure at the impact of the balls does not exceed the value of their hardness (7 GPa), since no traces of plastic deformation were found on steel balls when the mill worked without a sample. Using the general theory of elasticity in the case of a short rod striking a drummer with a much larger mass [20], it is possible to describe the stress state of a sample when struck with grinding balls. While the impactor speed has not significantly decreased and the stress is high, the front edge of the wave will go from one end to the other. Meanwhile, in the opposite direction from the second end of the rod, the reflected wave will go. The imposition of these waves will lead to large shear stresses. Thus, plastic deformations are possible already at relatively low impact speeds.

For several reasons, we used a detonation diamond powder with an average size of 10 nm produced by the SINTA company (Minsk, Republic of Belarus). First, this size allows you to directly explore a diamond particle in TEM without additional processing. Second, the average particle size is twice the value at which the quantum confinement effect begins to manifest itself [21]. Third, detonation nanodiamonds have a small concentration of defects [22].

3. Results and discussion

According to our estimates, the initial diamond powder (figure 1a) comprised no more than 3-4 % of particles containing planar lattice defects, namely, stacking faults and twins. To prevent iron from interacting with diamond, the diamond powder was mixed with silicon in a 1:4 weight ratio.

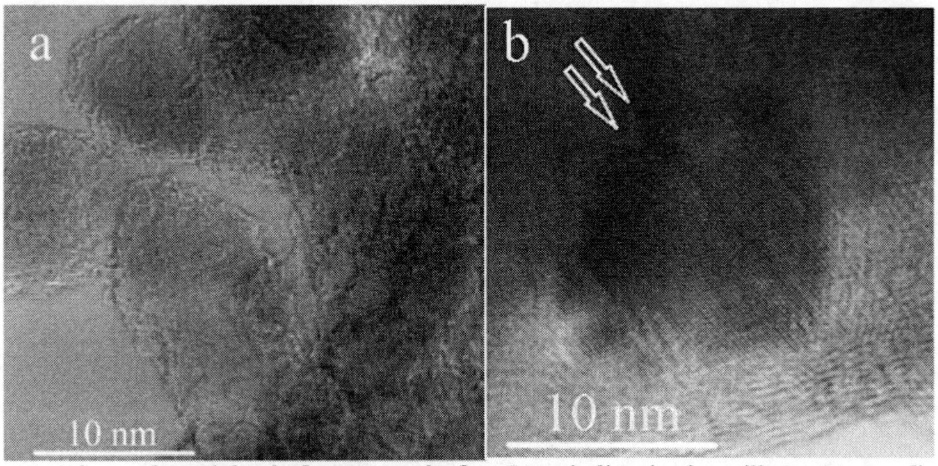

Figure 1. Diamond particles before **(a)** and after **(b)** grinding in the mill. In **(b)** – a diamond particle is shown surrounded by silicon particles. The arrows indicate twinning planes {111}.

According to the results of the TEM analysis, at least 30 % of the diamond particles after processing in the mill contain twins. Figure 1b shows a diamond particle surrounded by undissolved silicon, which encompasses the most common twin boundary in diamond. Such twin boundaries of the Σ= 3 type were observed in natural, artificial and thin film diamond layers [23], as well as in other materials that have the diamond lattice [24].

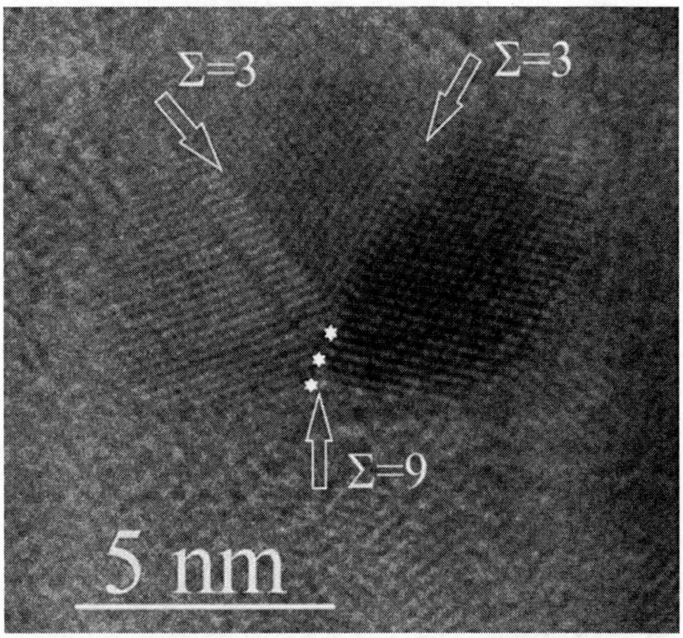

Figure 2. A diamond particle surrounded by silicon, after processing in a planetary mill. The asterisks indicate the incoherent part of the boundary.

In this work, we observed both coherent twin boundary and incoherent boundaries in different particles. In the case of the coherent boundaries, the lattices of twin components are mated along the boundaries, without causing macroscopic stresses in the crystal. The incoherent boundaries can be flat, stepped and curvilinear. Figure 2 shows the traditional twins of the first and second order (Σ = 3, 9). A similar twin (figure 2) structure of diamond films, grown by plasma-assisted CVD, was introduced and examined in [25].

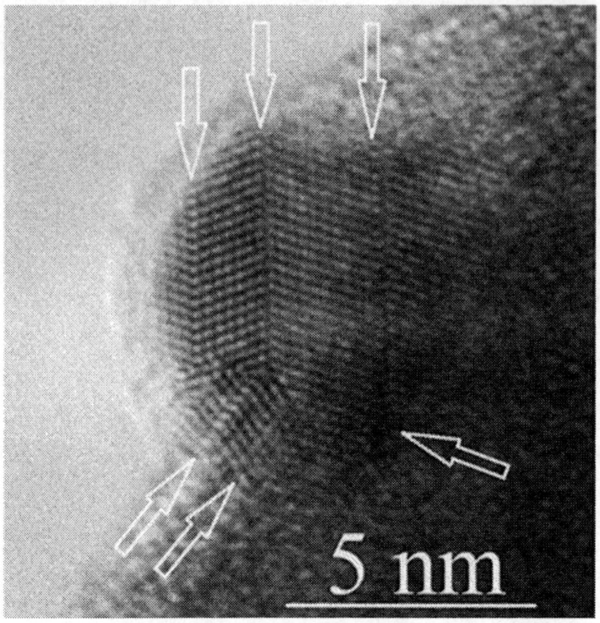

Figure 3. Multiple twinning in a diamond particle. The arrows indicate winning planes {111}.

Twins in diamond are shown in figure 3. The twinning plane {111} in this case is also the contact plane – *i.e.*, the lattices forming a twin have one common atomic layer located above the interface (in figure 3, the interfaces are indicated by arrows). Thus, when twinning of diamond crystals occurs along the {111} plane in the region of the lattice junction, the initial structure is not disturbed.

It was also found that the obtained material contains some relatively large (up to several nanometers) lonsdaleite fragments. Hexagonal diamond or lonsdaleite (2H) is a form of carbon found in meteorites in 1967. Lonsdaleite has a hexagonal crystal lattice (a=2.52, c=4.12A). The main difference of hexagonal from cubic (3C) diamond is in the packing of layers. The 3C crystal lattice in diamond is represented by the ABCABC sequence, whereas in lonsdaleite the 2H crystal lattice is represented by the ABAB sequence. Hence, a sequence of planes in diamond contains a lonsdaleite grid.

Some of the nanoparticles comprise lonsdaleite layers. Figure 4 shows a high-resolution image of a nanoparticle containing various defects that form such layers. In this figure, to the left, the sequence is ABCABCABC, whereas to the right, the monoatomic layer A removed from the sequence of ABCABCABC layers, which contains the mirror plane of symmetry, is demonstrated, thereby indicating the formation of a double structure in the lattice.

It is known that with increasing temperature, the lattice rearrangement is facilitated, which can lead to twinning. On the one hand, for twinning at an elevated temperature, less effort is required and larger residual twins can be obtained. But on the other hand, with increasing temperature, slip deformation begins to prevail, which impedes the twinning process [9]. Twin nucleation is not a heat-activated process, but it occurs in places with high stress concentration [10].

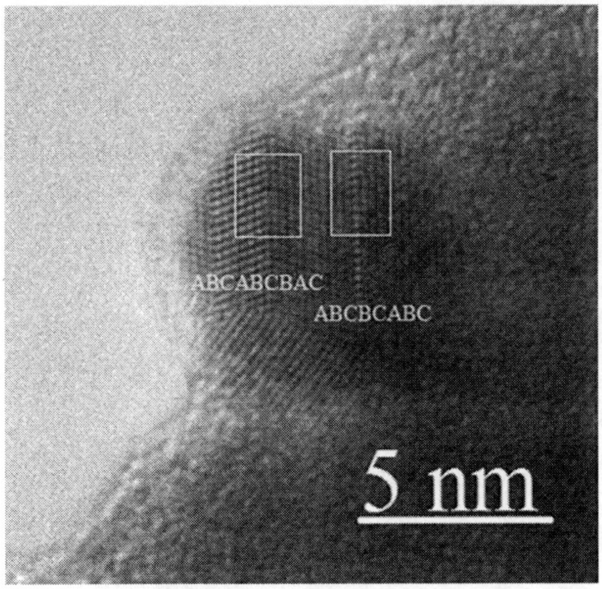

Figure 4. Diamond nanoparticle structure. Packing defects form lonsdaleite layers. To the left - the packaging layer ABCABCBAC. The stacking sequence is mirrored, which indicates the formation of a twin structure. To the right - one atomic layer A removed from the sequence of ABC ABCABC layers.

This can be explained by several reasons: blocking by the low mobility of atoms of partial or twin dislocations (in contrast to lattice dislocations); internal structure of the nucleus structure; or strain hardening. In general, the twinning process is more ordered than the slipping one. In the Pirouz model, the transition to twinning is due to the high Peierls force (*i.e.*, the force required to move dislocations in the plane of the atoms in the unit cell). This, in turn, leads to a very low mobility of lattice dislocations at low temperatures together with a large difference in the mobility of various partials dislocations. Therefore, the mechanism of twinning simultaneously with the concept of theoretical yield limit during machining was expected for diamond at a temperature ($T<T_{\vartheta}\approx 1200$ K). The results obtained are important for understanding the mechanisms of plastic deformation of diamondы and other covalent materials at room temperature.

4. Conclusions

The experiments considered herein demonstrated that in the diamond, the plastic deformation by mechanical twinning is possible under cyclic loads and stresses, as well as below the critical shear stresses of 55 GPa when the theoretical yield limit is achieved. Indeed, the maximum stresses in the diamond under the experimental conditions are limited by the hardness of steel balls (7 GPa). Various estimates also show that when machining in planetary mills, the mechanical stresses arising from the ball collision do not exceed 6GPa.

Thus, it was experimentally shown that for diamond at temperatures below 420 K (which is significantly below the Debye temperature) with cyclic stresses below 7 GPa, plastic deformation by mechanical twinning takes place.

5. Acknowledgment

The present work was funded by the Russian Foundation for Basic Research (RFBR) in the framework of Project No.18-29-19019.

References

[1] Wilks J and Wilks E 1991 *Properties and Applications of Diamond* (Oxford: Butterworth-Heinemann Oxford) p 525 https://doi.org/10.1002/adma.19920040421

[2] Gilman J J 2009 *Chemistry and Physics of Mechanical Hardness* (New York: Wiley Online Library) p 232 https://doi.org/10.1002/9780470446836

[3] Popov M, Kulnitskiy B and Blank V 2014 Superhard materials based on fullerenes and nanotubes *Compr. Hard Mater.* **3** 515 https://doi.org/10.1016/B978-0-08-096527-7.00057-X

[4] Popov M 2010 Stress-induced phase transitions in diamond *High Press. Res.* **30** 670 https://doi.org/10.1080/08957959.2010.525510

[5] Mao H K, Bell P M, Dunn K J Dunn, Chrenko R M and DeVries R C 1979 Absolute pressure measurements and analysis of diamonds subjected to maximum static pressures of 1.3–1.7 Mbar *Rev. Sci. Instrum.* **50** 1002 https://doi.org/10.1063/1.1135966

[6] Blank V, Popov M, Pivovarov G, Lvova N and Terentev S 1999 Mechanical properties of different types of diamond *Diam. Relat. Mater.* **8** 1531 https://doi.org/10.1016/S0925-9635(99)00079-5

[7] Sumiya H, Toda N and Satoh S 1997 Mechanical properties of synthetic type IIa diamond crystal *Diam. Relat. Mater.* **6** 1841 https://doi.org/https://doi.org/10.1016/S0925-9635(97)00149-0

[8] Brookes C A and Zhang L Y 1999 Cumulative deformation and fatigue of diamond—new developments *Diam. Relat. Mater.* **8** 1515 https://doi.org/https://doi.org/10.1016/S0925-9635(99)00076-X

[9] Klassen-Neklyudova M V 1964 *Mechanical Twinning of Crystals* (New York: Consultants Bureau) p 213

[10] Christian J W and Mahajan S 1995 Deformation twinning *Prog. Mater. Sci.* **39** 1 https://doi.org/10.1016/0079-6425(94)00007-7

[11] Sun L, He X and Lu J 2018 Understanding the mechanical characteristics of nanotwinned diamond by atomistic simulations *Carbon* **127** 320 https://doi.org/https://doi.org/10.1016/j.carbon.2017.11.021

[12] Tang J, Wang H-T, Lee D H, Fardy M, Huo Z, Russell T P and Yang P 2010 Holey silicon as an efficient thermoelectric material *Nano Lett.* **10** 4279 https://doi.org/10.1021/nl102931z

[13] Xu C, He G, Liu C and Wang H 2018 Twin-size effects on the hardness and plastic deformation mechanisms of nanotwinned diamond *Ceram. Int.* **44** 22121 https://doi.org/10.1016/j.ceramint.2018.08.324

[14] Huang C, Peng X, Yang B, Zhao Y, Xiang H, Chen X, Li Q and Fu T 2017 Molecular dynamics simulations for responses of nanotwinned diamond films under nanoindentation *Ceram. Int.* **43** 16888 https://doi.org/10.1016/j.ceramint.2017.09.089

[15] Titkov S V, Krivovichev S V and Organova N I 2012 Plastic deformation of natural diamonds by twinning: evidence from X-ray diffraction studies *Mineral. Mag.* **76** 143 https://doi.org/10.1180/minmag.2012.076.1.143

[16] Schoor M, Boulliard J-C, Gaillou E, Duparc O H, Estève I, Baptiste B, Rondeau B and Fritsch E 2016 Plastic deformation in natural diamonds: Rose channels associated to mechanical twinning *Diam. Relat. Mater.* **66** 102 https://doi.org/10.1016/j.diamond.2016.04.004

[17] Shiryaev A A, Frost D J and Langenhorst F 2007 Impurity diffusion and microstructure in diamonds deformed at high pressures and temperatures *Diam. Relat. Mater.* **16** 503 https://doi.org/10.1016/j.diamond.2006.10.001

[18] Kwon Y-S, Gerasimov K B and Yoon S-K 2002 Ball temperatures during mechanical alloying in planetary mills *J. Alloys Compd.* **346** 276 https://doi.org/10.1016/S0925-8388(02)00512-1

[19] Maurice D R and Courtney T H 1990 The physics of mechanical alloying: A first report *Metall. Trans. A* **21** 289 https://doi.org/10.1007/BF02782409

[20] Rabotnov Y N 1988 *Mechanics of Deformable Solids* [in Russian – Mekhanika

IOP Conf. Series: Materials Science and Engineering **693** (2019) 012022 doi:10.1088/1757-899X/693/1/012022

Deformiruemogo Tverdogo Tela] (Moscow: Nauka) p 711

[21] Popov M, Churkin V, Kirichenko A, Denisov V, Ovsyannikov D, Kulnitskiy B, Perezhogin I, Aksenenkov V and Blank V 2017 Raman spectra and bulk modulus of nanodiamond in a size interval of 2–5 nm *Nanoscale Res. Lett.* **12** 561 https://doi.org/10.1186/s11671-017-2333-0

[22] Daulton T L 2005 Nanodiamonds in the Cosmos *Synthesis, Properties and Applications of Ultrananocrystalline Diamond* vol 192 ed D M Gruen, O A Shenderova and A Y Vul' (Dordrecht: Springer) pp 49-62 https://doi.org/10.1007/1-4020-3322-2_5

[23] Kulnitskiy B, Perezhogin I, Dubitsky G and Blank V 2013 Polytypes and twins in the diamond–lonsdaleite system formed by high-pressure and high-temperature treatment of graphite *Acta Crystallogr. B.* **69** 474 https://doi.org/10.1107/S2052519213021234

[24] Kulnitskiy B, Annenkov M, Perezhogin I and Popov M 2016 Mutual transformation between crystalline phases in silicon after treatment in a planetary mill: HRTEM studies *Acta Crystallogr. B* **72** 733 https://doi.org/10.1107/S2052520616011422

[25] Shechtman D, Feldman A and Hutchison J L 1993 High-order twin boundaries in CVD diamond films *Mater. Lett.* **17** 211 https://doi.org/10.1016/0167-577X(93)90001-E

IOP Conf. Series: Materials Science and Engineering **693** (2019) 012023 doi:10.1088/1757-899X/693/1/012023

Production of few-layer and multilayer graphene by shearing exfoliation of graphite in liquids

V F Pershin[1,*], M N Krasnyanskiy[1], Z A A Alhilo[1], A M R Al-Mashhadani[1], A A Baranov[1] and A A Osipov[2]

[1]Department of Technology and Methods of Nanoproducts Manufacturing, *Tambov State Technical University,* 106 Sovetskaya Street, 392000, Tambov, Russian Federation
[2]*LLC "NanoTechCenter",* 51 Sovetskaya Street, 392000, Tambov, Russian Federation

[*]E-mail: pershin_home@mail.ru

Abstract. The article describes the process of liquid-phase shear exfoliation of graphite in a rotary apparatus with moving blades, in water and I-20A oil (Rosneft, Russia) used as a liquid. The linear dependence of the process intensity on the shear rate and graphite concentration in the initial suspension are established in the paper. To increase the share of graphene converted into graphene plates, it has been proposed to reduce periodically the concentration of graphene plates in suspension by adding pure liquid. It was found that the exfoliation rate of graphite is directly proportional to the concentration of graphite in the initial suspension and the rotor speed. The resulting graphene-containing oil-based suspension was used to modify greases.

1. Introduction

Nanomaterials have become a critical sector that can significantly push the boundaries of technology. The remarkable properties of carbon nanomaterials, such as high electrical (good electrical conductivity, as well as current density 108 A/cm^2 [1]) and thermal conductivities (thermal conductivity (5000 W/m·K) [2]), strength (Young's modulus (1 TPa) [3]), a large specific surface area (2630 m^2/g) [4,5], optical transmission (97.7%) [6,7].

One graphite layer is well known as monoatomic or monolayer graphene, and two and three graphite layers are known as bilayer and trilayer graphene. Graphene having more than 5 layers up to 10 layers is usually called few-layers graphene, and 20–30 layer graphene is usually called multilayer graphene or nanocrystalline thin graphite [8].

Any process for the manufacture or extraction of graphene can be attributed to the synthesis of graphene, depending on the desired size and purity of a particular product. In recent years, various methods have been developed for the synthesis of graphene, but mechanical splitting (peeling) [9], chemical peeling [10,11], chemical synthesis [12], and thermochemical vapor deposition (CVD) [13] are the most common.

Since this paper addresses the issue of modifying various materials (frost-resistant plastic lubricants, concrete, epoxy resin), i.e. materials used in industry in large volumes, the interest is represented by scalable methods for the production of graphene plates, providing improved performance characteristics of the modified material with a low production cost.

There are examples of the successful use of multilayer graphene plates in the modification of grease [14], concrete [15] and epoxy resin [16].

The aim of this work is to study the process of liquid-phase shear exfoliation of graphite and the preparation of graphene-containing suspensions that are used in the modification of materials, in particular, frost-resistant greases.

2. Critical analysis of liquid phase shear exfoliation methods

In liquid phase shear exfoliation, two main points were distinguished: particles of graphite were in a liquid, which weakens the bonds between the graphene layers and prevents agglomeration; forces were applied to the particles, which stratify the graphite. As a liquid, various solvents or water with the addition of surface-active substances was used.

High quality graphene was obtained in high yield through a solvothermic process [17]. Graphite flakes were dispersed in an organic solvent that can efficiently penetrate between the layers and was heated to 1000-2000°C. After heating to 1000°C in the atmosphere (95% Ar, and 5% H_2), graphite was mixed with a highly polar organic solvent (acetonitrile), in a teflon-coated autoclave. The mixture was then kept for 12 h at 180°C and treated with ultrasound. As a result, single-layer and two-layer graphene sheets were obtained. The disadvantage of this method was the difficulty of scaling and transition to industrial scale.

In [18], graphite was dispersed in an ionic liquid and irradiated with microwaves for 30 min. The installation is somewhat similar to ultrasonic treatment, but the processing time was significantly reduced, and an improved route for graphene exfoliation has been demonstrated. The authors reported high peeling efficiency, a yield of 93 wt.% And a selectivity of 95% with respect to single-layer graphene.

The use of supercritical fluids was also used for exfoliating graphite. Critical factors of this method were: high diffusion coefficient; extensibility and solvating ability of a supercritical fluid. Supercritical fluid can penetrate into the gap between the layers of graphite due to the high diffusion capacity and very low viscosity. When a rapid depressurization of the apparatus occurs, the supercritical fluid expands sharply, and forces arise that contribute to the process of graphite delamination [19, 20].

An attempt to increase the production of graphene using shear effects on particles of graphite created by a rapidly rotating blade, was made in [21]. In [22], hydrodynamics in supercritical CO_2 was investigated. This resulted in a yield of 63% monolayer graphene. The main problem of the supercritical fluid method is the need to provide a high-pressure reactor. For example, in [23], the pressure in the reactor was 12 MPa. In combination with temperature and the need to organize mechanical effects on graphite, even according to the authors, it is quite difficult to implement on an industrial scale.

Another shear-controlled process of exfoliation of graphite in an organic solvent is presented in [24]. In this approach, a rapidly rotating tube (diameter 16 mm, rotation speed 7000 rpm) was used to create an intense shear in a fluid containing graphite. In [25], a similar technology was used and multilayer with a thickness of not more than 20 nm and a small amount of flakes ~ 1 nm thick were obtained. This approach is less energy intensive than ultrasonic treatment, but industrial implementation is problematic.

Using a rotating disk (diameter 200 mm, rotation speed 2500 rpm) [26] generated the effects of transverse forces on graphite particles in dynamic thin films on the surface of the disk and this led to the exfoliation of graphite flakes and the production of graphene. In our opinion, the results of the studies confirm the possibility of obtaining graphene from graphite only due to shear effects, but has no prospects for industrial realization.

In [27], a mixer with a high rotor speed was used. A mixer with a diameter of 32 mm was used in the work, the rotation speed was 4500 rpm. The concentration of graphite in the initial suspension was 50 mg/ml, the volume of the suspension was 4.5 liters, and the processing time was 20 minutes. The treated suspension was centrifuged and the quality of the finished product was determined. As a result of the analysis, it was established that the thickness of the graphene particles obtained could be controlled by the gap between the stator and the rotor and the smaller this gap, the less layers in the graphene plates; the yield and quality of the product depend on the concentration of the initial suspension, the speed of rotation of the rotor, the processing time; rotor diameter. It is established that with a smaller stator diameter, the lateral dimensions of the graphene plates are smaller and vary from 0.35 to 0.9 μm, and the thickness of the plates from 0.9 to 1.3 nm. The analysis made by atomic force microscopy has shown that graphene sheets consist of 10 layers and less.

Raman spectroscopy confirmed the presence of single-layer graphene in the final suspension. The transition to a pilot plant with a rotor diameter of 110 mm was implemented. Analysis of the results showed that this setup can produce a suspension with a concentration of low-layer graphene 0.07 mg/ml. Plant capacity is 5.3 g/h. The authors carried out calculations and found that several such plants with a total volume of the treated suspension of 10 m^3 will be able to produce 100 g/hour. Judging by the fact that the pilot plant worked with a capacity of 0.3 m^3, 33 installations would be required. In our opinion, the latter method of obtaining suspensions is the most successful and can be used on an industrial scale.

3. Experimental setup

The liquid-phase exfoliation process was carried out in a laboratory setup, as in [28]. The main difference from the rotor apparatus, which was used in [27], is that we used moving blades. Under the action of centrifugal forces, these blades were pressed against the inner surface of the stator and slide along this surface without a gap. Thus, shear forces are transmitted to the particles directly by e-blades. As shown by the results of experiments, this method of transferring shear forces to particles increases the intensity of the exfoliation process.

Experiments were performed as follows. Preparing 5 L of a suspension of crystalline graphite fuels and lubricants-1 with a concentration with a concentration of 3% to 10%, poured it into a cylindrical container, installed the apparatus and started processing the suspension. Every 10 minutes, the process was stopped and 100 ml samples were taken. The suspension was centrifuged, the precipitate was removed and the concentration of graphene nanostructures in the remaining suspension was determined. After analysis, the sample was mixed with the residue, poured into a container and the exfoliation process continued. After the concentration of the next sample increased by less than 5% of the concentration of the previous sample, the exfoliation process was stopped. During the experiments, the rotor speed was changed: 5000; 10,000; 12,000; 15,000 rpm.

4. Results and discussion

In the first series of experiments, water was used as a liquid. Figure 1 (curve 1) shows the results of changes in the concentration of graphene structures over time, with a rotor speed of 10,000 rpm and a concentration of graphite in the initial suspension of 50 mg/ml. It can be seen that at the beginning of the process (up to a concentration of graphene nanoplastic of 1.7 mg/ml) the dependence is linear and this fully agrees with the results of [29]. With further processing of the suspension, the concentration grows more slowly and at a concentration of about 1.9-2.0 mg/ml, the exfoliation process stops. We decided to reduce the concentration of graphene nanoplates by adding pure water. The results of the experiments presented in figure 1 (curve 2) show that after reducing the concentration, the exfoliation process continues with the same intensity. Thus, periodically reducing the concentration of graphene nanoplastic materials can significantly increase the proportion of graphite, which is converted to graphene. The resulting suspension can be used to modify concrete [30].

With an increase in the concentration of graphite in the initial suspension from 5% to 10%, the maximum concentration of graphene structures in the suspension (1.79 mg/ml, figure 1, curve 2) was reached not in 25 minutes, but in 14 minutes. With an increase in the rotor speed, a similar situation was observed. For example, with an increase in speed from 10,000 rpm to 15,000 rpm, the time for which the maximum concentration was reached decreased from 25 minutes to 18 minutes, but the nature of the curve did not change.

In the second series of experiments, I-20A industrial oil (Rosneft, Russia) was used as a liquid. The concentration of graphite in the initial suspension was 5%, and the rotor speed was 10,000 rpm. The original suspension was passed through a rotary apparatus and after every five processing cycles, samples were taken for analysis. The intensity of the process of the formation of graphene structures was estimated from the change in the optical density coefficient (K), the numerical value of which characterizes the average integral number of layers in graphite particles [31].

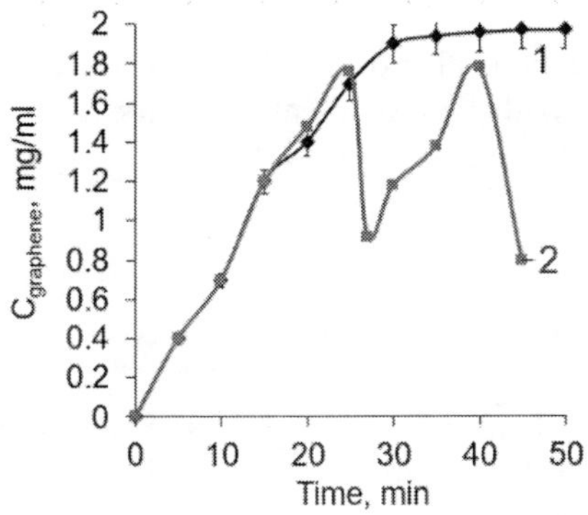

Figure 1. The dependence of the concentration of graphene nano-plates ($C_{graphene}$) on time: 1 – continuous mode; 2 – cyclic mode with a decrease the concentration of graphene structures.

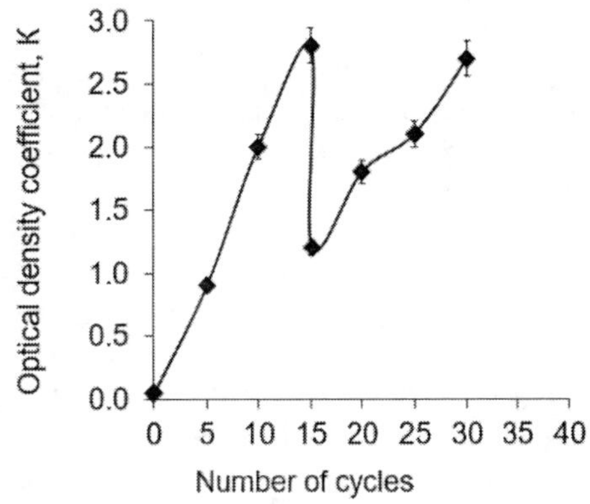

Figure 2. The dependence of the optical density coefficient on number of cycles.

Figure 2 shows the dependence of the coefficient K on the number of suspension processing cycles. The duration of each cycle is 3 minutes. As you can see, up to 15 cycles this dependence is linear, and then the intensity of the peeling process begins to decrease, so we used the same technological method as in the preparation of the aqueous suspension, i.e. decrease in concentration. The resulting dependence is similar to that shown in figure 2.

We carried out similar experiments in which the concentration of graphite in the initial suspension varied from 5% to 20%. It was found that with a decrease in concentration of graphite by 2 times, the intensity of exfoliation decreases by about 2 times, and with an increase in the concentration by 2 times, it increases by about 2 times. Thus, there is reason to believe that the intensity of the exfoliation process is directly proportional to the concentration of graphite in the initial suspension. At the same time, it should be noted that with a further increase in the concentration of graphite, the intensity of the process grows more slowly, so it makes no sense to use the initial suspension with a graphite concentration of more than 20%.

Graphene-containing suspensions are used in the modification of greases [31], including frost-resistant [32]. The results of determining the tribological characteristics showed that at a concentration of graphene structures obtained in this work of 0.15%, the diameter of the wear spot 50% reduction, attrition rate showed almost increased 2.7 times, carrying capacity increased 3 times. These results are in good agreement with previous ones. In the work [32], the concentration was slightly lower (0.1%), but the price of graphene structures obtained using long-term treatment of the suspension with ultrasound was at least 10 times higher. than using shear exfoliation. Thus, the proposed technology of shear exfoliation of graphite directly in oil has great prospects for industrial use.

5. Conclusion

As a result of experimental studies, it has been established that the concentration of graphene structures in the finished suspension is directly proportional to: the number of processing cycles; rotor speed; the concentration of graphite in the original suspension. In addition, it was found that to increase the efficiency of the exfoliation process, the fraction of graphite that could be converted into graphene structures can be achieved by periodically reducing the concentration of graphene structures in the suspension being processed, by adding pure liquid to this suspension. The technology for producing graphene structures directly in oil by shear exfoliation of graphite has good prospects for industrial implementation.

Acknowledgments
The work was funded by the Ministry of Science and Higher Education of the Russian Federation in the framework of the Federal Target Program "Research and Development in Priority Areas of Advancement of the Russian Scientific and Technological Complex for 2014–2020" (Agreement No. 14.577.21.0253, Unique Identification Number: RFMEFI 57717X0253).

References
[1] Moser J, Barreiro A and Bachtold A 2007 Current-induced cleaning of graphene. *Appl. Phys. Lett.* **91** 163513 doi.org/10.1063/1.2789673
[2] Balandin A, Ghosh S, Bao W, Calizo I, Teweldebrhan D, Miao F and Lau J 2008 Superior thermal conductivity of single-layer graphene. *Nano Lett.* **8**(3) 902 doi.org/10.1021/nl0731872
[3] Lee C, Wei X, Kysar J W and Hone J 2008 Measurement of the elastic properties and intrinsic strength of monolayer graphene. *Science.* **321** 385 doi.org/10.1126/science.1157996
[4] Bolotin K, Sikes K J, Jiang Zh, Klima M, Fudenberg G, Hone J, Kim Ph and Stormer H L 2008 Ultrahigh Electron Mobility in Suspended Graphene. *Solid State Communications.* **146.** 351 doi.org/10.1016/j.ssc.2008.02.024
[5] Morozov S V, Novoselov K S, Katsnelson M, Schedin F, Cunha D E, Jaszczak J and Geim A K 2008 Giant Intrinsic Carrier Mobilities in Graphene and Its Bilayer. *Physical review letters.* **100** 016602 https://doi.org/10.1103/PhysRevLett.100.016602
[6] Cai W W, Zhu Y W, Li X S, Piner R D and Ruoff R S 2009 Large area few-layer graphene/graphite films as transparent thin conducting electrodes. *Appl. Phys. Lett.* **95** 123115. doi.org/10.1063/1.3220807
[7] Li X, Zhu Y, Cai W, Borysiak M, Han B, Chen D, Piner R D, Colombo L and Ruoff R S 2009 Transfer of large-area grapheme films for high-performance transparent conductive electrodes. N*ano Lett.* **9** 4359 /dx.doi.org/10.1021/nl902623y
[8] Das S and Choi W 2011 Graphene Synthesis *Graphene: Synthesis and Applications* ed W Choi and J W Lee (NewYork: CRC Press, Taylor and Francis Group) chapter 2 pp 27-57
[9] Novoselov K S, Geim A K, Morozov S V, Jiang D, Zhang Y, Dubonos S V, Grigorieva I V and Firsov A A 2004 Electric fieldeffect in atomically thin carbon films. *Science.* **306** 666 doi.org/10.1126/science.1102896
[10] Allen M J, Tung V C and Kaner R 2009 Honeycomb Carbon: A Review of Graphene. *Chemical Rev.* **110** 132 https://doi.org/10.1021/cr900070d
[11] Viculis L M, Mack J J and Kaner R 2003 A Chemical Route to Carbon Nanoscrolls. *Science* **299** 1361 doi.org/10.1126/science.1078842
[12] Park S and Ruoff R 2010 Chemical Methods for the Production of Graphenes. *Nature Nanotechnology* **5** 309 doi.org/10.1038/nnano.2010.69
[13] Reina A, Jia X, Ho J, Nezich D, Son H, Bulovic V, Dresselhaus M S and Kong J 2009 Large area, few-layer graphene films on arbitrary substrates by chemical vapor deposition. *Nano Lett.* **9** 30 doi.org/110.1021/nl801827v
[14] Chernin I Z, Smekhov F M and Zherdin Yu V 1982 *Epoxy Polymers and Compositions* [in Russian] (Moscow: Khimiya) p 232
[15] Metcalf A 1978 *Surface Section in Metal Composites,* vol 1 [in Russian] (Moscow: Mir) p 437
[16] Vettegren V I, Bashkarev A Ya and Suslov M A 2007 Influence of the shape of filler particles on the strength of a polymer composite *Journal of Technical Physics* [in Russian] **77(6)** 135
[17] Qian W, Hao R, Hou Y, Tian Y, Shen C, Gao H and Liang X 2009 Solvothermal-assisted exfoliation process to produce graphene with high yield and high quality. *Nano Res.* **2** 706 doi.org/10.1007/s12274-009-9074-z
[18] Matsumoto M, Saito Y, Park C, Fukushima T and Aida T 2015 Ultrahigh through put exfoliation

of graphite into pristine 'single layer' graphene using microwaves and molecularly engineered ionic liquids. *Nat. Chem.* **7(9)** 730 doi.org/10.1038/nchem.2315

[19] Pu N W, Wang C A, Sung Y, Liu Y M and Ger M D 2009 Production of few-layer graphene by supercritical CO_2 exfoliation of graphite. *Mater Lett.* **63** 1987

[20] Rangappa D, Sone K, Wang M, Ujjal K G, Golberg D, Itoh H, Ichihara M and Honma I 2010 Rapid and direct conversion of graphite crystals into high-yielding, good-quality graphene by supercritical fluid exfoliation. *Chem A Eur J.* **16(22)** 6488

[21] Li L at al. 2015 Preparation of graphene nanosheets by shear-assisted supercritical CO_2 exfoliation. *Chem. Eng. J.* **284** 78 doi.org/10.1016/j.cej.2015.08.077

[22] Song N, Jia J, Wang W, Gao Y, Zhao Y and Chen Y 2016 Green production of pristine graphene using fluid dynamic force in supercritical CO_2. *Chem. Eng. J.* **298** 198 https://doi.org/10.1016/j.cej.2016.04.022

[23] Gao Y, Shi W, Wang W, Wang Y, Zhao Y, Lei Z and Miao R 2014 ultrasonic-assisted production of graphene with high yield in supercritical CO_2 and its high electrical conductivity film. *Ind. Eng. Chem. Res.* **53** 2839 doi.org/10.1021/ie402889s

[24] Chen X, Dobson J and Raston C 2012 Vortex fluidic exfoliation of graphite and boron nitride. *Chem. Commun.* **48** 3703 doi.org/10.1039/c2cc17611d.

[25] Wahid M, Eroglu E, Chen X, Smith S and Raston C 2013 Functional multi-layer graphene-algae hybrid material formed using vortex fluidics. *Green Chem.* **15** 650 doi.org/10.1039/C2GC36892G

[26] Chen X, Boulos R A, Dobson J F and Raston C L 2013 Shear induced formation of carbon and boron nitride nano-scrolls. *Nanoscale.* **5(2)** 498 doi.org/10.1039/C2NR33071G

[27] Paton K et al. 2014 Scalable production of large quantities, defect-free few-layer graphene by shear exfoliation in liquids. *Nature materials.* **13** 624 doi.org/10.1038/nmat3944

[28] Rosu D, Cascaval C, Mustaţă F and Ciobanu C 2002 Cure kinetics of epoxy resins studied by non-isothermal DSC data. *Thermochimica Acta* **383** 119 doi.org/10.1016/S0040-6031(01)00672-4

[29] Pershin V F, Al-Shiblawi K, Baranov A and PaskoT 2019 Production of low-layer graphene by liquid phase shear exfoliation. *St. Petersburg Polytechnic University Journal of Engineering Science and Technology* **25(1)** 143 doi.org/10.18721/JEST.25114

[30] Al-Shiblavi K A, Pershin V F and Pasko T V 2018 Modification of cement by few-layer graphene. *Science Vektor of Tolyatti State University* [in Russian] **4(46)** 6 https://doi.org/10.18323/2073-5073-2018-4-6-11

[31] Melezhyk A V, Kotov V A and Tkachev A G 2016 Optical properties and aggregation of graphene nanoplatelets. *Journal of Nanoscience and Nanotechnology* **16** 1067 doi.org/10.1166/jnn.2016.10496

[32] Pershin V F, Ovchinnikov K A, Al-Hilo Z A, Stolyarov R A, and Memetov N R 2018 Development of environmentally safe lubricants modified by graphene nanotechnologies in Russia *Nanotechnologies in Russia* **13**(5-6) 344 doi: 10.1134/S1995078018030138

IOP Conf. Series: Materials Science and Engineering **693** (2019) 012024 doi:10.1088/1757-899X/693/1/012024

Nitration and oxidation of graphite by mechanochemical treatment

V Chaika, V Savin, L Savina and I Zherebtsov

Laboratory of Physical Materials Science IRC, *Immanuel Kant Baltic Federal University,* 14A Nevskogo Street, 236016 Kaliningrad Russian Federation

E-mail: v_chayka8@rambler.ru

Abstract. When combining special mechanochemical processing of graphite in high-speed and planetary ball mills and heat treatment, it was revealed that condensed graphene oxide is formed, and a nitrogen-containing carbon phase is synthesized. Melamine was used as the material for introducing nitrogen into the system; during the heat treatment it releases ammonia. Besides, it improves the process of compaction of graphite powders into workpieces in the production of refractive X ray optics. The mechanochemical synthesis of graphite with melamine (not more than 1 wt.%) and two-stage annealing at 360 and 450 °C leads to the formation of some forms of they lead to the formation of graphene oxide and a nitrogen-containing carbon phase. X-ray studies revealed the appearance of peaks that can be attributed to peaks of graphene oxide and carbon nitride. The formed condensed carbon compounds with nitrogen and oxygen are characterized by natural aging.

1. Introduction

One of the known nitrogen-containing carbon materials is graphite carbon nitride (g-C_3N_4) [1]. The increased scientific interest to this compound is due to its promising semiconducting, catalytic and sorption propertiesattractive for practical use. Graphene oxide (GO) and a family of metal nitride nanostructures may be obtained from the g-C_3N_4. While its hardness and compressibility is a lower than those of diamond-like compounds, its stability under natural conditions is the highest in comparison with the other allotropic forms of nitrogenized condensed carbon compounds [1].

Methods for preparation of the g-C_3N_4 are based on different options of chemical, thermal and hydrothermal treatments at high pressure and high temperature [2,3], vapor deposition, and their various combinations [4].

The methods used to produce GO and graphene-like substances include: (a) chemical reduction using various reagents (*e.g.*, hydrazine) [5]; (b) thermal reduction performed in vacuum or in various gaseous media (*e.g.*, argon, hydrogen, hydrogen/argon mixtures) [6]; (c) electrochemical reduction [7]; and (d) combination of various methods such as chemical reduction and thermal annealing (TA) or thermal reduction and hydrogenation [6]. There are some similarities in the methods used for obtaining the g-C_3N_4 and GO. Particularly, material processing in disintegrators, including ball millings, commonly employed for synthesizing the g-C_3N_4 and GO, especially at the preparatory stages. In recent decades, mechanochemical treatment of powder materials in ball mills and the so-called mechanochemical synthesis (MCS) have shown advantageous results in developing new compound.

IOP Conf. Series: Materials Science and Engineering **693** (2019) 012024 doi:10.1088/1757-899X/693/1/012024

Considering the aforementioned, the present paper considers and discusses results of experimental investigations on nitration and oxidation of graphite from a powdered mixture of MPG-7 graphite and melamine by combing the MCS and TA. Melamine is a component that improves the technological process of compaction of graphite powder materials into blanks during the production of refractive X-ray optics [8].

2. Methods and materials

A graphite powder with a fraction of ≤ 20 μm was used at initial stage of the MCS processing. The powder of specified dispersion was obtained by grinding a lumpy MPG-7 powder (Graphite Service, Chelyabinsk, Russia) with a size of 1-5 mm in a PM-100 planetary mill (Retch Technology, Hamburg, Germany) for 1 h under fixed conditions [9]. Next, up to 1 wt.% of RN-M40 melamine powder (Roshal Group, Roshal, Russia) was added to the obtained material.

The MCS procedure was carried out for 1 h in various ball mills with different types of headsets: 1) in the PM-100 mill – in a steel glass with steel balls and in an agate glass with agate balls at 300 rpm; 2) in an Emax high-speed ball mill (Retch Technology) – with steel cups and balls with the processing mode of 1500 conventional units. The treatment was carried out in residual air atmosphere. The ratio of the mass of balls to the mass of loading was 9:1. After the MCS procedure, a two-stage annealing of the obtained powder was performed at: +360 °C for 1 h, and at +450 °C for 1 h.

The selected annealing conditions correspond to the decomposition temperature of melamine [10]. At the first annealing stage, the melamine decomposes with release of ammonia (NH_3), and melem is formed as a result of the following reaction: $C_3H_6N_6 \rightarrow NH_3 + C_6H_6N_{10}$. At the second annealing stage, melon $[C_6N_7(NH)(NH)]_n$ polymerization occurs with the release of ammonia (NH_3) [10]. After this two-stage annealing, the powder material was again subjected to the MCS for 1 h in the same headsets.

A DRON-3M X-ray diffractometer (Burevestnik, Yekaterinburg, Russia) with Cu-K_α-radiation was used for analysis of the materials obtained.

3. Results and discussion

The X-ray diffraction studies show (figure 1) that, the diffraction peaks of graphite weakening and broadening and their center of gravity are shifted towards smaller diffraction angles in comparison with the initial graphite materials for all the three powder samples. This is usual for the MCS processing of the powder due to its grinding, formation and accumulation of various defects, material doping by atoms of the residual atmosphere, and decomposition products of melamine.

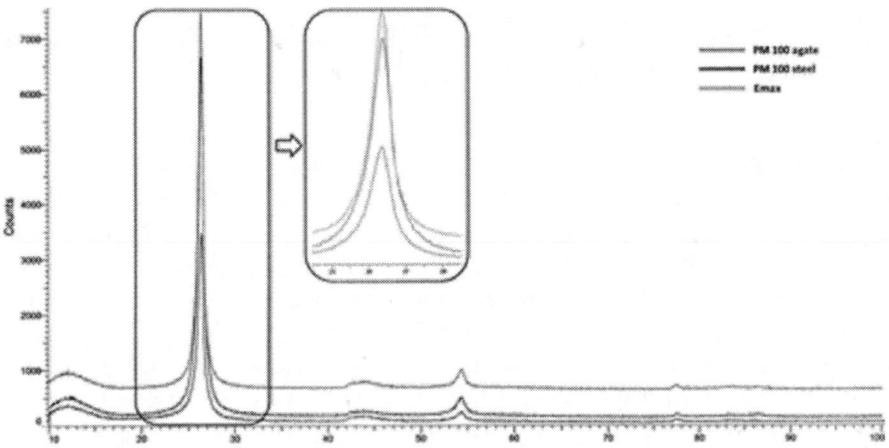

Figure 1. Diffractograms of the powder samples obtained through MCS processing in the Emax (steel set) and PM-100 (steel and agate headsets) ball mills.

The maximum attenuation of the intensity of the diffraction peaks is observed for the sample processed in the Emax mill. This result is consistent with its technological characteristics: more high-energy processing, complex interaction of the glass, grinding bodies and powder, accompanied by abrasive and impact effects on the material being processed. When the materials were processed in the PM-100 mill, the abrasive effects prevails [11].

The MCS-processed powder was compacted using a PP 25 manual hydraulic pellet press (Retch Technology) in a cylindrical mold with a diameter of 15 mm and a force of 872 MPa. The obtained samples in the form of tablets poorly kept their shape and showed a tendency to stratification during aging.

The X-ray studies were performed on the flat surface of the compacted powder samples (tablets) with residual mechanical strength. The obtained results demonstrate (figure 2) that, in addition to the main peaks corresponding to graphite, the diffraction patterns of the sample contain a diffusion halo, which is an X-ray scattering characteristic for a substance in an amorphous or glassy state. A diffuse halo with a center of gravity of $2\theta = (12.5 \pm 0.1)°$ (angular degrees) is clearly detected. Slightly diffus diffraction peaks (not related to graphite) are present at $2\theta = (43.5 \pm 0.1)°$, $(79.0 \pm 0.2)°$ and $(83.5 \pm 0.2)°$ (Table 1).

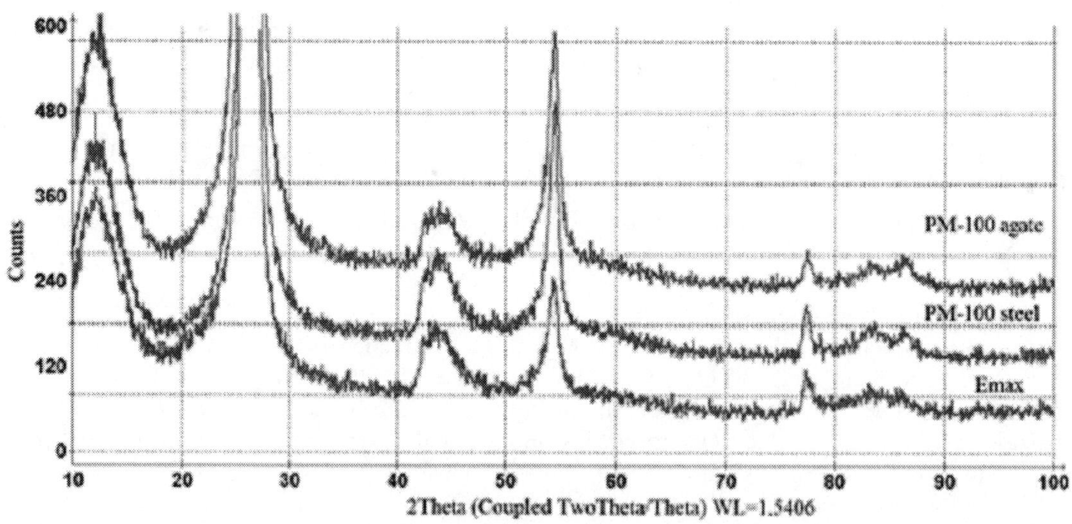

Figure 2. X-ray diffractograms of the samples compacted from the powder processed by the MCS in the Emax (steel set) and PM-100 (steel and agate headset) mills.

In [12], the GO obtained by the Hummers method was subjected to heat treatment. The oxygen content and structural phase changes were analyzed and according to their results, upon the graphite – to GO transition, the graphite peaks (002) at $2\theta = 26.5°$ (Cu-K_α-radiation) and (101) at $2\theta = 44.8°$ disappear, and the peak related to the GO (001) becomes apparent at $2\theta = 9.8$. Due to the introduction of the oxygen atoms into the interlayer space, the hexagonal graphite structure swells in the [001] direction (along the "c" axis). As a consequence, the interplanar distance increases from 0.336 nm for graphite to 0.867 nm for GO. The article [13] presents the results of X-ray diffraction studies on GO obtained by the Hummers method, and the peaks at $2\theta = 10.813°$ and $42.427°$ (Cu-K_α-radiation) corresponding to the interplanar spacing of 0.8176 and 0.2129 nm, respectively, were observed.

According to our results (figure 1 and table 1), the X-ray diffraction peaks of the graphite are shifted towards smaller angles, which is typical for the MCS. And the peaks, presumably corresponding to the GO, are shifted toward larger angles $2\theta = 12.5°$ and $43.4°$ with interplanar distances of 0.706 and 0.208 nm. Two other peaks at 2θ = 79° and 83.5° can also be attributed to the GO structure, since they have a similar appearance (blur). Smaller values of experimental interplanar distances can be explained not only by the imperfect form of the crystal structure due to numerous defects formed during the MCS, but also by the reduced (compared to the other synthesis methods) oxygen content interacting with graphite. The amount of oxygen entering the material is limited by its content in the residual atmosphere of the sealed glass. The MCS process is accompanied by the consumption of oxygen for the formation of volatile, stable oxides of CO_2 and CO, which reduces the number of oxygen atoms entering into the synthesis of GO.

The authors of [14] investigated the synthesis and bulk structures of carbon nitrides with different degrees of condensation using thermoanalytical methods. They also note the difference in results when processing in open and closed containers. During the polymerization of melamine in a nitrogen atmosphere at 350 °C, in closed containers, a diffraction peak in the region of 13° appears on diffractograms (Cu-K_α-radiation). This peak is assigned by the authors to tri-s-triazine blocks (C_6N_7). Melon chains are built from these blocks, and a variety of compounds with different structures of carbon nitride may be produced from them.

Table 1. Parameters of the X-ray diffractograms of the samples compacted from MCS powder processed in the Emax (steel set) and PM-100 (steel and agate headset).

hkl	PM-100 steel headset				PM-100 agate headset				E_{max}			
	<x>, degree	H, count	FWHM, degree	d, Å	<x>, degree	H, count	FWHM, degree	d, Å	<x>, degree	H, count	FWHM , degree	d, Å
001*	12.52	210	3.4	7.062	12.53	172	3.37	7.056	12.47	153	3.45	7.089
002	26.34	6457	0.68	3.379	26.33	6585	0.58	3.381	26.31	3181	0.71	3.383
100	42.48	32	0.49	2.125	42.56	63	0.89	2.122	42.4	31	0.37	2.129
100*	43.33	47	1.71	2.086	43.7	68	1.3	2.069	43.21	55	1.77	2.091
101	44.79	57	2.39	2.021	44.81	70	2.44	2.020	44.54	57	2.27	2.032
004	54.25	298	1.15	1.689	54.29	300	0.93	1.688	54.23	147	1.16	1.689
110	77.55	48	0.85	1.229	77.47	77	0.81	1.231	77.47	29	0.43	1.231
	79.1	10	1.29	1.209	78.68	24	1.03	1.215	79.24	35	1.61	1.227
	83.43	26	2.7	1.157	83.45	33	2.41	1.157	83.44	20	3.03	1.157
006	86.47	31	1.96	1.124	86.51	30	2.18	1.124	86.28	17	2.6	1.126

Note: hkl with an asterisk (*) refers to GO

Given that the synthesized MCS powders and compacts based on them are in a state of "undersaturation" ("underdevelopment") with the oxygen, then in a state of natural aging in air, these processes can continue in the direction of enrichment of the material with the oxygen. In this work, a repeated study of this material was carried out after being kept in vivo within two months. The MCS powderswere re-compacted in a cylindrical mold with a diameter of 10 mm and a force of 624.6 MPa.

The expected result, which has practical significance, was obtained: the pressed samples retain their shape well, do not stratify, and have sufficient mechanical strength for research. Figure 3 presents the diffraction patterns of such tablets.

All previously detected peaks significantly reduce their intensity and broadening. The peak of an amorphous halo at 2θ = 12.5°is shifted towards smaller angles by an amount beyond the experimental error (2θ≈ 11.5 ± 0.1°). The obtained results are in good agreement with the assumption that the process of saturation of the MCS powder with the oxygen and the physical-chemical processes of interaction of graphite with elements present in the bulk of the powder material or coming there from the atmosphere continue during aging under natural conditions. As a result, on the diffractograms of tablets from such MCS powders at small angles, four distinct diffraction peaks appear at angles of 2θ≈

6.6°, 7.1°, 8.2°, and 17.8° (figure 4). Moreover, if the first three peaks in the three samples are slightly shifted, then the latter has a constant location and intensity (Table 2).

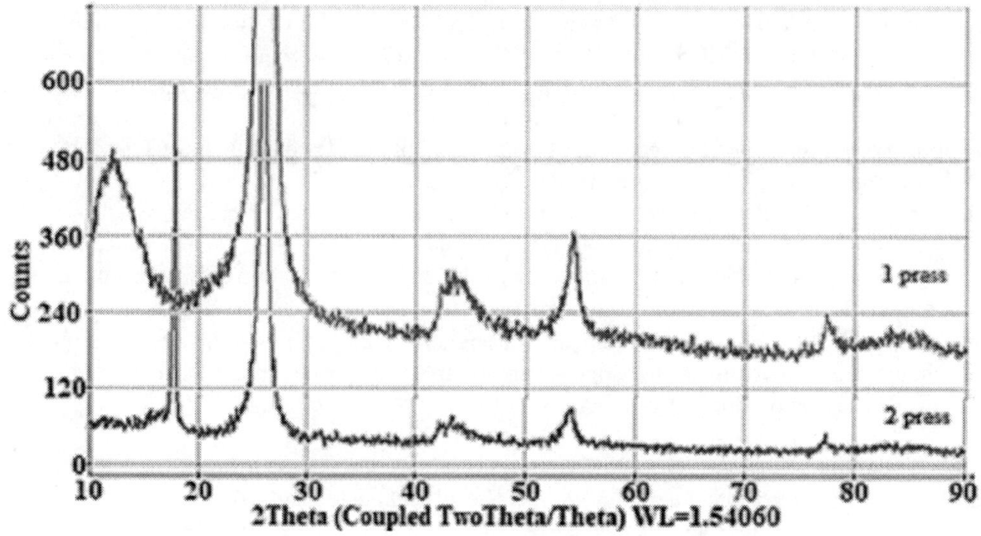

Figure 3. X-ray diffractograms of the samples compacted from the powder processed by the MCS in the Emax (steel set) mill after being kept for 60 days.

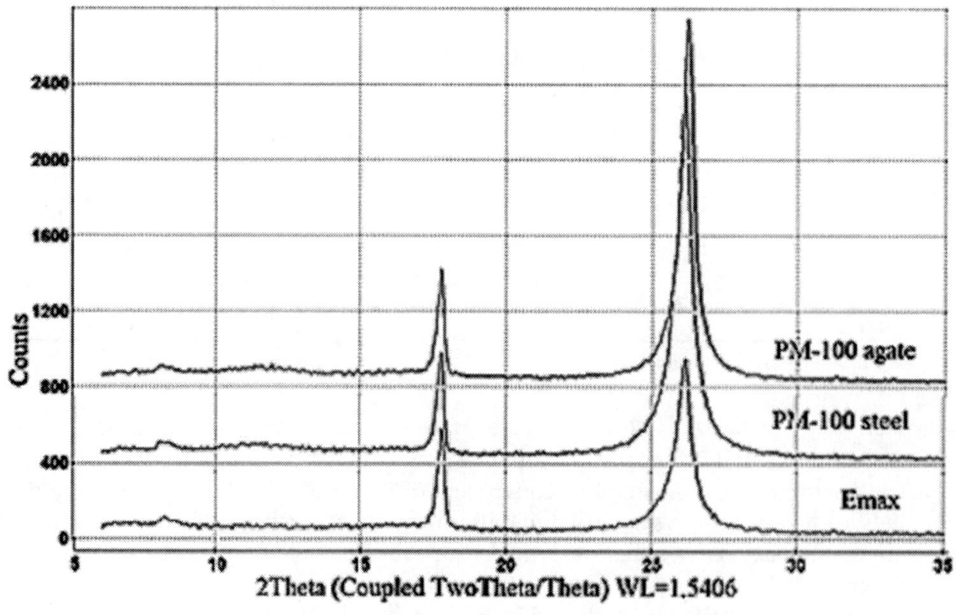

Figure 4. Fragment of X-ray diffractograms of tablets from the MCS powder after being kept for 60 days.

Table 2. Results of calculations of diffractograms of the samples obtained by compaction of MCS powder, which was kept under natural conditions for 60 days.

hkl	PM-100 steel headset				PM-100 agate headset				E_{max}			
	<x>, degree	H, count	FWHM, degree	d, Å	<x>, degree	H, count	FWHM, degree	d, Å	<x>, degree	H, count	FWHM, degree	d, Å
	6.64	19	0.54	13.296	6.62	17	0.43	13.336	6.41	17	0.19	13.774
	7.17	13	0.4	12.314	7.18	23	0.12	12.297	6.97	14	0.62	12.667
	8.2	43	0.64	10.769	8.19	41	0.88	10.783	8.16	42	0.59	10.822
	11.53	31	2.6	7.666	11.64	36	2.8	7.593	11.17	13	5.19	7.912
	16.61	28	2.56	5.331	-	-	-	-	16.44	24	2.49	5.386
	17.8	522	0.28	4.977	17.8	569	0.27	4.977	17.8	514	0.28	4.977
002	26.15	1866	0.73	3.404	26.23	1833	0.54	3.394	26.14	861	0.68	3.405

The appearance of similar peaks at low angles is noted by the authors of [15] when studying the melamine polycondensation in a nitrating medium at high pressure and temperature. The analysis of the chemical composition of the material obtained showed that graphite forms C_3N_4. However, the rhombic cell modeled by the authors for the structure obtained was not confirmed and requires further refinement and development.

4. Conclusion

A method of mechanochemical treatment of graphite in ball planetary and high-speed mills of MPG-7 graphite powder with additions of melamine powder in an amount of up to 1 wt.% in combination with heat treatment (annealing) and aging (under natural conditions) was proposed for providing synthesis of the condensed matter state by combining carbon with nitrogen and carbon with oxygen.

The effect of the following MCS technological parameters of the process was shown: ball mill type (PM-100 and Emax; headset (steel, agate); processing modes (load, energy, time, process sequence).

It was confirmed that, according to the proposed modes, during the MCS, the synthesis of condensed carbon compounds with nitrogen and carbon with oxygen proceeds in a new crystalline and/or amorphous-crystalline state.

For the formation of condensed carbon compounds with nitrogen and oxygen that are formed during the mechanochemical synthesis, the phenomenon of aging under natural conditions was found to be inherent.

The aging processes are associated with physicochemical regularities (absorption, diffusion, catalytic chemical reactions of atoms and molecules of absorbed elements with graphite and derivatives formed during its mechanochemical processing) occurring in the activated powder material upon contact with the atmosphere.

5. Acknowledgment

The study was supported by the subsidy allocated for the implementation of the Program for Enhancing the Competitiveness of the Immanuel Kant Baltic Federal University. The work was performed under Project No. 16.1419.2017/PCh, funded by the Ministry of Science and Higher Education of the Russian Federation.

References

[1] Horvath-Bordon E, Croc E, Svoboda I, Fuss H and Riedel R 2005 Potassium melonate, $K_3[C_6N_7(NCN)_3]\cdot 5H_2O$, and its potential use for the synthesis of graphite-like C_3N_4 materials. *New J. Chem.* **29** 693 https://doi.org/10.1039/B416390G

[2] Zhang Z, Leinenweber K, Bauer M, Garvie L A J, McMillan P F and Wolf G H 2001 High-pressure bulk synthesis of crystalline $C_6N_9H_3{\cdot}HCl$: a novel C_3N_4 graphitic derivative. *J. Am. Chem. Soc.* **123** 7788 https://doi.org/10.1007/978-94-010-0520-3_3

[3] Zuk A Z, Burdina K P, Milyavskii V V, Kravchanko O V, Borodina T I, Babaev V G and Guseva M B 2001 Shock-wave loading of amorphous carbon nitride in conservation ampoules. *High-Temperature Thermal Physics* [Teplofizika Vysokikh Temperatur – in Russian] **39** 154 https://doi.org/10.1023/A:1004199304109 160

[4] ChowdhuryA, Cameron D, Hashmi M and Gregg J 1999 Evidence for continuous areas of crystalline β–C_3N_4 in sputter-deposited thin films. *J. Mater Res.* **14** 2359 https://doi.org/10.1557/JMR.1999.0314

[5] Stankovich S, Dikin D A, Piner R D, Kohlhaas K A, Kleinhammes A, Jia Y, Wu Y, Nguyen S B T and Ruoff R S 2007 Synthesis of graphene-based nanosheets via chemical reduction of exfoliated graphite oxide. *Carbon* **45** 1558 https://doi.org/10.1016/j.carbon.2007.02.034

[6] Kaniyoor A, Baby T T, Arockiadoss T, Rajalakshmi N and Ramaprabhu S 2011 Wrinkled graphenes: A study on the effects of synthesis parameters on exfoliation-reduction of graphite oxide. *J. Phys. Chem. C* **115** 17660 1558 https://doi.org/10.1021/jp204039k

[7] Sundaram R S, Gomez-Navarro C, Balasubramanian K, Burghard M and Kern K 2008 Electrochemical modification of grapheme. *Adv. Mater.* **20** 3050 https://doi.org/10.1002/adma.200800198

[8] Roth T, Alianelli L, Lengeler D, Snigirev A and Seiboth F 2017 Materials for X-ray refractive lenses minimizing wave front distortions. *MRS Bulletin.* **42** 430 https://doi.org/10.1557/mrs.2017.117

[9] Chaika V A, Savin V V, Savina L A, Osadchy A V and Rybin M G 2019 Features of Raman spectra of mechanically activated graphite. *J. Phys. Conf. Ser.* **1172** 012027 https://doi.org/10.1088/1742-6596/1172/1/01202

[10] Lotsch B, Doeblinger M, Sehnert J, Seyfarth L, Senker J, Oeckler O and Schnick W 2007 Unmasking melon by a complementary approach employing electron diffraction, solid-state NMR spectroscopy, and theoretical calculations -structural characterization of a carbon nitride polymer. *Chem. Eur. J.* **13** 4969 https://doi.org/10.1002/chem.200601759

[11] Chaika V A, Savin V V, Savina L A, Osadchy A V, Zherebtsov I S and Medvedskaya P N 2018 The use of mechanochemical processing of graphite in planetary ball mills to obtain silicon-doped carbon composite pellets. *News of Higher Educational Institutions* [Izvestiya Vysshikh Uchebnykh Zavedeniy – in Russian] **61** 38 https://doi.org/10.6060/ivkkt.20186111.10y

[12] Botas C, Alvarez P, Blanco C, Santamara R, Granda M, Gutiyrrez M D, Rodrguez-Reinoso F and Menundez R 2013 Critical temperatures in the synthesis of graphene-like materials by thermal exfoliation-reduction of graphite oxide. *Carbon* **52** 476 https://doi.org/10.1016/j.carbon.2012.09.059

[13] Paulchamy B, Arthi G and Lignesh B D 2015 A simple approach to stepwise synthesis of graphene oxide nanomaterial. *J. Nanomed. Nanotechnol.* **6** 4 https://doi.org/10.4172/2157-7439.1000253

[14] Thomas A, Fischer A, Goettmann F, Antonietti M, Muller J-O, Schlogl R and Carlssonc J M 2008 Graphitic carbon nitride materials: variation of structure and morphology and their use as metal-free catalysts. *J. Mater. Chem.* **18** 4893 https://doi.org/10.1039/b800274f

[15] Alves I, Demazeau G, Tanguy B and Weill F 1999 On a new model of the graphitic form of C_3N_4. *Solid State Commun.* **109** 697

IOP Conf. Series: Materials Science and Engineering **693** (2019) 012025 doi:10.1088/1757-899X/693/1/012025

Analysis of the structural characteristics of graphene materials obtained by electrochemical exfoliation of graphite

Yu A Khan[*], T P Dyachkova, E A Burakova, E S Bakunin, A V Rukhov, N V Orlova and E Yu Obraztsova

Tambov State Technical University, 106 Sovetskaya street, Tambov, 392000, Russian Federation

[*]E-mail: khantermail@gmail.com

Abstract. The aim of the research is to study the structural characteristics of graphene materials obtained by electrochemical exfoliation of graphite. The objects of study are prepared in electrochemical cells using alkaline and acid electrolytes with various applied voltages. Scanning electron microscopy, Raman spectroscopy, and X-ray diffraction have been used as research methods. The possibility of a theoretical calculation of the structural characteristics of nanomaterials (the defectiveness parameter, the density of the defects, the average distance between defects and average crystallite size) obtained by electrochemical exfoliation, according to Raman spectroscopy, is discovered. The results of these calculations satisfactorily correlate with the data of X-ray analysis.

1. Introduction

Electrochemical exfoliation of graphite is a promising method for the synthesis of graphene nanostructures. The availability of the source material, ease of implementation and the possibility of varying the process conditions in a wide range are significant advantages over other synthesis processes. This method allows one to obtain graphene in a short time. The electrolytic cell is filled with electrolyte solutions, which can be presented as water mixtures of the acid [1-3], inorganic salts [4-6], ionic liquids [7-10], and alkalis [11]. The bath is immersed electrodes, at least one of which is made of graphite [12]. The use of different types of graphite as the electrode, the change in the composition of the electrolyte, voltage, flow direction (constant, variable) allows you to control many parameters of the process. For example, it can be the rate of individual stages of the electrochemical process and their mechanisms, the ability to stabilize and functionalize free fragments by different groups, to modify by metal atoms [13].

For the analysis of the obtained materials, a wide range of various analytical methods are used, the most obvious of which is electron microscopy. But it permits to analyze only a small part of their surface. Therefore, this method does not give a complete understanding of the properties of the material. The method of Raman spectroscopy makes it possible to significantly supplement and expand the understanding of the structure of layered graphene materials. The degree of surface defects and the average distance between them, the crystal composition of the sample, the approximate number of layers, and the presence of certain functional groups on the area are possible to determine on the basis of the obtained spectra [14-16].

Raman spectra of carbon nanomaterials contain several characteristic peaks. The peaks D (~ 1350 cm^{-1}) and G (~ 1550 cm^{-1}) are the main ones for most materials. Peak D presents the disorder of the crystal structure of the samples, which reduces the overall symmetry of the structure. It requires a structural defect to be activated and represents "breathing" vibrations of cyclic carbon structures. It is created by transverse optical phonons and can be associated with both interplanar and intraplanar double resonance. In the second case, a second branch appears at peak D - peak D\` (~ 1600 cm^{-1}). It is important to note that the intensity of peak D depends not only on the presence of defects in the structure of carbon hexagons. It is also based on their type, depending on the scattered electrons. Peak G is associated with the vibrational mode of sp2-carbon and demonstrates the presence of a defect-free crystal structure. The ratio of peak performance D/G is used as a relative indicator of defectiveness.

The spectra also contain second-order peaks: 2D (~ 2700 cm^{-1}), D+G (~ 2900 cm^{-1}). The 2D peak demonstrates the presence of a single graphene layer and is divided into several smaller bands with an increase in the number of layers. This peak is not related to the presence of defects since it occurs when a pulse is held by two phonons with the same antiparallel wave vectors. Defects are necessary for the excitation of the D + G peak. This peak appears as a result of the interaction of phonons with various momenta [14].

The purpose of this research is to determine the structural characteristics (density of the defects, average crystallite size, average distance between defects) of graphene nanomaterials obtained by the electrochemical method using theoretical calculations based on Raman spectroscopy data.

2. Materials and methods

The object of the study was multilayer graphene materials obtained by electrochemical separation of graphite foil Graflex (sample 1). In the first case (sample 2), electrochemical exfoliation was carried out in a two-electrode cell using pulsed current. The duration of the cathode and anode pulses was 1 s, and the current density was maintained at 0.1 A/cm^2. The process was carried out in a 0.1 M potassium hydroxide solution. The resulting product was centrifuged and repeatedly washed with distilled water to remove electrolyte residues to pH=7, and then subjected to ultrasonic treatment for 8 hours and dried in a vacuum oven for 24 hours at a temperature of 40 °C.

In the second case (samples 3 and 4), electrochemical exfoliation was carried out in a two-electrode cell at a constant current. The current density was maintained at a level of 0.125 A/cm^2 (5 V, sample 3) and 0.25 A/cm^2 (10 V, sample 4) was applied. The platinum electrode acted as a cathode, and the graphite foil served as the anode. The process was carried out in a 0.15 M sulfuric acid solution. The resulting product was centrifuged and washed repeatedly with distilled water to remove electrolyte residues to pH=7, and then subjected to ultrasonic treatment for 8 hours and dried in a vacuum oven for 24 hours at a temperature of 40℃. Direct current with a density of 0.25 A / cm2 and a voltage of 5 V and 10 V was applied to the element.

For preparing the material for diagnosis, parts of the mixtures were dried for 24 hours to obtain a dry nanographite film. Then, the film was subjected to diagnostics on a Thermo Scientific DXR Raman Microscope (USA) with a laser wavelength of 633 nm with an aperture of 50 μm of the "slit" type. The structural analysis of the samples was carried out on a X-ray diffractometer "Difrey-401" (XRD, "Scientific instruments", St. Petersburg, Russia) with a Bragg-Bretano focus. Images of the nanomaterials were taken using scanning electron microscopy (SEM) on a Carl Zeiss "Merlin" with a resolution of up to 0.8 nanometers.

To determine the average distance between defects, the equation (1) can be used [15]:

$$\frac{I_D}{I_G} = C_A \frac{(r_A^2 - r_S^2)}{(r_A^2 - 2r_S^2)} \left[e^{\frac{-\pi r_S^2}{L_D^2}} - e^{\frac{-\pi(r_A^2 - r_S^2)}{L_D^2}} \right] \tag{1}$$

where, L_D is the average distance between defects, r_A is the radius of the zone around the defect, produced by *D* peak, r_S is the radius of the structurally disordered space, C_A is the active surface parameter, which is taken depending on the area of the active zone, C_S is the unstructured surface parameter, which is depending on the area of the unstructured zone. In the study, $C_S = 0$ was taken. This is due to consideration of the ideal case without complete breaking of the C – C bond, as well as the conditions of low defect density (L_D > 10 nm), and is not considered in the formula. During the experiments, the values of r_A, r_S, and C_A were revealed, with the last parameter being inversely proportional to $E_L{}^4$. Formula (1) can be approximated to formula (2) with allowance for large L_D values and the dependence of the CA parameter on the laser energy. (2):

$$L_D^2 = \frac{(4.3 \pm 1.3) \cdot 10^3}{E_L^4} \left(\frac{I_D}{I_G}\right)^{-1} \tag{2}$$

In terms of the density of the defects $n_D = 10^{14}/(\pi L_D^2)$:

$$n_D = (7.3 \pm 2.2) \cdot 10^9 E_L^4 \left(\frac{I_D}{I_G}\right) \tag{3}$$

The possibility of applying this approach to the evaluation of the structural characteristics of products of electrochemical exfoliation of graphite is due to their common nature with the objects considered in [15-16].

Graphene crystallite size determination is possible with the assumption of a defect-free inner planar surface. Also in [17] separation of a polycrystalline zone into a structured crystalline and disordered around the crystallite, and the contribution of both zones to peak formation was introduced, equation (4). Crystallite is understood as a defect-free surface area. It was found that the dependence of the crystallite size is uniform for different values of the *D/G* ratio, independently of the laser excitation energy. This allows us to use the proposed calculation to predict the size of the flat surface in the obtained samples.

$$\frac{I_D}{I_G} = \frac{I_D^S + I_D^A}{I_G^S + I_G^A} \tag{4}$$

Formula (5) allows determining the intensity of the Raman scattering of the vibrational mode γ:

$$I_\gamma^C = \frac{\omega_s^4}{\varepsilon_0^2 c^4} \int_C \left|G(r;\ \omega_s)\chi_\gamma^C(r;\ \omega, \omega_s)E(r;\ \omega)\right|^2 d^2r, \tag{5}$$

where, C is a domain that includes the crystalline zone A and the disordered zone S; $G\ (r;\ \omega_s)$ is the outgoing Green`s function, which includes the whole system; $\chi_\gamma{}^C\ (r;\ \omega,\ \omega_S)$ is the Raman susceptibility of the vibrational mode $_\gamma$ over the domain C; $E\ (r;\ \omega)$ is the excitation of the electric field; r is the position of the sample plane; c and ε_0 are the speed of light and the dielectric constant of the medium in vacuum; ω and ω_s - the frequency of the incident and scattered light.

When equations (4) and (5) have been combined with considering the peculiarities of the D and G peak in the crystalline and disordered areas, the ratio (6) was obtained:

$$\frac{I_D}{I_G} = \frac{C_D^S 4 l_s (L_a - l_S) + C_D^A (L_a - 2l_S)\left[1 - e^{-2(L_a - 2l_S)/l_A}\right]}{C_G^S 4 l_S (L_a - l_S) + C_G^A (L_a - 2l_S)^2} \tag{6}$$

where, l_A and l_S are the coherence lengths of electrons/holes and optical phonons; C-parameters are constant factors of the intensity function of Raman oscillations. The coherence lengths were obtained by approximating the experimental data. According to them, formula (7) for the crystallite size calculation was derived, which is in agreement with Tuinstra-Koenig ratio:

$$L_a \approx \frac{(490 \pm 100)}{E_L^4} \left(\frac{I_D}{I_G}\right)^{-1} \tag{7}$$

The formula (7) was derived experimentally. Different lasers were used in the analysis of samples. This dependence persists in spite of significant changes in the laser energy. On the one hand, this permits the application of (7) to the other graphenes. On the other hand, the parameters of the functions, which assums to be constant for graphenes obtained by thermal treatment of diamond-like amorphous carbon, may differ for graphenes obtained by electrochemical exfoliation method. In any case, this approach can be used to identify qualitative patterns of changes in the structure of the product.

3. Results and discussion

Figure 1 shows the SEM image of the graphene material obtained by the method of electrochemical exfoliation of graphite. It consists of two-dimensional structures with a large surface length (figure 1, left). Defects of the graphene layers are presented as some perceptible irregularities and bends. Edge plots also represent part of a disordered structure. Methods of microscopy are the most obvious and powerful but localized around a small area of the sample under study. When zooming in to cover a larger surface the accuracy of this method decreases. Even in the image with a larger scale (figure 1, right), small flexural disorders can be hardly seen. At the same time, it is not possible to determine the size of defects and the defect-free areas between them. Method for a limited time allows us to analyze only a small part of the surface, and therefore it does not give a complete picture of the properties of the material.

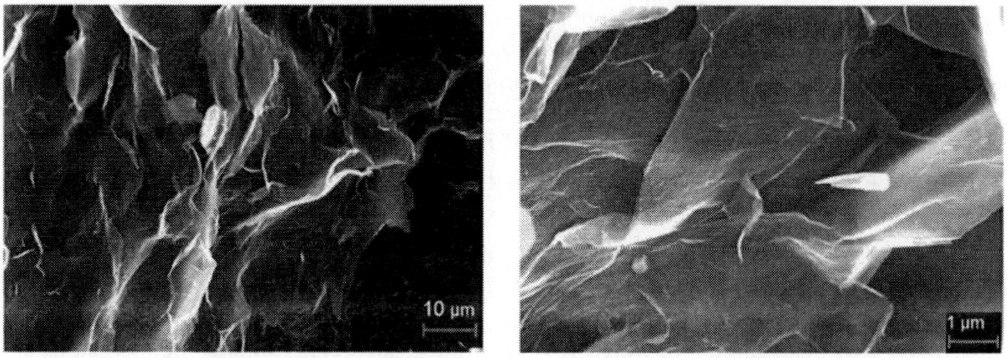

Figure 1. SEM images of the graphene nanostructured material obtained by the exfoliation method: general (left) and large (right) frames.

The Raman spectra of the samples (Figure 2) have typical peaks: *D* (~ 1350 cm^{-1}); *G* (~ 1550 cm^{-1}); *D'*(~ 1600 cm^{-1}); *2D* (~ 2700 cm^{-1}) and *D+G* (~ 2900 cm^{-1}), characterized by different intensity and width at half-height (FWHH) (table 1).

The spectrum of the starting material (sample 1) contains only peaks *2D* and *G*. For sample (2) obtained in an alkaline electrolyte, a low-intensity peak *D* appears. Graphite separation into fragments with the increasing of the percentage of atoms, which interacts with an environment, can be the reason. Also uprising of the surface defects, due to the aggressive influence of the alkali, contributes to peak growth. However, the value of the intensity ratio of the peaks *D/G*, which is often used as an indicator of the defectiveness of the material [18], for sample (2) is only slightly higher than for sample 1. In this case, it is possible that the formed fragments consist of a large number of layers. The use of sulfate electrolyte during exfoliation contributes to deeper structural changes in the material. On Raman spectra of samples (3) and (4), peak *D* becomes intense, *D'* and *D+G* appear, and *2D*, on the contrary, becomes less pronounced. The *D/G* ratio greatly grows. This is due to the increase in intraplanar flexural defects on the surface of graphene layers and attaching of the oxygen-included functional groups. Accordingly, crystallite sizes, which can be considered as flat areas without defects, are reduced. As a result, *D'* peak becomes intense. The increase in the intensity of the *D+G* peak indicates the validity of this assumption.

Increasing the voltage from 5 (sample 3) to 10 V (sample 4) during electrochemical exfoliation helps to reduce the intensity of typical peaks and the *D/G* ratio. This may indicate the forming of a less defective product.

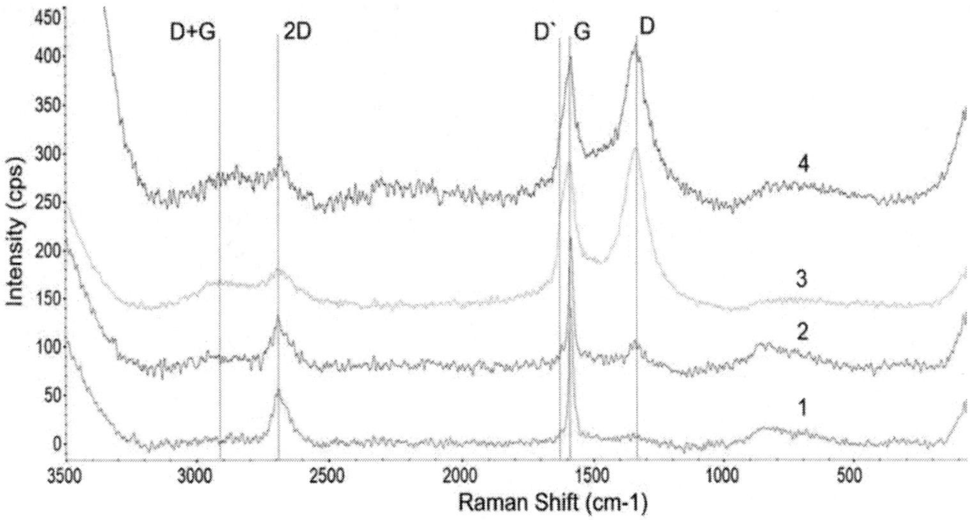

Figure 2. Raman spectra of the source material (1) and samples obtained by using KOH solution as electrolyte (2) and acidic solutions with direct current 5V (3) and 10V (4).

The XRD results of the initial graphite and materials obtained by the electrochemical exfoliation are shown in figure 3 and table 1. The diffractogram of the sample (1) (figure 3, a) contains 2 peaks of higher and lower intensity with 2θ values equal to 41.11 (100) and 85.42 (112), respectively. They indicate that the material consists of crystalline graphite [19]. For samples (2), (3) and (4) (figure 3b and 3d) the most intense peak on the diffractogram is shifted and has $2\theta = 36.52 - 36.61$. It is caused by the distance of the graphene planes from each other to a greater distance compared to graphite. On the diffractogram of sample (2), there is no second peak, which indicates that its volume is predominantly represented by low-layered fragments. The intensities of the peaks in the diffraction patterns of samples (3) and (4) are noticeably lower than in the first two cases, which indicates an increase in the fraction of the unstructured phase. For samples 2-3, a halo component is observed, also confirming the nanosize character of the objects.

Table 1. Characteristics of the Raman and main XRD (Intensity=100%) peaks of the source material (1) and nanographite samples obtained using KOH solution (2) and acidic electrolytes with 5V (3) and 10V (4).

D		***G***		***D`***	***2D***	***D+G***				
Int., cps	**FWHH, cm^{-1}**	**Int., cps**	**FWHH, cm^{-1}**	**Approx. int., cps**	**Int., cps**	**Int., cps**	***D/G***	**2θ**	**Int. (2θ)**	**d (Å)**
~13	-	119	19	-	48	-	0.11	41.11	50900	3.262
19	~56	99	18	-	40	-	0.19	36.52	19159	3.655
245	124	242	73	181	76	59	1.01	36.61	165	3.647
114	114	137	41	66	42	26	0.82	36.52	613	3.655

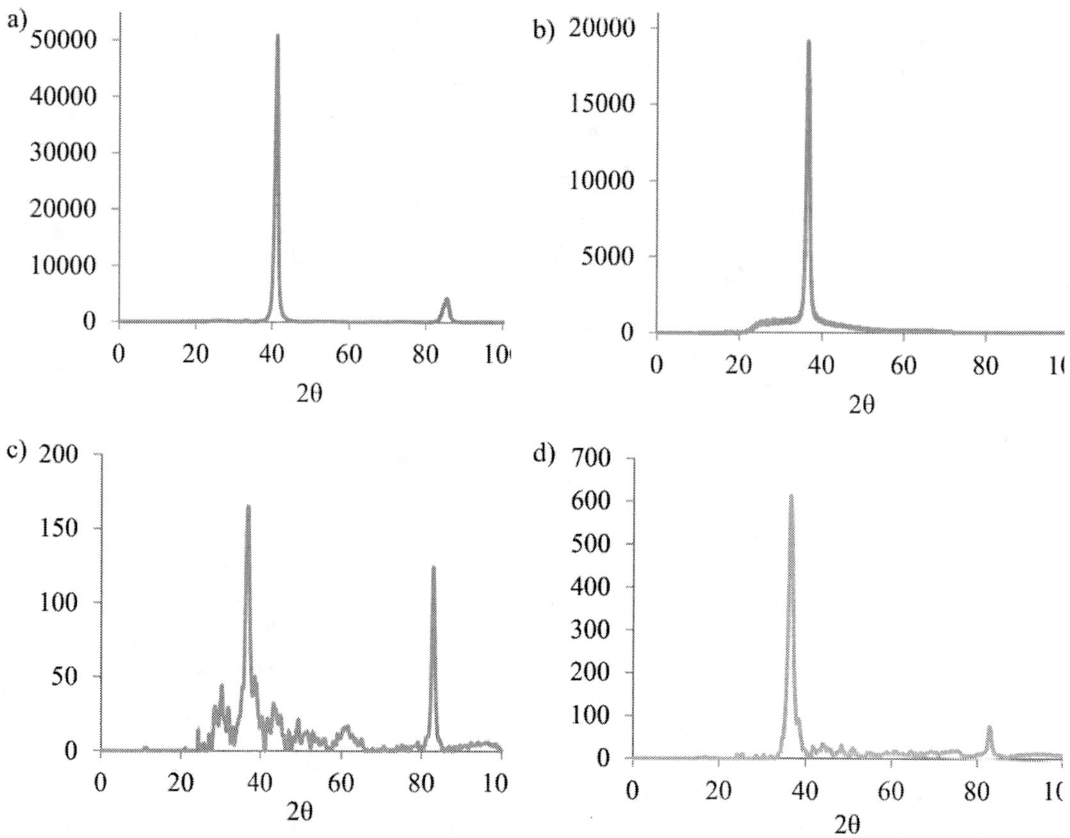

Figure 3. XRD spectra of the source material (a) and samples obtained by using KOH solution as electrolyte (b) and acidic solutions with direct current 5V (c) and 10V (d).

Table 2 presents the results of calculating the crystallite size (L_a), the interdefect distance (L_D) and the number of defects per surface area (n_D) of the samples of materials obtained by electrochemical exfoliation of graphite, based on Raman spectroscopy using formulas (2), (3) and (7).

Table 2. Calculated characteristics of the samples: crystallite size (L_a), inter-defect distance (L_D) and number of defects per unit surface area (n_D) for graphite electrochemical exfoliation products

Synthesis parameters	L_a, nm	L_D, nm	$n_D \cdot 10^9$, cm^{-2}
Source	240-363	43-59	8.28-15.42
KOH Solution	139-210	33-45	14.30-26.64
H_2SO_4 Solution, 5V	26-39	14-19	76.02-141.60
H_2SO_4 Solution, 10V	32-49	15-21	61.72-114.96

According to the calculations, areas of the defect-free surface of the largest size, the smallest defect density and inter-defect distance are formed when using an alkaline electrolyte. Taking into account the data of XRD and Raman spectroscopy, we can conclude that the resulting particles are nanoscale. In turn, the use of acid electrolytes permits to obtain a material with a less pronounced crystal structure and a large number of defects in graphene layers.

4. Summary

Raman spectroscopy data provide detailed characteristics of graphene nanomaterials obtained by electrochemical exfoliation of graphite: the defectiveness parameter, the average crystallite sizes, the defect density, and the average distance between them. The values of the parameters can be obtained by using D and G peaks information and experimentally determined equations. They were found by analysis of the Ar^{+}-bombarded graphene and polycrystalline graphene, which were described by the common theoretical data with experimental correction by Raman spectroscopy, XRD, transmission electron microscopy, scanning tunneling microscopy. The results of the analysis of samples obtained using sulfuric acid correlate with XRD data.

Acknowledgments

The research was supported by the Russian Foundation for Basic Research (Grant No. 18-29-19150 mk).

References

[1] Liu J *et al.* 2013 A green approach to the synthesis of high-quality graphene oxide flakes via electrochemical exfoliation of pencil core *RSC Adv.* **3**. 11745

[2] Su C Y, Lu A Y, Xu Y, Chen F R and Khlobystov A N 2011 High-quality thin graphene films from fast electrochemical exfoliation. *ACS Nano* **5 (3)** 2332

[3] Morales G M, Schifani P, Ellis G, Ballesteros C, Martinez G, Barbero C and Salavagione H J 2011 High-quality few layer graphene produced by electrochemical intercalation and microwave-assisted expansion of graphite. *Carbon* **49** 2809

[4] Parvez K, Wu Z S, Li R, Liu X, Graf R, Feng X and Mullen K 2014 Exfoliation of graphite into graphene in aqueous solutions of inorganic salts. *J. Am. Chem. Soc.* **136** 6083

[5] Yang S, Bruller S, Wu Z S, Liu Z, Parvez K, Dong R, Richard F, Samori P, Feng X and Mullen K 2015 Organic radical-assisted electrochemical exfoliation for the scalable production of high-quality graphene. *J. Am. Chem. Soc.* **137** 13927

[6] Hofmann M, Chiang W Y, Nguyen T D and Hsieh Y P 2015 Controlling the properties of graphene produced by electrochemical exfoliation. *Nanotechnology* **26** 335607

[7] Manzhos R A and Krivenko A G 2016 Synthesis of low-layer graphene structures by electrochemical splitting of graphite in aqueous solutions. *Proc. VIII Conf. Modern methods in Theoretical and Experimental Electrochemistry*, ed Parfenyuk V I (Ivanovo: Ivanovo State Chemical Engineering University) p 121

[8] Liu N, Luo F, Wu H, Liu Y, Zhang C and Chen J 2008 One-step ionic-liquid-assisted electrochemical synthesis of ionic-liquid-functionalized graphene sheets directly from graphite. *Adv. Funct. Mater.* **18** 1518

[9] Singh V V *et al.* 2012 Greener electrochemical synthesis of high quality graphene nanosheets directly from pencil and its SPR sensing application. *Adv. Funct. Mater.* **22** 2352

[10] Lu J, Yang J X, Wang J, Lim A, Wang S and Loh K P. 2009 One-Pot Synthesis of Fluorescent Carbon Nanoribbons, Nanoparticles, and Graphene by the Exfoliation of Graphite in Ionic Liquids. *ACS Nano* **3 (8)** 2332

[11] Lee J H, Shin D W, Makotchenko V G, Nazarov A S, Fedorov V E, Choi J Y, Kin J M and Yoo J B 2009 One-Step exfoliation synthesis of easily soluble graphite and transparent conducting graphene sheets. *Adv. Mater.* **21** 4383

[12] Wahab H S, Ali S H, Adi M and Hussein A 2015 Synthesis and characterization of graphene by Raman spectroscopy *Journal of Materials Sciences and Applications* **1 (3)** 130

[13] Ejigu A, Miller B, Kinloch I A and Dryfe R A W 2018 Optimisation of electrolytic solvents for simultaneous electrochemical exfoliation and functionalisation of graphene with metal nanostructures *Carbon* **128** 257

[14] Ghosh S, Ganesan K, Polaki S R, Ravindran T R, Krishna N G, Kamruddin M and Tyagi A K 2014 Evolution and defect analysis of vertical graphene nanosheets *J. Raman Spectrosc.* **45** (8) 642

[15] Cancado L G, Jorio A, Martins Ferreira E H, Stavale F, Achete C A, Capaz R B, Moutinho M V O, Lombardo A, Kulmala T S and Ferrari A C 2011 Quantifying defects in graphene via raman spectroscopy at different excitation energy *Nano Lett.* **11** 3190

[16] Lucchese M M, Stavale F, Martins Ferriera E H, Vilani C, Moutinho M V O, Capaz R B, Achete C A and Jorio A 2010 Quantifying ion-induced defects and Raman relaxation length in graphene *Carbon* 48. 1592

[17] Lazzarini A, Piovano A, Pellegrini R, Agostini G, Rudic S, Lamberti C and Groppo E Graphitization of activated carbons: a molecular-level investigation by INS, DRIFT, XRD and Raman techniques. *Physcs Proc.* **85** 20

[18] Ribeiro-Soares J et al. 2015 Structural analysis of polycrystalline graphene systems by Raman spectroscopy *Carbon* **95** 646

[19] Herrero-Latorre C, Alvarez-Mendez J, Barciela-Garcia J, Garcia-Martin S and Pena-Crecente R M 2014 Characterization of carbon nanotubes and analytical methods for their determination in environmental and biological samples: a review. *Anal. Chim. Acta.* **853** 77

IOP Conf. Series: Materials Science and Engineering **693** (2019) 012026 doi:10.1088/1757-899X/693/1/012026

Polyaniline / mesoporous carbon composites as promising materials for supercapacitors

I V Gutnik, T P Dyachkova, E A Burakova and I A Gavrilov

Tambov State Technical University, 106 Sovetskaya Street, 392000 Tambov Russian Federation

E-mail: anosowa_i_w@mail.ru

Abstract. The surface modification of mesoporous carbon with polyaniline was carried out by the method of oxidative polymerization of aniline. It was found that the initial conditions for the synthesis and the percentage of carbon template in the reaction mixture have little effect on the course of temperature dependencies. However, these parameters affect the yield of the target product and the electrically conductive characteristics of the resulting materials. Composites with a polyaniline / mesoporous carbon mass ratio of 1:1 have better electrical conductivity. The electrical capacity of these samples in the aqueous electrolyte is 299.6 F/g, which is 3.5 times higher than in the organic electrolyte. Information on the polyaniline in the modifying layer on the surface of the mesoporous carbon was obtained by the method of Raman spectroscopy.

1. Introduction

Recently, due to the fast-growing market of portable electronics, there is a growing demand for energy storage systems, including due to durability (100,000 charge-discharge cycles), a fast charge process and low self-discharge, ability to work at low temperatures without a significant reduction in performance are supercapacitors (SC) [1]. An important role in the operation of devices that serve for the accumulation and transformation of energy is played by electrode materials, in which carbon nanomaterials (CNM) and electrically conductive polymers can be used [2–5].

The use of CNMs (carbon nanotubes, graphene, graphene oxide, mesoporous carbon) in this area is due to their high electrical conductivity and electrochemical stability. A significant drawback of such materials is the relatively low specific capacity. In contrast, conductive polymers, for example, polyaniline (PANI), are characterized by higher capacitance values. However, the instability of the polymer during the charge-discharge cycle imposes a restriction on their use in the SC.

An effective approach to solve this problem is to obtain hybrid polyaniline / carbon nanostructure composites [6–8]. When a polymer is combined with carbon nanomaterials, an increase in the stability of PANI and an increase in the electrical capacity of carbon nanomaterials is observed. Consequently, supercapacitors with electrode materials of this type will have a longer service life.

To create electrode materials with the best performance (stability during charge / discharge, high electrical capacitance), it is advisable to modify the CNM polyaniline with the most developed surface. Mesoporous carbon (MC) satisfies this requirement, the specific surface area of which exceeds the analogous parameter of carbon nanotubes. This opens up broad prospects for obtaining a composite with a high specific electrical capacity.

For example, in [9] it is reported about an increase in the electrical capacity of polyaniline 2.5 times when applied to the surface of porous carbon with a specific surface equal to 422 m^2/g. The resulting composite is characterized by a high electrical capacity (up to 700 F/g) and retains up to 93.6% of the initial capacity after 10,000 charge-discharge cycles [10].

In connection with the promise of using polyaniline-based composites, the purpose of this work was to study the laws of the oxidative polymerization process of aniline on the surface of mesoporous carbon and to study the most important electrochemical properties of the composites obtained.

2. Materials and methods

Mesoporous carbon (manufactured by Nanotechcenter LLC, Tambov, Russia) was used with the following characteristics: specific surface area – 2640 m^2/g (according to BET); pore volume – 1.8 cm^3/g; pore diameter – 3.6 nm.

PANI/MC composites were obtained by the method of oxidative polymerization of aniline. For this purpose, the MC was introduced into an aqueous solution of H_2SO_4 with a given pH value (1 or 3). Aniline sulfate (0.04 M; 0.05 M) and ammonium persulfate (0.05 M; 0.07 M) were added to the reaction mixture with stirring. During the process, the temperature and pH of the reaction mixture were measured.

The obtained samples were washed with distilled water to a neutral pH of the filtrate and isopropyl alcohol until the color of the filtrate disappeared. The sample was dried in an inert atmosphere at a temperature of 80 °C to constant weight.

The content of PANI in the composite and its output from the theoretically possible was determined by the increase in the mass of the MC after the modification. Raman spectra of the samples were recorded using a DXR Raman Microscope instrument (Thermo Scientific, USA) at an excitation laser wavelength of 633 nm.

The specific electrical conductivity of the samples was also determined, and comparative electrochemical studies were carried in a standard three-electrode cell out in aqueous (3M H_2SO_4) and organic (1M tetraethylammonium tetrafluoroborate solution in acetonitrile) electrolyte solutions.

3. Results and discussion

The effect of the pH of the reaction mixture and the concentration of carbon material on the patterns of the oxidative polymerization of aniline on the surface of the MC was studied. The experimental data were used to plot the temperature (t) changing during the process (Figure 1). They turned out to be similar to dependences obtained earlier in the modification of carbon nanotubes with polyaniline [11,12].

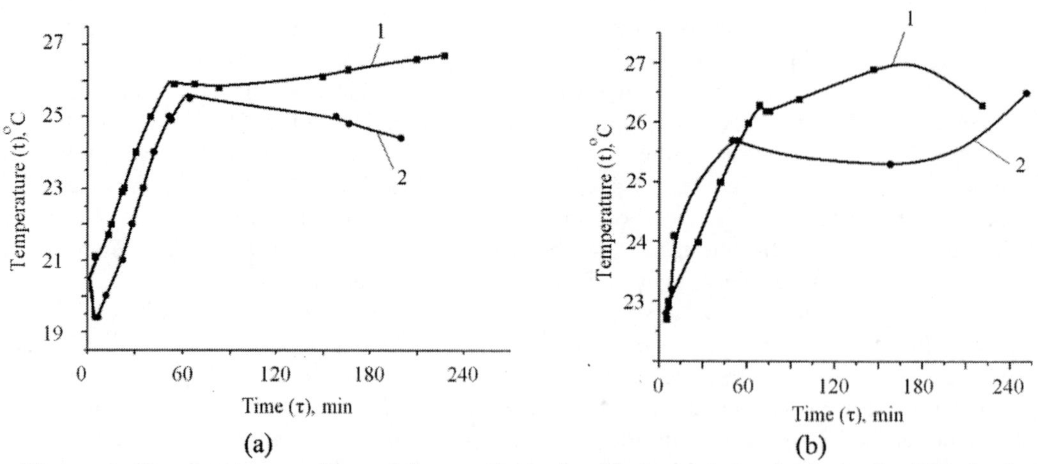

Figure 1. Temperature profiles of the reaction of oxidative polymerization of aniline in the presence of MC at initial pH = 1 (1) and initial pH = 3 (2). The content of PANI in the composite is: 75 wt.% (a), 50 wt.% (b).

Table 1 shows the PANI yield during the oxidative polymerization of aniline. The initial conditions of the reaction ambiguously affect the yield of the target product (η). During the synthesis of polyaniline without a dispersed template, the yield of PANI decreases with decreasing pH. The introduction of mesoporous carbon into the reaction mixture promotes an increase in η. However, during the synthesis of the PANI composite (75 wt.%) / MC (25 wt.%) (pH = 3), PANI yield decreases by 6%.

Table 1. Characterization of the reactions of oxidative polymerization of aniline.

Sample	pH (initial)	t_{max}, °C	PANI yield (η), %
PANI	1	29.5	78.3
	3	29.3	86.6
PANI (75 wt.%) / MC (25 wt.%)	1	26.3	98.6
	3	25.6	80
PANI (50 wt.%) / MC (50 wt.%)	1	27.1	88
	3	26.3	99

The Raman spectra of PANI and its composites with MC contain peaks at 1490 cm^{-1} (C = N stretching vibrations in quinoid ring) [13], 1389 cm^{-1} (C~N+ ring-stretching vibrations of N-phenylphenazine units) [14], 1161 cm^{-1} (C–H bending vibration of the semiquinonoid rings) [13], characteristic of the emeraldine salt of PANI (Figure 2). These peaks are particularly intense in the spectra of pure polyaniline and composites PANI (75 wt.%) / MC (25 wt.%) (Spectra 1-4, Figure 2). Consequently, under the chosen conditions a conductive form of polyaniline is formed.

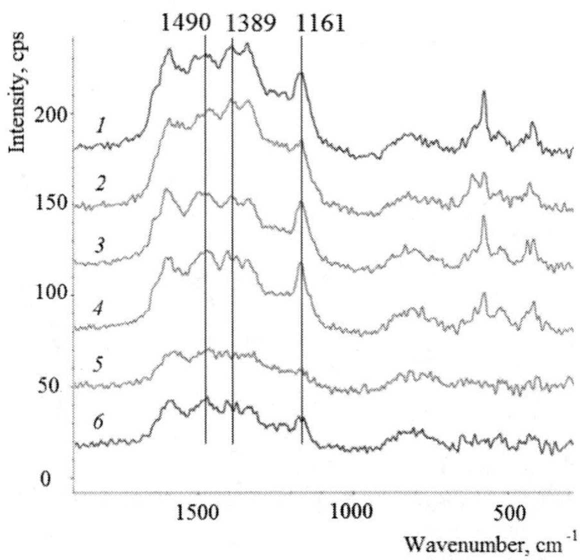

Figure 2. Raman spectra of PANI (*1*, *2*) and PANI / MC composites containing 75 (*3*, *4*) and 50 (*5*, *6*) wt.% of polyaniline. The initial pH is 1 (*1*, *3*, *5*) and 3 (*2*, *4*, *6*).

The value of electrical resistivity (ρ) depends on both the mass content of PANI in the composite and the initial pH value (Table 2). Polyaniline synthesized without a dispersed template has the highest electrical resistivity (ρ). The conductivity of a PANI sample synthesized in a more acidic medium (at pH = 1) is 1.5 times higher than that of the sample obtained at pH = 3. The combination of PANI with mesoporous carbon significantly reduces the electrical resistivity of the material. An increase in the electrical conductivity of the composite is promoted by synthesis in a more acidic medium and an increase in the content of MC. The fact that the substrates coated with conducting polymers have higher conductivity than the individual components can be explained by improved ordering of the polymer chains grown on substrates compared with random chain-aggregation produced in the absence of substrate [15].

Table 2. Electrical resistivity of PANI and PANI / MC composites.

Sample	pH (initial)	Specific electrical resistance, ohm·cm
PANI	1	29.74
	3	50.2
PANI (75 wt.%) / MC (25 wt.%)	1	7.6
	3	21.76
PANI (50 wt.%) / MC (50 wt.%)	1	1
	3	1.05

In the course of the study, the electrochemical properties of composites PANI (50 wt.%) / MC (50 wt.%) were studied. In Figure 3 shows cyclic current-voltage curves.

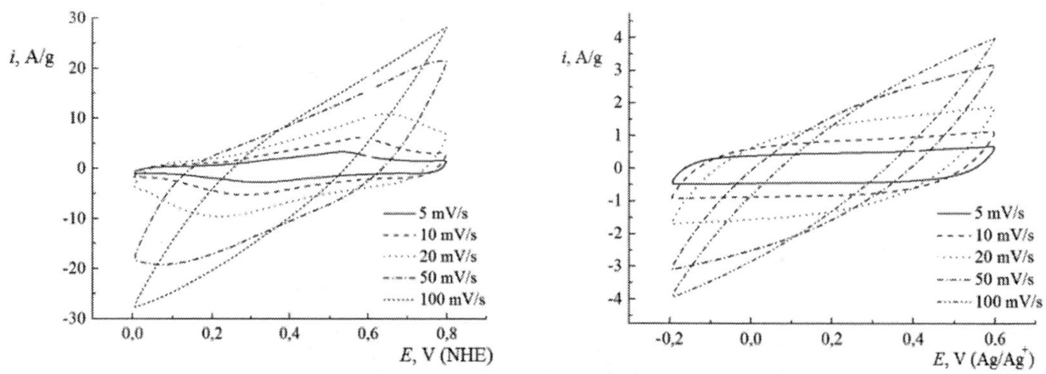

(a) (b)

Figure 3. Cyclic current-voltage curves of the composite PANI (50 wt.%) / MC (50 wt.%): electrolyte is: 3M H_2SO_4 (a); 1M solution of tetraethylammonium tetrafluoroborate in acetonitrile (b).

The cyclic current-voltage curves (CCVC) in the aqueous electrolyte have pronounced peaks due to the Faraday process (Figure 3). The current values for CCVC in the aqueous electrolyte are much higher than in acetonitrile. This is due to the fact that in the latter case only the charge of the electric double layer occurs without an additional flow of the Faraday process.

On the basis of the obtained CCVC curves, the specific capacity (designation) by formula (1) was found using the method of integration in the Origin 8 Pro program:

$$C_s = \frac{1}{mv(E_f - E_i)} \int_{E_i}^{E_f} I(E)dE \quad (1)$$

where m – electrode mass; v – potential sweep rate; E – electrode potential; $I(E)$ – current strength on the current-voltage curve; E_i and E_f – integration limits on the current-voltage curve.

The results of the calculations are presented in Table 3. A decrease in the capacitance of the studied samples by 79.7% (in the aqueous electrolyte) and by 86.6% (in the organic electrolyte) is observed with an increase in the sweep rate to 100 mV/s. Capacity decreases due to increased kinetic difficulties during the course of the electrochemical reaction.

Table 3. Specific capacitances of the studied electrode in aqueous and organic electrolytes at different potential scan rate.

Sample	Specific electrical capacitance C_s, F/g at potential sweep rate v, mV/s				
	5	10	20	50	100
3M aqueous solution H_2SO_4					
PANI (50 wt.%) / MC (50 wt.%)	299.6±4.8	281.6±7.9	256.6±16.7	146.8±32.3	60.9±16.9
1M solution of tetraethylammonium tetrafluoroborate in acetonitrile					
PANI (50 wt.%) / MC (50 wt.%)	80.8±5.8	68.7±8.7	51.1±10.8	22.3±11.4	10.8±5.0

In Figure 4 shows the dependences obtained by the method of electrochemical impedance. Comparing the impedance hodographs in an aqueous and organic electrolyte, one can see that the studied electrodes in an organic electrolyte have a much greater resistance compared to an aqueous electrolyte, therefore the specific capacitance in an aqueous electrolyte is much higher.

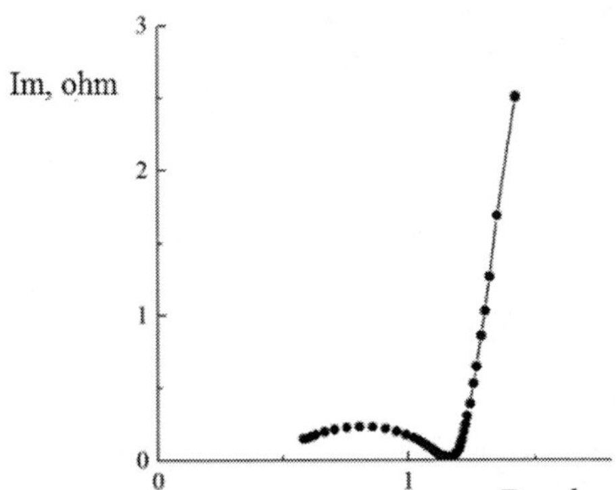

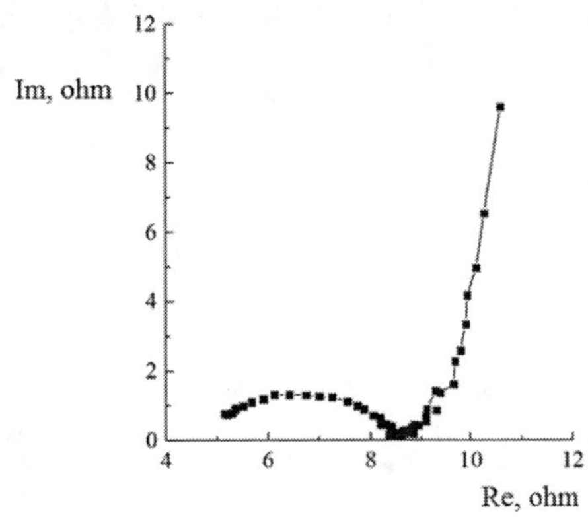

Figure 4. Dependences obtained using alternating current tests in 3M H_2SO_4 (a); in 1M tetraethylammonium tetrafluoroborate solution in acetonitrile (b).

4. Conclusion

The conditions of the oxidation polymerization of aniline in the presence of mesoporous carbon are slightly affected by the reaction conditions and the mass ratio of polymer and MC in the composite. However, these factors determine the electrically conductive properties of the samples. So the best electrically conductive properties have materials PANI (50 wt.%) / MC (50 wt.%) obtained at pH = 1. With a potential sweep rate of 5 mV/s in a 3M aqueous solution of sulfuric acid, this composite has a rather high specific electrical capacity (299 F/g).

On the basis of the experimental data obtained, it can be concluded that composites based on polyaniline and mesoporous carbon can be successfully used in the manufacture of supercapacitors based on aqueous electrolytes.

IOP Conf. Series: Materials Science and Engineering **693** (2019) 012026 doi:10.1088/1757-899X/693/1/012026

References

[1] Libich J, Máca J, Vondrák J, Čech O, Sedlaříková M 2018 Supercapacitors: Properties and applications. *J. of Energy Storage* **17** 22

[2] Liu L, Niu Z, Chen J 2018 Flexible supercapacitors based on carbon nanotubes *Chinese Chemical Letters* **29** 571 https://doi.org/10.1016/j.cclet.2018.01.013

[3] Ding B, Guo D, Wang Y, Wu X, Fan Z 2018 Porous carbon nanotube/graphene composites for high-performance supercapacitors. *J. of Power Sources* **398** 113 https://doi.org/10.1016/j.jpowsour.2018.07.063

[4] Li J, Tang J, Yuan J, Zhang K 2018 Porous carbon nanotube/graphene composites for high-performance supercapacitors. *Chemical Physics Letters* **693** 60 https://doi.org/10.1016/j.cplett.2017.12.052

[5] Li Z, Jiao X, Li C, Chen D 2018 Synthesis and application of nanocages in supercapacitors. *Chemical Eng. J.* **351** 135 https://doi.org/10.1016/j.cej.2018.06.108

[6] Liu J 2018 A high performance all-solid-state flexible supercapacitor based on carbon nanotube fiber/carbon nanotubes/polyaniline with a double core-sheathed structure. *Electrochimica Acta* **283** 366 https://doi.org/10.1016/j.electacta.2018.06.158

[7] Aydinli A, Yuksel R, Unalan H E 2018 Vertically aligned carbon nanotube – Polyaniline nanocomposite supercapacitor electrodes. *Int. J. of Hydrogen Energy* **43** 18617 https://doi.org/10.1016/j.ijhydene.2018.05.126

[8] Zhang L, Huang D, Hu N, Yang C, Li M, Wei H, Yang Z, Su Y, Zhang Y 2017 Three-dimensional structures of graphene/polyaniline hybrid films constructed by steamed water for high-performance supercapacitors. *J. Power Sources* **342** 1 https://doi.org/10.1016/j.jpowsour.2016.11.068

[9] Zhao W 2017 Design and fabrication of flexible supercapacitor devices by using mesoporous carbon/polyaniline. *Surface and Coatings Technology* **320** 595 https://doi.org/10.1016/j.surfcoat.2016.10.075

[10] Zhang X, Lin Q, Zhang X, Peng K 2018 A novel 3D conductive network-based polyaniline/graphitic mesoporous carbon composite electrode with excellent electrochemical performance. *J. of Power Sources* **401** 278 https://doi.org/10.1016/j.jpowsour.2018.08.091

[11] Dyachkova T P, Anosova I V, Tkachev A G et al. 2018 Synthesis of Composites from Functionalized Carbon Nanotubes and Polyaniline. *Inorg. Mater. Appl. Res.* **9** 305 https://doi.org/10.1134/S2075113318020089

[12] Stejskal J, Sapurina I, Trchová M 2010 Polyaniline nanostructures and the role of aniline oligomers in their formation. *Progress in Polymer Science* **35** 1420 https://doi.org/10.1016/j.progpolymsci.2010.07.006

[13] Šeděnková I, Trchová M, Stejskal J 2008 Thermal degradation of polyaniline films prepared in solutions of strong and weak acids and in water – FTIR and Raman spectroscopic studies. *Polymer Degradation and Stability* **93** 2147 https://doi.org/10.1016/j.polymdegradstab.2008.08.007

[14] Ćirić-Marjanović G, Trchová M, Stejskal J 2008 The chemical oxidative polymerization of aniline in water: Raman spectroscopy. *J. of Raman Spectroscopy* **39** 1375

[15] Stejskala J, Acharya U, Bober P 2019 Surface modification of tungsten disulfide with polypyrrole for enhancement of the conductivity and its impact on hydrogen evolution reaction. *Applied Surface Science* **492** 497 https://doi.org/10.1016/j.apsusc.2019.06.175

IOP Conf. Series: Materials Science and Engineering **693** (2019) 012027 doi:10.1088/1757-899X/693/1/012027

Microstructure of 1,4-diethynylbenzene frontal polymerization products

N N Volkova[1], A F Zholudev[1], M B Kislov[1], E V Suslova[2] and L S Yanovskiy[1,3]

[1]Laboratory of Combustion in High-speed Flows, *Institute of Problems of Chemical Physics,* 1 Academician Semenov Avenue, Chernogolovka 142432, Moscow region, Russian Federation
[2]Chemical Faculty, *Lomonosov Moscow State University,* 1/3 Leninskie Gory, Moscow 119991, Russian Federation
[3]Physico-Chemical Faculty, *Lomonosov Moscow State University,* 1 Leninskie Gory, Moscow 119991, Russian Federation

[1]E-mail: nnvolkova@rambler.ru

Abstract. A comparative study of two processes was carried out: frontal polymerization of 1,4-diethynylbenzene, proceeding as a result of high-temperature initiation under increased nitrogen pressure (P = 2–6 MPa), and ammonium perchlorate burning in the same conditions. It was shown that the linear rate of thermal transformations of 1,4-diethynylbenzene is close to the linear burning rate of the energetically saturated compound. The structure of the carbon material obtained as a result of thermal transformations of 1,4-diethynylbenzene was investigated using elemental analysis, Raman spectroscopy and scanning electron microscopy.

1. Introduction

The widespread use of acetylene and its derivatives in organic chemistry and in the field of materials science is associated with their unique reactivity caused by high degree of triple bond unsaturation [1]. Alkyne-containing molecules, because of their high reactivity, provide the possibility of producing carbon materials (CMs) of various shapes and sizes, and preparation of carbon nanomaterials (CNMs) is the focus of many research groups [2-5].

Two-step methods are usually proposed [2-5] for obtaining carbon materials: preliminary synthesis of precursors or oligomers and their pyrolysis at temperatures from 400 °C to 1000 °C. Another unique property of acetylene derivatives is the high heat of their formation, what makes them capable of reactions proceeding with high heat release and even explosion-like processes. This property allows these compounds to be used as high-energy fuels [6].

The purpose of the present work is to show the possibility of carbon materials producing by self-propagating frontal polymerization of the solid 1,4-diethynylbenzene (DEB) and to analyze the microstructure of the material obtained.

2. Methods and materials

The following reagents were used: 1,4-diethylbenzene (DEB), CAS 935-14-8, T_m = 368 K of "chemically pure " (reagent) grade was purified by recrystallization from hexane, the content of the main substance was 99.5%, density ρ = 1.150 kg/m^3; ammonium perchlorate (AP), CAS 7790-98-9, standard GJB617A-2003 99.5% series 1207 of 12.2017, density ρ = 1.950 kg/m^3.

A standard Crawford bomb (or a constant pressure bomb, figure 1) with a volume of $3{\cdot}10^{-3}$ m^3 was used to determine linear rates of thermal transformations of DEB and AP. Samples for testing were prepared by blind pressing in the form of a cylinder with a diameter of 8 mm and a height of 12 mm. The samples prepared this way were placed into the bomb in an asbestos-cement cup of the same internal size (figures 1) followed by filling of 2 or 6 MPa of nitrogen (99.999%, Logika Ltd). Thermal initiation of the process was carried out from the open side of the cup using a nichrome spiral when an alternating current with voltage of 8 V was applied to it. The initiation time was 2 sec; the temperature herewith did not exceed 1500 K. Temperature measurements were performed using a thermocouple mounted at a distance of 4 mm from the open side of a cup. The thermocouple was touching the side surface of the monomer pressed into this cup. The pressure in the system was monitored using inductive pressure sensors DD-10 and ID-2I.

The average time of the process t was determined from experimental dependencies of pressure on process time (figure 2) as an interval between the time of beginning of the growth and the time of pressure drop. The average linear rate of the process U was defined as $U = h/t$, where h is the height of the cylindrical sample. After the process, the solid residue was collected and weighed.

The morphology and microstructure of the carbon material were characterized using the JEOL JSM-6390LA scanning electron microscope operating at 25 kV and the SENTERRA Bruker Raman spectroscope with a laser wavelength of 532 nm.

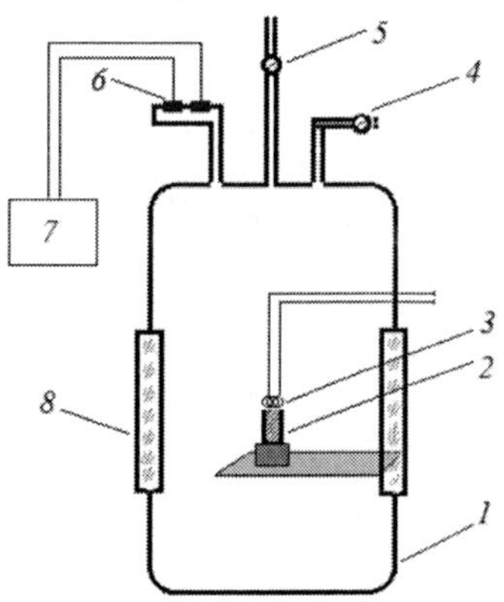

Figure 1. The Crawford bomb: 1 – power case; 2 – cup with a sample; 3 – nichrome spiral; 4 and 5 – faucets; 6 – pressure sensors; 7 – computer; 8 – windows.

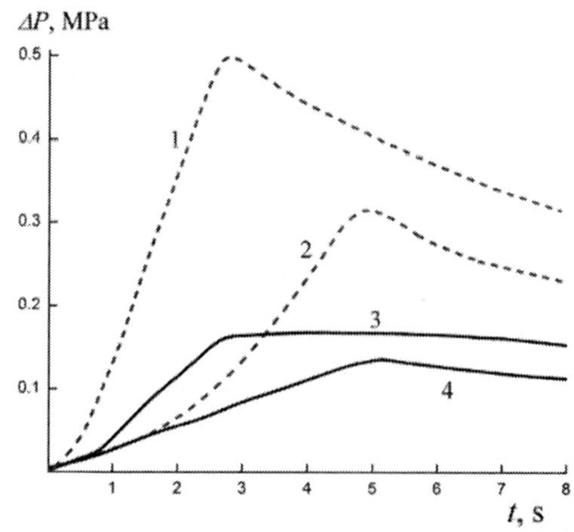

Figure 2. The pressure change during AP combustion (1, 2) and thermal conversion of DEB (3, 4) at initial pressure in the bomb of 6 (1, 3) and 2 MPa (2, 4).

3. Results and discussion

It is known that AP ignites after thermal initiation and burns due to internal oxygen with the release of gaseous products. The thermal conversion of acetylene derivative, DEB, in the bomb occurs as follows. The open upper layer of the sample is heated as a result of thermal initiation, after which a vigorous evaporation process begins, followed by fast polymerization of DEB in the gas phase. Due to high thermal effect of triple bond polymerization [1], the temperature in the reactor remains high even after the initiation heat impulse is turned off. Since DEB has low melting (95 °C) and boiling (183 °C) points and high volatility, after polymerization of the first upper layer, each subsequent layer of DEB also evaporates and polymerizes in the gas phase. This may explain the unusual result of polymerization: an insoluble crosslinked polymer "extends" from the cup in the form of a black cylindrical rod. Its diameter is equal to 8 mm, which is consistent with the internal diameter of the cup.

Figure 3 shows the photo of the carbon material obtained as a result of the DEB transformations under the conditions described above. In this process, a cylindrical rod was formed, apparently, directly in the front of the reaction wave, moving out of the reactor due to the large volume of porous carbon material. The rod length reaches 50 cm. During growth, this rod can break under the influence of its gravity, since its structure is porous and fragile.

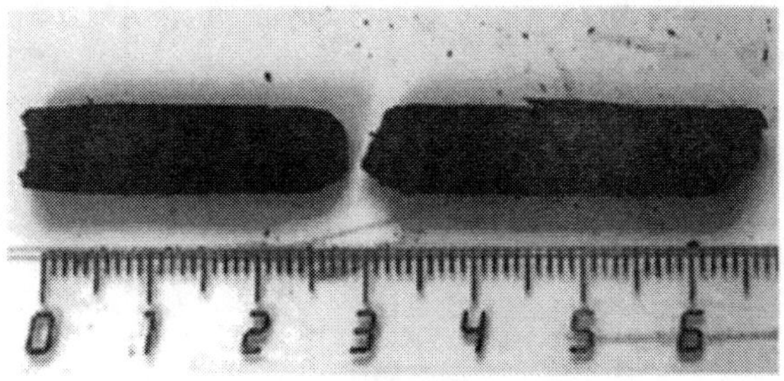

Figure 3. Carbon material obtained as a result of the DEB thermal FP in the Crawford bomb.

The process described above can be attributed to the thermal frontal polymerization – an autowave, self-propagating mode of polymerization reaction. This process begins after a local heat impulse initiation and heat impulse – local initiation. It differs from other polymerization modes by the presence of a moving narrow, usually flat spatial boundary, where the reaction takes place (the reaction front) separating the initial monomer from the final product, the polymer [7]. By the formal signs (presence of the reaction front, heat generation, dynamics), this process is similar to burning and to self-propagating high temperature synthesis (SHS).

It should be noted that during the DEB polymerization under conditions of stationary propagation of the reaction front, the temperature in the reactor reached 1540 K and remained high until the end of the process. In addition, the reaction was accompanied by periodic fluctuations in temperature around the mean value of 1500 K and by appearance of weak flashes of dark red flame.

Figure 2 presents the results obtained by studying the dynamics of AP combustion and of DEB thermal transformations and demonstrates that duration of the thermal DEB transformation (the time of the pressure increase to the maximum point) coincides with the burning time of AP. In other words, the linear rate of the thermal FP of diacetylene is almost equal to the linear burning rate of the high-energy substance. The linear rate of these processes, or reaction front advance velocity, depends on the pressure of the inert gas: during the DEB polymerization, its value increases from 2.3 mm/s to 3.4 mm/s with pressure increasing from 2 to 6 MPa.

The combustion of AP is accompanied by a greater increase in pressure (ΔP_{max} = 0.50 MPa, figure 2, curves 1) compared with the maximum value of the pressure increase in the case of the DEB frontal polymerization (ΔP_{max} = 0.16 MPa, figure 2, curves 3). This is quite natural, since the combustion of AP is accompanied by a strong gas evolution, and the increase in pressure during the DEB polymerization is apparently due mainly to the evaporation of the monomer. In addition, it should be considered that the carbonization process involves aromatization and condensation with evolution of H_2, which also contributes to the pressure increase in the bomb. Besides that, the mass burning rate of AP is higher than the mass rate of the DEB transformation, since the AP density ρ is 1.950 kg/m^3, while that of DEB is 1.150 kg/m^3.

A high temperature in the reactor may lead to a partial degradation (pyrolysis) of both initial and polymerizing molecules of DEB. Nevertheless, the yield of the solid residue of the process is almost 100%, and elemental analysis shows that the resulting product contains 96.49% C, 0.822% H and 0.45% N atoms by weight (nitrogen atoms are apparently absorbed by the carbon material during the

IOP Conf. Series: Materials Science and Engineering **693** (2019) 012027 doi:10.1088/1757-899X/693/1/012027

experiment in the bomb). Hence, the pressure in the system is growing mainly as a result of the vapor formation of DEB.

Figure 4 shows Raman spectrum of the carbon material obtained as a result of the DEB thermal FP. Raman spectroscopy is one of the techniques which can complement the previous analysis and contribute to better understanding of carbon materials (CMs) structure. In Raman spectra, carbon materials typically exhibit two broad bands, called D (disordered) and G (graphitic). In highly oriented pyrolytic graphite, only a G-peak appears in Raman spectra at 1582 cm^{-1}, while in disordered carbons, besides the G-peak, a D-peak appears at lower wavenumbers 1350 cm^{-1} due to the asymmetric vibrations of the carbon bonds near defects in the structure of CMs. The information reflected in Raman spectra depends on both ordering and hybridization [8].

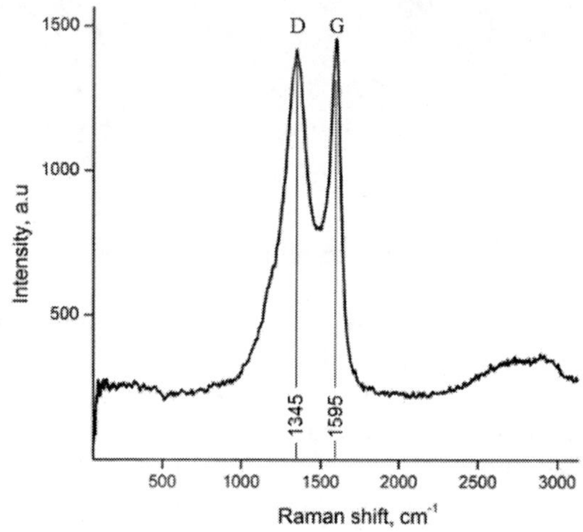

Figure 4. Raman spectrum of the carbon material obtained as a result of the DEB thermal FP at the initial pressure of 6 MPa.

The *D*- and *G*-peaks of the completely amorphous carbons spectra normally are distorted. The spectrum of the carbon material obtained in our experiments demonstrates the presence of well-defined *D* (1345 cm^{-1}) and *G* (1595 cm^{-1}) modes, which indicates that a graphite structure was developed in the carbonized DEB. In other words, the sample contains some amounts of sp^2 bonded carbon. Such atoms can be located on the surface layers and in nanoflakes of graphite. The height of the *D* peak is almost equal to the intensity of the *G* peak, which means a high degree of structure disorder. However, since the high *D*-peak intensity directly depends on the presence of sixfold aromatic rings [8], the significant height of the peak in the Raman spectrum (Figure 4) indicates the presence of a large number of aromatic rings in the carbon material obtained.

Figure 5 shows SEM images of the carbon material obtained at the initial pressure of 6 MPa. The most significant feature of the microstructure of these samples, in our opinion, is the presence of micro spheres with a rather narrow size distribution. The average diameter of the micro spheres is 100-200 nm, and only a small fraction of the micro spheres has size of 500 nm (figure 5 *c*, *d*). Apparently, such narrow size distribution is a consequence of the frontal method of polymerization. This happens because the thermal wave of the stationary polymerization front creates uniform conditions for the formation of micro particles resulting from the polymerization. The fact that the polymerization proceeds in the gas phase creates conditions for preventing the agglomeration of the particles during the process.

IOP Conf. Series: Materials Science and Engineering **693** (2019) 012027 doi:10.1088/1757-899X/693/1/012027

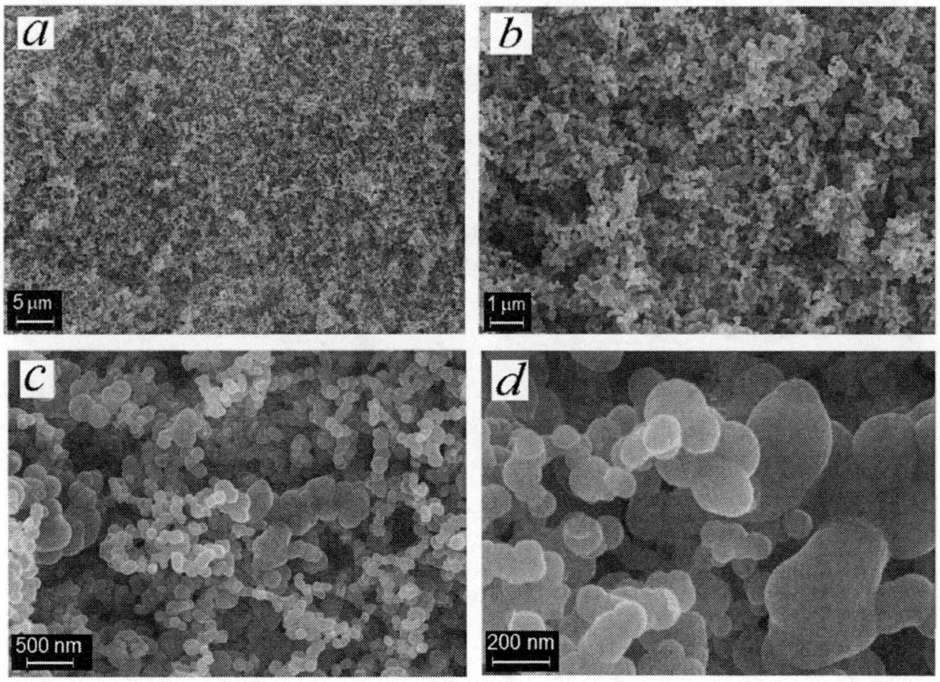

Figure 5. SEM images of the carbon material obtained as a result of the DEB frontal polymerization at a pressure of 6 MPa.

4. Conclusion

The regularities of thermally initiated frontal polymerization of 1,4-diethynylbenzene proceeding under conditions of elevated pressure (2-6 MPa) of an inert gas are investigated. It was shown that the temperature in the reactor could rise during this process up to 1540 K.

The carbon material obtained consists of microspheres with a narrow size distribution. The average diameter of the particles is 100-200 nm. Frontal polymerization of 1,4-diethynylbenzene is an ecologically clean (since no solvent is used in this process), fast, and energy-saving approach to preparing structured carbon and composite materials. Moreover, it should be noted that the most advantageous factor of frontal polymerization, from the perspective of its implementation, is that the process can be performed in continuous reactors [7].

5. Acknowledgments

This work was financially supported by IPCP RAS from the resources of the government assignment 0089-2014-0019 “Creating of high-energy materials and technologies for developing and future systems”, with financial support from the program #56 of the Presidium of the Russian Academy of Sciences. The study was carried out using the equipment bought on account of the Moscow State University Development Programme fund.

References

[1] Diederich F, Stang P J and Tykwinski R 2005 *Acetylene chemistry: Chemistry, Biology and Material Science* (Weinheim: Wiley-VCH) p 508

[2] Rondeau-Gagnéa S and Morin J F 2014 Preparation of carbon nanomaterials from molecular precursors. *Soc Rev* **43(1)** 85 DOI:10.1039/C3CS60210A

[3] Oishi S S, Botelho E, Luscombe C K and Rezende M C 2014 Synthesis and characterization of polyarylacetylene for use in the monolithic vitreous carbon processing. *Polímeros* **24(9)** 541

http://dx.doi.org/10.1590/0104-1428.1623
[4] Tseng W C, Chen Y and Chang G W 2009 Curing conditions of polyarylacetylene prepolymers to obtain thermally resistant materials. *Polymer Degradation and Stability* **94** 2149 https://doi.org/10.1016/j.polymdegradstab.2009.09.008
[5] Herghel L, Kübel C, Lieser G, Räder H J and Müllen K 2002 Pyrolysis in the mesophase: A chemist's approach toward preparing carbon nano- and microparticles. *J. Am Chem Soc.* **124** 13130 https://doi.org/10.1021/ja020363u
[6] Lempert D, Kazakov A and Yanoskiy L 2019 Potential capability of some ethynyl derivatives as fuel dispersants for solid fuel ductedrockets. *Proc. 22nd International Seminar New Trends in Research of Energetic Materials* (Pardubice, Czech Rep) pp 527–532
[7] Davtyan S P, Berlin A A and Tonoyan A O 2011 Advances and problems of frontal polymerization processes. *Review J. Chem.* **1** 56 https://doi.org/10.1134/S207997801101002X
[8] Urbonaite S, Hälldahl L and Svensson G 2008 Raman spectroscopy studies of carbide derived carbons. *Carbon* **46(14)** 1942 https://doi.org/10.1016/j.carbon.2008.08.004

Development of thermo-electrochemical cells based on flexible nanocomposite electrodes with oxidized multi-walled carbon nanotubes coating

V Shpekina[1,*], I Burmistrov[1,2], N Gorshkov[1], D Artyukhov[1], N Kiselev[1], N Kovyneva[1] and Y Smirnova[3,4]

[1]Solid State Ionics Laboratory, *Yuri Gagarin State Technical University of Saratov*, 77 Politechnicheskaya Street, 410054, Saratov, Russian Federation
[2] Laboratory of Materials Science, *Tambov State Technical University*, 106 Sovetskaya Street, 392000, Tambov, Russian Federation
[3]*Tyumen Industrial University*, 38 Volodarskogo Street, 625000, Tyumen, Russian Federation
[4]*Tobolsk Industrial Institute (Branch of Tyumen Industrial University),* 5 Zona Vuzov, 626158, Tobolsk, Russian Federation

*E-mail: vshpekina@mail.ru

Abstract. While conversion of low-grade waste heat into electricity is an important energy harvesting strategy, the thermo-electrochemical cells are one of the most promising devices for this application. This paper shows the study on thermo-electrochemical cell with polymer electrodes coated by oxidized multi-walled carbon nanotubes and the $Fe(CN)_6^{3-}/Fe(CN)_6^{4-}$ based electrolyte. The developed cell demonstrates the current density values of more than 13 A/m^2 and a specific power of 140 mW/m^2. Hypothetical Seebeck coefficient was found equal to 0.7 mV/K by the calculation based on the temperature dependencies of the open circuit voltage.

1. Introduction

Conversion of low-grade waste heat (i.e. at temperatures < 150 °C) into electricity is a promising energy harvesting strategy. Low-grade waste heat gained from industrial or geothermal processes, solar heaters or collectors, biomass fermentation, human body, etc. by being a significant source of energy can be effectively harvested for electricity production. One of the key challenges regarding low-grade waste heat harvesting is an improvement of the devices efficiency while decreasing their cost [1-3]. Majority of studies shows the application of solid-state thermoelectric and Stirling engines for low-grade heat harvesting [2]. However semiconductors based thermoelectric have the insufficient efficiency at low temperatures due to low voltage range (μV/K). Meantime, Stirling engine technology demonstrates problems with long-term reliability while having high initial cost [1-3].

When comparing the devices' efficiency in Wh/dollars, thermo-electrochemical cells shows advantages over Stirling engines or solid-state thermoelectric devices [1,2] due to their low cost. However, their low energy conversion efficiency and low real output power limit their commercial applications [3-6].

A typical thermo-electrochemical cell consists of two identical electrodes working at different temperatures in contact with an electrolyte containing a redox pair [7,8]. Temperature gradient between electrodes causes oxidation of the redox pair on the anode (hot electrode) and reduction on the cathode (cold electrode) [6-8].

Aqueous ferri/ferrocyanide solution is an archetypical electrolyte system due to its large reaction entropy, high Seebeck coefficient (theoretical limit for aqueous solution is 1.4 mV/K) and high exchange current density [3].

Temperature difference between electrodes causes the oxidation of $[Fe(CN)_6]^{4-}$ ions on the hot electrode:

$$[Fe(CN)_6]^{4-} \rightarrow [Fe(CN)_6]^{3-} \tag{1}$$

and the reduction of $[Fe(CN)_6]^{3-}$ ions on cold electrode

$$[Fe(CN)_6]^{3-} \rightarrow [Fe(CN)_6]^{4-} \tag{2}$$

The potential difference is an important factor determining the device's power output and characterized by the Seebeck coefficient:

$$S = \frac{\Delta s_{B,A}}{nF} = \frac{\partial V}{\partial T} \tag{3}$$

The production of flexible devices is one of the promising trends in modern electronics. The thermocells flexibility can allow them to be used for heat harvesting from various non-planar heat sources such as heated pipeline, reactor or vessel, human body heat, etc. Flexible thermocells can be made of flexible body materials such as polyethylene terephthalate [9] and silicone [10] and flexible electrodes, typically based on carbon nanotubes. However, the device's performance indicators (open-circuit voltage, short-circuit current and power output) is significantly limited by the low thermal conductivity of the body materials and low wettability of native multi-walled carbon nanotubes (MWCNTs) with aqueous ferri/ferrocyanide based electrolyte. This study shows the use of MWCNTs-based polymer composite with improved thermal [11,12] and electrical [13-15] conductivity allowing to increase the real temperature difference between internal electrodes surface (in the MWCNTs cover) in the cell and thus to improve the output open circuit voltage.

The MWCNTs show significant advantages including high electrical conductivity, flexibility, thermal stability, high surface area (i.e. 300-350 m^2/g for "Taunit-M" MWCNTs, used in current study) and ultra-low areal density [16-18]. Thus, MWCNTs become the most common electrode for modern thermal cells [1-3]. However, native CNTs have low wettability while interaction with aqueous electrolytes. Meantime, high wettability of oxidized MWCNTs allows increasing the charge transfer between electrolyte and MWCNTs electrode.

This study aims at performance indicators (open-circuit voltage, short-circuit current and power output) of thermo-electrochemical cell based on a flexible and heat conducting polymer electrodes coated by oxidized MWCNTs. Flexible cell body with well performance indicators provide have a potential applications as commercial battery based on thermo-electrochemical cells for waste heat gained from various sources such as industrial or geothermal processes, solar heaters or collectors, biomass fermentation, human body heat, heated reactor, pipeline or vessel.

2. Materials and methods

2.1. Materials

A flexible polymer composite based on MWCNTs ("Taunit-M" [18] produced by NanoTechCenter LLC, Russia) and ethylene-1-octene copolymer (Lucene LC370 TM produced by LG Chem Ltd, South Korea) was used as an flexible electrode materials. A laboratory roller mixer UB-6175 (UGNLab, Russia) was used for sample processing. Processing took place at the temperature of 120±3°C and treatment time of 5-30 min. The details on polymer preparation methodology can be found in [14].

The electrode was covered with MWCNTs in order to get a greater specific surface area. The following oxidation of MWCNTs allowed increasing their wettability.

MWCNTs oxidation was carried out in two-necked flask with backflow condenser. The reaction temperature was maintained at 100-105°C and controlled with thermometer. Concentrated nitric acid (HNO_3) was used as an oxidizing agent in MWCNT/acid ratio equal to 1/20. Oxidation process continued for 2 h. After the oxidation completion, the products were rinsed with distilled water till the neutral pH. Separation of reaction products was performed by decantation with the following vacuum filtering and drying at 60 °C for 48 h. The details on oxidation technique can be found in [19].

Mixture of aqueous 0.3 M solution of the mixture of two components: potassium ferricyanide (III) $K_3[Fe(CN)_6]$ and potassium ferrocyanide (II) $K_4[Fe(CN)_6]$ with a molar ratio of 1/1 was used as an electrolyte.

2.2. Measurement and characterization technique

Figure 1 shows the schematic representation of the studied thermo-electrochemical cell with the electrodes made of composite sheets (1x1 mm size and 0.7-0.8 mm thickness). The composite sheets were coated with a MWCNTs by spreading their ethanol-based suspension (1 wt.%) and the following drying for 1 hour at 60 °C. Separator was placed between the electrode sheets. Electrolyte was introduced into the interelectrode space and sealed with a rubber gasket on perimeter.

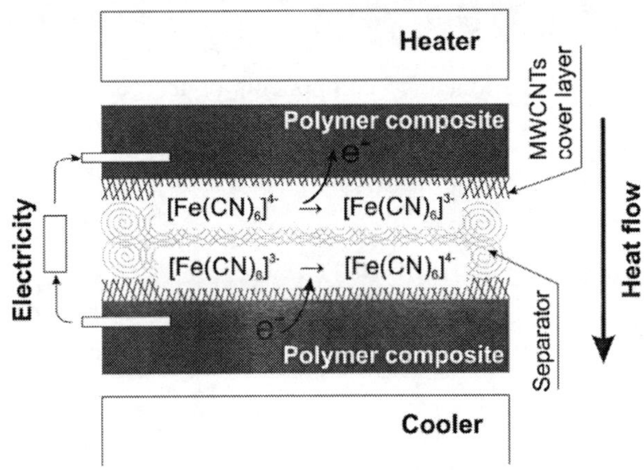

Figure 1: Schematic representation of thermo-electrochemical cell operation.

The measurement of current-voltage characteristic of thermo-electrochemical cell was carried out by potentiostat P-30i (Elins, Russia) in potentiostatic mode. The scanning electron microscopy (SEM) images were obtained by scanning electron microscope "EXplorer" ASPEX (USA).

3. Results and discussion

A SEM study of electrodes (figure 2a and 2b) proved the presence of 1 μm thick oxidized MWCNTs layer over polymer composite. The used separator (figure 2c) consisted of 10-20 μm thick polypropylene fibers. The MWCNTs distribution in the composite was shown as uniform (figure 2d).

Figure 3 shows the output power-voltage and current-voltage characteristics of the developed cell The open-circuit voltage (U) and short-circuit current (j) increases with temperature difference increase, resulting in a parabolic power output (P) curves (figure 3a and 3b). The Seebeck coefficient is found to be ~ 0.7 mV/K (figure 3c). The output power of a thermo-cell also increases with an increase in the temperature gradient (figure 3d).

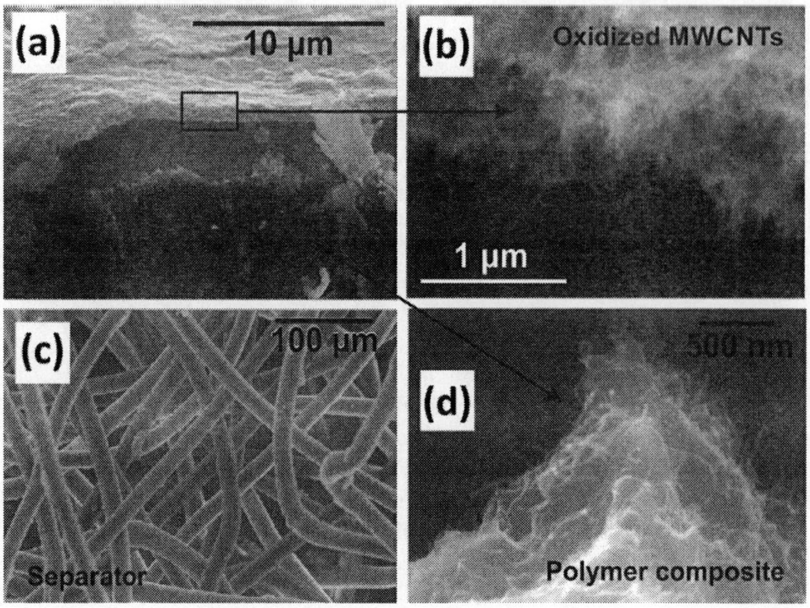

Figure 2: SEM image of components of the cell: (a) – electrode structure; (b) – cover oxidized MWCNTs layer; (c) – separator; (d) – fracture of polymer nanocomposite.

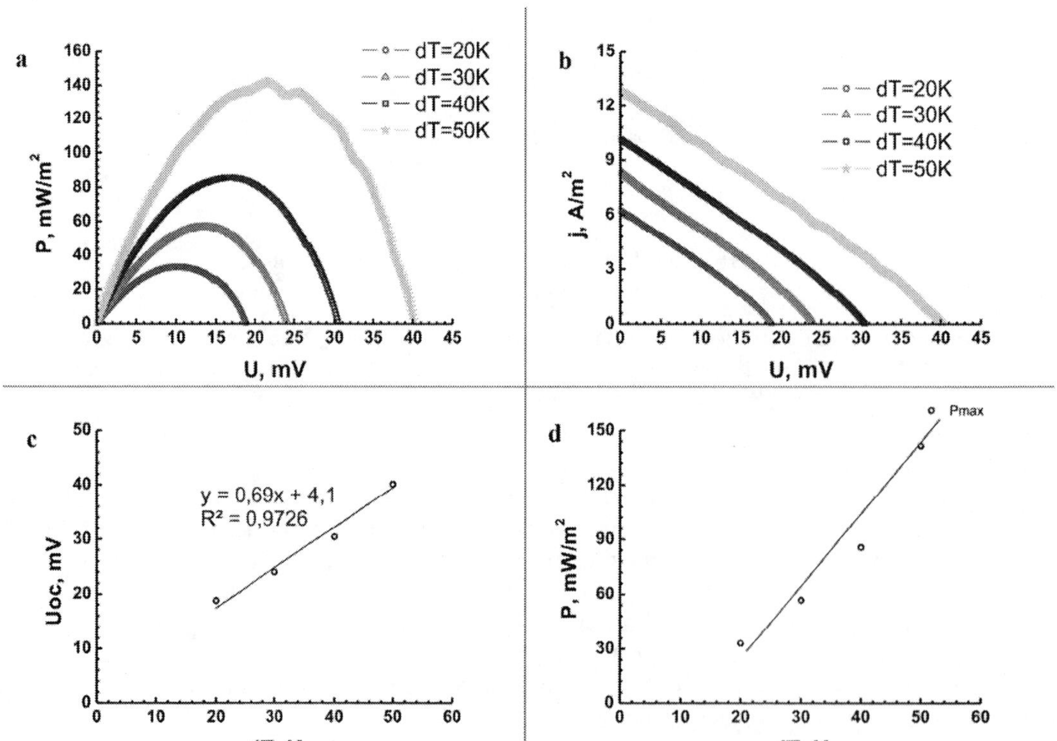

Figure 3. Electrophysical properties of thermocells: (a) – power output – open-circuit voltage and (b) – short-circuit current – open-circuit voltage dependenses for different temperature; (c) and (d) – temperature dependenses of open-circuit voltage and power output respectively.

Maximum output power increases from 26 mW/m^2 (ΔT = 20°C) to 140 mW/m^2 (ΔT = 50°C) (figure 3a). At the maximum temperature difference used in the experiment (hot electrode temperature T_{hot} = 75°C, ΔT = 50°C), the cell showed a current density of 13 A/m^2 and a voltage of 40 mV.

The obtained result (output power) is not as high as has been shown in literature for the best performers (i.e. non-flexible carbon nanotube aerogel sheet onto metal electrodes demonstrated the maximum power output of 6.6 W/m^2 [1]), however high enough for polymer flexible cells. The difference in results can be explained by the high resistance of the polymer electrode. The best values of output power for flexible thermo-electrochemical cell have been shown about 76 – 230 mW/m^2, as per [10] while current study shows the output power of 140 mW/m^2. However worth noting that in [10] these values have been obtained for tube-shaped flexible thermo-cell. The maximum output power for the flat thermo-electrochemical cells did not exceed 1 mW/m^2 as per [9].

Literature review shows the short-circuit current density of 0.39 A/m^2 and open-circuit voltage of 5 mV [10] for flexible thermo-cells compared to 13 A/m^2 and 40 mV, respectively, gained in current study. The different result can be explained by a higher real temperature gradient between the internal surfaces of the electrodes.

4. Conclusion

This paper shows the results of the study on the performance indicators (open-circuit voltage, short-circuit current and power output) of thermo-electrochemical cell with the flexible and heat- and electricity-conducting polymer electrodes coated with oxidized MWCNTs. The study showed that the polymer composite electrode with high thermal conductivity can improve thermo-electrochemical cell performance. Flat flexible thermo-electrochemical cell with polymer electrodes coated with oxidized MWCNTs showed the maximum output power of 140 mW/m^2, a current density of 13 A/m^2 and a voltage of 40 mV. Thus, the developed thermocells can be potentially applied for economically convenient harvesting untapped supplies of low-grade heat.

Acknowledgments

This research was financially supported by Russian Foundation for Basic Research (RFBR) according to the research project No. 18-29-19121\1 and by the Ministry of Education and Science of the Russian Federation (project 4.6197.2017/8.9).

References

[1] Im H and Kim T 2016 High-efficiency electrochemical thermal energy harvester using carbon nanotube aerogel sheet electrodes. *Nat. Commun.* **7** 10600 doi.org/10.1038/ncomms10600

[2] Hu R, *et al.* 2010 Harvesting waste thermal energy using a carbon-nanotube-based thermo-electrochemical cell. *Nano Lett.* **10** 838 doi.org/10.1021/nl903267n

[3] Dupont M F, MacFarlane D R and Pringle J M 2017 Thermo-electrochemical cells for waste heat harvesting – progress and perspectives. *Chem. Commun.* **53** 6288 doi.org/10.1039/C7CC02160G

[4] Quickenden T I and Mua Y 1995 A review of power generation in aqueous thermogalvanic cells. *J. Electrochem. Soc.* **142** 3985 doi.org/10.1149/1.2048446

[5] Gunawan A, Lin C H and Buttry D A 2013 Liquid thermoelectrics: review of recent and limited new data of thermogalvanic cell experiments. *Nanosc. Microsc. Therm.* **17** 304 doi.org/10.1080/15567265.2013.776149

[6] Burmistrov I, Kovyneva N, Gorshkov N, Gorokhovsky A, Durakov A, Artyukhov D and Kiselev N 2019 Development of new electrode materials for thermo-electrochemical cells for waste heat harvesting. *Renew. Energy Focus* **29** 42 doi.org/10.1016/j.ref.2019.02.003

[7] Kim K and Lee H 2019 Diglyme-based thermoelectrochemical cells operable at both subzero and elevated temperatures. *Energy Technol.* Online doi.org/10.1002/ente.201900857

[8] Alzahrani H A H, Buckingham M A, Marken F and Aldous L 2019 Success and failure in the incorporation of gold nanoparticles inside ferri/ferrocyanide thermogalvanic cells. *Electrochem. commun.* **102** 41 doi.org/10.1016/j.elecom.2019.03.007

[9] Im H, Moon H G, Lee J S, Chung I Y, Kang T J and Kim Y H 2014 Flexible thermocells for utilization of body heat. *Nano Res.* **7** 443 doi.org/10.1007/s12274-014-0410-6

[10] Yang H D, Tufa L T, Bae R M and Kang T J 2015 A tubing shaped, flexible thermal energy harvester based on a carbon nanotube sheet electrode. *Carbon* **86** 118 doi.org/10.1016/j.carbon.2015.01.037

[11] Muratov D, Kuznetsov D, Ilinykh I, Burmistrov I and Mazov I 2015 Thermal conductivity of polypropylene composites filled with silane-modified hexagonal BN. *Composites Science and Technology* **111** 40 doi.org/10.1016/j.compscitech.2015.03.003

[12] Mazov I, Burmistrov I, Il'inykh I, Stepashkin A, Kuznetsov D and Issi J-P 2014 Anisotropic thermal conductivity of polypropylene composites filled with carbon fibers and multiwall carbon nanotubes. *Polym. Compos.* **36** 1951 doi/abs/10.1002/pc.23104

[13] Burmistrov I, Gorshkov N, Ilinykh I, Muratov D, Kolesnikov E, Anshin S and Mazov I 2016 Improvement of carbon black based polymer composite electrical conductivity with additions of MWCNT. *Compos. Sci. Technol.* **129** 79 doi.org/10.1016/j.compscitech.2016.03.032

[14] Burmistrov I, Gorshkov N, Ilinykh I, Muratov D, Kolesnikov E, Yakovlev E and Mazov I 2017 Mechanical and electrical properties of ethylene-1-octene and polypropylene composites filled with carbon nanotubes. *Compos. Sci. Technol.* **147** 71 doi.org/10.1016/j.compscitech.2017.05.005

[15] Burmistrov I, Gorshkov N, Anshin S, Kolesnikov E, Kuskov K, Ilinykh I, Issi J-P, Vikulova M and Kuznetsov D 2019 Enhancement of percolation threshold by controlling the structure of composites based on nanostructured carbon filler. *J. Electron. Mater.* **48** 5111 doi.org/10.1007/s11664-019-07287-3

[16] Gong K, Yan Y, Zhang M, Su L, Xiong S and Mao L 2005 Electrochemistry and electroanalytical applications of carbon nanotubes: a review. *Anal. Sci.* **21** 1383 doi.org/10.2116/analsci.21.1383

[17] Zhang M, Fang S, Zakhidov A, Lee S B, Aliev A and Williams C 2005 Strong, transparent, multifunctional, carbon nanotube sheets. *Science* **309** 1215 doi.org/10.1126/science.1115311

[18] Tkachev A, Mikhaleva Z and Burakova E 2009 Investigation of methods for improving the activity of catalysts for producing carbon nanostructural materials. *Theor. Found. Chem. Eng.* **43** 739 doi.org/10.1134/S0040579509050212

[19] Burmistrov I, Muratov D, Ilinykh I, Kolesnikov E, Godymchuk A and Kuznetsov D 2016 The effects of liquid-phase oxidation of multiwall carbon nanotubes on their surface characteristics *IOP Conference Series: Materials and Engineering.* **112** 012004 doi.org/10.1088/1757-899X/112/1/012004

Removal of Cu^{2+}, Zn^{2+} and Pb^{2+} ions using a graphene-containing nanocomposite: a kinetic study

A Babkin[1], I Burakova[1], A Burakov[1], D Kurnosov[1], E Galunin[1], A Tkachev[1] and I Ali[2]

[1]Department of Technology and Methods of Nanoproducts Manufacturing, *Tambov State Technical University,* 106 Sovetskaya Str., Tambov 392000, Russian Federation
[2]Department of Chemistry, *Jamia Millia Islamia (Central University)*, Jamia Nagar, New Delhi 110025 India

[1]E-mail: iris_tamb68@mail.ru

Abstract. In the present paper, the urgent issue of removing heavy metal ions from aqueous solutions is considered by the example of extraction of Cu^{2+}, Zn^{2+} and Pb^{2+} ions using an innovative sorbent - polyhydroquinone/graphene nanocomposite. Kinetics of the metal sorption on this material was studied. Moreover, the effect of solution pH on the sorption capacity of the nanocomposite was estimated. Besides, to elucidate mechanisms of kinetic removal and determine process parameters, experimental data were fitted to kinetic (pseudo-first- and pseudo-second-order, and Elovich) and diffusion (intraparticle and Boyd) models. The kinetic experiments showed that equilibrium is reached within 60 min. The effective diffusion coefficient values calculated according to the Boyd model were found to increase in the following sequence: Pb^{2+} ($1.26\ 10^{-13}$ m^2 s^{-1}) < Zn^{2+} ($3.74\ 10^{-13}$ m^2 s^{-1}) < Cu^{2+} ($4.24\ 10^{-13}$ m^2 s^{-1}). Finally, it was established that the metal sorption process is controlled by the internal diffusion of the sorbate into the bulk (volume) of the sorbent.

1. Introduction

In the field of highly efficient purification of aquatic systems, many scientific papers prove that carbon nanostructured materials are more effective compared to currently used commercially available sorbents [1-4]. Moreover, they are non-toxic and thermally stable. The prospect of using such nanomaterials in sorption-desorption processes is justified, on the one hand, by their unique properties (such as specific surface area, presence of surface chemically active sites, high chemical affinity for components being extracted), and, on the other hand, by the similarity of the chemical nature (allotropy) and a set of mechanical characteristics to those of the most widely used industrial sorbents [5, 6]. This will allow them to be employed without doing any harm to the microelemental and mineral composition of soils and aquatic media. In the near future, the proposed carbon nanostructured sorbents will become an alternative to conventional sorption materials based on rocks, clays, hydrated aluminosilicates, activated carbons, low-cost agro-waste biosorbents, *etc.*

Among the existing physical and chemical processes of removal and subsequent isolation of harmful contaminants, sorption-desorption cycles play a significant role. Only their practical implementation makes it possible to extract, transfer and accumulate toxic materials for appropriate storage conditions. In this regard, elucidating fundamentals and empirical dependencies of selective extraction of various harmful contaminants from target aquatic and terrestrial systems using nanostructured sorbents is an important task.

Although a number of scientific studies (table 1) have shown the promising use of carbon graphene-containing structures (multi- (MWCNTs) and single-walled (SWCNTs) carbon nanotubes, graphene oxide (GO)) as sorbents for the extraction of such heavy metals as Fe, Cu, Ni, Zn, Cr, Pb, *etc* [6-8], to the best of our knowledge, mechanisms of metal removal processes occurring through sorption-desorption cycles have not completely and reliably studied yet so far.

Table 1. Comparative characteristics of the adsorption of Cu^{2+}, Pb^{2+}, Zn^{2+} on various carbon nanomaterials.

Adsorbent	Experimental conditions					Q_e, mg·g^{-1}
	pH	Temperature, K	Initial concentration, mg L^{-1}	Adsorbent weight, g	Time, min	
Pb^{2+}						
MWCNTs (oxidized) [9]	5	room temperature	40	1	150	50
GO [10]	5	room temperature	-	0.02	90	44.56
rGONF (ferrite-reduced GO) Nanocomposite [10]	5	room temperature	-	0.02	30	25.78
Mn_3O_4-coated activated carbon [11]	5	303	50	1 g·L^{-1}	70	49.82
Cu^{2+}						
Mn_3O_4-coated activated carbon [11]	5	303	50	1 g·L^{-1}	70	25.66
GO/MWCNTs/Fe_3O_4 nanocomposite [12]	7	room temperature	25	25	-	20.83
GO [13]	6	room temperature	100	0.03	20	33
Zn^{2+}						
SWCNTs [14]	7	room temperature	60	0.05	360	39.52
Oxidized MWCNTs [15]	5	room temperature	3 mmol L^{-1}	0.005	120	0.2 mmol g^{-1}
GO [16]	6	room temperature	100	0.03	20	26

Considering the aforementioned, the aim of the present work was to study sorption regularities using the kinetic approach, which allows estimating the effect of system parameters, varying over time, on the degree of extraction of target products from model systems and their retention. A novel polyhydroquinone/graphene (PHQ/G) nanocomposite was employed as a sorbent. The results of the work will contribute to the expansion of the application of the developed technologies in wastewater treatment programs in populated areas, where aquatic systems are contaminated with heavy metals and toxic organic compounds due to the abundance of industrial effluents.

2. Experimental part

2.1. Material and reagents

The PHQ/G nanocomposite was produced by NanoTechCenter Ltd (Tambov, Russia). The technology of its manufacturing is described elsewhere [17]. Heavy metal salts – $Pb(NO_3)_2$, $Cu(NO_3)_2$ and $Zn(NO_3)_2$ (all – analytical grade)– were acquired from Laverna Lab Ltd (Moscow, Russia). Heavy metal stock solutions were obtained by diluting certain amounts of the corresponding salts in deionized water. Before using in tests, the solutions diluted to the required concentrations.

2.2. pH effect on the sorption capacity

pH is an important parameter, since it determines the number of acid-base groups, which, in turn, affects the degree of interaction between the contaminant and the surface groups of the sorbent. To study its effect on the removal sorption (removal) capacity of the PHQ/G nanocomposite for Pb^{2+}, Cu^{2+} and Zn^{2+} ions, the following experiment was performed at room temperature.

0.03 g of the PHQ/G nanocomposite were placed into 50-mL conical tubes. Then, 30 mL of the metal solution at a preset concentration and pH value was poured into the tubes. Prior to that, the solution pH was fixed at the predetermined value using a buffer system obtained by adding corresponding acids (hydrochloric, acetic, and aminoacetic) and alkali (sodium hydroxide) in certain proportions. The pH of each solution was measured using an HI 2210 pH meter (Hanna Instruments, Woonsocket, RI, USA). The tubes were shaken for 60 min on a Multi Bio RS-24 programmable rotator (Biosan, Riga, Latvia), and then filtered to separate the solid phase. The equilibrium metal concentration in the selected aliquots was measured using an MGA-915 MD atomic absorption spectrometer (Atompribor, St. Petersburg, Russia).

The equilibrium sorption capacity (Q_e, mg g^{-1}) of the material was determined according to equation (1):

$$Q_e = \frac{V(Co-C)}{m}, \quad (1)$$

where C_0 – initial metal concentration in solution (before sorption), mg L^{-1}; C – equilibrium metal concentration in solution (after sorption), mg L^{-1}; V – solution volume, mL; m – sorbent weight, g [18].

2.3. Kinetic study

To identify the features of the Cu^{2+}, Pb^{2+} and Zn^{2+} sorption on the PHQ/G nanocomposite, batch kinetic tests were carried out. The purpose was to establish the time to reach equilibrium (at which the sorption and the desorption rates are equal) in the sorbate-sorbent system under specific process conditions, and to describe the mechanism of metal ion sorption.

0.01 g of the PHQ/G nanomaterial and 30 mL of 100-mg L^{-1} were put into 50-mL tubes, and then shaken on the rotator (Biosan) for 60 min. The average number of experimental samples to be analyzed was 13 (*i.e.*, it corresponded to sampling time intervals). Due to the high sorption activity found for the material during the initial period of the process, it was necessary to pick the maximum possible number of samples – after 10, 20, 30 and 60 s. Further aliquot sampling intervals were 1, 2, 5, 7, 10, and 15 min. After this time, the sorption activity of the nanocomposite usually decreases significantly, and thus, the sampling was carried out after 30, 45, and 60 min.

To estimate parameters and describe possible kinetic mechanisms of the metal sorption on the PHQ/G nanocomposite, the following models were implemented (Table 2): kinetic – pseudo-first- and pseudo-second-order, and Elovich, and diffusion – intraparticle and Boyd [18].

Table 2. Kinetic and diffusion models implemented [18].

Model	Equation
Pseudo-first-order	$log(Q_e - Q_t) = log(Q_e) - \frac{k_1 t}{2.303}$ (2)
Pseudo-second-order	$\frac{t}{Q_t} = \frac{1}{k_2 \cdot Q_e^2} + \frac{1}{Q_e} \cdot t$ (3)
Elovich	$Q_t = \frac{1}{\beta} ln(\alpha\beta) + \frac{1}{\beta} ln(t)$ (4)
Intraparticle diffusion	$Q_t = k_{id} \cdot t^{0.5} + C$ (5)
Boyd	$F = 1 - \frac{6}{\pi^2} \sum_{n=1}^{n=\infty} \frac{1}{n^2} exp(-Bn^2)$, (6) where $B = \frac{D\pi^2}{r_0^2}$ (7)

Parameters of equations (2)-(7): t – sorption time, min; Q_e – equilibrium sorption capacity, mg g^{-1}; Q_t – sorption capacity at time t, mg g^{-1}; k_1 – pseudo-first-order reaction rate constant (min^{-1}); k_2 – pseudo-second-order reaction rate constant, g mg^{-1} min^{-1}; α – initial flow rate, min^{-1} mg g^{-1}; β - desorption constant (degree of surface coverage and activation energy of chemisorptions), g mg^{-1};k_{id} – intraparticle diffusion coefficient, min $^{-1/2}$; c – boundary layer thickness constant, mg g^{-1}; F – degree of process completion; n – number of sum items; B – time-dependent parameter ; D – effective diffusion coefficient (m^2 s^{-1}); r_0 - average radius of the sorbent grain, m.

3. Results and discussion

The results of the pH effect experiments are presented in figure 1.

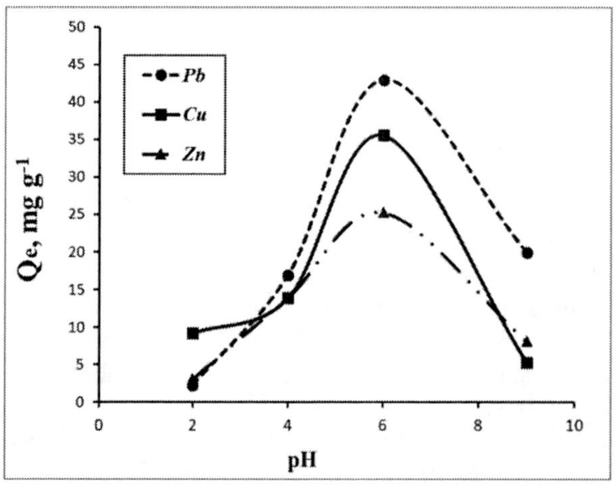

Figure 1. The aqueous solution pH effect on the Pb^{2+}, Cu^{2+} and Zn^{2+} sorption capacity of the PHQ/G nanocomposite (conditions: sorbent weight – 0.03 g, temperature – 298 K, and time – 60 min).

From this figure, it can be seen that the maximum sorption characteristics of the nanocomposite being studied are achieved in the solutions at pH=6 (the dependences have extrema at this point). An increase in the degree of extraction of Pb^{2+}, Cu^{2+} and Zn^{2+} ions with an increase in pH indicates that these ions are in optimal interaction condition. Relatively low sorption capacity values in the acidic medium (at pH = 2 and 4) are probably due to the protonation of the sorbent, which leads to the displacement of Pb^{2+}, Cu^{2+} and Zn^{2} ions (they may be available in solution as aqua complexes [16]). In the pH range of 3-6, the metal sorption capacity increases and reaches the maximum value at pH=6. In this area, the functional groups of the nanocomposite are ionized, and the existing aqua complexes may be dissociated. Further increasing the solution pH leads to a decrease in the sorption capacity of the material. Based on the results of this test, the other experiments were carried out in buffer solutions at the optimum pH value (*i.e.*, pH = 6).

Figure 2 shows the graphical dependences of the sorption capacity of the PHQ/G nanocomposite and the reference materials in their original form – carbon nanotubes (CNTs), graphene oxide (GO), and activated carbon (AC) on the phase contact time.

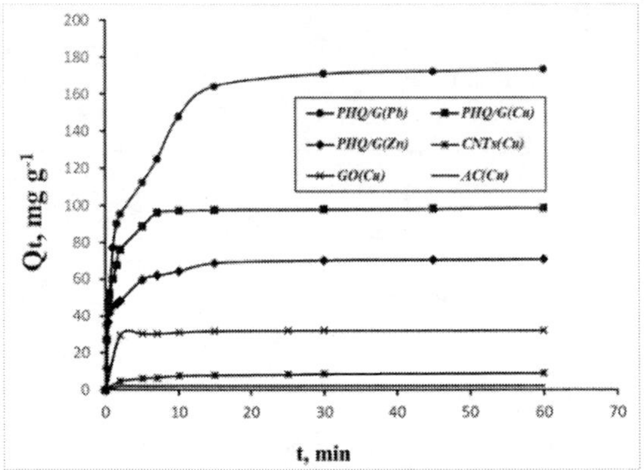

(a)

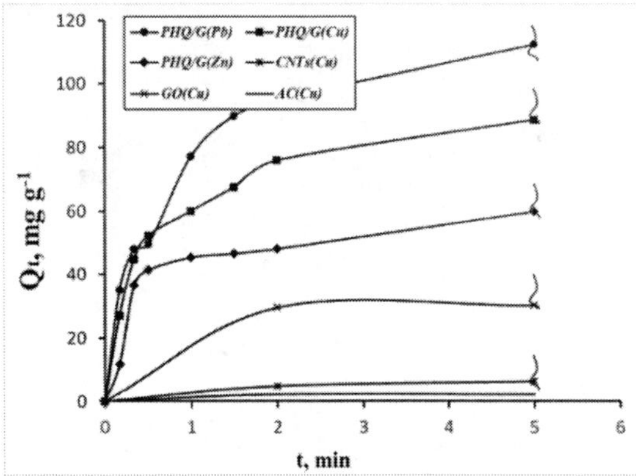

(b)

Figure 2. Kinetic curves constructed for the metal sorption on the PHQ/G nanocomposite in comparison with the reference materials (CNTs, GO and AC) in the time ranges of 0-60 min **(a)** and 0-5 min **(b)** (conditions: sorbent weight – 0.01 g, initial metal concentration in solution – 100 mg L^{-1}, temperature – 298 K, and pH=6).

It can be seen that the sorption capacity of the PHQ/G material is 2-25 times higher than that of the reference materials. It should be noted that the nature of the metal sorption on the CNTs, GO and AC differs from process taking place on the nanocomposite, which is characterized by a sharp increase in the sorption capacity during the initial period of time – about 70 % of the sorbate is sorbed within the first 5 min of the process (Figure 2b). The sorbate/sorbent system reaches equilibrium 60 min after the sorption starts.

Besides, it was experimentally found that the maximum sorption capacity ($Q_{e,\exp}$) of the PHQ/G nanocomposite is as follows: ~174, ~97 and ~64 mg g^{-1} regarding Pb^{2+}, Cu^{2+} and Zn^{2+}, respectively.

Figures 3-5 and Tables 3 and 4 demonstrate the fitting of the experimental data, obtained for the sorption kinetics of all the metal ions, to the kinetic (pseudo-first-order, pseudo-second-order, and Elovich) and diffusion (intraparticle) models (for the Boyd diffusion model, the graphs are not shown; the parameters of this model were obtained using the computer algebra system Maxima).

IOP Conf. Series: Materials Science and Engineering **693** (2019) 012029 doi:10.1088/1757-899X/693/1/012029

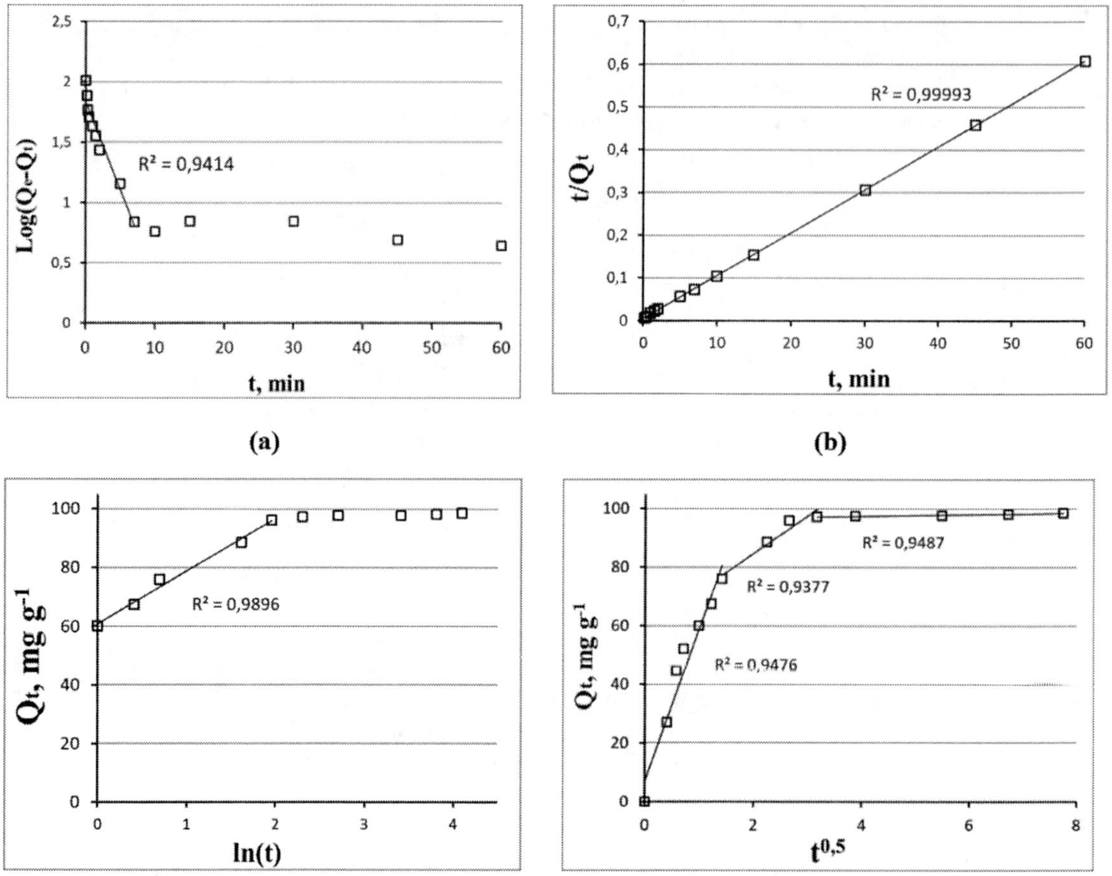

(c) (d)

Figure 3. Cu^{2+} sorption data fitted to the pseudo-first-order (a), pseudo-second-order (b), Elovich (c) and intraparticle diffusion (d) models.

IOP Conf. Series: Materials Science and Engineering **693** (2019) 012029 doi:10.1088/1757-899X/693/1/012029

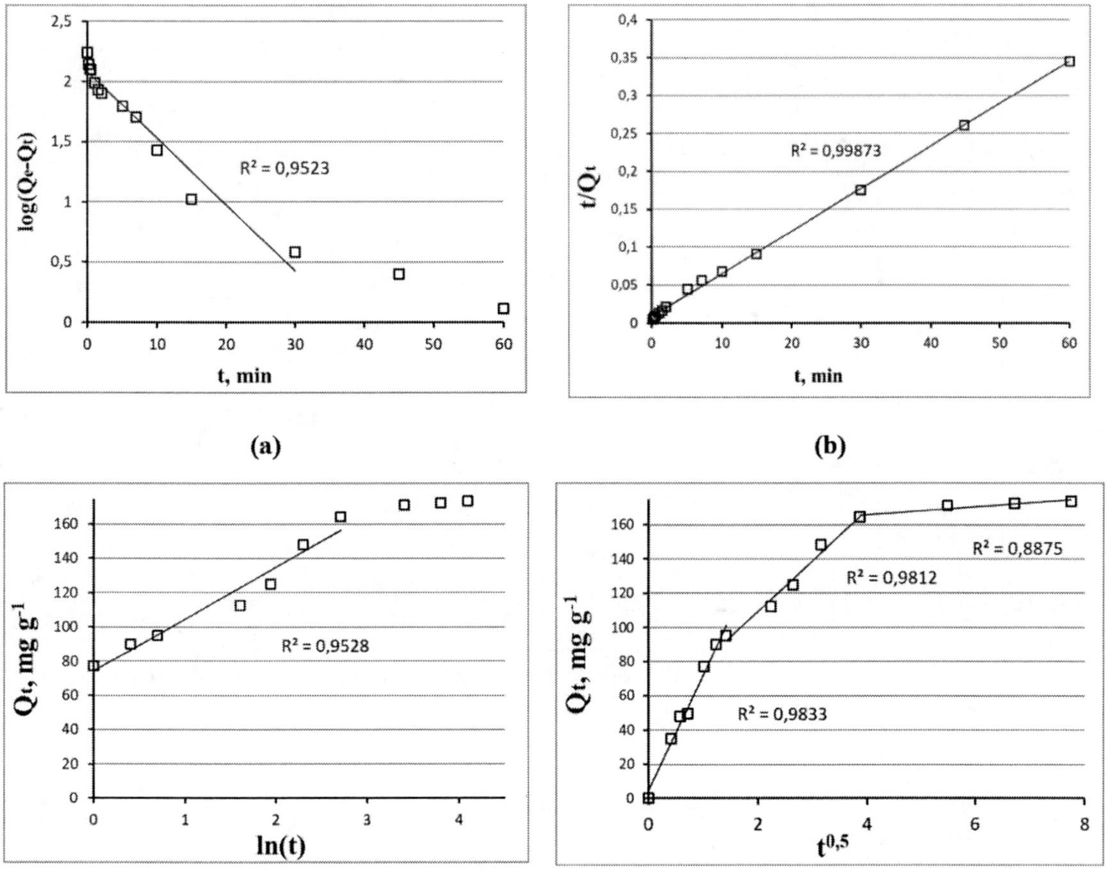

(a) (b)

(c) (d)

Figure 4. Pb^{2+} sorption data fitted to the pseudo-first-order (a), pseudo-second-order (b), Elovich (c) and intraparticle diffusion (d) models.

IOP Conf. Series: Materials Science and Engineering **693** (2019) 012029 doi:10.1088/1757-899X/693/1/012029

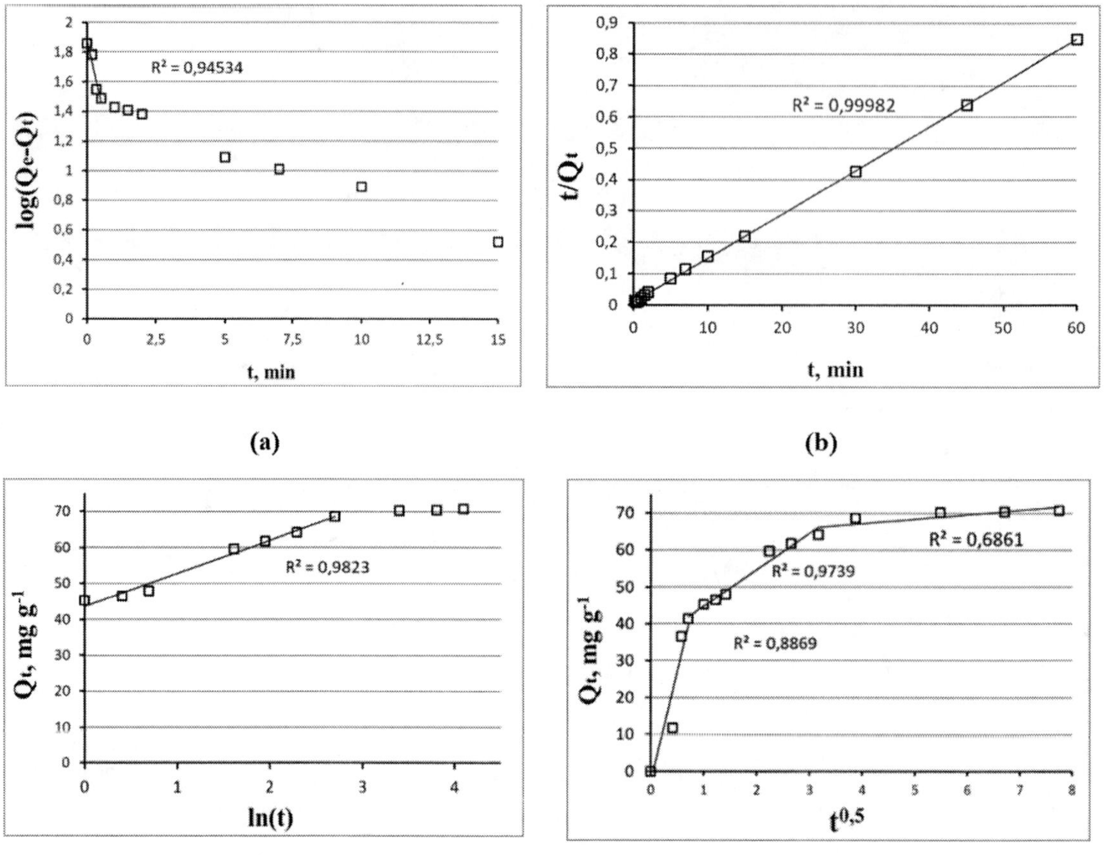

Figure 5. Zn^{2+} sorption data fitted to the pseudo-first-order (a), pseudo-second-order (b), Elovich (c) and intraparticle diffusion (d) models.

The high correlation of the experimental data to the pseudo-first-order model, judged by the coefficients of determination (R^2) being equal to 0.941 (Cu^{2+}), 0.945 (Zn^{2+}) and 0.952 (Pb^{2+}), mainly in the initial period of time (up to 7 min – for Cu^{2+}, up to 30 min – for Pb^{2+}, and up to 2 min – for Zn^{2+}), evidences the occurrence of chemical interactions between the active functional groups of the sorbent and the heavy metal ions.

The pseudo-second-order model satisfactorily describes the removal process for all the ions (R^2= 0.999 (Cu^{2+} and Zn^{2+}) and 0.998 (Pb^{2+})). The correspondence of the experimental data to this model indicates that the sorbate–sorbent chemical reaction (*i.e.*, pseudo-second-order reaction) takes place. It should be noted that the calculated Q_e values coincide with experimental $Q_{e,\exp}$ values for all the heavy metal ions. At the same time, the pseudo-second-order reaction rate constant is on average 10-50 times lower than during the interaction between the sorption sites of the material and the metal ions.

The correspondence of the data to the Elovich model (R^2= 0.989 (Cu^{2+}), 0.983 (Zn^{2+}) and 0.953 (Pb^{2+})) suggests that the sorbent has high-energy heterogeneous surfaces [18]. This leads to very rapid sorption of the solute by the sorbent until the system stabilizes.

Table 3. Parameters of the kinetic models obtained for the metals sorption on the PHQ/G nanocomposite.

Sorbate	Pseudo-first-order			Pseudo-second-order			Elovich		
	Q_e, mg g^{-1}	k_1, min^{-1}	R^2	Q_e, mg g^{-1}	k_2, g mg^{-1} min^{-1}	R^2	α, min^{-1} mg g^{-1}	β, g mg^{-1}	R^2
Cu^{2+}	68.6	0.336	0.941	99.4	0.0210	0.999	1809.1	0.083	0.989
Zn^{2+}	74.13	0.808	0.945	71.6	0.0183	0.999	1005.4	0.109	0.983
Pb^{2+}	119.6	0.165	0.952	178.6	0.0034	0.998	337.3	0.033	0.953

Table 4. Parameters of the diffusion models obtained for the metals sorption on the PHQ/G nanocomposite.

Sorbate	Intraparticle diffusion									Boyd		
	Line 1			Line 2			Line 3			B	D	R^2
	k_{id1}, min$^{-1/2}$	C_1, mg g^{-1}	R^2	k_{id1}, min$^{-1/2}$	C_2, mg g^{-1}	R^2	k_{id3}, min$^{-1/2}$	C_3, mg g^{-1}	R^2		·10^{-13} m^2 s^{-1}	
Cu^{2+}	51.8	7.348	0.948	12.9	59.05	0.938	0.27	96.41	0.949	0.00865	4.24	0.980
Zn^{2+}	60.7	3.267	0.887	9.86	34.95	0.974	1.17	62.5	0.686	0.00763	3.74	0.880
Pb^{2+}	68.2	4.448	0.983	29.5	50.36	0.981	2.35	156.5	0.886	0.00264	1.29	0.980

The metal diffusion penetration effect on the overall process rate was elucidated by implementing the intraparticle diffusion model (R^2=0.938-0.949 (Cu^{2+}), 0.686-0.974 (Zn^{2+}) and 0.886-0.983 (Pb^{2+})). It is clear that the graphs are non-linear over the entire time interval, which means that the sorption process is limited by more than one stage. The intraparticle diffusion sorption model has three intersecting lines for each ion. The first and the second lines are associated with surface sorption and intraparticle diffusion, whereas the third line represents the final stage of equilibrium, at which the intraparticle diffusion begins to slow down due to the extremely low concentration of the sorbate remaining in the solution. However, for all the ions, the graphs leave the origin, therefore, the internal diffusion limits the total time of the sorption process.

To confirm this assumption, the Boyd model (R^2=0.980 (Cu^{2+} and Pb^{2+}) and 0.880 (Zn^{2+})) was implemented in the initial period of the process. Due to the mathematical processing of the experimental data, the effective diffusion coefficients characterizing the rate of sorbate ion penetration inside the sorbent grains were calculated. They were found to increase in the following sequence: Pb^{2+} (1.29 10^{-13} m^2 s^{-1}) < Zn^{2+} (3.74 10^{-13} m^2 s^{-1}) < Cu^{2+} (4.24 10^{-13} m^2 s^{-1}).

Thus, the values obtained for the adsorption capacity of the PHQ/G nanocomposite regarding the Cu^{2+}, Pb^{2+} and Zn^{2+} ions are significantly higher in comparison with the ones obtained for the other materials listed in table 1 (CNTs, GO, modified activated carbon). This fact makes it possible to judge on the feasibility of using the PHQ/G nanocomposite in the processes of liquid-phase sorption of heavy metal ions.

4. Conclusion

As a result of the experimental studies on the sorption of Cu^{2+}, Zn^{2+} and Pb^{2+} ions onto the PHQ/G nanocomposite material, it was found that the equilibrium is reached within 60 min. The following maximum sorption capacity of the sorbent was achieved: ~174, ~97 and ~64 mg g^{-1} regarding Pb^{2+}, Cu^{2+} and Zn^{2+}, respectively.

Using the pseudo-first and second-order, Elovich and intra-diffusion models, it was found that the sorption is limited by the internal diffusion of the metal ions into the porous space of the PHQ/G nanocomposite, taking into account the contribution of the chemical interaction between the metal ions and the functional groups of the sorbent during the initial period of the process (according to the pseudo-second order model).

Acknowledgment
The research was supported by the Ministry of Science and Higher Education of the Russian Federation under Project No. 16.1384.2017/PCh.

References

[1] Hua M, Zhang S, Pan B, Zhang W, Lv L and Zhang Q 2012 Heavy metal removal from water/wastewater by nanosized metal oxides, a review *J. Hazard. Mater.* **211-212** 317–331
[2] Ashour R W, Abdelhamid H N, Abdel-Magied A F, Abdel-Khalek A A, Ali M M, Uheida A, Muhammed M, Zhou X and Dutta J 2017 Rare earth ions adsorption onto graphene oxide nanosheets *Solvent Extr. Ion Exch.* **35** 91–103
[3] Harvey P J, Handley H K and Taylor M P 2015 Identification of the sources of metal (lead) contamination in drinking waters in north-eastern Tasmania using lead isotopic compositions *Environ. Sci. Pollut. Res.* **22** 12276–12288
[4] Fu F and Wang Q 2011d Removal of heavy metal ions from wastewaters, a review *J. Environ. Manag.* **92** 407–418
[5] Gu D and Fein J B 2015 Adsorption of metals onto graphene oxide, surface complexation modeling and linear free energy relationships *Colloids Surf. A: Physicochem Eng. Aspects* **481** 319–327
[6] Dreyer D R, Park S, Bielawski C W and Ruoff R S 2010b The chemistry of graphene oxide *Chem. Soc. Rev.* **39** 228–240
[7] Kumar R, Chawla J and Kaur I 2015 Removal of cadmium ion from wastewater by carbon-based nanosorbents, a review *J. Water Health* **13** 19–33
[8] Duru I, Ege D and Kamali A R 2016 Graphene oxides for removal of heavy and precious metals from wastewater. *J. Mater. Sci.* **51** 6097–6116
[9] Wang H J, Zhou A L, Peng F, Yu H and Chen L F 2007 Adsorption characteristic of acidified carbon nanotubes for heavy metal Pb(II) in aqueous solution *Mater. Sci. Engin. A.* **466** 201-206
[10] Lingamdinne L P, Kim I S, Ha J H, Chang Y Y, Koduru J R and Yang J K 2017 Enhanced Adsorption Removal of Pb(II) and Cr(III) by Using Nickel Ferrite-Reduced Graphene Oxide Nanocomposite *Metals* **7** 225-237
[11] Lee M E, Park J H, Chung J W, Lee C Y and Kang S 2015 Removal of Pb and Cu ions from aqueous solution by Mn_3O_4-coated activated carbon *J. Indus. Engin. Chem.* **21** 470-475
[12] Long Z, Zhan Y, Li F, Wan X, He Y, Hou C and Hu H 2017 Hydrothermal synthesis of graphene oxide/multiwalled carbon nanotube/Fe_3O_4 ternary nanocomposite for removal of Cu (II) and methylene blue *J. Nanoparticle Res.* **19(9)** 318-329
[13] Babkin A V, Burakova I V, Burakov A E, Neskoromnaya E A, Kucherova A E and Kurnosov D A 2018 Kinetics of the Cu(II) sorption from aqueous solutions by carbon nanomaterials *Nanosystems: physics, chemistry, mathematics* **9(1)** 117-119
[14] Lu C and Chiu H 2006 Adsorption of zinc (II) from water with purified carbon nanotubes *Chem. Engin. Sci.* **61** 1138-1145
[15] Cho H H, Wepasnick K, Smith B A, Bangash F K, Fairbrother D H and Ball W P 2010 Sorption of aqueous Zn[II] and Cd[II] by multiwall carbon nanotubes: the relative roles of oxygen-containing functional groups and graphenic carbon *Langmuir* **26** 967–981
[16] Galunin E V, Burakova I V, Neskoromnaya E A, Babkin A N, Melezhik A V, Burakov A E and Tkachev A G 2018 Adsorption of heavy metals from aqueous media on graphene-based nanomaterials *AIP Conference Proceedings* **2041** 020007
[17] Babkin A V, Melezhik A V, Kurnosov D A, Mkrtchyan E S, Burakova I V, Burakov A E and Galunin E V 2018 Kinetics of the Adsorption of Synthetic Dyes on a Polyhydroquinone/Graphene Carbon Nanocomposite *J. Phys.: Conf. Ser.* **1124** 081030
[18] Gautam R K and Chattopadhyaya M Ch 2016 *Nanomaterials for Wastewater Remediation* (Oxford: Elsevier) p 353

Investigation of the properties of two-dimensional molybdenum disulfide films synthesized by the CVD method

S Smagulova, A Semenova, E Zakharkina and P Vinokurov

Graphene Nanotechnology Laboratory, *North-Eastern Federal University*, 58 Belinsky Street, 677027 Yakutsk Russian Federation

E-mail: smagulova@mail.ru

Abstract. Molybdenum disulfide films were synthesized through chemical vapor deposition. Optimal synthesis parameters (temperature, time, gas pressure, gas flow rates, and geometrical arrangement of substrates and crucibles with precursors), allowing for the growth of single-layer domains of molybdenum disulfide with maximum domain sizes (250 μm), were found. The technology for processing substrates, including mechanical, chemical and plasma-chemical processing, was developed. The results of studies on the structural, optical, electrical properties of the synthesized and transferred films using optical, scanning electron and atomic force microscopes, elemental analysis, measurement of Raman scattering spectra, luminescent and electrical characteristics are presented.

1. Introduction

After the discovery of graphene and its unique properties [1], the investigation of new two-dimensional materials became one of the most popular research topics. Among such materials, the class of transition metal dichalcogenides, in particular, molybdenum disulfide (MoS_2), has gained much research attention. As the other transition metal dichalcogenide family, MoS_2 is widely used in many areas, such as a hydrodesulfurization catalyst, photocatalyst, photocell, and as a dry lubricant, due to its distinctive electronic, optical, and catalytic properties. Multi-layer MoS_2 is a semiconductor with an indirect 1.2 eV band gap [2]. However, most interesting properties appear in single-layer molybdenum disulfide, which is a direct-gap semiconductor with a 1.8 eV band gap [3]. Investigating two-dimensional MoS_2 opens new opportunities in the development of new nano- and opto-electronic devices.

A number of methods for obtaining two-dimensional materials have been developed: mechanical splitting, electrochemical intercalation with lithium ions, magnetron sputtering, pulsed laser sputtering, *etc* [4]. The existing techniques for producing MoS_2 do not allow controlling the number of atomic layers and lateral sizes of domains. Nowadays, the most promising and relatively inexpensive method for synthesizing two-dimensional MoS_2 films of high quality is chemical vapor deposition (CVD) [5-7]. This method makes possible to obtain homogeneous two-dimensional MoS_2 films with lateral domain sizes of up to 100 μm. The advantage of this technique is the high quality of samples grown and the possibility of material synthesis in large quantities. Raman spectroscopy is one of the main methods for characterizing two-dimensional structures that allows to obtain information on the number of atomic layers and the quality of the MoS_2 structure.

MoS_2 films synthesized by the CVD method can be used as active components for nanoelectronic devices or building blocks for complex layered heterostructures. Vertical heterostructures are currently considered as the main direction of thin-film electronics development. For such heterostructures, a special name has been introduced – van der Waals heterostructures, since they are expected to be formed by sequential growth of layers, and in a single structure, the layers will be held by van der Waals forces. Currently, several types of these heterostructures, in particular, the ones based on transition metal dichalcogenides (MoS_2, WS_2) and graphene, have been developed and studied. An important achievement of the van der Waals heterostructures based on two-dimensional materials is the demonstration of the possibilities of new physical approaches and new principles for constructing electronic structures. The formation of such structures opens up many opportunities for developing various kinds of electronic devices, ranging from transistor structures and optoelectronic elements, to various gas sensors and biological sensors.

Considering the aforementioned, in this work, the effect of temperature, gas feed rate, synthesis time and location of substrates on the kinetics of formation of two-dimensional MoS_2 was studied.

2. Experimental part

Silicon with a 285 nm thick SiO_2 oxide film was used as substrate. Before the process, the substrate surface was pretreated in an ultrasonic bath with acetone and isopropyl alcohol for 5 min to remove contaminants. To remove organic bonds from the surface, the substrate was treated in a sulfur-peroxide solution in a 1:3 ratio, followed by washing in deionized water. Then, the substrate was processed in an oxygen plasma for 5 min at a power of 150 W. MoS_2 was synthesized by the CVD method in a RS 80/750/ three-zone tube furnace (Nabertherm, Lilienthal, Germany), shown schematically in figure 1.

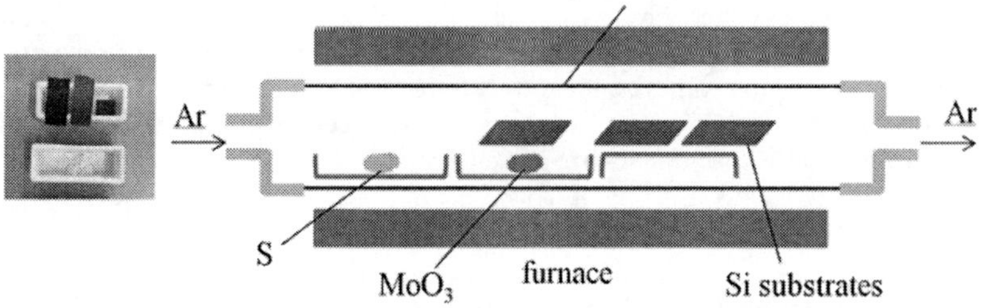

Figure 1. Schematic illustration of a setup for CVD growth of MoS_2.

MoO_3 and sulfur powders (Sigma Aldrich, St_Louis, USA) in a weight ratio of 1:13 were used as initial precursors for the synthesis. The substrates, on which MoS_2 was grown, were located above the crucible with the MoO_3 powder. To control the synthesis, the following parameters were selected: temperature, time, pressure, argon feed rate, and precursor concentration. The whole process can be divided into five stages [8], as shown in figure 2:

1. The reaction chamber is pumped out to a pressure of 10^{-5} bar for 1 h.
2. When the temperature reaches 300 °C, inert gas (Ar) is fed into the chamber at a flow rate of 100 cm^3/min for 1 h to a pressure of 1 atm.
3. The chamber is heated to a synthesis temperature of 700 °C.
4. The growth of MoS_2 on SiO_2 occurs in the Ar flow at a rate of 10 cm^3/min for 30 min at 700 °C.
5. After synthesis, the chamber wis naturally cooled to room temperature in the Ar stream at a flow rate of 300 cm^3/min.

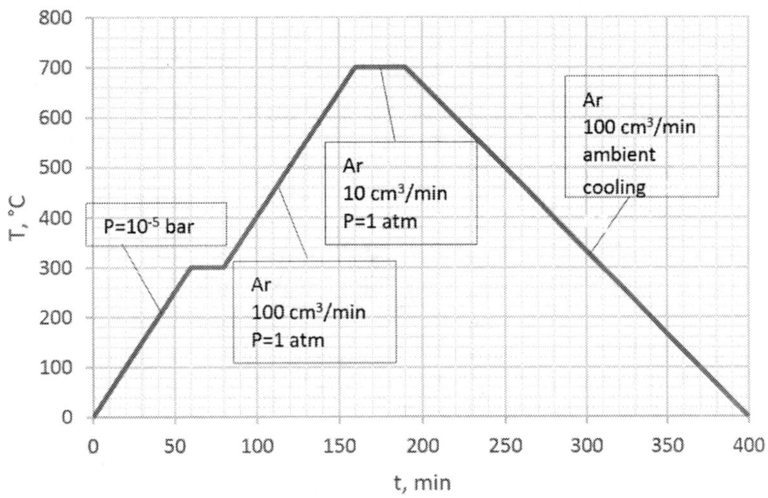

Figure 2. Temperature conditions during synthesis

The obtained films were studied by Raman spectroscopy using an INTEGRA SPECTRA system (NT-MDT, Zelenograd, Russia) equipped with a green laser at λ = 532 nm. The morphology of the surface of the MoS_2 films grown by the CVD over the SiO_2 substrate was studied by an AFM/STM "SolverNext" scanning probe microscope (NT-MDT). The morphology and elemental composition of the films were determined by scanning electron microscopy using a JSM-7800F instrument (JEOL, Tokyo Japan). Optical images were taken by an Eclipse LV100 optical microscope (Nikon, Tokyo Japan).

3. Results and discussion

It is known that in the Raman spectra of MoS_2, two most intense peaks related to the vibration modes in the plane and outside the plane in the region of 400 cm^{-1} can be observed [9]. The vibration modes of atoms of both molybdenum and sulfur are limited in the horizontal plane; therefore, the vibrations of atoms will be located along the vertical plane. With an increase in the number of layers, a decrease in the frequency of vertical vibrations mode will be observed, because with an increase in the number of layers, the van der Waals forces will also increase, which will suppress atomic vibrations along the vertical plane. In [9], it is shown that in the domains of the MOS_2 grown by the CVD method on the SiO_2, a decrease in the Raman peak can observed with a decrease in the film thickness (the number of atomic layers). The frequency difference between these modes for the single-layered MoS_2 was found to be $\Delta v = 19$ cm^{-1}, for the double-layered MoS_2 - $\Delta v = 21$ cm^{-1}, and for the multi-layered MOS_2 - $\Delta v = 25$ cm^{-1}. Figure 3 presents the Raman spectra of the studied samples.

The analysis of the Raman spectra showed that the samples represent films of the single-layered MoS_2. In addition, measurements of the Raman spectra of the double- and multi-layered MoS_2 were carried out (figure 3).

The optical measurements showed the growth of the MoS_2 domains on the SiO_2 surface in the form of triangles (figure 4) with lateral sizes of up to 80 μm. The domain size varied depending on the location of the substrate relative to the molybdenum-containing precursor. In the samples that were located 5-7 cm from the precursor MoO_3, the domain size was 10-20 μm. With decreasing the distance between the samples and the precursor, the domain size increased up to 40 μm. It was experimentally established that the maximum domain size is obtained directly above the crucible with the MoO_3. It is possible that larger flakes were grown near the Mo precursor due the higher $_{MoO3}$ concentration and different gas flow through the substrate surfaces.

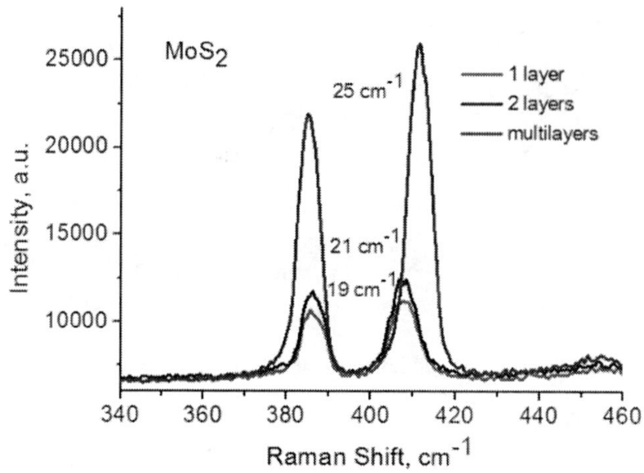

Figure 3. Raman spectra of the synthesized MoS_2.

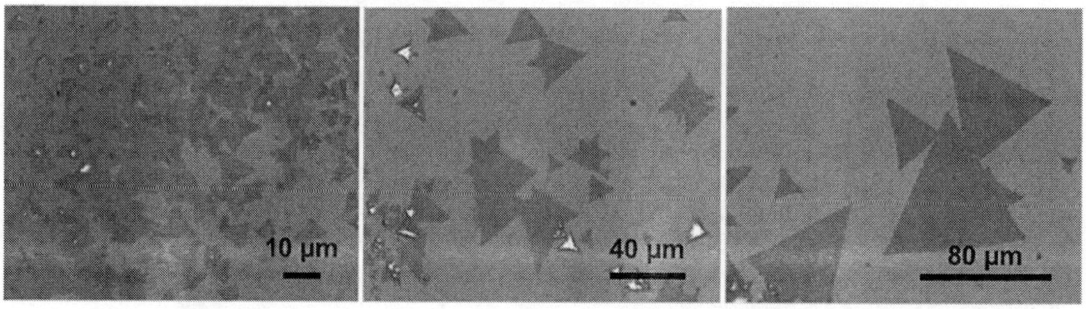

Figure 4. Optical images of the domains grown at different distances from the MoO_3 precursor: more than 5 cm (a), more than 2 cm (b), and directly above the precursor (c).

The AFM measurements showed that the thickness of a single MOS_2 layer is 0.9 nm (figure 5). Besides, at the center and the edges of some domains, defects in the form of overgrown fragments of the same material were found.

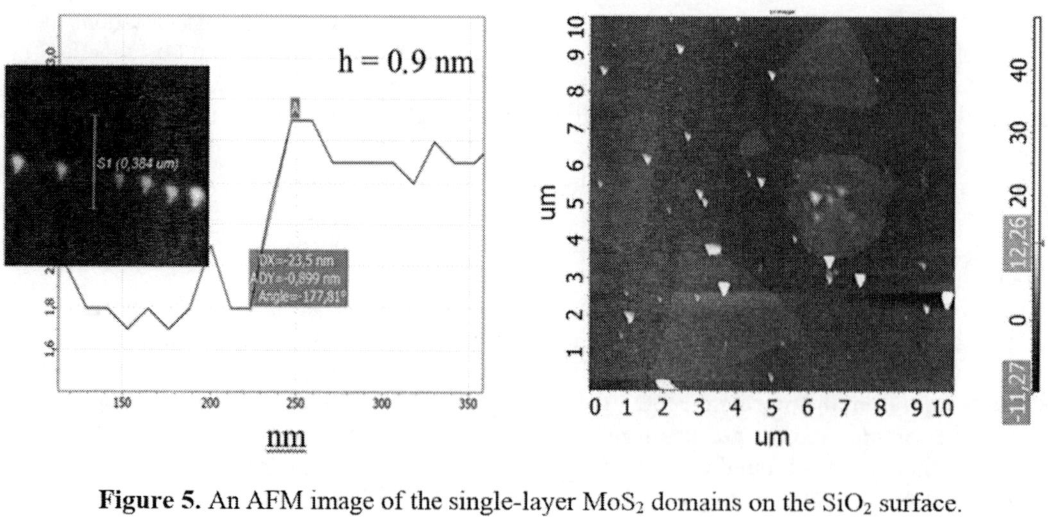

Figure 5. An AFM image of the single-layer MoS_2 domains on the SiO_2 surface.

The data on the energy dispersive X-ray spectral microanalysis (figure 6), presented in table 1, made it possible to determine the elemental composition of the samples - molybdenum (0.19 at.%), sulfur (0.48 at.%), silicon (28.41 at.%), and oxygen (70.91 at.%). It should be noted that the spectral peaks of molybdenum and sulfur are located at the values of about 2.3 keV and have only minor deviations with respect to each other, which made it difficult to identify these elements (figure 6). Nevertheless, it was found that the atomic sulfur content is twice the atomic molybdenum content.

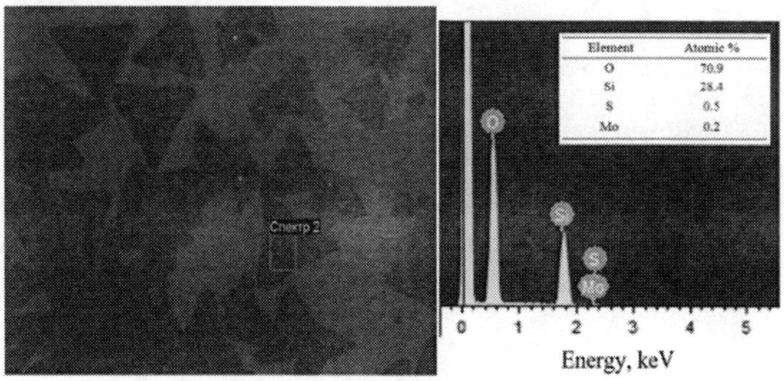

Figure 6. The line-scan EDX spectrum of a selected region from the STEM image (marked by a pink color rectangle).

4. Conclusion

A method for the synthesis of two-dimensional MoS_2 films by the CVD was developed. The AFM and optical measurements showed the growth of the MoS_2 domains on the SiO_2 surface in the form of triangles with lateral sizes of up to 80 μm. The domain size varied depending on the location of the substrates relative to the molybdenum-containing precursor. It was experimentally established that the maximum domain size is achieved directly above the crucible with the MoO_3. The analysis of the Raman spectra demonstrated that the grown films mostly consist of MoS_2 monolayers. Besides, the Raman spectra of the double- and multi-layered MoS_2 were measured. A decrease in the peaks of the Raman spectra peaks was observed with a decrease in the thickness of the MoS_2 film grown through the CVD method over the SiO_2 surface. The frequency difference between these modes was found to be $\Delta v = 19$, 21 and 25 cm^{-1} for the single-, tdouble- and multi-layered MoS_2, respectively

5. Acknowledgments

The work was supported by the Russian Foundation for Basic Research, Grant No. 18-42-140005 p_a, and by the Ministry of Science and Higher Education of Russia, State Assignment No. FSRG-2017-0017.

References

[1] Novoselov K, Geim A, Morozov S, Jiang D, Zhang Y, Dubonos S, Grigorieva I and Firsov A 2004 Electric field effect in atomically thin carbon films. *Science* **306** 666 https://doi.org/10.1126/science.1102896

[2] Gordon R, Yang D, Crozier E, Jiang D and Frind R 2002 Structures of exfoliated single layers of WS_2, MoS_2, and $MoSe_2$ in aqueous suspension. *Phys. Rev. B* **65** 125407 https://journals.aps.org/prb/pdf/10.1103/PhysRevB.65.125407

[3] Zhao W, Ribeiro R, Toh M, Carvalho A, Kloc C, Castro Neto A and Eda G 2013 Origin of indirect opticfl transitions in few-layer MoS_2, WS_2 and WSe_2. *Nano Lett.* **13** 5627 https://arxiv.org/ftp/arxiv/papers/1309/1309.0923.pdf

[4] Wong S, Liu H and Chi D 2016 Recent progress in chemical vapor deposition growth of two dimensional transition metal dichalcogenides. *Prog. Cryst. Growth. Ch.* **62** 9

https://doi.org/10.1016/j.pcrysgrow.2016.06.002
[5] Zhan Y, Liu Z, Najmaei S, Ajayan P M and Lou J 2012 Large-area vapor-phase growth and characterization of MoS2 atomic layers on a SiO2 substrate. *Small* **8** 7 966 https://doi.org/10.1002/smll.201102654
[6] Vollath D and Szabo D 1998 Synthesis of nanocrystalline MoS_2 and WS_2 in a microwave plasma *Mater. Lett.* **35** 236 https://doi.org/10.1016/S0167- 577X(97)00247-4
[7] Gong Y, Lin J, Wang X, Shi G, Lei S, Lin Z, Zou X, Ye G, Vajtai R, Yakobson B, Terrones H, Terrones M, Tay B, Lou J, Pantelides S, Liu Z, Zhou W and Ajayan P 2014 Vertical and in-plane heterostructures from WS_2/MoS_2 monolayers. *Nat. Mater.* **13** 1135 https://doi.org/10.1038/nmat4091
[8] Zakharkina E I, Semenova A A, Boiakinov E F, Vinokurov P V and Smagulova S A 2018 CVD growth of two-dimensional MoS_2 depending on the location of the SiO_2 substrates. *AIP Conf. Proc.* **2041** 1 040001 https://doi.org/10.1063/1.5079366
[9] Li H, Zhang Q, Chong Ray Yap C, Kang Tay B, Hang Tong Edwin T, Olivier A and Baillargeat D 2012 From bulk to monolayer MoS_2: Evolution of Raman scattering. *Adv. Funct. Mater.***22** 1385 https://doi.org/10.1002/adfm.201102111

IOP Conf. Series: Materials Science and Engineering **693** (2019) 012031 doi:10.1088/1757-899X/693/1/012031

Investigation of the properties of carbon quantum dots synthesized by the hydrothermal method

S Smagulova, M Egorova and A Tomskaya

Graphene Nanotechnology Laboratory, *North-Eastern Federal University,* 58 Belinsky Street, 677027 Yakutsk Russian Federation

E-mail: smagulova@mail.ru

Abstract. Carbon quantum dots have a number of advantages and can be widely used as a material for photodetectors, chemical element detectors, biosensors, and for various optoelectronic applications. The optimal conditions for the hydrothermal synthesis of carbon quantum dots from glucose are considered. Carbon quantum dots, having an amorphous-crystalline structure with functional surface groups, were obtained. Luminescence spectra of these dots showed the presence of bright luminescence in the green region. The dependences of the luminescence spectra as a function of the excitation length were measured. Experiments on the functionalization of the carbon quantum dots with chromophores were carried out in order to study their effect on the luminescence intensity.

1. Introduction

In the recent years, quantum dots have gained increasing attention. They are of great interest in the field of bioimaging and biomedicine due to their high quantum yield and unique optical properties. However, their wider use is limited due to some of their drawbacks such as high toxicity, low biocompatibility, high cost, and low chemical inertness. Moreover, they show poor solubility in aqueous solutions [1]. Considering this, carbon-based quantum dots look increasingly advantageous. Carbon quantum dots (CQDs) belong a class of zero-dimensional carbon nanomaterials that contain sp^2 carbon structures and functional groups. They are superior to the conventional semiconductor quantum dots in terms of low toxicity, high biocompatibility, good chemical inertness and solubility. [2,3] Can also work as effective drug carriers due to their small size and biocompatibility [4, 5]. In addition, their excellent photo- and electro-catalytic characteristics also allow their use in the field of energy saving and environmental protection [6, 7].

CQDs exhibit unique photoluminescent properties that depend on several factors – first, on the particle size and the uniformity of the particle size distribution, second, on the surface passivating compounds which affect not only spectra of luminescence, but also its intensity, and third, on the functionalization of the surface with organic molecules, which can be used to control luminescence spectra. The possibility of controlling the photoluminescence of CQDs (spectra and intensity of luminescence, quantum yield, band gap, and the other parameters) with simple changes in the synthesis conditions makes them a promising material in various areas [8].

The aim of this work was to synthesize CQDs by the hydrothermal method and to study the effect of chromophore functionalization on their luminescence properties.

2. Experimental

For CQD synthesis, 1 g of glucose ($C_6H_{12}O_6$, "Ruskhim", Russia) was dissolved in 15 mL of deionized water. 6 ml of an aqueous solution of ammonia (NH_3, 25.6%, "Vostokreaktiv", Russia) were added and mixed. The solution was transferred to a polytetrafluoroethylene autoclave ("TOPTION Laboratory Store", China). The autoclave was placed into a universal drying oven (ShSP-0.25-60-C, Russia) for 10 min, 30 min, 1 h, 4 h, 8 h, 10 h, and 18.5 h at a temperature of 180 °C. The resulting brown suspension was filtered through a 100-nm track membrane, then dialyzed in a dialysis bag ("Aksioma-Bio", Russia) with a pore size of 3.5 kDa for 12 h. For further characterization, a solution from the dialysis bag was used. CQD particles were size-separated by electrophoresis. The electrophoresis equipment was self-made from a 10 mL syringe body (vertical positioning of a syringe), with gold contacts. The CQD suspension from a dialysis bag was poured into the syringe body and a constant voltage of 30 V from a voltage source was applied (APS-7151, "AKTAKOM", Russia) for 20 min. During this time, the solution was separated by size and charge: in the lower part of the syringe (positive contact), the dark part of the suspension (larger particles) was collected, whereas in the upper part (negative contact), the lighter part (smaller particles) was remained.

Chromophore is a covalently saturated group causing electron absorption (for instance, C=C, C=O and NO_2). For synthesizing them, 1 g of glucose was dissolved in 10 mL of deionized water and subjected to pyrolysis on a hot plate until evaporation and browning. The resulting black precipitate was dispersed with deionized water. Then, the solution was filtered through a 100-nm track membrane and dialyzed (3.5 kDa) for 12 h. This solution contained glucose chromophores. Similarly, chromophores of citric acid were obtained.

For functionalization of the CQD with chromophores, 3 mL of the chromophores from the glucose or citric acid were added to 7 mL of the CQD solution. As a result, two CQD mixtures with chromophores were obtained. Then, the mixtures were transferred to the polytetrafluoroethylene autoclave and heated for 1 h in a drying oven at 90 °C.

The luminescent properties were determined using a Perkin Elmer LS 50B spectrometer (USA). The absorption properties were measured using a UNICO UV-VIS 2804 spectrophotometer (Russia). Structural studies were performed on a Varian 7000 FT-IR Fourier transform infrared (IR) spectrometer (USA).

3. Results and discussion

The resulting carbon dot solutions had brown color under room lighting, exhibited green luminescence upon excitation with a laser (λ = 420 nm) and had greenish blue glow under UV light (λ = 325 nm) (figure 1).

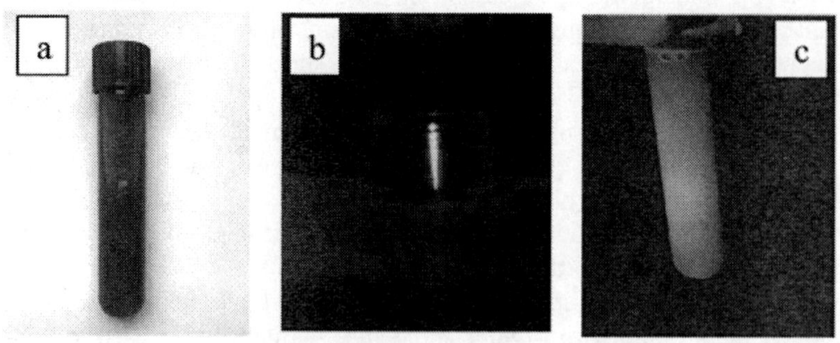

Figure 1. The CQD solution from the glucose exposed to daylight **(a)**, under the laser irradiation (λ = 420 nm) **(b)**, and under UV light (λ = 325 nm) **(c)**.

Figures 2(a) and 2(b) show the photoluminescence spectra of the CQD aqueous suspensions from the glucose after 4 and 8 h of hydrothermal treatment at different excitation wavelengths (300, 350, and 400 nm), whereas figure 2(c) demonstrates the IR spectra of the glucose+ammonia solution at different duration of hydrothermal treatment. From the luminescence spectra, it can be observed that the positions of the maxima of the photoluminescence peaks is shifted from 400 to 473 nm, when the excitation wavelength changes from 300 to 400 nm. The position of the fluorescence emission peak depends on the excitation wavelength. This is due to the inhomogeneous size distribution of the CQDs, chemical composition of the surface, presence of various emission traps, or characteristics of the fluorescence mechanism which is currently not explained [9].

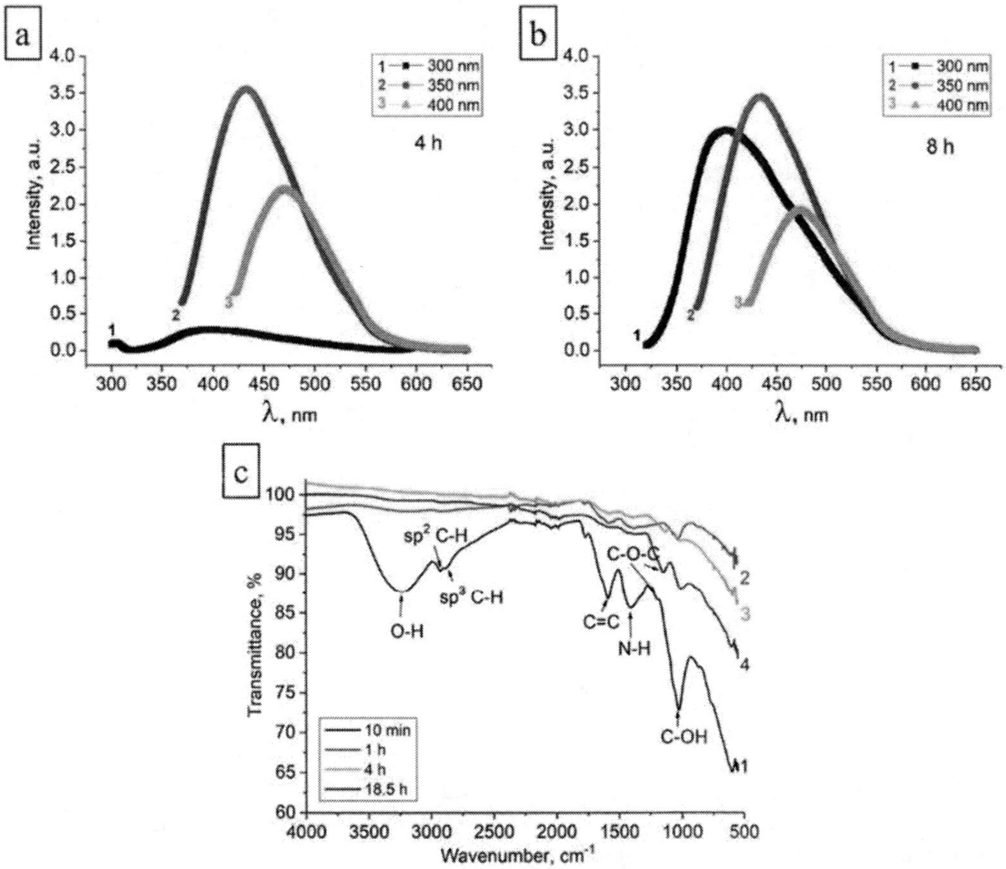

Figure 2. Luminescence spectra of the CQD solutions from the glucose after hydrothermal treatment for: 4 h **(a)** and 8 h **(b)** at different excitation wavelengths. IR spectra of the glucose+ammonia mixture after hydrothermal treatment for: 10 m (1), 18.5 h (2), 4 h (3), and 1 h (4) **(c)**.

The IR spectra show that most of the obtained CQDs consists of both sp^2-carbon structures and oxygen-containing functional groups, including carboxyl, carbonyl, hydroxyl and epoxy groups.

During the hydrothermal treatment, the glucose is dehydrated to form a C=C bond, which is the elementary unit of the obtained CQDs. The absorption peak at 1641 cm^{-1} is related to the C=C stretching, a broad peak at 3392 cm^{-1} is assigned to the O-H groups, the absorption at 1027, 2927 and 1380 cm^{-1} refers to the C-OH and C-H stretching vibrations and the N-H bond, respectively (figure 2(c)). In addition to the C-OH, COC, and C-H vibrations, the C=O stretching vibrations (~1610 cm^{-1}) were also detected, in combination with asymmetric and symmetric stretching vibrations of C-O-C

(~1300 and ~1200 cm^{-1}) in the carboxyl group. This may be due to the oxidation of the C–OH bonds by residual or decomposed oxygen atoms under hydrothermal conditions [10,11].

The IR spectra of the CQD solution from the glucose after 10 min of hydrothermal treatment indicates the presence of wide absorption bands in the areas of 2450-3100 and 3200-3600 cm^{-1} related to the vibrations of the bonds of aliphatic C-H and O-H groups, respectively. A sharp and intense peak with a center in the region of 1100 cm^{-1}corresponds to the C-OH vibrations [12]. The detected functional groups located on the CQD surface act as a passivating layer that facilitates the solubility of the CQDs in water, and improve their photoluminescence properties [13].

The presence of a large number of residual hydroxyl groups increases the hydrophilicity and stability of the CQDs in polar solvents [9,12]. These hydrophilic groups not only provide luminescence, but also contribute to the safety of the product, allowing them to be used in biochemical studies. As the synthesis time increases, the intensity of the absorption bands decreases, which can be explained by the decrease in the number of functional groups on the CQD surface as a result of intramolecular changes in the composition of the functional groups during the hydrothermal processing.

Investigation of UV-VIS spectra showed (figure 3a) that both glucose and the glucose+ammonia mixture have no absorption above 350 nm without hydrothermal treatment. However, after the hydrothermal treatment, significant absorption appears in the wavelength region above 300 nm (figure 3(b)), which is consistent with the presence of visible luminescence. The positions of the absorption maxima shift in different ways with the time of the hydrothermal treatment. The reason for this may be a change in the composition of the surface functional groups during the hydrothermal treatment.

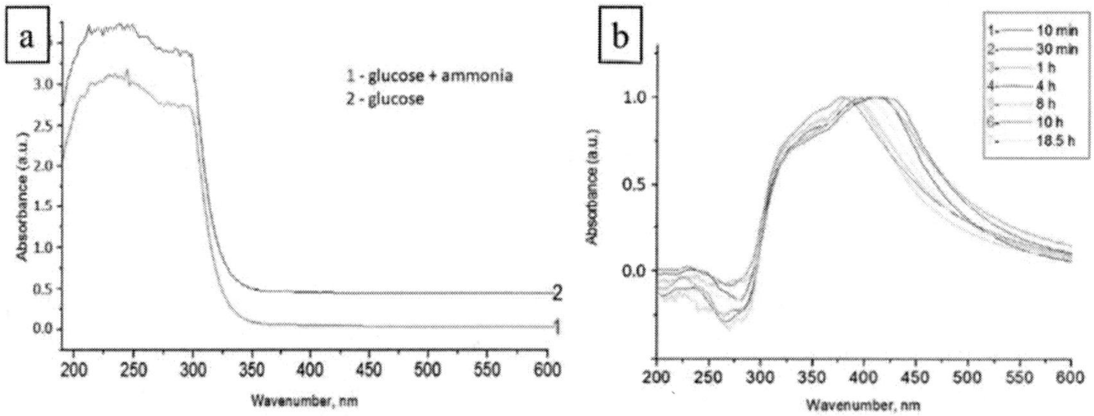

Figure 3. UV-VIS absorption spectra of the initial solutions of the glucose (1) and the glucose+ammonia mixture (2) **(a)**, and the CQD solutions from the glucose obtained after hydrothermal treatment at different duration **(b)**.

For the separation of particles by size, the method of electrophoresis was used. Further, to adjust the luminescence spectra of the CQDs, the chromophores from the glucose or the citric acid were added to the solution obtained after the electrophoresis. After the hydrothermal treatment, the samples were characterized.

Figure 4 shows the luminescence spectra of the CQDs from the glucose after the electrophoresis and chromophore functionalization. After analyzing the luminescence spectra, it was found that the chromophores shift the spectra to the long-wavelength region and increase the luminescence intensity of the CQD solution. The shift in the position of the maxima of the luminescence peaks may be due to the different composition of the chromophore functional groups (figure 5).

IOP Conf. Series: Materials Science and Engineering **693** (2019) 012031 doi:10.1088/1757-899X/693/1/012031

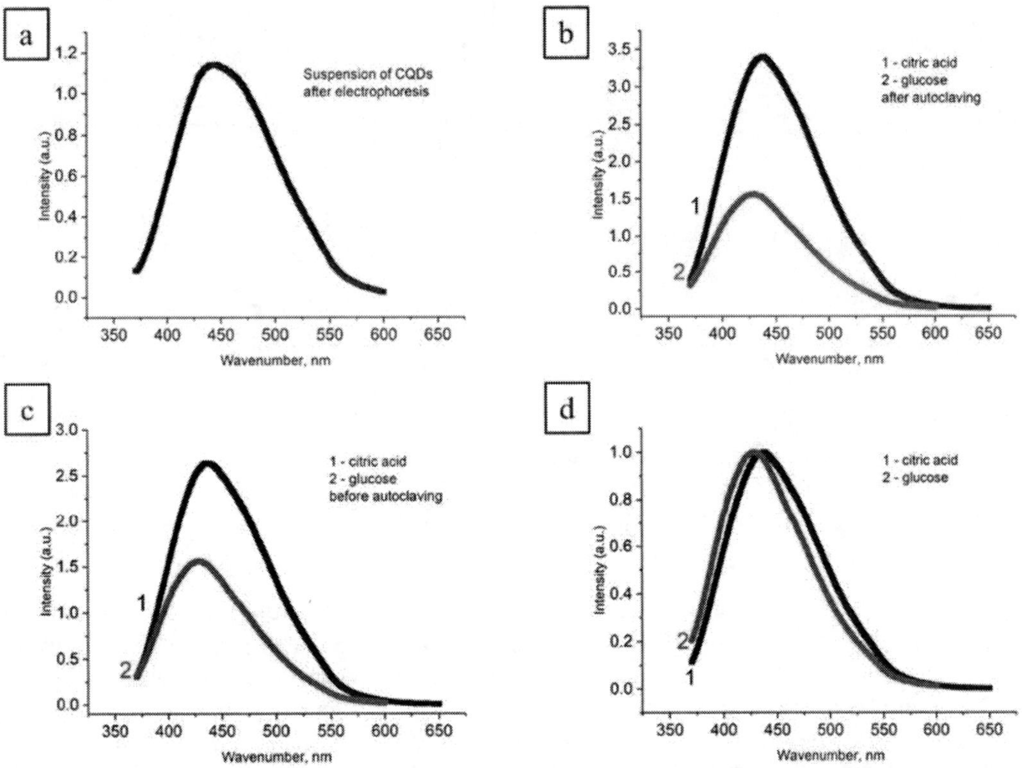

Figure 4. Photoluminescence spectra recorded for: the CQD suspension from glucose (λ = 350 nm): after electrophoresis **(a)**; the CQD solutions after the introduction of the chromophores from the citric acid (1) and the glucose (2) before autoclaving **(b)**; the CQD solutions with the chromophores from the citric acid (1) and the glucose (2) after autoclaving **(c)**; and normalized spectra of the CQD solutions with the chromophores from the citric acid (1) and the glucose (2) **(d).**

Figure 5(a) shows the IR spectra of the chromophores from the citric acid and the glucose and the CQD solutions from the glucose after the functionalization by the chromophores and heating in an autoclave. It can be seen that after the hydrothermal treatment, the intensity of the OH and C–O–C functional groups decreases. Besides, from the spectra, it can be concluded that the composition of the functional groups for the CQDs from the citric acid (figure 5(a) - (1)) and the glucose (figure 5(a) - (2)) is different.

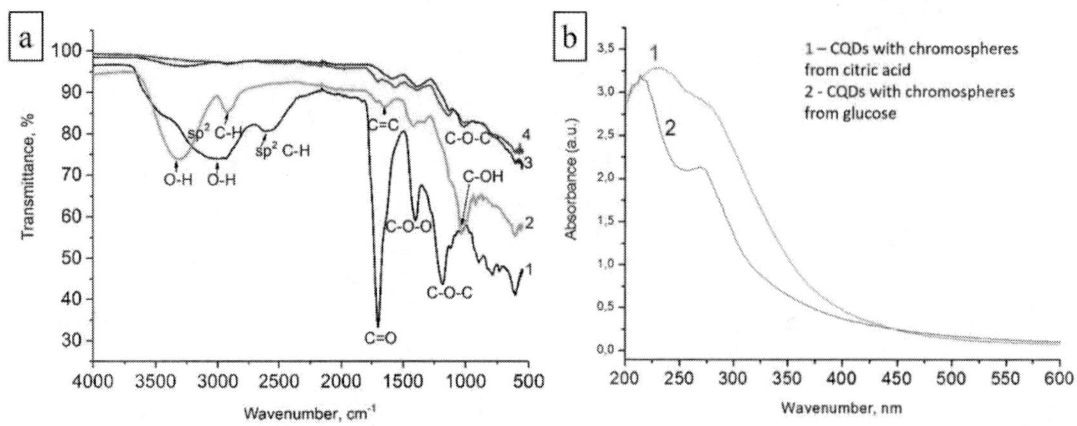

Figure 5. IR spectra of the chromophores from citric acid (1) and from glucose (2), the CQD solutions functionalized with the chromophores from the glucose (3) and the citric acid (4) **(a)**; and UV-VIS spectra of the CQD solutions from the glucose after functionalization with the chromophores from the citric acid (1) and the glucose (2) **(b)**.

Figure 5(b)presents the UV-VIS spectra of the CQD solutions from the glucose after additional functionalization with the chromophores from the citric acid and the glucose. It can be concluded that the functionalization with different chromophores can be used to control the luminescence spectra of the CQDs. Absorption peaks can be observed in the regions of 215 and 268 nm. For the samples with the citric acid chromophores, an absorption peak is observed in the region of 230 nm, and a weak absorption shoulder at 270 nm caused by the $\pi - \pi^*$ and $n - \pi^*$ transitions associated with the C=C and carbonyl/hydroxyl groups [14].

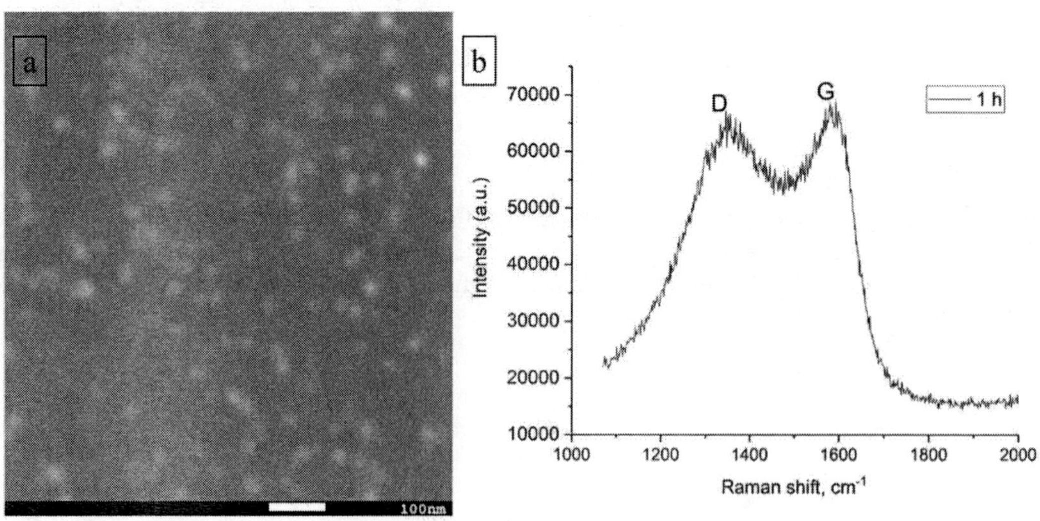

Figure 6. SEM image **(a)** and Raman spectra **(b)** of the CQDs from the glucose.

Figure 6 shows the SEM image of the CQDs from the glucose and its Raman spectra. The average size of the CQDs was found to be 23 nm.

4. Conclusion

Hydrothermal methods were used to synthesize CQDs, which have bright luminescence in the green spectral region. It was shown that the luminescence spectra are shifted to the long-wavelength region, depending on the excitation wavelength. It was suggested that this phenomenon is caused by energy states associated with surface defects of the CQDs. It was found that the functionalization of CQDs with chromophores changes the luminescence spectra and their intensity. The presence of the chromophores on the CQD surface gives a bright, stable luminescence and provide good solubility and stability in water. Luminescent CQDs synthesized from glucose are environmentally friendly, biocompatible and have great prospects for applications in the field of biology and medicine.

Acknowledgments

The work was supported by the Russian Foundation for Basic Research, grant No.18-02-00449, and by the Ministry of Science and Higher Education of Russia, State Assignment No.FSRG-2017-0017.

References

[1] MerianJ, Gravier J, Navarro D and Texier I 2012 Fluorescent nanoprobes dedicated to in vivo imaging: from preclinical validations to clinical translation *Molecules* **17** 5564 https://doi.org/10.3390/molecules17055564

[2] Yuan F, Li S, Fan Z, Meng X, Fan L and Yang S 2016 Shining carbon dots: Synthesis and biomedical and optoelectronic applications *Nano Today* **11** 565 https://doi.org/10.1016/j.nantod.2016.08.006

[3] Qu S, Zhou D, Li D Ji W, Jing P, Han D, Liu L, Zeng H and Shen D 2016 Toward efficient orange emissive carbon nanodots through conjugated sp^2-domain controlling and surface charges engineering *Adv. Mater* **28** 3516 https://doi.org/10.1002/adma.201504891

[4] HassanM, GomesV G, DehghaniA and Ardekani S M 2018 Engineering carbon quantum dots for photomediated theranostics *Nano Res.* **11** 1 https://doi.org/10.1007/s12274-017-1616-1

[5] Molaei M J 2019 Carbon quantum dots and their biomedical and therapeutic applications: A review *RSC Adv.* **9** 6460 https://doi.org/10.1039/C8RA08088G

[6] Yuan T, Meng T, He P, Shi Y X, Li Y, Li X, Fan L and Yang S 2019 Carbon quantum dots: an emerging material for optoelectronic applications *J. Mater. Chem.* C **7** 6820 https://doi.org/10.1039/c9tc01730e

[7] Dimos K 2016 Carbon quantum dots: surface passivation and functionalization *Curr. Org. Chem.* **20** 682 https://doi.org/10.2174/1385272819666150730220948

[8] Chen J, Liu J, Li J, Xu L andQiao Y 2017 One-pot synthesis of nitrogen and sulfur co-doped carbon dots and its application for sensor and multicolor cellular imaging *J. Colloid Interface Sci.* **485** 167https://doi.org/10.1016/j.jcis.2016.09.040

[9] Bao L, Liu C, Zhang Z L and Pang D W 2015 Photoluminescence tunable carbon nanodots: Surface-state energy-gap tuning *Adv. Mater.* **27** 1663 https://doi.org/10.1002/adma.201405070

[10] Wu Z, Zhang P, Gao M, Liu C, Wang W, Leng F and Huang C 2013 One-pot hydrothermal synthesis of highly luminescent nitrogen-doped amphoteric carbon dots for bioimaging from Bombyxmori silk–natural proteins*J. Mater. Chem. B* **1** 222868 https://doi.org/10.1039/C3TB20418A

[11] Du F, Jin X, Chen J, Hua Y, Cao M, Zhang L, Li J, Zhang L, Jin J, Wu C, Gong A, Xu W, Shao Q amd Zhang M 2014 Nitrogen-doped carbon dots as multifunctional fluorescent probes *J. Nanoparticle Res.* **16** 2720 https://doi.org/10.1007/s11051-014-2720-8

[12] Eliseev A A and Lukashin A V 2010 *Functional Nanomaterials* [in Russian – Funktsionalnye Nanomaterialy] (Moscow: Fizmatlit) p 456

[13] Zhu H, Wang X, Li Y, Wang Z, Yang F and Yang X 2009 Microwave synthesis of fluorescent carbon nanoparticles with electrochemiluminescence properties *ChemComm.* **34** 5118

https://doi.org/10.1039/b907612c
[14] Yu J, Xu C, Tian Z, Lin Y and Shi Z 2016 Facilely synthesized N-doped carbom dots with high fluorescent yield for sensimg Fe^{3+} *New J. Chem.* **40** 2083 https://doi.org/10.1039/C5NJ03252K

Plasma-chemical synthesis of carbon nanotubes and graphene with mesoporous architecture for energy applications

M Shavelkina*, E Shkolnikov, S Kochanova, and E Sidorova

Joint Institute for High Temperatures, *Russian Academy of Sciences,* 13/2 Izhorskaya Street, Moscow, 125412, Russian Federation

*E-mail: mshavelkina@gmail.com

Abstract. We develop the universal technique of producing the mesoporous carbon nanomaterials. We manage to synthesize, in one and the same reactor, in the different regimes, carbon nanotubes and graphene. The technique consists in conversion of the carbon containing substances in the helium plasma flow generated by the plasma torch at low pressure. We investigate the samples using the scanning electron microscope and the synchronous thermal analysis. The studies, by the limited evaporation method, of the porous structure show that the graphene flakes and the balled nanotubes are mesoporous and thus provide the specific capacity value rather high for the electrochemical electrodes of the supercapacitor (125–133 F/g) at the charging/discharging resistance of 121/100 Ω.

1. Introduction

Supercapacitors (SC) are the promising energy storage means: they provide both high voltage and excellent capacity, that is, high specific energy values. In the industrial-manufactured SC, the high-disperse carbon materials with developed specific surface, such as the activated coals, serve as the active elements. Graphene might substitute the activated coal – it has both high (maximal for the pure carbon materials) specific surface and high electrical conductivity. Yet, the graphene sheets tendency to aggregation is the main obstacle in its application as the SC electrodes. This tendency results in a decrease of the specific surface accessible for the charged particles and thus in the SC capacity decrease [1]. To overcome the problem, the 3D-materials should be created [2]: application of the thermal plasma jet systems might provide this opportunity [3]. For example, graphene synthesized via the CVD-method [4] or obtained via the mechanical [5] or the liquid-phase [6] graphite exfoliation represents the flat sheet whereas in the plasma jet, it becomes cupola-shaped [3]. For example, graphene synthesized via the CVD-method [4] or obtained via the mechanical [5] or the liquid-phase [6] graphite exfoliation represents the flat sheet whereas in the plasma jet, it becomes cupola-shaped [3]. Then, the material porosity also influences on the specific capacity value [7]. Presence of the mesopores, 20–100 Å in radii, makes it possible to increase essentially the capacitance and the kinetic parameters of the capacitors with the double electric layer as against those with the microporous carbon material [8].

The present work is aimed at determination of possibility to apply, as the supercapacitor electrode component, the carbon nanostructures with the mesoporous architecture, synthesized in the volume of the helium plasma jet generated by the Plasma torch.

2. Experimental section

To synthesize graphene and carbon nanotubes (CNT), we used the high-power direct current plasma torch with the expanding anode channel [3]. Figure 1 shows the reactor layout. The precursor is introduced into the plasma torch simultaneously with the plasma-forming gas. We managed to decompose the hydrocarbons both in the arc discharge domain and to the plasma jet itself to form a vapor-gas mixture; when cools, it condenses to form the flakes in the collector. The pressure in the reactor was equal to 77–710 Torr; we maintained the dynamic pumping by means of the water pump. We kept the arc current constant and equal to 300–350 A. The plasma torch power reached 35 kW. We took methane, propane-butane mixture, and acetylene as the hydrocarbons and helium as the plasma-forming gas. The mass flow rate of the hydrocarbons and nitrogen varied independently.

For each experiment, we added the hydrocarbons after the temperatures in the collector reached 350–400°C. We monitored the collector temperature by means of the chromel-alumel and chromel-copel thermocouples.

We examined the synthesis products by the traditional methods of studying the doped graphene [8]. To directly register the dimension parameters and the images of the plasma torch pyrolysis products, we engaged the Hitachi S5500 scanning electron microscope (SEM) and applied the in-lens technology to obtain both ultra-high resolution and high sensitivity of the EDS (Energy Dispersive X-ray Spectrometry) analysis.

To evaluate the efficiency of the synthesis, the phase composition and the thermal stability of the carbon products, we used the DSC (Differential Scanning Calorimetry) and the TG (Thermogravimetry) thermal analysis methods (STA 449 F3 Jupiter platform from Netzsch) applying the linear heating in air at 10 K/min within the temperature range of 20–1000°C.

To determine the characteristics of the porous structure (the pore volume, the pore radius, and the surface area), we used a relatively new method of adsorption: "the limited evaporation" (LE) based on the analysis of kinetics of the adsorbate evaporation from the test material [9]. To evaluate the specific surface area of the materials, we applied the classic BET (Brunauer, Emmett, and Teller) method using the Quadrasorb SI low-temperature nitrogen adsorption analyzer.

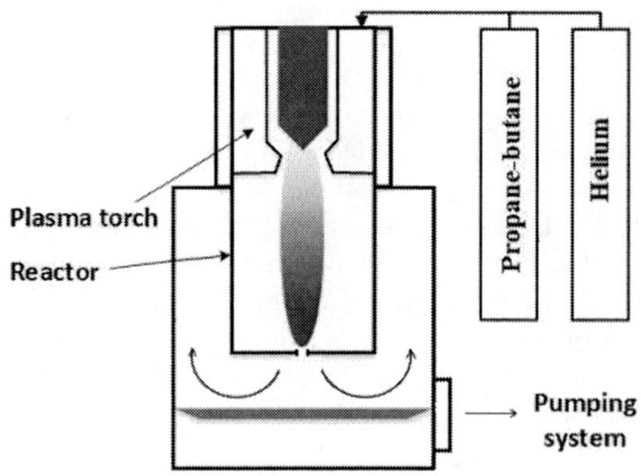

Figure 1. The reactor layout.

As the electrode material, we applied the composition of graphene or CNT, commercial activated coal and fluoroplastic. We introduced the carbon nanomaterials in the form of suspension (solid carbon/

IOP Conf. Series: Materials Science and Engineering **693** (2019) 012032 doi:10.1088/1757-899X/693/1/012032

alcohol = 0.004/1 and 0.05/1). At the first stage of the electrode preparation, we performed the water-coal mixture homogenization with addition of the necessary quantity of the fluoroplastic suspension. Then, we flatted the paste on the jeweler rolls with the subsequent drying at 200°C. At the second stage, we milled the dried electrode blank until the grit state and pressed it into the necessary thick plate. We obtained the end product by flatting the plate on the roll, under the gradual clearance decrease, up to the necessary thickness (90 to 300 μm). The final electrode area was within 9 cm^2. We established the treated electrodes mirror-like on the foil with the adhesive and soldered the construction from both sides. Then, we filled up the obtained pocket with the electrolyte: 1-mol Et_4NBF_4 solution in the acetonitrile – and vacuumed and pressurized the construction.

To study the pore structure of the graphene and the CNT structure, we analyze the cumulative and the differential distributions of the pores over the sizes. When measuring by the LE method, we performed the vaporization using benzol with the critical diameter of 0.6 nm as the adsorbate. We calculated the pore radii from the obtained isotherms using the Kelvin equation (1):

$$ln\left(\frac{p}{p_o}\right) = -\frac{kV\sigma}{rRT}\cos\varphi, \qquad (1)$$

where, σ is the surface tension, V – the molar volume of the liquid, φ– the limiting wetting angle, k – the pore form-factor, r – the pore radius, Å.

For the electrochemical measurements, we fixed the ready plane-parallel cell between the clamping plates of the Analyzer of the chemical current sources (Russia). Before the measurements, we established the potentio-dynamic mode of the potential scanning (5 mV/s). The process finished when we reached the data reiteration regime per cycle. Then, to obtain the capacitance characteristic, in the galvanic-static mode, we charged the cell up to the voltage of 2.5 V and discharged with the preset currents; and then cycled the cell in the charging-discharging mode by the current of 5.5 mA/cm^2 within the voltage range of 0–2.5 V and 1.25–2.5 V. We determined the specific (per unit SC volume) SC capacitance at the discharge, according to the formula (2):

$$\mathrm{C} = (\mathrm{I} * \mathrm{t})/((\mathrm{Uinit} - \mathrm{Uend})2 * \mathrm{l} * \mathrm{S}), \qquad (2)$$

where, I is the discharge current, t – the discharge duration, Uinit – the initial discharge voltage, Uend – the final discharge voltage, l – the electrode thickness, S – its area.

3. Results and discussion

The experiments on the propane-butane mixture decomposition show that the maximal yield of the graphene materials is obtained when helium, at the pressure of 350–710 Torr, is engaged as the plasma forming gas and at the hydrocarbon mass flow rate of 0.3 g/s. Depending on the medium pressure, the plasma producing gas flow rate, and the composition, the graphene morphology varies from the crumpled structure up to the large-scale flakes with the transverse size from 50 nm up to 1000 nm. We synthesized the 1000 nm flakes at the helium pressure of 350 Torr and the mass flow rate of 0.75 g/s. Figure 2 shows an example of the SEM image of the synthesized products. In those conditions, the quantity of the graphytized particles is within 4 mass %. That conclusion is based on the thermo-gravimetric analysis of the synthesized carbon structures.

The carbon nanotubes, 20 nm in diameter (figure 3), occur at 710 Torr of helium and the propane-butane mass flow rate of 0.4 g/s. According to the thermogravimetry, at the specimen heating, the maximal mass loss takes place at 400–600°C; the graphitized particle content is above 20 mass %.

Figure 2. SEM image of crumpled graphene morphology at 500 Torr in helium / propane-butane atmosphere.

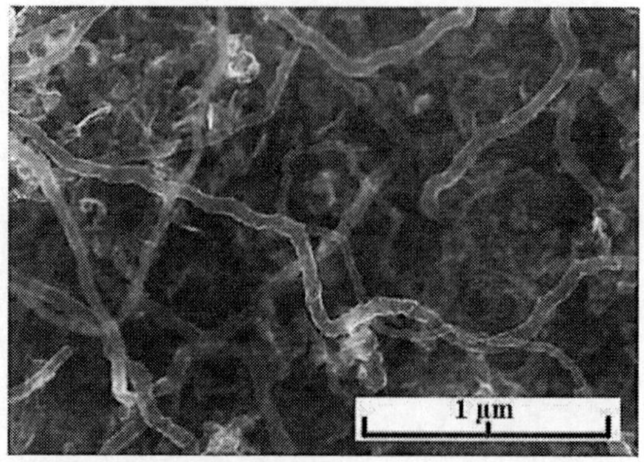

Figure 3. SEM-image of the carbon nanotube morphology at 710 Torr, helium propane-butane atmosphere.

Figure 4 shows the data indicating that the pore distribution domains in graphene and in the CNT differ. For the CNT (1), the main peak is located within the range of r_p = 70–1100 Å and is divided into two peaks – this is possibly caused by the sub-layer presence. The peak respecting to the sub-layer is within the range of r_p = 200–1020 Å. The pores with the r_p = 200–300 Å respect to the main layer. Also present is the peak within the lower micropore domain of r_p = 6.8 Å. For graphene (2), the main peak is within the range of r_p = 200–1020 Å and consists of three peaks. The peak of r_p = 200–400 Å respects to the sub-layer domain. The pores within the range of r_p = 1050 Å are those of the substrate. We calculated the pore radii from the desorption isotherms with account for the thickness of the adsorption film on the surface of the pores, as well as according to the Broekhoff-de Boer equations [9].

From the cumulative distribution of the pores over the radii (Figure 5), we see that the pore volume curve exhibits the typical rise in the mesopore domain. This effect is caused by the desorption process from the initial mesopores. The measured pore volume for graphene is higher than the same for the CNT.

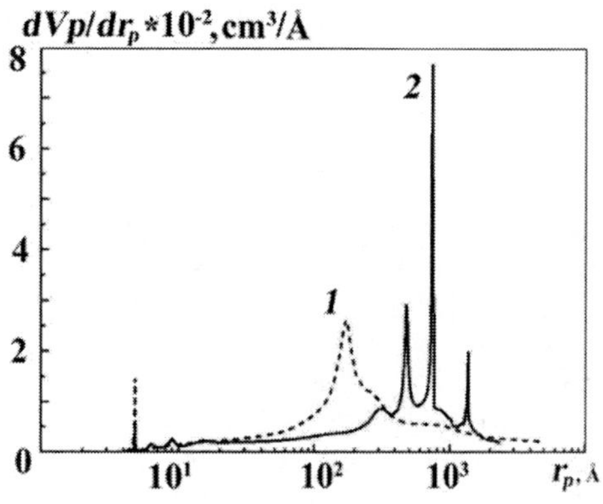

Figure 4. Differential pore distribution over the Kelvin radii.

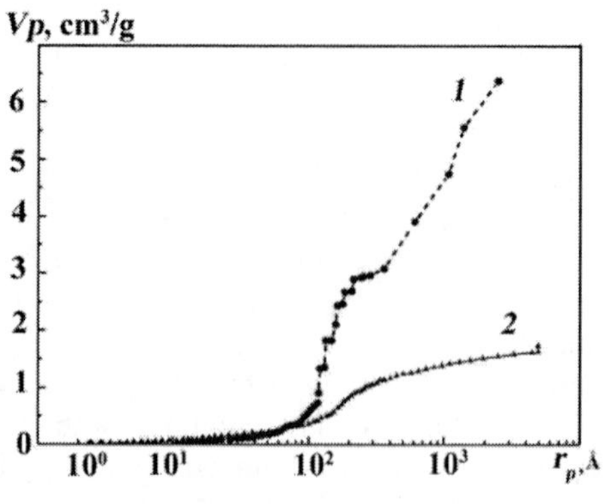

Figure 5. Cumulative pore distribution over the Kelvin radii (the adsorbate – benzol).

The plot of the specific surface distribution over the pore radii (Figure 6) shows that the predominant surface of the graphene specimen is located in the lower mesopore domain (8–20 Å) and equals to

IOP Conf. Series: Materials Science and Engineering **693** (2019) 012032 doi:10.1088/1757-899X/693/1/012032

450 cm^3/g. For the CNT, within the same domain, the specific surface equals to 350 cm^3/g.

Electrical measurements of the SC parameters with the electrode made of the active carbon material (the F-4D fluoroplast suspension) and the metal current feeder (the A 99 foil, 20 μm thick) confirm these suppositions. We added the CNT and graphene, with the different concentration, under the normal conditions, at the room temperature and the atmospheric pressure, into the active carbon material (the activated coal). Figure 7 shows the histograms reflecting the influence of the carbon nanotubes and graphene presence in the electrode mass composition on the SC characteristics (the volumetric specific capacity). Note that graphene exhibits the maximal effect of the specimens taken.

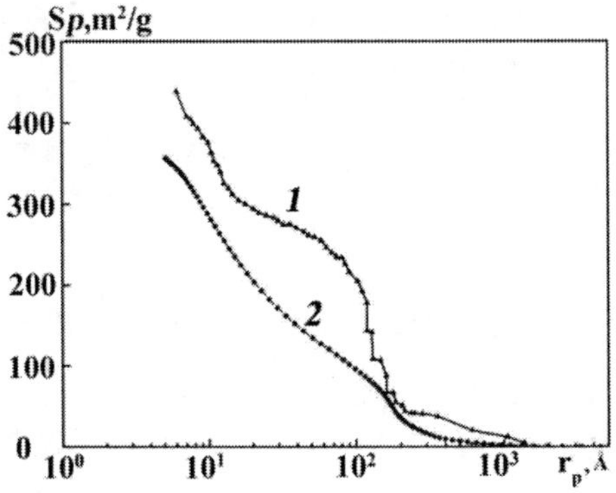

Figure 6. Distribution of the specific surface over the pore radii.

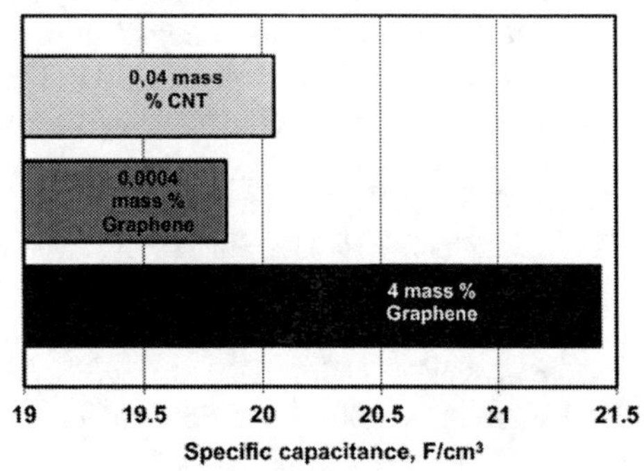

Figure 7. Histograms reflecting the effect of the presence of carbon nanotubes and graphene in the composition of the electrode mass on the characteristics of the SC (a volumetric specific capacity).

The results of SC cell parameter measurements show that with the CNT adding, the values of the specific SC capacitance equal to 125 F/g whereas with the graphene adding – to 133–132 F/g. Note that if we varied the graphene content in the electrode from 0.004 to 0.05 mass %, then the values of the resistance during charging decreased strictly from 165 to 100 Ω. We suppose that the graphene morphology facilitates the molecule orientation under the electric field application.

In general, the results of this work are consistent with the data in [10] that graphene is a more efficient material for chemical current sources than carbon nanotubes.

4. Conclusion

We investigate the electrochemical properties of the double-layered supercapacitor: its combined cell includes graphene and carbon nanotubes. We show that the high-disperse carbon nanomaterials synthesized in the plasma jet are mesoporous, with the specific surface up to 500 m^2/g. We develop the technique of preparation of the electrode mass with those materials. We added graphene and the carbon nanotubes into the electrodes as the ultrasound treated solutions. We investigate the electrical parameters of the prototype of the double-layered supercapacitor with the non-water electrolyte on the acetonitrile base. We show the possibility of increase of the supercapacitor power via adding the carbon nanomaterials. We state that by varying the graphene adding concentration in the electrode, we might reduce the supercapacitor resistance.

Acknowledgments
The work is supported by the Russian Foundation for Basic Research, project # 18-08-00165 and # 19-08-00145.

References
[1] Eletsky AV, Zycerman V Yu and Kobzev G A 2015 Nanocarbon materials: Physicochemical and exploitation properties, synthesis methods, and enegretic applications. *High Temperature.* **53** 130 doi.org/10.7868/S0040364415010032
[2] El-Kady M F, Ihns M, Li M, Hwang J, Mousavi M, Chaney L, Lech A and Kaner R 2015 Engineering three-dimensional hybrid supercapacitors and microsupercapacitors for high-performance integrated energy storage *Proceedings of the National Academy of Sciences.* **112** 4233 doi.org/ 10.1073/pnas.1420398112
[3] Shavelkina M B, Filimonova E A, Amirov R Kh and Isakaev E Kh 2018 Methane/nitrogen plasma-assisted synthesis of graphene and carbon nanotubes. *J. Phys. D: Appl. Phys.* **51** 294005 doi: 10.1088/1361-6463/aacc3d
[4] Pan Y, Zhang H G, Shi D X, Sun J T, Du S X, Liu F, and Gao H-J 2009 Highly Ordered,Millimeter-scale, continuous, single-crystalline graphene monolayer formed on Ru (0001) *Adv Mater.* **21** 2777 doi.org/10.1002/adma.200800761
[5] Novoselov KS, Geim AK, Morozov SV, Jiang D, Zhang Y, Dubonos SV, Grigorieva I V, and Firsov A A 2004 Electric field effect in atomically thin carbon films. *Science* **306** 666 doi: 10.1126/science.1102896
[6] Cooper A J, Wilson N R, Kinloch I A and Dryfe R A W 2014 Single stage electrochemical exfoliation methodfor the production of few-layer Graphene via intercalation of tetraalkylammonium cations *Carbon* **66** 173 doi.org/10.1016/j.carbon.2013.09.009
[7] Kötz R and Carlen M 2000 Principles and applications of electrochemical capacitors. *Electrochimica Acta* **45** 2483 doi.org/10.1016/S0013-4686(00)00354-6
[8] Wang H, Varghese J and Pilon L 2011 Simulation of electric double layer capacitors with mesoporous electrodes: Effects of morphology and electrolyte permittivity *Electrochimica Acta* **56** 6189 doi: 10.1016/j.electacta.2011.03.140
[9] Shkolnikov E I, Sidorova E V, Malakhov A O, Volkov V V, Julbe A and Ayral A 2011 Estimation of pore size distribution in MCM-41-type silica using a simple desorption technique. *Adsorption* **17** 911 doi: 10.1007/s10450-011-9368-9
[10] Karimov KS, Fatima N, Siddiqui K J and Khan M I 2019 Flexible thermoelectric cells fabricated by rubbing-in technology withrubber-carbon nanotubes/graphene composites *Mater. Sci. Energy Technol.* **2** 551 doi.org/10.1016/j.mset.2019.06.001

Fireproof intumescent systems modification by means of phosphorus-containing metal/carbon nanocomposites for construction applications

R. Mustakimov, V. Kodolov and V. Kodolova-Chukhontseva

M. Kalashnikov Izhevsk State Technical University, 7 Studencheskaya Street, 426069 Izhevsk, Russia

E-mail: rostmust@mail.ru

Abstract. In the present paper, a method for modifying a copper/carbon nanocomposite with ammonium polyphosphate is considered. The optimum ratio of reagents at the modification of composite was found using infrared and X-ray photoelectron spectroscopy. Experimental tests were carried out for samples of fireproof intumescent coatings based on epoxy resins modified with the phosphorus-containing copper/carbon nanocomposites. An increase of more than 20 % in such properties as adhesive durability and fire-retardant performance for the coatings modified with the above nanocomposites was determined.

1. Introduction

The protection of buildings against destructive effects of high temperatures and fire is the main part in the designing of any structures. The application of incombustible materials as well as the use of special fire resistant coatings seem to be the most widespread methods to ensure fire resistance of structures.

At present, fireproof intumescent coatings can be successfully introduced into construction materials because their volume increases several times while forming an incombustible low-conductivity layer. The basic efficient characteristics of such coatings are as follows:

1) multiple intumescences;
2) intumescent layer durability;
3) intumescent layer structure (pore forms and sizes) [1,2].

It should be noted that recently modification of intumescent coatings with nanosystems has become the most commonly used method for improving foam coke characteristics [3-5]. However, in turn, the corresponding nanosystems should be modified before using them for developing fireproof coatings.

For example, in work [6], novel flame retardant system for wood polymer composites was developed based on the combination of ammonium polyphosphate with phytic acid modified layered double hydroxide. In another examples [2] mechanical, thermal, fire and morphological properties of thermoplastic composites filled with fire retardants and microcrystalline cellulose were investigated in detail. In work [7] fire resistant properties of nano-structured intumescent coatings were investigated.

The aim of this work was to develop a method for modifying copper/carbon nanocomposites with ammonium polyphosphate for using in epoxy-based intumescent flame-retardant coatings.

2. Materials and methods

2.1. Materials
A copper/carbon nanocomposite (Cu/C NC) was synthesized at Izhevsk Electromechanical Plant "Kupol" JSC (Izhevsk, Russia) through a mechanochemical method. The image of the nanocomposites is presented in figure 1.

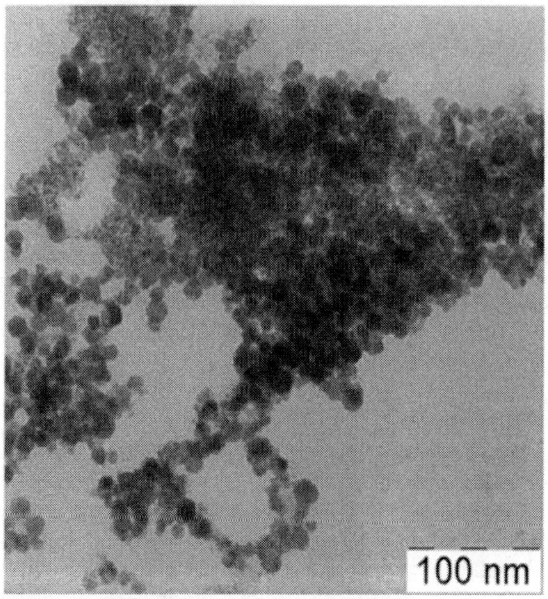

Figure 1. TEM image of the Cu/C NC.

Ammonium polyphosphate (APP; Wellchem, Quzhou City, China), having crystalloid phase II (with extent of polymerization n > 1000) was used for modification of this nanocomposite.

The polymeric base of the coating composition included the following components: ED-20 epoxy resin ED-20 (FSE Plant named after Y.M. Sverdlov, Dzerzhinsk, Russia), polyethylene polyamine (PEPA) as cross-linking agent (Uralchimplast, Nizhny Tagil, Russia), and APP (Wellchem). To achieve uniform dispersion with phosphorus, the nanocomposite was introduced as a fine-dispersed suspension in the PEPA.

2.2. Nanocomposite modification and fine-dispersed suspension production
The modification of the Cu/C NC with the APP was performed through mechanochemical interaction of the reagents using a mechanical mortar by adding a polar medium and, further, by heating the reactive mass for 3 min.

The process of fine-dispersed PEPA-based nanocomposite suspension production consists in preliminary grinding of the initial reagents in the mechanical mortar and further diluting (accompanied by mixing) to the required concentration. After these procedures, the obtained suspension is sonicated using a Volna-M USTA-1/22 dispenser (U-SONIC, Biysk, Russia) for 3 min for uniform volume-based size distribution.

2.3. Spectroscopic studies
Infrared (IR) studies were performed on a FSM 1201 (Infraspek Ltd, Saint-Petersburg, Russia) Fourier-transform IR spectrometer in the wave number region corresponding to 4000–650 cm^{-1}[5].

X-ray photoelectron spectroscopy (XPS) studies were carried out on an XPS magnetic spectrometer (PTI of Ural Branch of RAS, Izhevsk, Russia) in a way described in [8].

2.4. Studying the fireproof coating properties

2.4.1. Coating sample preparation for fire-retardant performance determination. Fireproof (or fire-retardant) parameters were determined according to the ASTM E-14-10 method. The tests were carried out with plate-form samples of the following sizes: length – 50-55 mm, width – 10-15 mm, and thickness – 1-1.3 mm. The fireproof composition was deposited over the correspondent plate. The layer thickness on each plate was 1 mm. The samples were cross-linked for 24 h under normal conditions.

2.4.2. Sample preparation for adhesive durability determination. The adhesion durability was determined to Russian National Standard (GOST) No. 14753-73. The experiments were carried out on steel plates under displacement efforts.

2.4.3. Fire-retardant performance determination. The tests were performed on the steel plates coated with the epoxy resins modified by the APP with the phosphorus-containing Cu/C NC [Cu/C↔P] ($3 \cdot 10^{-3}$ %). For these tests, the flame of a gas burner was used.

3. Results and discussion

3.1. Cu/C NC modification conditions

The studies on the initial Cu/C NC show that it represents metal clusters associated with carbon fibers or metal clusters within the carbon shell consisting of 3-4 layers of carbon fibers. Each carbon fiber includes acetylene and carbine fragments with delocalized electrons on fragments joints. In other words, the nanocomposite obtained has active surface, which contains double bonds and delocalized electrons [3].

For the studies of the APP layers thickness influence on the metal cluster magnetic moment and electron structure at the modification, the following ratios of Cu/C NC and APP were used: 1 (NC): 0.5 (APP); 1(NC) : 1(APP); 1 (NC) : 1.5 (APP)

3.1.1. IR investigations. The fragments of the IR spectra depending on the Cu/C NC : APP ratios are presented in figure 2. The characteristics of spectra peaks are set in table 1. The peaks at 1072 cm^{-1} attributed to P-O-C groups are noted on all the spectra. In this spectrum the higher intensity of the above peak is found when the ratio of reagents is equaled to 1 (NC) : 0.5 (APP). In this case there are decreased intensities for peaks corresponding to P-O-P group (904 cm^{-1}) and P=O bond (close to 1254 cm^{-1}). These data can evidence of a more extensive reaction between the Cu/C NC and the APP. In spectra of samples obtained at the ratio of 1 (NC) :1.5 (APP) the peak in region 1006 cm^{-1} attributed to P-O bond appeared that may be indicative of the presence of APP groups.

In table 1, the comparison of peaks intensities for P-O-P and P-O-C groups as well as peaks intensities for P=O bond is given. The increased peak intensity for P=O bond at 1253 cm^{-1} in the spectrum of APP-modified Cu/C NC (*i.e.*, Cu/C↔P) can be explained by the availability of oxygen atoms close to the phosphorus atom. These results correspond to the data obtained in the XPS studies.

IOP Conf. Series: Materials Science and Engineering **693** (2019) 012033 doi:10.1088/1757-899X/693/1/012033

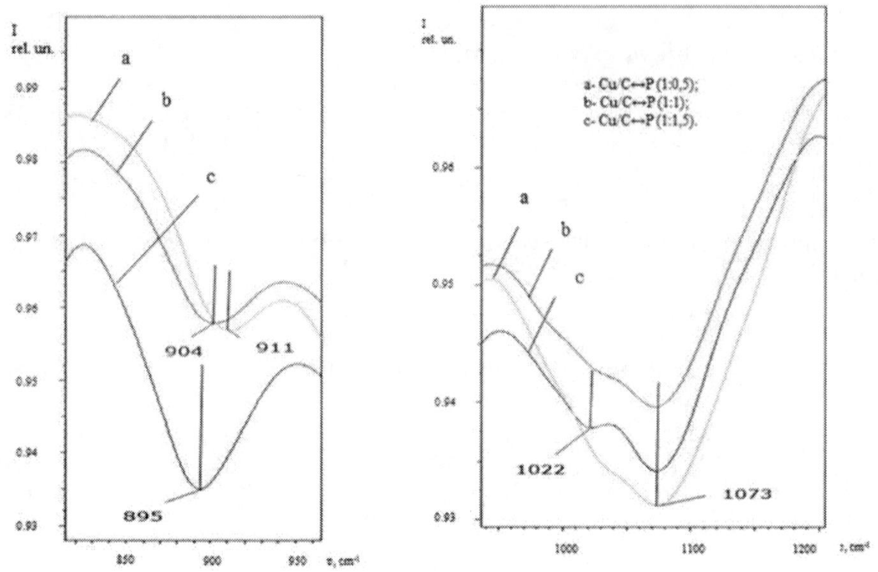

Figure 2. Fragments of the IR spectra recorded for the the Cu/C↔P NC at the NC : APP ratios: 1:0.5 **(a)**, 1:1 **(b)**, and 1:1.5 **(c)**

Table 1. Characteristics of IR spectra peaks for the Cu/C↔P NC samples obtained at the NC : APP ratios: 1:0.5; 1:1; and 1:1.5.

N (peak)	Sample	Spectrum peaks characteristics			Groups and bonds attributed to peaks [9]
		Peak wave number, cm^{-1}	Intensity	Half width	
	1:0.5	911	0.0105	42.32	P-O-P
1	1:1	904	0.0118	51.67	P-O-P
	1:1.5	895	0.0245	57.45	P-O-P
	1:0.5	1073	0.0547	149.26	P-O-C
2	1:1	1072	0.0399	135.31	P-O-C
	1:1.5	1073	0.0194	71.08	P-O-C
	1:0.5	1254	0.0057	50.48	P=O
3	1:1	1254	0.00904	48.37	P=O
	1:1.5	1253	0.02225	53.03	P=O

IOP Conf. Series: Materials Science and Engineering **693** (2019) 012033 doi:10.1088/1757-899X/693/1/012033

3.1.2. X-ray photoelectron spectroscopic investigations. For the chemical bonds (bonds between metal atoms, carbon and phosphorus) formation mechanism in above systems, the spectra of inner levels such as C1s, O1s, Cu3s, P2p are investigated (figures 3 and 4, table 2).

According to figure 3, the phosphorus reduction process proceeds more extensively at a ratio 1:0.5 (with a decrease of the APP layer) that is explained by the increasing of electron quantization.

The peak corresponding to energy EP2p equaled to 132.5 eV is attributed to C=N bond [10]. This fact can be explained by the interference which is increased because of the direct electromagnetic field formed at the annihilation. It is possible that the reduced The phosphorus is found between the carbon fibers of the mesoparticle shell.

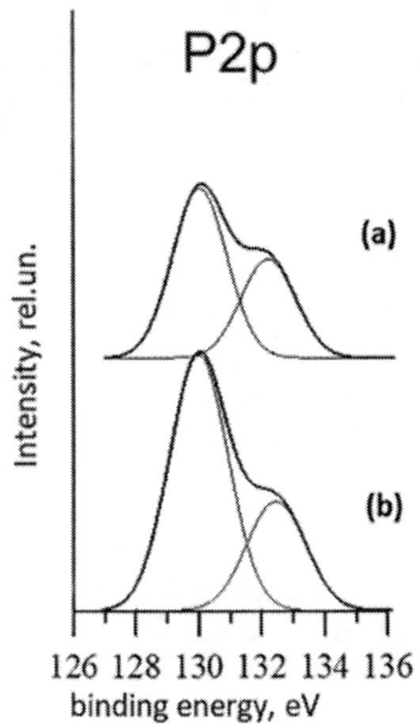

Figure 3. P2p spectra recorded for the Cu/C↔P NC. NC : APP ratios – 1:1 **(a)** and 1:0.5 **(b)**.

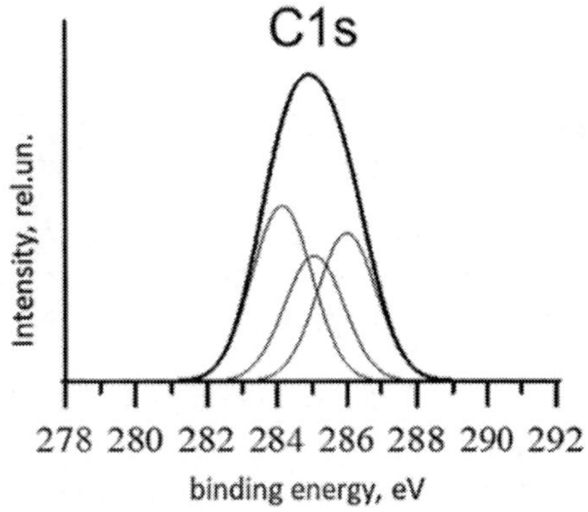

Figure 4. C1s spectra recorded for the Cu/C↔P NC. NC : APP ratio – 1:0.5.

In the investigation of the C1s spectra of the Cu/C↔P NC (figure 3) obtained at theratio of Cu/C NC : APP ratio of 2 (or 1:0.5) the spectra consist of three components: sp^2 hybridization for C – C bond (284 eV); sp^3 hybridization for the same bond (286 eV) and also for C–H bond (285 eV). The third component corresponds to C–H bond in poly acetylene fragment of nanocomposite carbon shell. The intensities of the first and second components are close. Therefore, it is possible that the nanogranules of the Cu/C↔P NC has the spherical form. The intensity of C–H peak at the ratio being equal to 2 is three times higher than the intensities of peaks at other ratios used in this investigation. That effect is possible when the carbon shell deformation occurs with the increasing polyacetylene fragments in the surface layer.

It follows from the comparison of copper atomic magnetic moments (table 2) that the higher increase of the copper atomic magnetic moment (M_m, μB)is achieved for the Cu/C↔P NC obtained at the reagent ratio being equal to 2 (M_m = 4.5 μB). At a ratio of reagents being equal to 1, the copper atomic magnetic moment decreases more than twice. This result can be explained by the decreasing electron quantization velocity at the Ammonium Polyphosphate layer thickness growth.

Table 2. Parameters of multiple splintering of Cu3s spectra in the nanosystems.

Sample	I_2/I_1	Δ, eV	M_m, μB
Cu/C NC	0.2	3.5	1.3
Cu/C↔P NC (1:1)	0.4	3.5	2.0
Cu/C↔P NC (1:1.5)	0.4	3.5	2.0
Cu/C↔P NC (1:0.5)	0.85	3.5	4.5

I_2/I_1 – the ratio of multiple splintering lines maximums intensities;
Δ - energetic distance between the multiple splintering maximums in Me3s spectra.

On the basis of above investigations, the optimum ratio of initial reagents is the ratio equal to 2. According to IR spectra and XPS studies, the reaction between the Cu/C NC and the APP proceeds more completely with the maximum growth of copper atomic magnetic moment.

3.2. Studying fireproof intumescent coatings modified with the phosphorus containing copper/carbon nanocomposites

A set of laboratory experiments on comparing the modified coatings with the initial ones were carried out, and the effect of the phosphorus-containing copper/carbon nanocomposite on the physical-mechanical characteristics and fireproof efficiency of the obtained coatings was assessed.

3.2.1. Determining the coating adhesive durability.

For the determination of the Cu/C NC introduction efficiency the following compounds were prepared:

1) ED-20+APP (20 %) with PEPA (reference sample);
2) ED-20+APP (20 %) with fine-dispersed suspension of Cu/C NC;
3) ED-20+APP (20 %) with fine-dispersed suspension of Cu/C↔P NC.

The quantity of Nanocomposites introduced to compositions equaled to 0.1 %. This quantity of nanocomposites is considered as the maximum for achieving a positive effect without negative consequences. The diagram of the adhesive durability dependence on the composition and forms of the coatings is presented in figure 5.

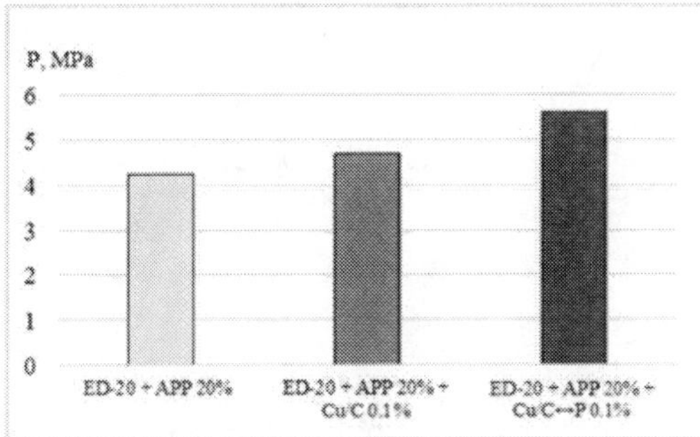

Figure 5. A dependence of the adhesive durability of the samples on the coating content.

According to above diagram, the introduction of the Cu/C↔P NC into the coating composition increases the adhesive durability by 22 % as compared to the fireproof coatings without this nanocomposite.

3.2.2. Fire tests of a fireproof coating. The photos of the fireproof coating samples before and after the fire tests are presented in figures 6(a) and 6(b), respectively. The height of foam coke was measured from the steel plate surface to the upper level of the foam coke layer. The dependence of the foam coke height on the coating content is given in figure 7.

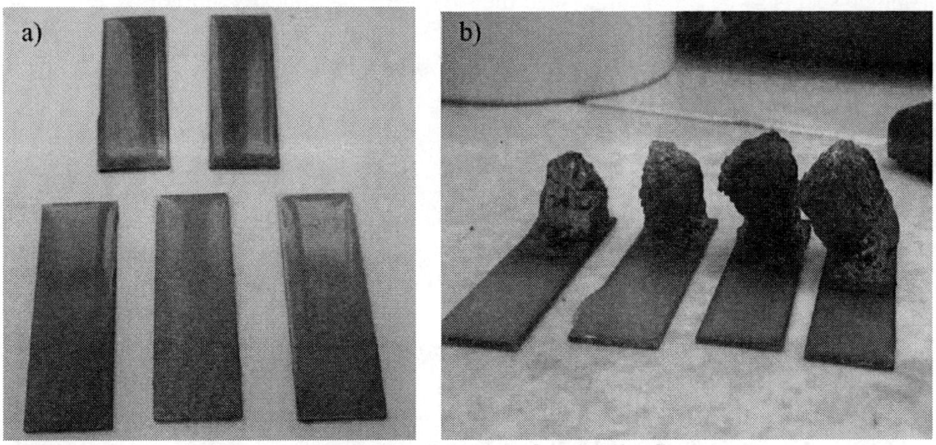

Figure 6. Photos of the samples before **(a)** and after **(b)** the fire tests.

As can be seen, the height of the Cu/C↔P NC-modified foam coke sample was found to be 28 % greater than that of the initial sample. Thus, one can assume that the introduction of phosphorus in the form of ammonium polyphosphate into the nanocomposite increases the activity of the latter.

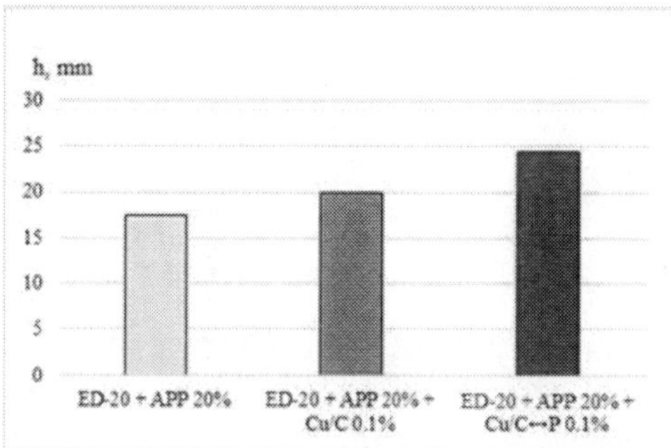

Figure 7. A dependence of the sample foam coke height on the fireproof coating content.

4. Conclusions

For the first time, phosphorus-containing copper/carbon nanocomposites were synthesized through activation and mechanochemical modification. In this case, the change of element oxidation states, as well as an increase of the copper atomic magnetic moment, takes place. At the same time, the elements and functional groups with them appear in the carbon shell of the nanocomposites. These facts open a new era for further investigations and development of metal/carbon nanocomposites as well as searching for their possible application fields. For instance, the application of these active nanosystems to obtain fireproof intumescent coatings containing epoxy resins and ammonium polyphosphate seems to be promising, since they increase the adhesive durability by 22 % compared to the fireproof coatings without nanocomposite introduction, furthermore, the height of foam coke samples, which are modified by means of the phosphorus-containing copper/carbon nanocomposite, is 28 % greater than that of the unmodified samples.

References

[1] Gravit M., Gumenyuk V., Sychov M. and Nedryshkin O. 2015 Estimation of the pores dimensions of intumescent coating for increase the fire resistance of building structures. *Procedia Eng.* **117** 119 https://doi.org/10.1016/j.proeng.2015.08.132

[2] Cavdar A.D., Torun S.B., Ertas M. and Mengeloglu F. 2019 Ammonium zeolite and ammonium phosphate applied as fire retardants for microcrystalline cellulose filled thermoplastic composites. *Fire Saf. J.* **107** 202 https://doi.org/10.1016/j.firesaf.2018.11.008

[3] Kodolov V.I., Trineeva V.V. and Kopylova A.A. 2017 Mechanochemical modification of metal/carbon nanocomposites. *Chemical Physics & Mesoscopics* [Khimicheskaya Fizika i Mezoskopiya – in Russian] **19** 569

[4] Kodolov V.I. and Trineeva V.V. 2017 New scientific trend – chemical mesoscopics. *Chemical Physics & Mesoscopics* [Khimicheskaya Fizika i Mezoskopiya – in Russian] **19** 454

[5] Mustakimov R.V., Kodolov V.I., Shabanova I.N. and Terebova N.S. 2017 Modification of copper/carbon nanocomposites by means of ammonium polyphosphate for the application as modifier of epoxy resins. *Chemical Physics & Mesoscopics* [Khimicheskaya Fizika i Mezoskopiya – in Russian] **19** 50

[6] Kalali E.N., Zhang L., Shabestari M.E., Croyal J. and Wang D-Y 2019 Flame-retardant wood polymer composites as potentials fire safe bio-based materials for building products: preparation, flammability and mechanical properties. *Fire Saf. J.* **107** 210 https://doi.org/10.1016/j.firesaf.2017.11.001

[7] Han Zh, Fina A. and Malucelli G. 2015 Thermal shielding performances of nano-structured intumescent coatings containing organomodified layered double hydroxides. *Progr. Org. Coat.* **78** 504 https://doi.org/10.1016/j.porgcoat.2014.06.011

[8] Shabanova I.N., Terebova N.S., Sapozhnikov G.V. and Kodolov V.I. 2017 X-ray photoelectron spectroscopy of the mechanism of surface functionalization of metal/carbon nanostructures

with sp-elements. *Physics of the Solid State* [Fizika Tverdogo Tela – in Russian] **59** 174

[9] Pretsch E., Buhlmann P. and Affolter C. 2000. *Structure Determination of Organic Compounds* (Berlin-Heidelberg: Springer-Verlag Berlin Heideberg) p. 439

[10] Wang J.Q., Wu W.M. and Feng D.M. 1992. *The Introduction to Electron Spectroscopy (XPS/XAES/UPS)* (Beijing: National Definite Industry Press) p. 640

Rheological properties of "Vaseline – Carbon nanoparticles" model systems under conditions of non-destructive deformations"

M Shilov[1,5,7], A Smirnova[1], A Gvozdev[2], N Rozhkova[3], T Dyachkova[4], A Burkov[5], D Stolbov[1,6], S Savilov[6] and N Usol'tseva[1]

[1]Nanomaterials Research Institute, *Ivanovo State University,* 39 Ermaka Street, 153025 Ivanovo Russian Federation
[2]Faculty of Engineer, *Ivanovo State Agricultural Academy*, 45 Sovetsjaya Street, 153012 Ivanovo Russian Federation
[3]Institute of Geology, *Karelian Research Center, Russian Academy of Sciences*, 11 Pushkina Street, 185910 Petrozavodsk Russian Federation
[4]Institute of Technology, *Tambov State Technical University*, 106 Sovetskaya Street, 392000 Tambov Russian Federation
[5]Institute of Chemistry and Ecology, *Vyatka State University*, 36 Moskovskaya Street, 610000 Kirov Russian Federation
[6]Faculty of Chemistry, *Lomonosov Moscow State University*, GSP-1 Leninskie Gory, 119991 Moscow Russian Federation
[7]Faculty of Electromechanics, *Ivanovo State Power University*, 34 Rabfakovskaya Street, 153003 Ivanovo Russian Federation

[1,5,7]E-mail: Mshilov@yandex.ru

Abstract. The paper presents experimental data on the rheological properties of "Vaseline – Carbon Nanoparticles" systems, which were obtained through oscillation tests. The nanoparticles used are multi-layered oxidized graphene (Gr), single- (SWCNTs) and multi-walled (MWCNTs) carbon nanotubes, as well as shungite nanocarbon (Sh). It was demonstrated that when carbon nanoparticles were added to the Vaseline (VM) at concentrations from 0.5 to 1.5 wt.%, the elasticity of the systems increased as compared with the pure VM, wherein the system containing 1.0 wt. % of MWCNTs was characterized by the greatest elasticity. All the studied carbon nanoparticles proved to be rheomodifiers. However, Gr, SWCNTs and MWCNTs acted essentially depending on their concentration in the vaseline system, resulting in both shortening and lengthening of the relaxation time. The natural shungite nanocarbon always led to the decrease of the relaxation time within the studied concentration range. The obtained results form an experimental basis for the development of scientific approaches to the production of lubricants operating in various friction modes

1. Introduction

One of the most effective ways of ensuring the reliability and durability of machine moving parts is the use of lubricating materials (LMs) as components of these parts. If designers try to reduce the LM specific input into the friction unit to extend the entire service life, it is necessary to use plastic lubricating materials (PLMs). They consist of at least two components – base oil (dispersion medium) and thickener (dispersed phase). The thickener forms a structural lattice and determines the PLM properties. Industry-grade PLMs contain modifiers that make the optimal LM lattice and additives (poorly soluble compounds) that impart the lubricant the required functional properties. Defining the functionality of specific additives in PLMs is complicated by their possible interaction with surface molecules of a solid object as well as with other lubricant components [1].

Modern PLMs contain additives of both organic and inorganic nature [2,3]. It is known that addition of different carbon nanoparticles (CNPs) can improve the tribological properties of LMs [4,5]. However, all those studies were carried out on commercially produced PLMs, which makes it impossible to isolate the effect of certain CNP additives on tribological properties of PLMs, regardless of other components. Rheological characteristics are among important operational properties of LMs. Managing the viscoelastic properties of LMs allows to optimize the LM composition scientifically. Flow curves seem to be a useful tool for studying the lubricant internal lattice and registering its changes.

Considering the aforementioned, the objective of the present research was to determine the dependence of the rheological properties of "Vaseline-CNPs" model systems on the type and concentration of CNPs in non-destructive (oscillation) tests. The main goal was to develop a scientific selection of additives to LMs operating under various loads. It is known that different viscosity of LMs is required for heavy- and light-load machines [1]. The secondary goal was to study the effect of the shear frequency on the characteristics of the viscoelastic flow of the analyzed model systems.

2. Materials and methods

Medical vaseline (VM) (Pharmaceutical Warehouse Himfarmprodukt LLC, St.-Petersburg, Russia), manufactured according to the Federal standard requirements LSP-005886/08-230708, was used as the base for VM/CNPs model systems.

The following CNPs, representing various allotropic carbon modifications,were introduced as additives (at concentrations ranging from 0.5 to 1.5 wt.%):

- multi-layered oxidized graphene (Gr) (NanoTechCenter LLC, Tambov, Russia) representing 3-5-nm thick nano-sized layered platelets. The number of graphene layers ranges from 5 to 10, on the periphery of which hydroxyl and carboxyl groups are available;
- single-walled (SWCNTs) and multi-walled (MWCNTs) carbon nanotubes (Arry GmbH, Weiher, Germany): SWCNTs are 1–2 nm in diameter, ~20 μm in length, their purity exceeds 90 wt.%; whereas MWCNTs are 3–10 nm in diameter, ~20 μm in length, the purity exceeds 95 wt.%);
- shungite carbon nanoparticles (Sh), obtained from high-carbon Shungit I rock (Shunga deposit), Russia). The particles have multi-level fractal structure resulting from the sequential aggregation of ~ 1 nm graphene fragments. Turbostratic stacks of ~1.5–2.5 nm and globular stack formations averaging ~6 nm in linear size determine the secondary and tertiary levels of the structure. Globule aggregates of tens of nanometers in size complete the structure [6-8]. It was demonstrated that the Sh basic element is a reduced graphene oxide [9]. In our tests, we used a condensate of water dispersion of Sh nanoparticles [10], which was placed into chloroform and then dried.

The VM/CNPs systems were obtained by mechanical mixing. The VM and CNPs components were weighed with 0.1 mg accuracy using analytical balance.

To determine the rheological characteristics of the samples, a StressTech rheometer by REOLOGICA Instruments AB (Lund, Sweden) was used. It included a digital interface rotational viscosimeter and a personal computer with a preinstalled control- and experiment-data-processing software suite. As a sensor element, we used a plane-plane system of the following geometry: 2r – 20 mm, the in-plane gap was 0.4 mm. The shaft was rotated by a compressed-air low-friction turbine. We used a precision automatic micrometer to set the gap between the sensor elements. To maintain constant temperature, a 0.1 °C accuracy thermostat was used. Oscillation tests were done at constant temperature of 30 °C and constant torque $M_{trq} = 2.094$ N·μm. The viscoelastic properties were estimated by varying the shear frequency in the 0.1 - 12 Hz frequency range. To preserve the system structure at the beginning of the test, the stress and strain values were significantly lower than the yield strength and amounted to $\tau_0 = 1$ Pa and $\gamma_0 = 2.5 \cdot 10^{-2}$ Pa .

In non-destructive tests, we measured the following parameters: the complex shear modulus (G^*) as well as its components – elastic (or storage) modulus (G') and viscous (or loss) modulus (G'') components, the phase angle (δ) and the phase angle tangent ($tan\ \delta$). A decrease in the phase angle δ over the entire range (0 - 90°), depending on the frequency, indicates a decrease in the viscous and, accordingly, an increase in the elastic inputs into the sample characteristics.

3. Results and discussion

In [1], nanoparticles are considered as rheo-improvers and specifically viscosity modifiers. According to their Working Site Category, they relate to bulk agents, whereas, according to their Working Mechanism Category, they belong to physical additives. By functionality, the rheo-improvers could be categorized as "indirect" tribo-improvers. Therefore, we began our study with such values as the phase angle and the complex shear modulus, which together determine the viscoelasticity map of the VM/CNPs systems. The map is shown in figure 1. In this map, the shift from left to right represents the increase in viscosity, whereas the shift from bottom to top represents the solid-to-liquid transition.

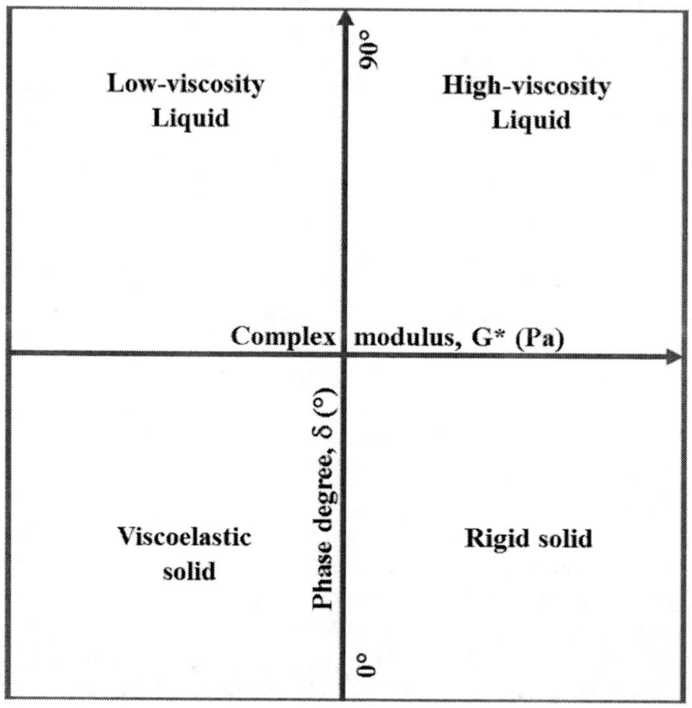

Figure 1. A viscoelasticity map.

The performed tests demonstrated that the general trend for all the VM/CNPs systems studied in the considered range of phase angles δ ($0° < \delta < 90°$) and frequencies f ($0.1\ Hz < f < 12\ Hz$) is a gradual decrease of δ, which leads to an increase of the elastic component G' and a decrease of the viscous component G'' of the complex shear modulus G^* as compared to the pure VM. The corresponding formula for calculating G^* is given below:

$$G^* = G' + iG''. \tag{1}$$

The experimental data on the oscillation tests for the pure VM and VM/CNP systems are summarized in the viscoelasticity map in figure 2. When drawing the map, we used numerical data for the complex shear modulus G^* and phase angle δ at the test end time $t = 65$ s.

IOP Conf. Series: Materials Science and Engineering **693** (2019) 012034 doi:10.1088/1757-899X/693/1/012034

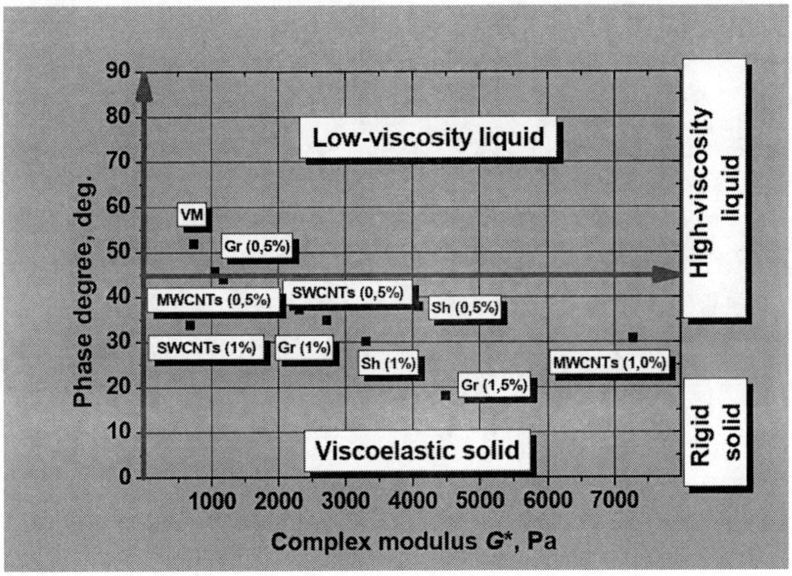

Figure 2. A viscoelasticity map of the VM and the VM/CNPs systems.

When the studied carbon allotropes were added to VM, the G^* value increased over time in all the samples, except the VM/SWCNTs (1.0 wt %) system. The G^* value for the latter system was 682 Pa, *i.e.* close to that of the pure VM – 731 Pa. In general, in the VM/CNP systems, we observed a tendency for increasing G' and G^* and decreasing G'' as compared with the pure VM (figure 2). However, the G'' value in the systems turned out to exceed that of the pure VM. The greatest increase in the G^* value (figure 2) was caused by addition of the MWCNTs (1.0 wt.%).

All the tested systems demonstrated a decrease in the phase angle δ values. Moreover, this decrease depended on the CNPs concentration and lay within the limits of $\delta \approx$ 46-38° when adding 0.5 wt.% of the CNPs and 35-30° – for the systems containing 1.0 wt.% of the CNPs. The increase in the additive concentration up to 1.5 wt.% in the VM/Gr system reduced δ value to 18°. Thus, according to the G'' and δ values, an increase in the system elasticity with the increase of the CNP concentration took place.

As examples given in figures 3 and 4, we show the dependence of the VM and VM/MWCNTs (1.0 wt.%) viscoelastic properties on the oscillation time. Evidently, once the values of the elastic G' and viscous G'' components of the complex shear modulus G^* crossed the equilibrium point, a sharp increase of the G^* value was recorded, which for the VM/MWCNTs (1.0 wt.%) system reached 7300 Pa (δ= 31°, figure 4), whereas the corresponding value for the pure VM was 731 Pa (δ = 52°, figure 3).

In his classic monograph, Schramm [11] demonstrated that it was possible to register structural rearrangements within the system at low frequencies of the 1 Hz order by the relaxation times λ_i. That is why we took this indicator to analyze the effect of the structure and concentration of CNPs in the studied systems. We established certain correlations between the relaxation time and the angular velocity of the systems. In these systems, time is one of the most important factors signaling that the system returns to its original state once the external load is removed (figure 5). As it is shown in Figure 5, within the entire angular velocity range, a decrease of the λ_i values was recorded. However, the steady exponential decrease observed for the VM was characteristic only of the VM/CNPs systems in the following sequence of addtitive: SWCNTs (1.0 wt.%) > Gr (0.5 wt.%) > MWCNTs (0.5 wt.%) > SWCNTs (0.5 wt. %).

IOP Conf. Series: Materials Science and Engineering **693** (2019) 012034 doi:10.1088/1757-899X/693/1/012034

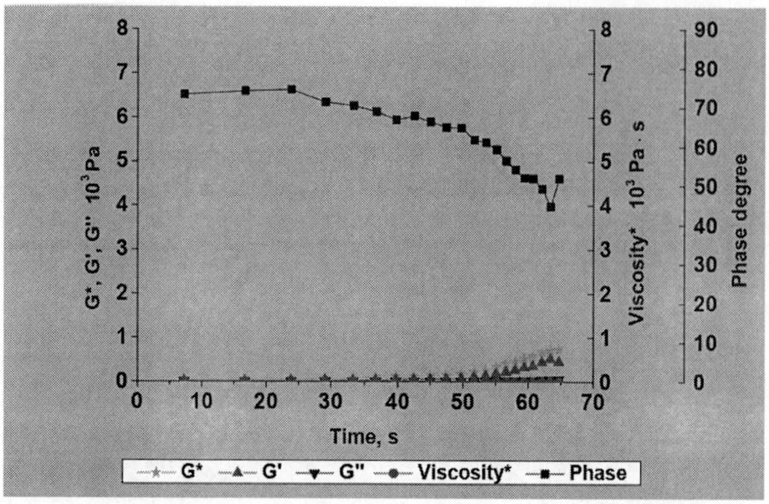

Figure 3. Changes in the viscoelastic properties of the pure VM at T = 30 °C, where G* – complex shear modulus, G'– elastic and G'' – viscous components of the complex shear modulus, *Phase* – phase degree, *Viscosity** – complex viscosity.

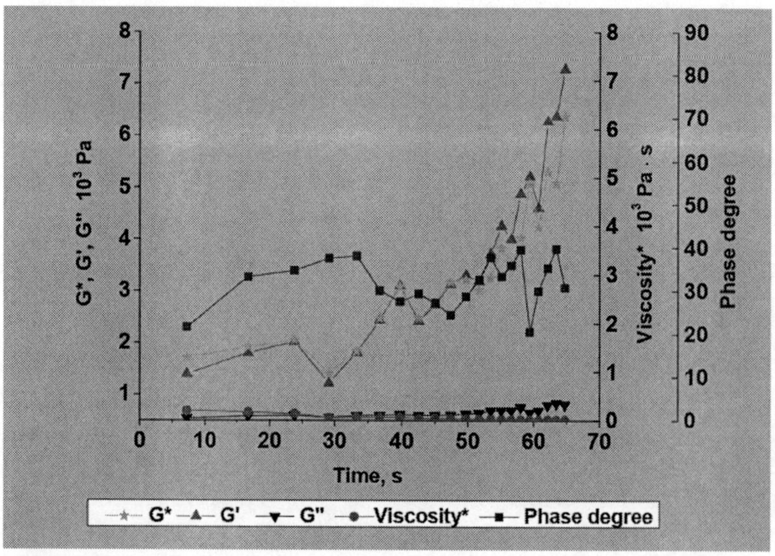

Figure 4. Changes in the viscoelastic properties of the VM/MWCNTs (1.0 wt.%) model system at T=30 °C, designations – see the legend for figure 3

For the other systems, the curves (figure 5) demonstrated a definite knee-bend in the 30–55 rad/s angular velocity range. The introduction of the MWCNTs (1.0 wt.%) into the VM caused the greatest λ_i decrease, making the relaxation time half the value of that for the pure VM. The maximum increase in the relaxation time was observed when 1.0 wt. % of the SWCNTs was introduced: the relaxation time increased by 2.6 times as compared to the pure (pristine) VM.

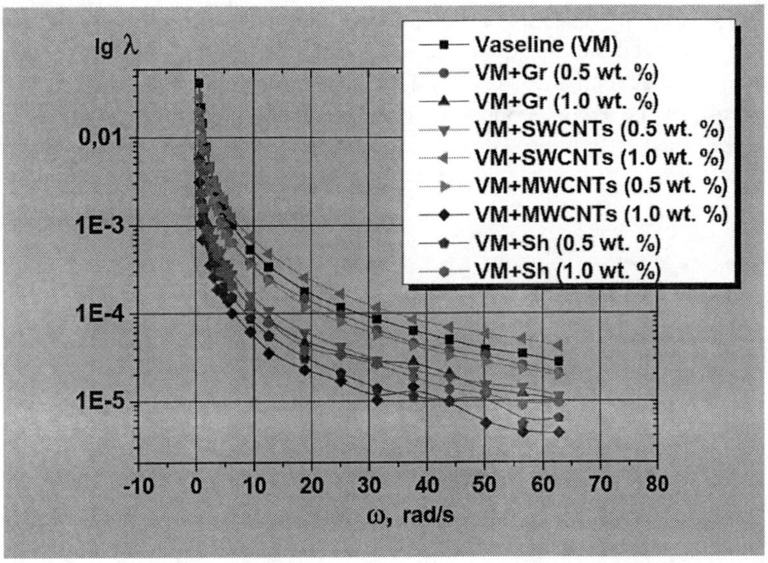

Figure 5. Dependence of the relaxation time logarithm λ_i. on the angular velocity ω for the VM/CNPs systems at T = 30 °C.

Based on Schramm's evaluation of the relaxation time λ_i given the $G' = G''$ equation, the quantitative values for λ_i and the corresponding frequencies satisfying this equation were obtained (table 1). The oscillation frequency was measured in the tests.

It was found that the CNPs presence in vaseline in any of the studied concentrations resulted in a shift of the intersection point of the curves, representing the elastic and viscous components ($G' = G''$) of the system and the change of λ_i. In all the systems, that point was reached earlier than in the pure VM.

The relaxation time at a low oscillation frequency (about 1 Hz) was calculated according to equation (2):

$$\lambda_i = \frac{1}{\omega G_i''}, \tag{2}$$

where i is the ordinal number of relaxation time value, ω is the angular velocity (rad/s), and G_i'' is the loss modulus.

Table 1. Dependence of the relaxation time λ on the CNPs concentration and structure.

Vaseline/CNPs system		Relaxation time, λ (μs)	Oscillation frequency, f (Hz) at G' = G''
CNPs	CNPs concentration (wt.%)		
–	0	540	1.5
Gr	0.5	830	0.9
	1.0	500	0.5
	1.5	340	0,4
SWCNTs	0.5	500	0.7
	1.0	1420	0.9
MWCNTs	0.5	640	1.0
	1.0	260	0.5
Sh	0.5	310	0.9
	1.0	450	0.6

The decrease in the system relaxation time corresponded to the destruction of the VM internal organization, whereas the increase of the relaxation time meant the formation of a new structure, the dimensions of which exceed those of the basic one. Analyzing the data in table 1, we could establish the following correlations between the structure-forming and structure-destructive CNPs effect and the CNPs type as well as the CNPs concentrations (for the concentrations of 0.5 and 1.0 wt.% the correlations are presented in figure 6).

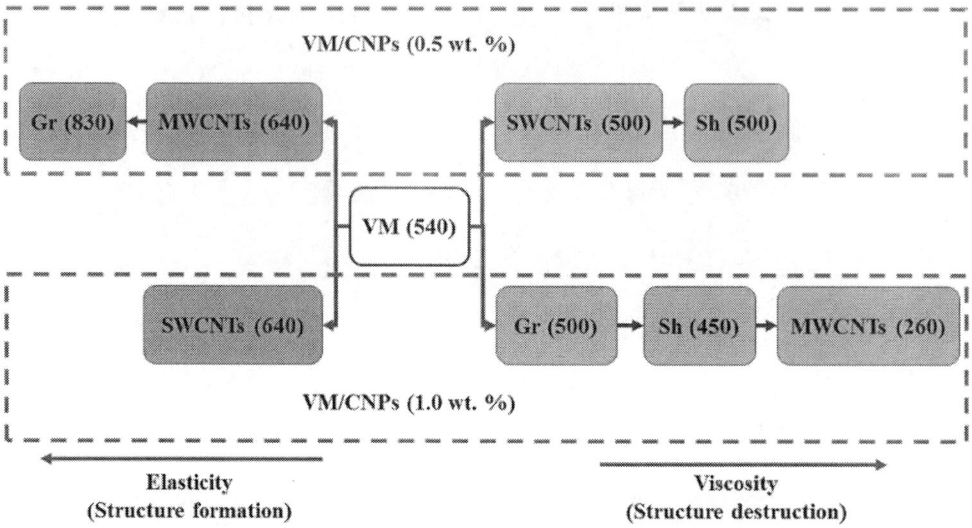

Figure 6. Viscoelastic properties of the VM/CNPs systems depending on the type and concentration of added nanoparticles. The relaxation time λ, μs is given in brackets.

Figure 6 illustrates that only Sh at both studied concentrations behaves as a structure-destructive factor, whereas the other CNPs additives can act in two ways depending on their concentrations: both as structure-destructive and as structure-forming agents.

4. Conclusion

The experimental study of the viscoelastic properties of Vaseline and its CNP-containing systems through oscillatory deformation showed that with the introduction of the studied carbon nanoparticles the model systems shift from the low-viscosity liquid zone to that of a viscoelastic solid. At the same time, the elasticity of the systems increases over the entire range of the frequencies studied. The maximum increase in the elastic properties was observed in the VM/Gr (1.5 wt.%) system. The least effect on the increase of the G' component of the G^* complex shear modulus was found after introducing of SWCNTs (1.0 wt.%).

The analysis of the relaxation times λ_i of the VM/CNP model systems demonstrated a steady decrease of λ_i with an increase of the angular velocity ω. Comparing λ_i of the G^* complex shear modulus at the equilibrium point of the elastic G' and viscous G'' components, it was found that the introduction of MWCNTs (1.0 wt.%) has the greatest effect upon diminishing the relaxation time.

All the studied nanoparticles acted as rheo-modifiers, however, the multi-layered oxidized graphene, single- and multi-walled carbon nanotubes acted essentially depending on their concentration in the vaseline system, which resulted in both shortening and lengthening of the relaxation time. In contrast, the natural shungite nanocarbon added in all the concentrations studied always led to the decrease in the relaxation time. The obtained results form the experimental basis for the development of scientific approaches to the production of lubricants operating in various friction modes.

5. Acknowledgments
The work was supported in the framework of the program of the Ministry of Science and Higher Education of the Russian Federation [Grant No. 16.1037.2017/4.6] and the Russian Foundation for Basic Research [Grant No. 18-29-19150_mk].

References
[1] Minami I 2017 Molecular science of lubricant additives. *Appl. Sci.* **7** 445 https://doi.org/10.3390/app7050445
[2] Lopes T, Gonsales A, Reguero A, Matos M, Díaz-García M and Badía-Laíño R 2015 Engineered silica nanoparticles as additives in lunbricant oils. *Sci. Technol. Adv Mater.* **16** 055005 https://doi.org/10.1088/1468-6996/16/5/055005
[3] Ilie F and Covaliu C 2016 Tribological properties of the lubricant containing titanium dioxide nanoparticles as an additive. *Lubricants* **4** 12 https://doi.org/10.3390/lubricants4020012
[4] Gvozdev A A, Smirnova A I, Berezina Ye V, Dunayev A V, Tkachev A G and Usol'tseva N V 2018 Study of the tribological characteristics of promising carbon nanoparticles containing lubricants. *Liquid Crystals and their Practical Appications* [Zhidkie Kristally i ikh Prakticheskoe Primenenie – in Russian] **18** 66 https://doi.org/10.18083/LCAppl.,2018.1.66
[5] Usol'tseva N V, Smirnova M V, Kazak A V, Smirnova A I, Bumbina N V, Il'in S O and Rozhkova N N 2015 Rheological characteristics of different carbon nanoparticles in cholesteric masogen dispersions as lubricant coolant additives. *Frict. Wear* **36** 380 https://doi.org/10.3103/S1068366615050165
[6] Sychov M M, Mjakin S V, Ogurtsov K A, Rozhkova N N, Matveychikova P V, Belyaev V V, Vysikailo F I and Nakanishi Y 2017 Effect of shungite nanocarbon deposition on the luminescent properties of ZnS:Cu particles. *Proc. 15th Int. Conf. on Global Research and Education (Warsaw)* vol 519 (Cham: Springer) p 19
[7] Sadovnichiy R V and Rozhkova NN 2014 Mineral associations of high-carbon shungite rocks of the Maksov deposit (Onega structure) Transacions of the Karelian Scientific Center of RAS [Trudy Karelskogo Nauchnogo Tsentra RAN – in Russian] **1** 148
[8] Rozhkova N N 2013 *Nanocarbon of Shungites: Structural and Physical-Chemical Properties, Activation Mechanisms* [in Russian – Nanouglerod Shungitov: Strukturnye i Fiziko-Khimicheskie Svoystva, Mekhanizmy Aktivatsii – in Russian], DSc thesis, Saint Petersburg State Polytechnic Institute (Technical University)
[9] Sheka E F and Rozhkova N N 2014 Shungite as the natural pantry of nanoscale reduced graphene oxide. *Int. J. Smart Nano Mater.* **5** 1 https://doi.org/10.1080/19475411.2014.885913
[10] Rozhkova N N, Rozhkov S S and Loshchilov A S, Russian Federation Patent No. 2642632 (25 January 2018)
[11] Shramm G 2003 *Fundamentals of Practical Rheology* [in Russian – Osnovy Prakticheskoy Reologii] (Moscow: Koloss) p 312

Production of graphene concentrates based on synthetic oils in rod drum mills

V F Pershin[1], **G B Zhumagaliyeva**[2], **A G Tkachev**[1], **A A Pasko**[1] and **A M Vorobyev**[3]

[1]Department of Technology and Methods of Nanoproducts Manufacturing, *Tambov State Technical University,* 106 Sovetskaya Street, 392000 Tambov Russian Federation
[2]*Zhangir Khan West Kazakhstan Agrarian-Technical University,* 51 Zhangir Khan Street, 090009 Uralsk, Republic of Kazakhstan
[3]Department of Operation of road transport and auto repair, *Tambov State Technical University,* 106 Sovetskaya Street, 392000 Tambov Russian Federation

[1]E-mail: pershin_home@mail.ru

Abstract. The article presents the results of the refinement of the mathematical model of the process of liquid-phase shear exfoliation of graphite and experiments on the production of graphene concentrates based on synthetic oils in a rod drum mill. The model parameters were identified and its adequacy to the real process was experimentally confirmed. It is given the dependence of the coefficient of light absorption, which characterizes the average integral value of graphene layers in suspension particles and their concentration on the processing time.

1. Introduction

Beginning in the second half of the nineteenth century, grinding in ball drum mills has been widely used in various industries, and now grinding by balls is one of the largest industries in the world [1]. A little later, planetary mills were invented, but the rotating drum and spherical grinding bodies play the main roles in the grinding process. The intensity of grinding and the quality of the finished product depend on the mode of movement of the grinding balls and the material in the cross section of the rotating drum. Modes of movement of the material in the cross section of the drum are well defined [2]: avalanching, slumping, rolling, cascading, cataracting, and centrifuging. In practice, most often use two modes cascading [3, 4] and cataracting [5, 6] and these two modes are the most well studied, both theoretically and experimentally [7].

In recent decades, a number of methods have been developed for producing graphene plates containing several layers using ball mills. For example, in the work [8] a top-down method was developed for producing colloidal dispersions of graphene sheets. After 30 hours of the shear-force-dominated grinding and centrifugation, single- and few-layer graphene sheets were readily prepared and homogeneously and stably suspended in the solvent medium N,N-dimethylformamide, at a concentration up to 0.08 mg/ml, achieving a yield higher than 32.0wt %. The graphene materials in the colloidal suspension were characterized using scanning and transmission electron microscopy and atomic force microscopy.

Graphene nanoplatelets (GNPs), one of the most widely used forms of graphene, consist of stacks of several graphene monolayers and may be of different thicknesses. Graphene plates, which consist of 7-10 monoatomic layers, are called multilayer grapheme [9].

In-Yup Jeon et al [10] reported that graphene nanofilters selectively functionalized at the edges (EFGnP) with various functional groups were efficiently prepared simply by using ball-grinding of graphite in the presence of hydrogen, carbon, a mixture of dioxide, sulfur trioxide or carbon dioxide. The authors believe that due to the universality of mechanochemical reactions that occur during the processing of graphite in a ball mill, various functional groups can be introduced into the broken edges of graphite with the presence of appropriate chemical vapors, liquids or solids in a ball crusher.

As it is known, grinding is the process of repeated destruction of the body under the action of external loads [11]. As a result of this process, the specific surface area of the material to be ground increases. The main grinding mechanisms: blow; abrasion; cutting; cracking. Since abrasion occurs mainly from shear stresses, this grinding mechanism is closest to the delamination of graphite. It should be noted that the destruction of particles by shear is considered the most economical method of grinding [11]. The ratio of average particle diameters before and after grinding is called the degree of grinding [12]:

$$i = d / d_0, \qquad (1)$$

where d is the average particle diameter after grinding, d_0 is the average diameter before grinding.

By analogy, you can enter the parameter degree or degree of exfoliation i_E and given that the specific surface depends on the number of layers, define it as follows:

$$i_E = n / n_0, \qquad (2)$$

where n is the average number of layers after exfoliation, n_0 is the average number of layers in the particles of the original graphite.

In [13, 14], we reported that graphene plates produced in a rod drum mill increase the tribological characteristics of lubricants, including frost-resistant ones. To increase the quantity and quality of graphene plates that are produced in a rod mill, it is necessary to optimize the process parameters and this can be done with the help of mathematical modeling.

To describe this process, a mathematical approach of random Markov processes, discrete in space and time, was implemented [13]. In the present work, we continue to improve this model and conduct experiments on the production of graphene concentrates in a rod drum mill.

2. Mathematical model

Earlier, we used the mathematical apparatus of random Markov processes in simulating the process of liquid-phase shear exfoliation of graphite in a rotary-type apparatus [15]. In fact, the separation of graphite occurred as a result of the simultaneous hard contact of the particle with the fixed and moving surfaces. In our case, the same exfoliation mechanism takes place and in [16] we proposed a mathematical model of the process, but did not consider the issues of identifying model parameters and checking its adequacy.

The object of modeling is the system, which is a suspension of graphite particles of a specific particle size distribution in oil with a specific concentration of graphite.

The state of the system can be characterized by the vector S (k):

$$S(k) = [c(1, k), c(2, k), \cdot \cdot \cdot , c(i, k), \cdot \cdot \cdot , c(N, k)], \qquad (3)$$

where k is the transition number, i is the cell number, c is the amount of particles with a certain number of graphene layers in the cell, and N is the total amount of cells.

The values of the elements of the initial state vector $S(0)$ of the system are determined by the characteristic of the original graphite and its concentration. The transition of the system from one state to another occurs suddenly during a time interval Δt. The change in the system over time can be expressed in the following relationships:

$$\begin{aligned} &S(1) = S(0)P, \\ &S(2) = S(1)P, \\ &\ldots\ldots\ldots\ldots \\ &S(i) = S(i-1)P, \\ &\ldots\ldots\ldots\ldots \\ &S(k) = S(k-1)P, \end{aligned} \quad (4)$$

where P is the matrix of transition probabilities.

The matrix P has the following form:

$$P = \begin{vmatrix} p_{1,1} & p_{1,2} & \cdots & p_{1,n} \\ p_{2,1} & p_{2,2} & \cdots & p_{2,n} \\ \cdots & \cdots & \cdots & \cdots \\ p_{n,1} & p_{n,2} & \cdots & p_{nn} \end{vmatrix} \quad (5)$$

where $p_{i,j}$ is the probability of transition of particles from the site i to the site j.

In this model, it is necessary to identify the parameters Δt and $p_{i,j}$. It is logical to assume that the parameter Δt is directly proportional to the angular velocity of rotation of the drum (ω, s^{-1}), because the greater this speed, the more often the particles will fall into the zone between the grinding rods and the inner surface of the drum, where shear exfoliation of these particles is realized.

If c is the number of particles with a certain number of graphene layers in a cell, then after one delamination cycle, two particles with a smaller number of layers are formed from one particle. Let us assume that the distribution of layers in two particles is equally probable, i.e. if the particle before stratification had 6 layers, then the following combinations are possible with the formation of two particles: 1 layer + 5 layers; 2 layers +4 layers; 3 layers + 3 layers. Since (to correct P_{ii}) is the probability that a particle does not fall on a given cycle in the zone of delaminating or the process of delaminating does not occur, the probability can be determined as follows:

$$p_{i,j} = p_{i0}/(N-i), \quad (6)$$

where p_{i0} is the probability of the participation of particles of the i-th cell in the exfoliation process at this transition.

It should be noted that the following condition is always fulfilled:

$$p_{i0} + p_{ii} = 1, \quad (7)$$

because at this transition the separation of one particle into two will occur or will not occur and there are no other options.

This model allows you to calculate the concentration of graphene structures in suspension. First of all, it is necessary to decide which particles will be taken into account, in other words, particles with how many layers will be taken into account. For example, it is necessary to calculate the concentration of graphene plates with B layers or less. Next, you need to determine the number N_B of the cell in which there are particles whose B graphene layers. After that, it is necessary to sum the masses of all the particles that are in the cells from N_B to the N:

$$M_{BN} = \sum_{i=B}^{N} i \cdot m \cdot c(i), \quad (8)$$

where M_{BN} - the mass of particles with the number of layers from B to N, m - the mass of one layer of grapheme.

The mass concentration of graphene in the suspension C_{1B} is equal to:

$$C_{1B} = M_{BN} / M_S, \tag{9}$$

where M_S - the mass of the suspension.

The identification of the parameter of the mathematical model was carried out as follows. When the concentration of graphite in the initial suspension was 10%, the rotational speed of the drum was equal to the critical speed (ω_{CR}):

$$\omega_{CR} = \sqrt{g/R}, \tag{10}$$

g – acceleration of gravity, R – inner radius of the drum.

The concentration of graphene plates with the number of layers from 1 to 15 was determined at a treatment time of 3, 6, 9, 12 and 15 hours. Further, by changing the value of the parameter p_{i0} from 0.1 d to 0.9 with a step of 0.1, we found such a value p_{i0} at which the deviations of the calculated values calculated by the formula (9) from the experimental values were minimal.

Since the concentration of graphite in the initial mixture is taken into account in the vector of the initial state, we assumed that when this concentration changes, the numerical value of the parameter should not change. In order for the mathematical model to take into account the speed of rotation of the drum during one transition of the system from one state to another, we adopted the time of one revolution of the drum. Thus, with an increase in the speed of rotation of the drum, the time required for the system to transition from one state to another, that is, the time for transformation of the state vector *S (i-1)*, in accordance with the relations (4) to the state vector S *(i)* decreases S *(i)*.

So, the model allows a quantitative assessment of the exfoliation process, but the main purpose of creating this model is to study the qualitative patterns of the influence of different parameters on the distribution of particles by the number of graphene layers. Thus, conducting numerical experiments using a mathematical model, we can better understand the process of graphite shear stratification and find new technical solutions to increase the intensity of this process.

Questions about the effect of the diameter of the drum and the coefficient of its filling with the suspension on the intensity of the exfoliation process will be discussed below.

3. Experiment

First of all, we investigated the movement of rods in the cross section of a smooth rotating drum with a diameter of 400 mm. Earlier [13], we found that the rods periodically rise up with the rotating drum, and then slide down to the starting position. Shift effects on particles of graphite can occur only when the rods slide down. High-speed video showed that the rods do not always slide on the inner surface of a rotating drum, but sometimes rotate. When the rods rotate about their own axis, shear forces on the particles do not act and there is no exfoliation process. Thus, during one revolution of the drum, only for a very short period of time, the separation of particles of graphite occurs and low-layer graphene is formed. The result is a low intensity shear exfoliation process. We have developed a rod drum mill design in which the rods are interconnected by a flexible cable and they cannot rotate about their own axes [13]. So the rods periodically rise up and then slide down. A new problem was that the time for lifting the rods up and the time they were sliding down were approximately the same. Thus, half the residence time of the rods in the rotating drum of the process of exfoliation of graphite does not occur. We solved this problem by connecting the rods with the base, as shown in figure 1. In this embodiment, the rods all the time slide relative to the inner surface of the rotating drum. It should be noted that the sliding of the rods occurs even when the rotational speed of the drum is greater than the critical speed (ω_{CR}).

In our opinion, at the best option, the speed of rotation of the drum should be maximum, and the number of rods should be such that they occupy one-fourth of the circumference of the drum, as shown in figure 1. We found that at the speed of rotation of the drum 1.5 from the critical speed, rods slide well on the inner surface of the drum. We used these recommendations during the process of exfoliating graphite in a laboratory mill.

Experiments with real exfoliation of graphite were carried out on a laboratory ball mill with rods (figure 2), performed in accordance with [13].

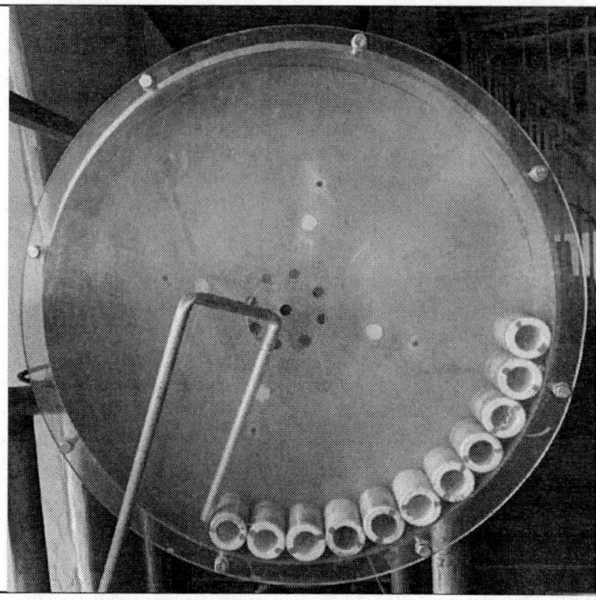

Figure 1. Rods connected by a flexible cable in a rotating drum.

Figure 2. Laboratory ball mill (a) and grinding rods (b).

Experiments were performed in the following sequence. Previously, using the paddle mixer, the initial suspension was prepared from graphite powder and synthetic oil. In the experiments used crystalline graphite brand GS-2. The mass concentration of graphite was varied from 5 mg / ml to 30 mg / ml. This suspension was loaded into the drum, where there were grinding rods. The number of rods varied from 3 to 12. The rods were used in three diameters: 8, 10, and 12 mm. The speed of rotation of the drum was changed from 28 to 140 rpm.

Every three hours the drum was stopped, three samples of the suspension were taken. These samples were centrifuged, the precipitate was removed and the concentration of graphene structures was determined. Since there are no proven methods for determining graphene concentration in oil, we developed a new method. This method is based on the difference in the density of oil and graphene: the density of oil is about 800 kg / m; The density of graphene is about 2000 kg / m. The concentration was determined as follows. We measured 25 ml of the suspension with an accuracy of 0.04 ml and weighed this suspension with an accuracy of 0.0001 g. The percentage concentration of graphene was determined as follows:

$$c_G = (G_S - \gamma_O V_S)100 / G_S \tag{11}$$

where G_S is the weight of the sample suspension, mg; γ_O - real oil density, mg / ml; V_S- sample volume, ml.

If the concentration value is multiplied by the total volume of the suspension, which we get after centrifuging and removing the sediment and divide by the processing time in the drum, then we obtain the performance value

4. Results and discussion

Figure 3 shows the characteristic dependence of productivity on the concentration of graphite in the initial suspension. In this case, the speed of rotation of the drum was equal to the critical, and the processing time of the suspension is 15 hours. We used the performance values at graphite concentration in the initial suspension 10% to determine the numerical value of the parameter (p_{i0}) and then calculate using the formula (6) the remaining elements of the transition probability matrix (5). These values were used in calculating the performance of a rod drum mill at other values of its rotational speed. Figure 4 shows the characteristic dependencies of performance on the relative speed of rotation of the drum. In this case, straight lines are the calculation according to the mathematical model for a laboratory drum with a diameter of 160 mm and relative speeds of rotation (ω / ω_{CR}) 1, 1.25 0.5, respectively. The concentration of graphite in the initial mixture was 10%. It can be seen that the calculated dependences 1, 2, and 3 are in satisfactory agreement with the experimental data, those. confirm the assumptions we made in deriving the formula.

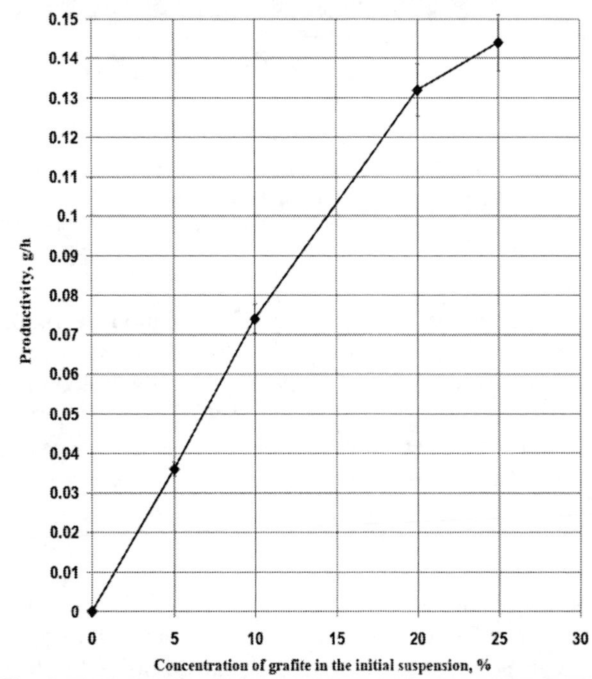

Figure 3. The dependence of the performance of the mill on the concentration of graphite in the initial suspension.

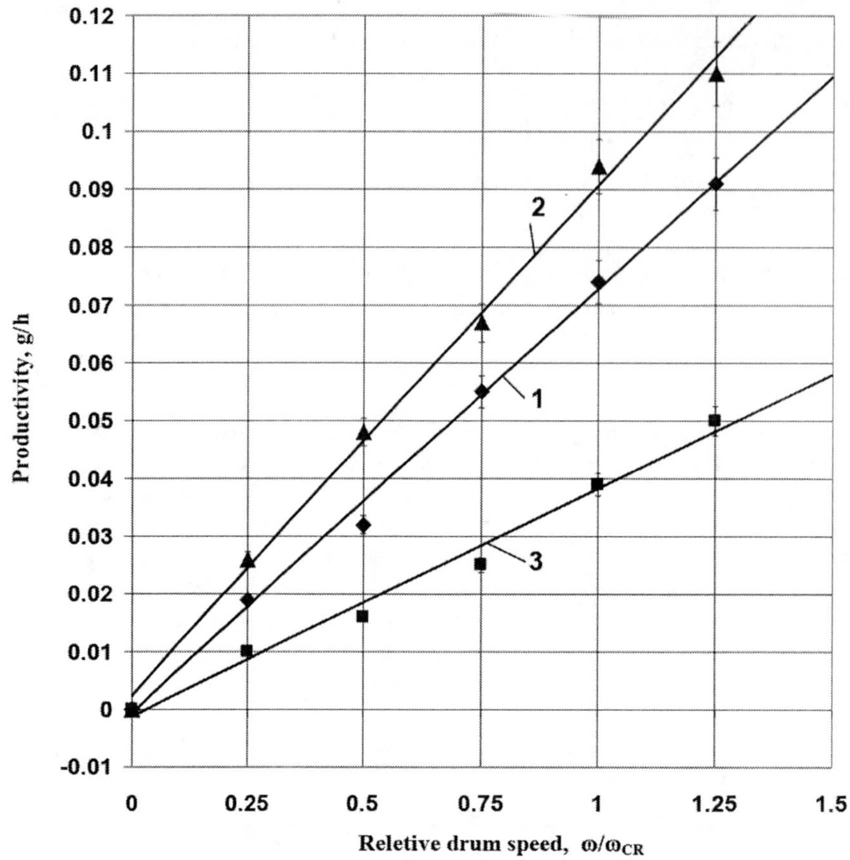

Figure 4. The dependence of the mill performance on the relative speed of its rotation.

The results of the experiments that we conducted with different numbers of rods showed that with an increase in the number of rods, the intensity of the exfoliation process increases, but we have not yet managed to find a relationship that adequately describes the effect of the number of rods on the intensity of graphene concentration in the finished suspension.

Conclusion

A mathematical model of the process of liquid-phase exfoliation of graphite is considered and the problem of identifying its parameters is solved. Conducted experimental studies to determine the effect of concentration graphite in the original suspension and the speed of rotation of the rotor on the performance of the rod drum mill. It has been established that the proposed mathematical model adequately describes the actual process of exfoliation of graphite and can be used to determine the parameters of the mill and the modes of its operation, which are necessary for ensuring a given performance.

Acknowledgments

The work was funded by the Ministry of Science and Higher Education of the Russian Federation in the framework of the Federal Target Program "Research and Development in Priority Areas of Advancement of the Russian Scientific and Technological Complex for 2014–2020" (Agreement No. 14.577.21.0253, Unique Identification Number: RFMEFI 57717X0253).

IOP Conf. Series: Materials Science and Engineering **693** (2019) 012035 doi:10.1088/1757-899X/693/1/012035

References

[1] Bodurov P and Genchev V 2015 New and More Effective Grinding Bodies for Drum Mills. Alternative of the Spherical Grinding Bodies *Journal of Multidisciplinary Engineering Science and Technology (JMEST)* **2**(9) 2516-2520

[2] Henein H, Brimacobe J K and Watkison A P 1983 The Modeling of Transverse Solids Motion in Rotary Kils. *Metallurgical Transactions B* **14**(2) 207-220

[3] Taberlet N, Richard P and Hinch E J 2006 S shape of a granular pile in a rotating drum *Physical Review E* **73** 050301

[4] Brewster R Grest G S and Levine A J 2009 Levine Effects of cohesion on the surface angle and velocity profiles of granular material in a rotating drum *Physical Review E* **79** 011305

[5] Rogachev A S Moskovskikh D O, Nepapushev A A, Sviridova T A and Vadchenko S G 2015 Experimental investigation of milling regimes in planetary ball mill and their influence on structure and reactivity of gasless powder exothermic mixtures *Powder Technology* **274** 44-52

[6] Fullenwider B, Schoenung J M, Kiani P, Ma K 2019 Two-stage ball milling of recycled machining chips to create an alternative feedstock powder for metal additive manufacturing. *Powder Technology* **342** 562-571

[7] Chattopadhyay P P, Manna I, Talapatra S and Pabi S K 2001 A mathematical analysis of milling mechanics in a planetary ball mill *Materials Chemistry and Physics* **68** 85-94

[8] Zhao W, Wu F, Wu H, and Chen G 2010 Preparation of Colloidal Dispersions of Graphene Sheets in Organic Solvents by Using BallMilling *Journal of Nanomaterials* http://dx.doi.org/10.1155/2010/528235

[9] Shahil K M F, Balandin A A 2012 Thermal Properties of Graphene: Applications in thermal interface materials *Solid State Communications* **152** (2012) 1331–1340

[10] In-Yup Jeon, Sun-Min Jung, Choi Hyun-Jung, Jeong-Min Seo 2013 Large-Scale Production of Edge-Selectively Functionalized Graphene Nanoplatelets via Ball Milling and Their Use as Metal-Free Electrocatalysts for Oxygen Reduction Reaction *J. Am. Chem. Soc.* **135** (4) 1386-1393

[11] Baranov D A et al *The processes and apparatus of chemical technology T. 2 Mechanical and hydromechanical processes* (Moscow, 2001) p 600

[12] Gusev Yu I et al *Design and calculation of chemical production machines* (M.-Mashinostroenie, 1985) p 408

[13] Zhumagalieva G, Pershin V, Tkachev A, Vorobiev A, Pasko A, and Galunin E 2018 Using a rod drum mill for graphene masterbatch production *AIP Conference Proceedings* **2041** 020010

[14] Tkachev A, Zhumagalieva G, Al-Hilo Z, Memetov N, Galunin E, Pershin V 2019 Modification of Frost-Resistant Plastic Lubricants Using Few- and Multi-Layered Graphene *Proceedings of the 4th World Congress on Recent Advances in Nanotechnology (RAN'19)* https://www.iso.org/10.11159/icnnfc19.105

[15] Al-Shiblavi K A, Pasko A A, Pershin V F. 2018 Modeling of the Process of Obtaining Graphene Structures Using Liquid Phase Graphite Shift Exfoliation *Transactions TSTU* **24**(4) 717-726

[16] Pershin V F et al., RU. Patent No. 2,670,495 (26 December 2017)

Synthesis and characterization of polyethylene terephthalate-reduced graphene oxide composites

P A Mikhaylov[1], M I Vinogradov[1], I S Levin[1], G A Shandryuk[1], A V Lubenchenko[2] and V G Kulichikhin[1]

[1]*A V Topchiev Institute of Petrochemical Synthesis RAS*, 29 Leninsky Avenue, Moscow 119991, Russia
[2]*National Research University Moscow Power Engineering Institute*, 14 Krasnokazarmennaya Street, Moscow 111250, Russia

E-mail: pmih@ips.ac.ru

Abstract. By means of *in situ* melt polycondensation of polyethylenterephathalate accompanied with simultaneous reduction of introduced graphene oxide, the composite of polyethylenterephathalate filled with reduced graphene oxide was prepared. Melting and crystallization of the reduced graphene oxide – polyethyleneterephthalate composite was studied by the differential scanning calorimetry method, and the rheological test was carried out by rotational rheometry and compared to the pristine polyethylenterephathalate. The morphology of the reduced graphene oxide – polyethyleneterephthalate composite was studied using optical microscopy. The reduced graphene oxide was isolated from the reduced graphene oxide – polyethyleneterephthalate composite and characterized by thermogravimetry, X-ray photoelectron spectroscopy and X-Ray diffraction methods. Since a part of polymer could not be removed from reduced graphene oxide particles by trifluoroacetic acid, the hypothesis about probable grafting of polyethyleneterephthalate on reduced graphene oxide sheets was suggested. Rheological behaviour of the reduced graphene oxide – polyethyleneterephthalate composite melt confirms this suggestion. According to the calculations based on thermogravimetry, the reduced graphene oxide, isolated from the reduced graphene oxide – polyethyleneterephthalate composite, consists of about 80% polyethyleneterephthalate.

1. Introduction

Graphene is a two-dimensional material composed from sp^2-carbon honeycombs with unique mechanical and electrical properties attracting a lot of attention last year as a filler for polymer composites. However, graphene tends to aggregate due to strong π-π stacking interactions [1]. As a result, dispersibility of graphene is very poor. It is a great challenge because the mechanical and electrical properties of composites strongly depend on the particles size, and qualitative distribution of graphene plates in polymer matrix is required to achieve outstanding properties of polymer-graphene composites. The most commonly used method of producing graphene is the reduction of exfoliated graphene oxide (GO). GO is typically produced by Hummers method [2] and its modifications from graphite by strong oxidative reagents such as a mixture of KMO_4, H_2SO_4 and $NaNO_3$. During oxidation, a lot of oxygen containing functional groups are attached to basal planes and edges of graphene sheets. The interplanar space between graphene sheets in GO increases in comparison with graphite. GO is hydrophilic and can be dispersed easily in water and polar organic solvents on few layers or even single layer structure by simple sonication [3]. The reduction of GO produces a material which is called "reduced graphene oxide" (RGO). The chemical, thermal, solvothermal, electrochemical reduction methods were commonly used to produce RGO. Hydrazine, ascorbic acid, sodium borohydride are often used as chemical reducers of GO [3]. The reduction of GO dispersion in appropriate solvent at elevated temperatures is a solvothermal process [4].

Non-functionalized RGO possesses poor dispersibility in polymer matrix because of loss of the most of the oxygen containing functional groups which provide interaction of graphene nanosheets and solvent or polymer matrix. This drawback was overcome by various *in situ* reduction approaches. Chemical reduction of GO in polymer solution or dispersion is widely used to prepare RGO-polymer composites [1,5]. The reduction of GO in polymer-GO composites by chemical [3] or thermal [6,7] post-treatment was also described. The process of polymerization in presence of GO dispersion accompanied with *in situ* GO reduction is a promising approach for large-scale production. These methods were successfully used to prepare polyimide [8,9], polyamide-6 [10], polyethylene terephthalate (PET) [11] composites during polycondensation. Melt processing to prepare composites is more economically preferable and ecologically friendly in comparison with solution-based process, because no solvent recovery is required.

Herein, RGO-polyethylene terephthalate (RGO-PET) composites prepared via *in situ* polycondensation with simultaneous reduction of GO are studied. RGO was isolated from polymer matrix and analyzed by thermogravimetry (TGA), X-Ray photoelectron spectroscopy (XPS) and X-ray diffraction (XRD) methods. In addition, thermal properties of the RGO-PET composite and rhelogical properties of the RGO-PET melts were tested.

2. Experimental part

2.1. Materials

Sulfuric acid, hydrochloric acid, trifluoroacetic acid ("Sigma-Aldrich"), sodium permanganate ("Reachem", Russia), dimethylformamide (DMF), potassium hydroxide, sodium, hydrogen peroxide, stibium oxide (III) ("Reachem", Russia) and ethanol of reagent grade purity were used without additional purification. All solvents, acids and alkalis were purchased from "Ekos-1" company (Russia) if other is not specified. Thionyl chloride (reagent grade) purchased from "Chimmed" (Russia) was distilled before application. Toluene (reagent grade) was refluxed for 3 h with Dean-Stark apparatus, for 2 h over sodium and distilled. Pyridine (reagent grade) was refluxed for 6 h with KOH and distilled. The purified graphite for GO preparation (Zaval'e deposit, Ukraine) was ground by a mill and sifted through standard 0.08 mm mesh sieves. Deionized water (DI) with specific electro conductivity less than 1 μSm/cm was used. Dichloracetic acid (DCA) purchased from "BRI LLC" company (Russia) for intrinsic and specific viscosity measurements has analytical grade purity. Terephthalic acid (TA), ethylene glycol (EG) and industrial polyethylene terephthalate containing 0.2 % isophthalic acid were kindly provided by ZAO "New Polymers Plant [*Zavod Novikh Polymerov*]" (Russia, Moscow region, Solnechnogorsk). TA and EG were "commercial grade purity". Ethylene glycol for synthesis of bis-(2-hydroxyethyl) terephthalate was distilled over sodium.

2.2. Synthesis and exfoliation of GO

Oxidation of graphite was carried out according to modified Hummer's method, using a mixture of $KMnO_4/H_2SO_4/H_3PO_4$ [12]. Detailed synthesis and characterization of graphite oxide were described in our previous work [13]. GO aqueous dispersion was prepared by mixing of 0.199 g GO in 25 ml DI water with IKA T10 Basic Ultra-Turrax disperser for 30 min. The viscous dark-brown dispersion was filtered through 0.45 μm PTFE filter.

2.3. Synthesis of terephthaloyl dichloride

TA (0.352 mol), $SOCl_2$ (1.52 mol) were refluxed in presence of few drops of DMF during 8 h. Excess of $SOCl_2$ was distilled out, and a residue was recrystallized from anhydrous toluene (m.p. 82-83 °C).

2.4. *Synthesis of bis(2-hydroxyethyl) terephthalate (BHET)*
A flask with a magnetic stirrer, containing EG (110 g) was cooled by ice bath for 30 min, and 0.235 mol of terephthaloyl dichloride was charged into the flask. Anhydrous pyridine (0.47 mol) was then added dropwise to the reaction mixture for 10 min. The ice bath was removed, the reaction mixture was heated and kept for 2 h at 40 °C. White precipitate was filtered on Büchner funnel, repeatedly washed with water and dried.

2.5. *Polycondensation of RGO-PET*
BHET (39.5 g), Sb_2O_3 catalyst (10 mg) were charged into a 250 ml three-neck round bottom flask equipped with a mechanical glass paddle stirrer, an argon inlet and a vacuum outlet. Aqueous GO dispersion (25 ml) containing 0.199 g of GO was mixed with EG (25 ml) and poured into the flask. The flask was immersed into the wood bath preheated to 130-150 °C and kept there for about 30-40 min until most of the water was distilled out. In order to remove the excess of EG, the wood bath was heated to 230 °C for about 40 min in slow flow argon. Then, the temperature of the wood bath was raised to 270-280 °C and the vacuum was applied (~1 Torr). After about 11 h of polycondensation, the RGO-PET was evacuated from the flask under the stream of argon and cooled on air. The intrinsic viscosity of the RGO-PET was 1.0 dl/g in DCA at 25 °C.

2.6. *Isolation of RGO from RGO-PET*
The RGO-PET (1.00 g) was dissolved in about 20 ml of TFA, filtered through a 0.22 μm polypropylene membrane filter and dried. The filtrate was thoroughly washed with fresh TFA and air dried. The yield was 25 mg (2.5%).

2.7. *Thermal reduction of GO (RGO-300)*
The sample of GO (100 mg) was placed in a round bottom flask. The flask was immersed into the wood bath and kept for 1 h at 300 °C under slow flow of argon.

2.8. *Characterization*
Intrinsic and specific viscosity were measured in DCA at 25 °C using Ubbelode viscometer.

X-ray diffraction patterns were obtained with Rigaku Rotaflex RU-200 X-ray source with a rotating copper anode (Cu K_α–radiation with characteristic wavelength λ=1.542 Å, a secondary curved graphite monochromator was used) at apparatus operating conditions 50 kV – 160 mA. The unit was equipped with a horizontal wide-angle goniometer Rigaku D/MAX-RC using Bragg-Brentano scheme in θ-2θ geometry. Scanning was carried out in the 3-40 angle range at 2θ scale with a speed of 1°/min and 0.04° increment. A scintillation counter was used as a detector of diffracted X-ray emission. XRD studies were carried out at room temperature.

XPS of the samples surface was carried out with the electron-ion spectroscopy module based on Nanofab 25 (NT-MDT) platform in ultrahigh oil-free vacuum (10^{-6} Pa). The X-ray source SPECS XR 50 without a monochromator with Mg anode as the X-ray source (1253.6 eV photon energy) was used. The spectra were registered with an electrostatic hemispherical energy analyzer SPECS Phoibos 225. The energy resolution based on the full width at half maximum (FWHM) of the spectrometer at the Ag3d5/2 line (peak) was 0.78 eV for non-monochromatic X-radiation Mg Kα. The energy positions of the spectra were calibrated with reference to the Cu2p3/2 (binding energy 932.62 eV), Ag3d5/2 (368.21 eV) and Au4f7/2 (83.95 eV) peaks. All survey spectra scans were recorded at the pass energy of 80 eV.

The TGA measurements were carried out using TGA / DSC 1 combined thermal analysis instrument, Mettler Toledo (Switzerland), in 70 μl alumina crucible at the heating rate of 10 °C / min from 30 to 1000 °C under 10 cm^3/min flow rate of argon.

The differential scanning calorimetry (DCS) measurement was performed by DSC 823e Mettler Toledo (Switzerland) in 40 μl aluminum crucibles without perforation at the heating/cooling rate of 10 °C/min (30-300 °C) under 70 ml/min flow rate of argon. The crystallization was studied at first cooling from 300 °C to 30 °C. The melting point and enthalpy were determined at second heating.

The degree of crystallinity X_c of PET and RGO-PET was measured using the following equation:

$$X_c = \Delta H_f(PET)/(\Delta H_f^0(PET)\times(W(PET))^{-1}\cdot 100\%, \quad (1)$$

where $\Delta H_f^0(PET) = 140$ J/g and $W(PET)$ is weight fraction of PET [13].

The graphene concentration X_g (TGA) in RGO was calculated from TGA using the following equation:

$$X_g(TGA) = \frac{R_{RGO} - R_{PET}}{R_{GO} - R_{PET}} \cdot 100\% \quad (2)$$

where R_{RGO}, R_{GO}, R_{PET} are the ratio of weight at 500 °C to weight at 300 °C for RGO, GO, and PET, respectively.

The graphene concentration X_g (XPS) in RGO was calculated from XPS measurements using the following equation:

$$X_g(XPS) = \frac{\omega(C)_G - \omega(C)_{RGO}}{\omega(C)_G - \omega(C)_{PET}} \cdot 100\% \quad (3)$$

where $\omega(C)_{RGO}$, $\omega(C)_G = 78\%$, $\omega(C)_{PET} = 65.2\%$ are weight fractions of carbon in RGO, graphene, and PET, respectively.

The rheological properties of melts at steady state shear flow were studied on a RS600 rotational rheometer (Thermo Haake, Germany) using a plate–plate operation unit at 270 °C in the shear rate range of 10^1-10^3 s^{-1}.

Morphological study of the RGO-PET thin film (about 10 μm) was performed using a Biomed 6 PO polarization optical microscope (Russia). The thin film was prepared by drying of the RGO-PET solution in TFA on air.

3. Results and discussion

As was mentioned above, GO forms stable dispersion in a variety of polar solvents. Nevertheless, the preparation of the GO dispersions in water-miscible solvents were often proceeded via solvent-exchange technique [1] because of the exceptional dispersibility of GO in water. We also used the solvent-exchange method *in situ* in a polycondensation flask. A mixture containing exfoliated GO, water, excess of EG and BHET was employed as a starting material for polycondensation. The concentration of GO was 0.5% w/w regarding the weight of the monomer (BHET). The synthesis of RGO-PET represents a several stage process. Firstly, the water was removed from the polycondensation flask by distillation. Secondly, the excess of EG was distilled out with a slow stream of argon at higher temperature. The EG boiling point (197 °C) is significantly higher than the water boiling point, and the water was easily removed by distillation at 100 °C. During the second step, the reaction media became black because of the GO reducing. After removing most of EG, the mixture was heated to 270-280 °C, under reduced pressure and continuous stirring to remove volatile by-product of polycondensation. The polycondensation was stopped when the reaction media became too viscous to stir.

In order to study RGO from the RGO-PET composite, the weighted RGO-PET sample was dissolved in TFA and filtered thought 0.22 μm membrane filter. The isolated RGO (2.5% w/w) was studied by TGA, XRD, and XPS methods. In addition, thermally reduced GO at 300 °C (RGO-300) was prepared and studied by XPS for chemical composition.

TGA data of GO, PET, RGO-PET, and RGO are presented in figure 1 and Table 1. GO is thermally unstable and demonstrates weight loss even below 100 °C (3%). The main weigh loss of GO was observed at the range of 160-225 °C (43%). These values are different from the values described in [14,15]. Zhou et al. reported about 10% weight loss at 100 °C and 20% – at 200 °C. Evidently, GO thermal behaviour depends on the graphite oxidation method and post-treatment such as drying. RGO-PET and neat PET did not demonstrate mass loss below 300 °C, while GO and RGO showed 49.5% and 11.4% mass loss, respectively. The observed weight loss of GO takes place due to the removal of some oxygen-containing functional groups, and the weight loss of RGO is attributed to the removal of the absorbed TFA, which was used for isolation of RGO. However, TFA was not detected in RGO by the XPS method which was carried out in ultrahigh vacuum (10^{-6} Pa). Hence, we concluded that TFA was bonded to RGO primarily by Van der Waals or hydrogen bonds and it was removed in vacuum. A drastic mass loss was observed at 400-500 °C for RGO, neat PET and RGO-PET. The unusual thermal behavior of RGO demonstrates the presence of PET in RGO in spite of the thorough purification. Hence, the RGO sheets were functionalized by PET during the polycondensation. In order to calculate the weight fraction of grafted polymer in RGO, it was assumed that RGO is a mixture of PET and graphene. The second assumption is that the thermal behavior of graphene in RGO is the same as the thermal behavior of GO at temperatures above 300 °C. This assumption is reasonable because the polycondensation was carried out at 280 °C.

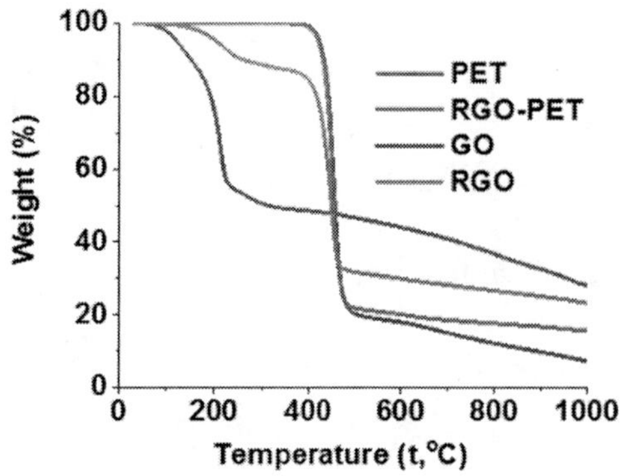

Figure 1. TGA curves of GO, RGO, neat PET, RGO-PET.

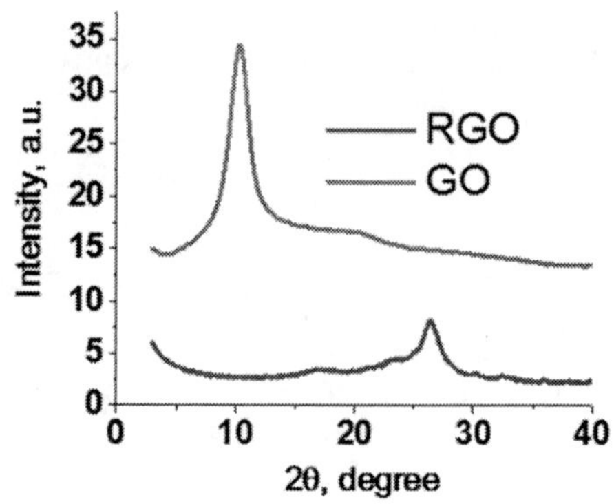

Figure 2. XRD pattern of GO and RGO.

Table 1. Mass loss of GO, RGO, RGO-PET and PET after heating from 300 °C to 500 °C and from 300 °C to 1000 °C.

Temperature range, °C	Residual weight, %			
	GO	RGO	RGO-PET	PET
300-500	92.8	35.8	21.9	20.4
300-1000	55.6	26.3	15.6	7.2

The XRD pattern of RGO and GO is shown in figure 2. The main reflex of GO appears at 10.3° corresponding to the interplanar distance between the graphene sheets equal to 0.858 nm. The diffraction peak of RGO is located at 26.4°. The presence of PET in RGO was revealed as a wide diffraction peak at about 20°, which was previously reported in [16].

The elemental analysis of GO, RGO, and RGO-300 was performed by the XPS method as it was described in [17]. The results are presented in Table 2.

Table 2. Relative atomic concentration of C and O in GO, RGO-300, and RGO.

Elements	GO, %	RGO, %	RGO-300
C	67.9	76.4	82.2
O	27.4	23.6	17.8
S	2.2	–	–
N	2.5	–	–

According to the PET formula, C/O ratio in PET is 2.5. Thermally reduced GO (RGO-300) contains 82.2% of carbon. Assuming that RGO sheets should consist of 82.2% carbon, the calculated by equation (3) weight fraction of PET in RGO is 55%, which is lower compared to the results based on TGA (80%). The reduction of GO in polycondensation melt is different from thermal reduction of GO due to the presence of a reducing reagent such as EG and acetaldehyde. We suppose RGO is more reduced than RGO-300 and that the XPS method underestimates PET content in RGO. As mentioned above, the weight concentration of RGO in RGO-PET is 2.5%. Relying on the TGA results, the true concentration of graphene in RGO-PET is about 0.5%.

Grafting of RGO with PET also was proved by swelling in two different solvents. There was swelling of RGO in TFA which is a solvent of PET and no swelling in $CHCl_3$ (figure 3).

Figure 3. Swelling of RGO in $CHCl_3$ (a) and in TFA (b).

DSC thermograms are presented in figure 4 and summarized in Table 3. The degree of crystallinity X_c and crystallization temperature T_c of RGO-PET are higher than X_c and T_c of neat PET. The increasing of X_c and T_c was also observed for PET/RGO composites prepared by mixing PET and RGO [16,18]. Glass transition temperature T_g and melting temperature T_m of RGO-PET are lower than that of PET.

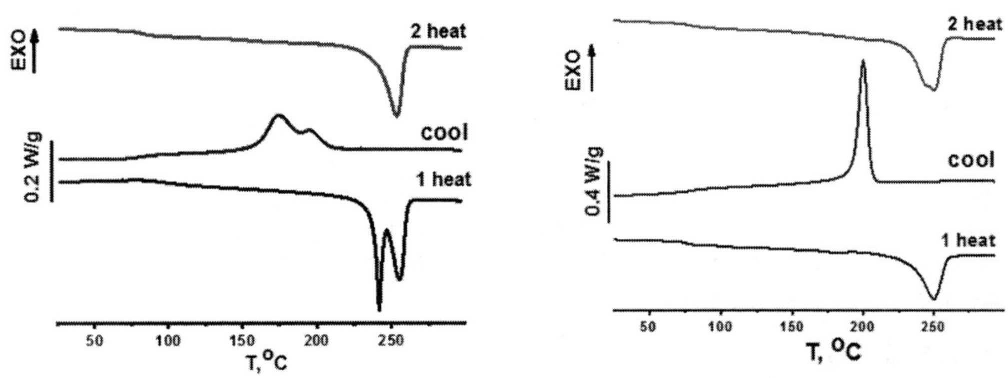

(a) (b)

Figure 4. DSC thermogram of pristine PET (a) and RGO-PET (b).

Table 3. DSC results for RGO-PET and PET.

Sample	T_g, °C	T_c, °C	T_m, °C	ΔH_f, J/g	X_c, %
PET	81.2	174.9	253.5	41.15	29.5
RGO-PET	77.4	205.6	250.8	42.78	30.7

The morphological features of the polymer composite were studied by optical microscopy (figure 5). Thanks to the obtained photographs, we observe a fairly uniform distribution of particles over the volume of the polymer. The sizes of agglomerates do not exceed 50 μm.

Traditionally, PET is a fiber forming polymer. That is why, to test the spinnability of melt filled with RGO, its rheological properties were measured. The flow curves of the neat PET melt and melt filled with RGO are shown in figure 6.

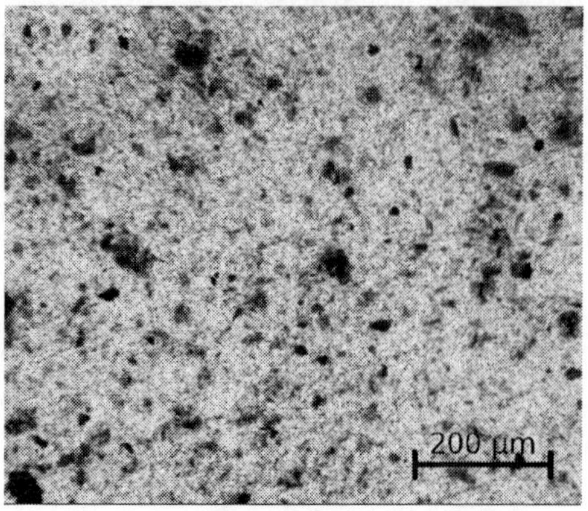

Figure 5. Microphotograph of RGO-PET.

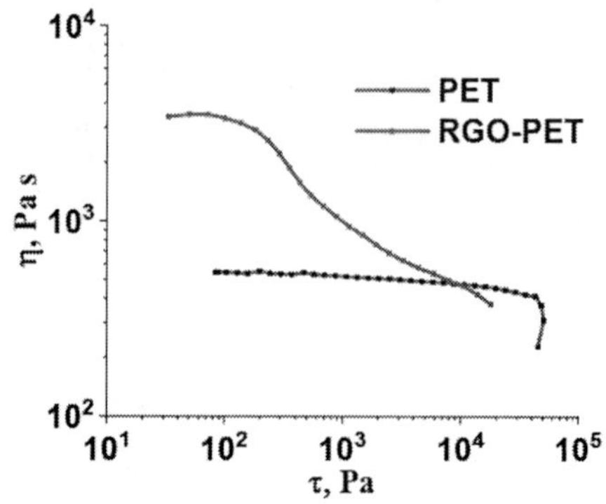

Figure 6. Flow curves of PET and RGO-PET.

As is seen, the PET melt has almost Newtonian behavior, and only at high shear stresses, its flow becomes unstable. The rheological properties of PET in the presence of 0.5% change drastically. The viscosity of RGO-PET is significantly higher than that of native PET at low stresses. Increasing the shear stress resulted in decreasing of viscosity after the short branch of constancy. Presumably, the mobility of PET macromolecules is restricted due to the presence of RGO particles. The most probable explanation is the connection of macromolecules with the RGO surface by contact forces of chemical nature due to grafting PET macromolecules on RGO sheets. In addition, partial physical contacts between solid particles may occur, though percolation threshold at 0.5% of filler content hardly could be achieved.

4. Conclusion

The RGO-PET composite was successfully synthesized from the monomer and GO dispersion by *in situ* melt polycondensation and simultaneous reduction of GO. The thermal behavior of RGO-PET is almost the same as that of the previously described RGO-PET composites prepared by mixing PET and graphene. The addition of RGO increases the degree of crystallinity and crystallization temperature and decreases melt temperature. The glass transition temperature of RGO-PET is lower as that of pristine PET. RGO isolated from the RGO-PET composite by filtration represents the graphene sheets with strongly bonded macromolecules of PET, which is proved by TGA, XPS, and XRD methods. The rheological behavior of RGO-PET was studied. The viscosity of RGO-PET is significantly higher than neat PET at low stresses and decreases with increase shear stress.

5. Acknowledgment

This work was supported by the Russian Science Foundation (Grant No. 17-79-30108).

References

[1] Xiao S-P and Huang H-X 2018 In situ vitamin C reduction of graphene oxide for preparing flexible TPU nanocomposites with high dielectric permittivity and low dielectric loss. *Polymer Testing* **66** 334 https://doi.org/10.1016/j.polymertesting.2018.01.026

[2] Hummers Jr W S and Offeman R E 1958 Preparation of graphitic oxide. *J. of the American Chemical Society* **80** 1339 https://doi.org/10.1021/ja01539a017

[3] Cobos M, González B, Fernández M J and Fernández M D 2018 Study on the effect of graphene and glycerol plasticizer on the properties of chitosan-graphene nanocomposites via in situ

green chemical reduction of graphene oxide. *Int. J. of Biological Macromolec.* **114** 599 https://doi.org/10/gdrqvz

[4] Soares C P P, Baptista R de L and Cesar D V 2017 Solvothermal Reduction of Graphite Oxide Using Alcohols. *Mater. Res.* **21** https://doi.org/10.1590/1980-5373-mr-2017-0726

[5] Wan T and Chen D 2019 In Situ Reduction of Graphene Oxide in Waterborne Polyurethane Matrix and the Healing Behavior of Nanocomposites by Multiple Ways. *J. of Polymer Sci. Part B* **57** 202 https://doi.org/10/gfw35h

[6] Zhang J, He Y, Zhu P, Lin S, Ju S and Jiang D 2019 In situ reduction of graphene oxide in the poly (vinyl alcohol) matrix via microwave irradiation. *Polymer Composites* **40** 170 https://doi.org/10/gfw35g

[7] Glover A J, Cai M, Overdeep K R, Kranbuehl D E and Schniepp H C 2011 In Situ Reduction of Graphene Oxide in Polymers. *Macromolec.* **44** 9821 https://doi.org/10.1021/ma2008783

[8] Yang H, Yu Z, Wu P, Zou H and Liu P 2018 Electromagnetic interference shielding effectiveness of microcellular polyimide/in situ thermally reduced graphene oxide/carbon nanotubes nanocomposites. *Appl. Surf. Sci.* **434** 318 https://doi.org/10.1016/j.apsusc.2017.10.191

[9] Yang H, Li Z, Zou H and Liu P 2017 Preparation of porous polyimide/in-situ reduced graphene oxide composite films for electromagnetic interference shielding. *Polymers for Adv. Technologies* **28** 233 https://doi.org/10.1002/pat.3879

[10] Zheng D, Tang G, Zhang H-B, Yu Z-Z, Yavari F, Koratkar N, Lim S-H and Lee M-W 2012 In situ thermal reduction of graphene oxide for high electrical conductivity and low percolation threshold in polyamide 6 nanocomposites. *Composite Sci. and Technol.* **72** 284 https://doi.org/10.1016/j.compscitech.2011.11.014

[11] Liu K, Chen L, Chen Y, Wu J, Zhang W, Chen F and Fu Q 2011 Preparation of polyester/reduced graphene oxide composites via in situ melt polycondensation and simultaneous thermo-reduction of graphene oxide *J. of Material Chemistry* **21** 8612 https://doi.org/10.1039/c1jm10717h

[12] Marcano D C, Kosynkin D V, Berlin J M, Sinitskii A, Sun Z, Slesarev A, Alemany L B, Lu W and Tour J M 2010 Improved Synthesis of Graphene Oxide. *ACS Nano* **4** 4806 https://doi.org/10.1021/nn1006368

[13] Mikhaylov P A, Levin I S, Lubenchenko A V, Bondarenko G N and Kulichihin V G 2019 Ultrasonication-assisted ultrafast solvothermal reduction of graphene oxide. *Mater. Sci. and Eng.* 10 https://doi.org/10.1088/1757-899X/525/1/012039

[14] Zhou T, Chen F, Liu K, Deng H, Zhang Q, Feng J and Fu Q 2011 A simple and efficient method to prepare graphene by reduction of graphite oxide with sodium hydrosulfite. *Nanotechnology* **22** 045704 https://doi.org/10/dr9jpj

[15] Kim D, Yang S J, Kim Y S, Jung H and Park C R 2012 Simple and cost-effective reduction of graphite oxide by sulfuric acid. *Carbon* **50** 3229 https://doi.org/10.1016/j.carbon.2011.11.014

[16] Awad S A and Khalaf E M 2018 Improvement of the chemical, thermal, mechanical and morphological properties of polyethylene terephthalate-graphene particle composites. *Bulletin of Mater. Sci.* **41** https://doi.org/10.1007/s12034-018-1587-1

[17] Lubenchenko A V, Batrakov A A, Pavolotsky A B, Lubenchenko O I, Ivanov D A, 2018 XPS study of multilayer multicomponent films. *Applied Surface Sci. Part A* **427** 711 https://doi.org/10.1016/j.apsusc.2017.07.256

[18] Aoyama S, Park Y T, Ougizawa T and Macosko C W 2014 Melt crystallization of poly(ethylene terephthalate): Comparing addition of graphene vs. carbon nanotubes. *Polymer* **55** 2077 https://doi.org/10.1016/j.polymer.2014.02.055

IOP Conf. Series: Materials Science and Engineering **693** (2019) 012037 doi:10.1088/1757-899X/693/1/012037

Effects of graphene oxide on white poplar X ASPEN (*Populus alba* x *Populus tremula*) hybrid microsprouts at various growth stages

O Zakharova[1,2], E Kolesnikova[3], D S Muratov[2], I Il'inikh[2], E Tsukanova[4], N Yevtushenko[5], N Strekalova[1] and A Gusev[1,2*]

[1]*Derzhavin Tambov State University*, 33 Internatsionalnaya Street, Tambov, 392000, Russia
[2]*National University of Science and Technology "MISIS"*, 4 Lenisnky Avenue, Moscow, 119991, Russia
[3]*All-Russian Scientific Research Institute of Sugar Beet and Sugar named after A.L. Mazlumov*, 86, VNIISS, Ramon district, Voronezh region, 396030, Russia
[4]*I.V. Michurin Federal Scientific Center*, 30, Michurin Street, Michurinsk, Tambov Region, 393774, Russia
[5]*Voronezh State Forestry University named after G.F. Morozov*, 8, Timiryazeva, Voronezh, 394087, Russia

[*]E-mail: nanosecurity@mail.ru

Abstract. Graphene-like materials are promising in biotechnology as antimicrobial agents and plant growth stimulants. We investigated the effects of graphene oxide nanoparticles in the composition of the MS0 medium on microclonal seedlings of the poplar and aspen hybrid. In the early stages of cultivation, graphene oxide improved microclone survival by 6–20%. However, then, along with some stimulation, negative effects were observed, manifested in a decrease in the development of the root system and a decrease in the thickness of the leaf lamina and the diameter of the stem. Our results indicate the promise of using graphene oxide in the early stages of the development of microclones of white poplar and aspen hybrids.

1. Introduction

Plant cell cultures are widely used both in the general and applied areas of plant biology such as cytology, embryogenesis, morphogenesis, nutrition, pathology and germplasm conservation, genetic manipulation, large scale microcloning and production of phytopathogen-free plants. Success of in vitro manipulations with plant cultures depends on the methods for surface disinfection, on the selected growth medium and growth-regulating chemicals as well as on donor plant genotype and physiological status and on cell cultures maintenance conditions. Culture media usually consist of macro- and microelements, amino acids, organic additives, vitamins, nitrogen sources, plant growth regulators and solidifiers and their compositions have a pronounced effect on morphogenetic potential of explants. Optimization of plant mineral nutrition promotes explants growth and morphogenesis. It also induces cell proliferation, organogenesis, somatic embryogenesis, improves stems quality and content of bioactive substances in cell and organ cultures [1]. Both chemical composition and physical properties of growth media should conform to the demands of each stage of microsprout development. At the first stage various antioxidants, such as cysteine, glutathione or ascorbic acid are added to the media in order

IOP Conf. Series: Materials Science and Engineering **693** (2019) 012037 doi:10.1088/1757-899X/693/1/012037

to inhibit damage from hydrolyzing enzymes thus preventing death of explants. Use of carbon materials fosters offshoots growth due to absorption toxic substances produced by explants. Khodakovskaya et al. demonstrate that multiwalled carbon nanotubes have the ability to enhance the growth of tobacco cell culture (55–64% increase over control) in a wide range of concentrations (5–500 μg/mL). Activated carbon (AC) stimulates cell growth (16% increase) only at low concentrations (5 μg/mL) while dramatically inhibits the cellular growth at higher concentrations (100–500 μg/mL) [2].

Authors of some studies report metallic nanoparticles (NPs) have positive effects when used during microcloning of plants. For instance, in the process of tobacco explants cultivation on modified medium containing TiO_2 NPs, high antibacterial activity of the studied NPs was observed [3, 4]. The best results for potato cells cultivation were obtained in the Murashige and Skoog medium (MC0) containing 1% (wt./wt.) of TiO_2 NPs [5], while no adverse effects on further growth and development of the culture cells was observed. Addition of 60 mg/L of TiO_2 NPs to the MC0 medium eliminated bacterial contamination in barley callus cultures [6].

Microbial contamination is a grave issue for plant cell cultures cultivation as it can lead to total destruction of all the experimental specimens. Both the explants and the laboratory environment act as sources of contamination. According to the accepted procedure for creating in vitro cultures various organs collected from plants undergo surface sterilization. Disinfection of explants is the most important procedure preceding the actual process of in vitro cultivation because in the growth media microorganisms develop quicker than explants and the former can seriously affect the culture initiation. Nevertheless, antimicrobial treatments are often ineffective. Various chemical agents can be used for obtaining sterilized explants, for example, bromine water, sodium hypochlorite, mercuric chloride, silver nitrate, antibiotics and fungicides, but the concentrations of the chemicals and exposure time may affect the quality of explants. Besides, some of the sterilizing agents show phytotoxic action [7, 8].

In the study [9] graphene was found to penetrate tomato seed husks that might facilitate water uptake, resulting in faster germination and higher germination rates. The authors of [10] investigated the effects of unfunctionalized graphene oxide and amine-functionalized graphene oxide on wheat (*Triticum aestivum*) after 9 days of hydroponic culture. They found that the incubation with both nanomaterials did not affect the final seed germination rate. The plant growth was enhanced under amine-functionalized graphene oxide exposure, and the root and stem lengths were increased. Several studies are focused on the phytotoxic effects of grapheme based materials. Authors of [11] reports graphene significantly inhibited plant growth and biomass of cabbage, tomato, and red spinach. Graphene cytotoxicity has been reported in the experiment on Arabidopsis thaliana cell culture [12], as well as amplification of the phytotoxicity of arsenic in wheat by graphene oxide [13].

However, the influence of graphene-like materials on microclonal propagation of tree cultures lacks thorough study. Thus, the aim of this work was to reveal the effects of graphene oxide in the cultural medium on the development indicators of white poplar and aspen (*Populus alba* L. x *Populus tremula* L.) hybrid microclonal sprouts.

2. Materials and methods

Graphene oxide (GO) NPs were prepared by the chemical exfoliation method (i.e. the Hummers method) [14], where graphite is oxidized with sulfuric acid and potassium permanganate. Concentrated H_2SO_4 (69 ml) was added to the mixture of graphite flakes (3 g, 1 wt. equivalent) with $NaNO_3$ (3 g, 0.5 wt equivalent), the obtained mixture was cooled down to 0°C. Then $KMnO_4$ (9 g, 3 wt equivalent) was added slowly, in order to keep the temperature of the reaction mixture below 20°C. After that the reaction mixture was heated up to 35°C and stirred for 30 min, at the same time 138 ml of water was added slowly to it. The reaction temperature was kept at 98°C by heating the mixture for 15 min on an electric heater. After this the heating was switched off and the reaction mixture was cooled down in a water-bath for 10 min. Then 420 ml of water and 3 ml of 30% H_2O_2 were added. The resulting mixture was run through a sieve (U.S. Sieve Size (WS Tyler, 300_m)) and then filtered through polyester fiber (Carpenter, Co). The obtained product was centrifuged (for 4 h at 4000 rpm), the supernatant fluid was decanted. The remaining solid material was rinsed successively with 200 ml of water, 200 ml of

30% HCl and 200 ml of ethanol (2×); after each rinsing cycle the mixture was run through a sieve (U.S. Sieve Size Standard), filtered through polyester fiber and centrifuged (for 4 h at 4000 rpm) while the supernatant fluid was decanted. The resulting material was coagulated with 200 ml of ether and the obtained suspension was filtered through a PTFE membrane with the pore size of 0.45 μm.

The obtained material was studied by atomic-force microscopy (AIST-NT instrument (AIST-NT, Russia) in semi-contact tapping mode with NTMDT AFM tips), Raman scattering spectroscopy (Thermo DXR Raman microscope (Thermo Scientific, USA) with 532 nm laser at 1mW power through 100× objective), optical (Biolam M-1 microscope, LOMO, Russia) and scanning electron microscopy (Vega3 microscope, Tescan, Czech Republic).

In order to prepare the samples for studying by Raman scattering spectroscopy, optical microscopy and scanning electron microscopy the colloidal solution of NPs was applied onto a silicon wafer and dried at 90 °C for 1 h.

For the survival rate study stem cuttings from the white poplar and aspen hybrid containing apical and axillary buds were cultured in vitro. Softwood shoots were taken in June from field grown plants. The 20-cm-long cuttings were stored at a low positive temperature for the period from 3 to 5 days. The cut surfaces were renewed and water was changed every 3 days. Then the shoots were washed in running water, cut into 3-5 cm segments and thoroughly washed in warm water with a surface-active agent. After that the cuttings were washed for 35 min in solution consisting of 200 μL of 2% sodium hypochlorite and 200 ml of distilled water with subsequent rinsing for 10 min in distilled water being thus prepared for sterilization. Sodium hypochlorite solutions were used as main sterilizing agents.

After sterilization the explants were planted into agar-solidified micro- and macro-salts Murashige and Skood medium (MS0) [15, 16]. The parts damaged by the sterilizing agent were removed before the cuttings were placed into the agarized growing medium.

The test tubes containing explants were placed on the light box shelves and cultivated at +24°C and 70% relative air humidity with 16 h photoperiod (5000 lx).

The newly developed microshoots were removed from the initial explants and were planter into the growing medium containing micro- and macro-elements and enriched with hormones and NPs promoting further growth and propagation.

The white poplar and aspen hybrid explants and microclones were cultivated on MS growth medium with addition of 0.75, 1.5, 3, 6 or 15 μg of GO per liter of the medium. In each variant, 10 seedlings were cultivated. The experiment was carried out in triplicate.

Throughout the cultivation process the dynamics of the morphometric parameters of the test tube plants was analyzed every second day, the following parameters were measured: the plant height, number of green leaves, number of yellow leaves, number of roots, condition on a 1 to 5 scale.

When the developed regenerants reached the height of 2 to 3 cm they were removed into culture tubes with the rooting medium, ½ woody plant medium (WPM) was used for this purpose [17]. GO NPs were added to the growth medium at the concentration of 1.5 μg of GO per liter of the medium in order to stimulate rhizogenesis.

Biochemical evaluation was carried out by measuring the photosynthesis activity which is the integral diagnostic indicator of the general plant resistance against stressors and by studying the oxidoreductases class enzymes, catalase in particular. It is established that the degree of catalase activity can be considered a practicable indicator for evaluating the activity of protection systems in organisms [18].

The photosynthesis activity in the chlorophyll-containing tissues was estimated by means of a fluorimetric indicator of physiological state IFSR-2 according to the method suggested in [19].

The catalase activity assay was carried out by the permanganatometric method - H_2O_2 titration in sulphuric acid. The enzyme activity was expressed as standard units (act. units or 1 mg of H_2O_2 per min) while taking into account that 1 ml of 0.1 μ $KMnO_4$ corresponds to 1.7 mg of hydrogen peroxide.

The histological and cytological studies of leaf lamina tissue measuring such criteria as stomatal density (i.e. the number of stomata per mm2) and stomatal pore size were carried out by means of a Video Test Morphology 4.0 hardware-software complex.

In order to study the effects of NPs on the white poplar and aspen hybrid microclones the plantlets of the uniform height with 4-5 leaves and at least 2 cm-long roots cultivated in the protected environment were used. The plantlets with well developed stem and root system were removed from the tubes with long forceps and with a special hook. The remaining agar was rinsed from the roots with 1 % potassium permanganate solution, the sprouts were planted into technological cassettes filled with moist cultivation soil and then placed into the greenhouse propagating frame.

The cultivation soil consisted of pH-neutral peat (Lama Torf, Russia) and perlite at the ratio of 3:1. The peat contained the following elements: nitrogen (N), phosphorus (P_2O_5), potassium (K_2O). The plants were kept in the controlled atmosphere at 22-24 °C and humidity of 85-90 %.

The planting cassettes willed with the soil were treated with solution of the reference growth regulator 2,4-epibrassinolide at 667 µg per liter and with GO solution at 3 µg per liter. Both solutions were prepared in distilled water.

3. Results and discussion

Figure 1 presents the results of atomic-force microscopy of the samples. The average size of the flakes varies from 0.1 to 3 µm (figure 1a, 1c) while their mean thickness is less than 1 nm (figure 1b, 1d), which indicates single layer GO.

(a) (b)

(c) (d)

Figure 1. The results of atomic-force microscopy of GO samples. Individual flakes marked by arrows.

In the Raman spectrum (figure 2a) one can observe the main lines characteristic for GO D (at 1338 cm^{-1}) and G (at 1590 cm^{-1}). Figure 2b represents a micrograph of the sample in the same spot where the Raman spectrum was taken. One can see folds and creases on the surface of individual flakes.

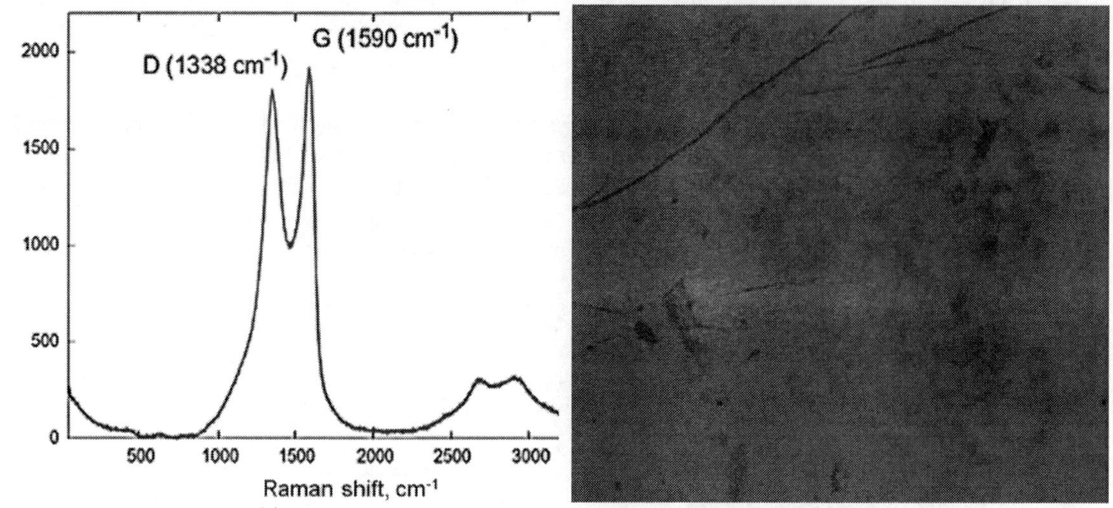

(a) (b)

Figure 2. The results of Raman scattering spectroscopy (a) and optical micrograph of the GO film on a silicon wafer (b). Folds and creases on the surface of individual flakes marked by arrows.

Figure 3 represents a scanning electron micrograph of the GO film on a silicon wafer after air-drying. Folds and creases on the surface of individual flakes can be clearly seen.

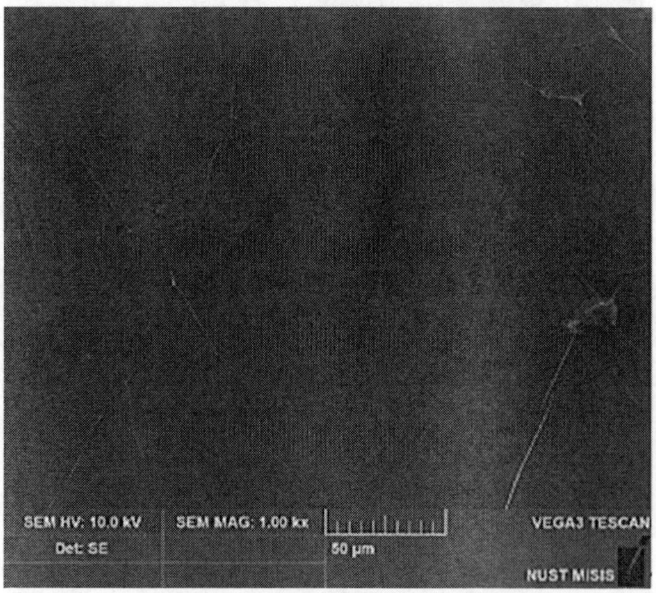

Figure 3. Scanning electron micrograph of the GO film on a silicon wafer after air-drying at 90 °C. Folds and creases on the surface of individual flakes marked by arrows.

After 1.5 months of cultivation a number of well-developed and free of any morphological anomalies microclones of the white poplar x aspen hybrid were obtained from the apical and axillary meristems (figure 4).

Figure 4. Microclones of the white poplar x aspen hybrid after 1.5 months of the tissue culture cultivation.

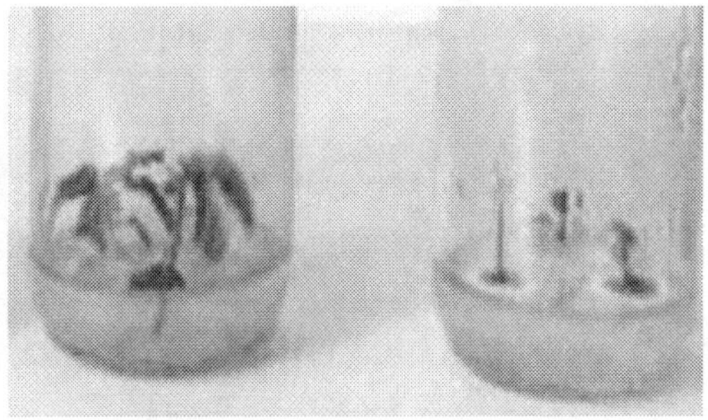

(a) (b)

Figure 5. Sterile and infected clones of woody plants on the cultivation medium with (a) and without (b) GO NPs.

During the cultivation process a positive effect of the NPs solutions on the growth processes in the tissue culture of the white poplar x aspen hybrid meristems was observed. The results presented in table 1 show the positive effect of GO in the cultivation medium on the analyzed parameters. The number of uninfected microclones was 100% at a concentration of GO 1.5 - 15 µg / L (figure 5), against a background of 80% in the control. However, it is worth noting that a dose of 15 µg / L inhibited the further development of seedlings, according to other indicators. The number of surviving passed microclones exceeded the control by 6-20% at 0.75 - 6 µg / L. The maximum stem length and the largest number of additional shoots were noted in groups 1.5 and 3 µg / L, root occurrence was observed here.

Table 1. Biomorphological parameters of the white poplar x aspen hybrid microclones cultivated on soils containing various concentrations of GO NPs.

NPs in cultivation medium, µg/L	Number of sterile microclones, pcs/%	Number of surviving microclones, pcs/%	Stem height, cm	Number of leaves, pcs	Number of additional shoots, pcs	Root occurrence	Microclones condition, on a 1 to 5 scale
Control	80.0	80.0	3.5	2	0	0	5
GO 0.75	86.0	86.0	3.5	2	0	0	5
GO 1.5	100.0	100.0	3.8	5	1	+	5
GO 3	100.0	100.0	4.0	3	2	+	5
GO 6	100.0	90.0	3.8	2	2	0	3
GO 15	100.0	70.0	3.4	2	0	0	2

Cuttings were taken from the normally developed microclones and transplanted into growing media enriched with hormonal components.

After 1-month cultivation on the growing media containing the following hormonal components: 0.2 mg/L of 6-benzylaminopurine (BAP) + 0.1 mg/L of indoleacetic acid (IA) + 0.3 mg/L of gibberellic acid (GA) we observed height reduction of the studied plants, while on the media containing both hormones and GO the number of additional shoots increased to up to 4 shoots per plant (table 2). The maximal stem height of 3.8 cm was registered in the plants cultivated on the hormone-free medium containing GO, in the control group the stem height was 3.5 cm. Introduction of GO NPs into the growing media also increased the number of surviving microclones up to 100%. One should note that roots were developed only by the microplants cultivated on GO-containing medium. The maximal number of leaves was observed in the same case - 5 leaves per plant.

Further studies of the influence of GO NPs at a concentration of 1.5 µg/L on the white poplar x aspen hybrid microclones at the rooting stage prior to transplanting into the soil showed adverse effect of GO NPs on the rooting process as only 30% of the microclones took root, which is by 20% less than in the control (table 3).

Table 2. Biomorphological parameters of the white poplar x aspen hybrid microclones cultivated for 1 month on media containing hormonal components and GO NPs at a concentration of 1.5 μg/L.

Version	Number of surviving clones, %	Stem height, cm	Number of leaves, pcs	Number of additional shoots, pcs	Root occurrence	Microclones condition, on a 1 to 5 scale
Control	80.0	3.5	4	0	0	5
0.2BAP+0.1IA+0.3GA	80.0	1.5	4	3	0	5
GO1.5	100.0	3.8	5	1	+	5
GO1.5+0.2BAP+0.1IA +0.3GA	100.0	2.4	4	4	0	5

Table 3. Morphometric parameters of the white poplar x aspen hybrid microclones cultivated at the rooting stage under the influence of GO NPs.

Version	Number of surviving microclones, %	Stem height, cm	Number of leaves, pcs	Number of additional shoots, pcs	Number of micro-clones with roots,%	Number of roots, pcs	Micro-clones condition, on a 1 to 5 scale
Control	100.0	5.0	7	0	50.0	1.0	4
GO 1.5	100.0	5.1	6	0	30.0	1.0	4

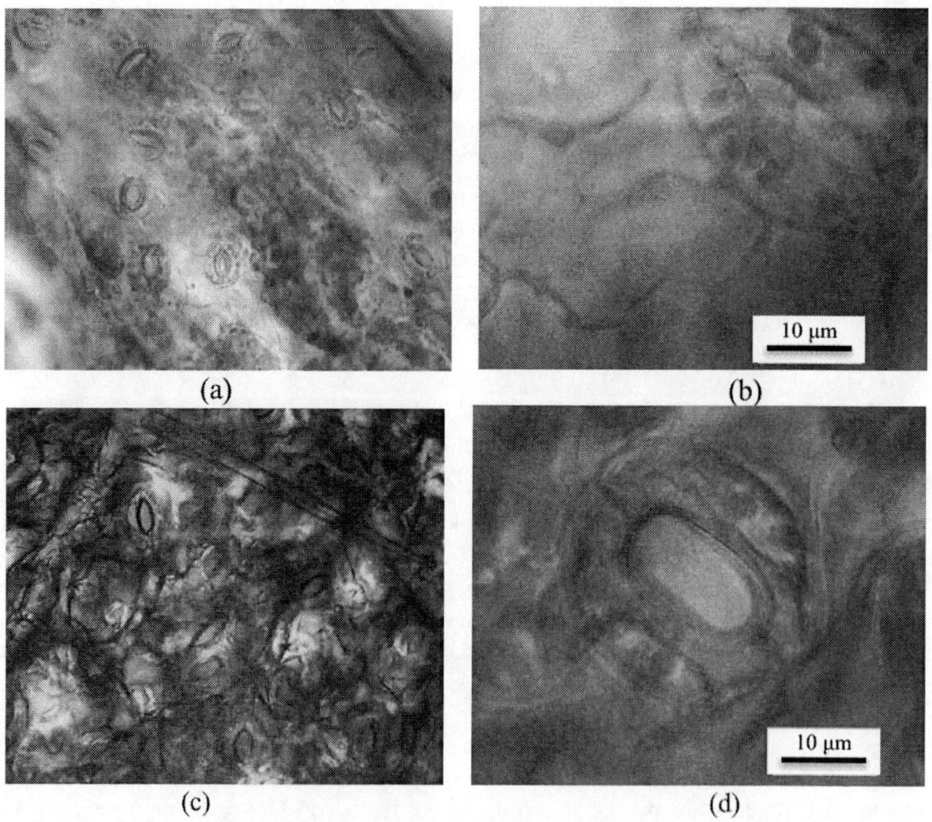

Figure 6. A leaf of the white poplar x aspen hybrid microclone: (a) - control, (c) - in the growth media with GO; a stoma of the white poplar x aspen hybrid microclone: (b) - control, (d) - in the growth media with GO.

The histological analysis showed that on the GO-containing media the plants developed higher stomatal density per mm2 and the stomata had larger openings (figures 6, 7a, 7b), these factors may result in improved heat tolerance of these plants.

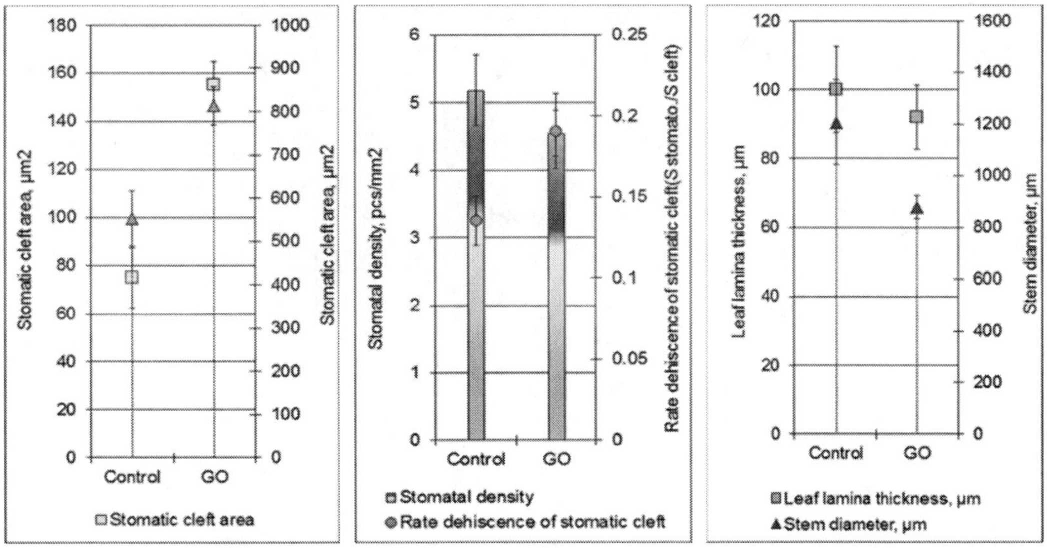

(a) (b) (c)

Figure 7. Biomorphological parameters of: (a), (b) stomata; (c) leaf and stem.

The leaf lamina thickness and stem diameter in the white poplar x aspen hybrid treated with GO were smaller than those in the control (figure 7c). This fact may be attributed to the nanomaterial phytotoxic effects which may manifest themselves in suppression of some morphometric parameters [10, 13].

Biochemical analysis of the sprouts showed slight increase in photosynthetic activity under the GO influence (figure 8) while the parameter dispersion within an individual plant was lower compared to the control (0.01580 vs 0.00323 relative units) prompting suggestions that the treatment not only increases the resistance potential in plants but also promotes stability of their state.

Figures 6-8 demonstrate positive effect of GO on the photosynthetic and antioxidant systems as well as on stomata number and function was noted, while at the same time some decrease in leaf lamina thickness and stem diameter was observed.

The study of the white poplar x aspen hybrid microclones during their adaptation to the protected environment showed that soil treatment with the GO solution had no observable effect on the sprouts during the adaptation period. At the same time, treatment with the reference growth regulating agent increased the number of surviving microclones by 10% while the number of leaves and additional shoots also increased (table 4). The noted small increase in the proportion of adapted plants in the GO group is within the statistical error.

Thus, during the cultivation we have observed positive effects produced by GO solutions on the growth processes of the white poplar x aspen hybrid meristems in the tissue culture. In every case the number of surviving microclones exceeded that in the control by 6-20%. The number of sterile microclones in the media containing more than 0.75 μg/L of GO reached 100%, thus exceeding the same parameter in the control group by 20%. Integrally, the maximal effect was observed at the GO NPs concentration of 3 μg/L.

Further cultivation of the microsprouts on the growth media enriched with hormones showed stem height reduction, while on the media containing both hormones and GO NPs the number of additional shoots increased up to 4 shoots per plant. The maximal stem height of 3.8 cm was registered in the plants cultivated on the hormone-free medium containing GO compared to 3.5 cm stems in the control group. Introduction of GO NPs into the growing media also increased the number of surviving microclones up to 100%. One should note that only the microplants cultivated on GO-containing medium developed roots. The maximal number of leaves was observed in the same case - 5 leaves per plant.

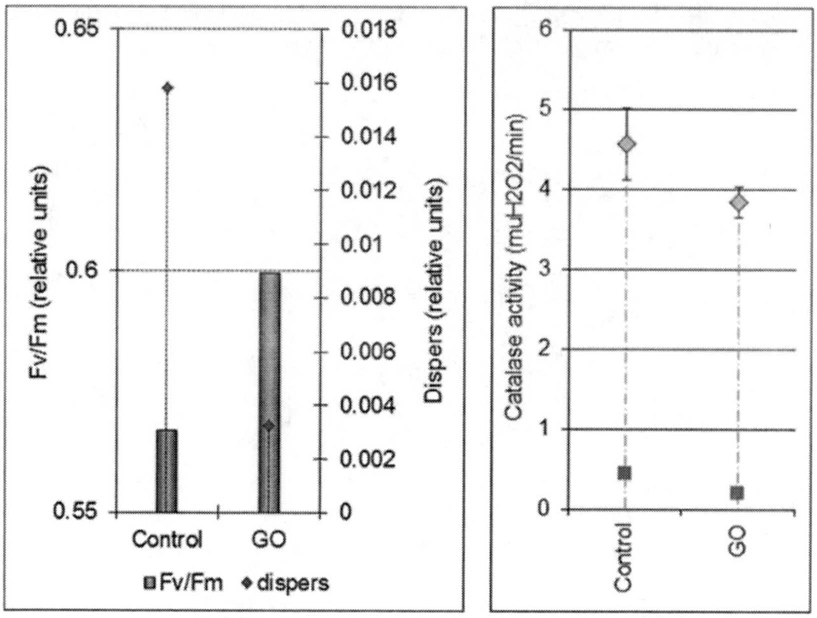

(a) (b)

Figure 8. Biochemical parameters of the sprouts: (a) photosynthetic activity, (b) catalase activity.

Table 4. Indicators of in vivo adaptation effectiveness of the white poplar x aspen hybrid microclones.

Version	Number of surviving microclones, %	Regenerated plant height, cm	Number of leaves, pcs	Number of wilted leaves, pcs	Number of adapted plants, %	Number of additional shoots, pcs	Microclones condition, on a 1 to 5 scale
Control	60.0	8.0	8	2	50,0	0	4
2,4-epibrassinolide	70.0	8.5	10	2	60.0	0	5
GO	60.0	8.4	7	4	51.0	0	4

Further studies of the influence of GO NPs showed adverse effect of GO NPs on the rooting process as only 30% of the microclones took root, which is by 20% less than in the control. The effect of soil treatment with the GO solution during adaptation of the plants to the protected environment was negligible.

The histological and biochemical assays showed that treatment with GO not only increased the resistance potential in plants but also promoted stability of their state.

On the one hand, the obtained results are connected with antibacterial and fungicide properties of GO described in a number of research papers [20, 21]. The material improves microsprouts survival rate by suppressing pathogenic microorganisms. On the other hand, some researchers mention phytotoxicity of graphene-like materials [10, 13, 22]. We observed similar effects at further stages of plant development. Besides, some of the described effects can be attributed to high sorption capacity of GO [23]. In particular, phytohormones and other active molecules can be aggregated on its surface and thus their biological activity can be extended.

Conclusion

Thus, we studied the effects of GO nanoparticles in the composition of the MS0 medium on microclonal seedlings of the poplar and aspen hybrid. In the early stages of cultivation, GO improved microclone survival by 6–20%. However, then, along with some stimulation, negative effects were observed, manifested in a decrease in the development of the root system and a decrease in the thickness of the leaf lamina and the diameter of the stem.

Our results prove that GO is a promising nanomaterial for application at the early stages of cultivation of white poplar and aspen hybrid microclones.

Acknowledgments

This study was supported by the Ministry of Education and Science of the Russian Federation, project no. RFMEFI57417X0159.

References

[1] Doo H K, Judy G and Iyyakkannu S 2017 Nanomaterials in plant tissue culture: the disclosed and undisclosed. *RSC Advances* 7 36492 doi: 10.1039/C7RA07025J

[2] Khodakovskaya M V., De Silva K, Biris A S, Dervishi E and Villagarcia H 2012 Carbon nanotubes induce growth enhancement of tobacco cells. *ACS Nano* **6** 2128 doi.org/10.1021/nn204643g

[3] Safavi K, Mortazaeinezahad F, Esfahanizadeh M and Asgari M J 2011 In vitro antibacterial activity of nanomaterial for using in tobacco plants tissue culture. *World Academy of Science, Engineering and Technology* **55** 372 doi: 10.13140/2.1.1236.8007

[4] Safavi K and Mortezaeinezhad F 2012 Role of modified tissue culture media by nanomaterial in potato propagation. *Res. Crop.* **13** 1 254

[5] Safavi K 2014 Effect of titanium dioxide nanoparticles in plant tissue culture media for enhance resistance to bacterial activity. *Bulletin of Environment, Pharmacology and Life Sciences* **3** 163

[6] Mandeh M, Omidi M and Rahaie M 2012 In Vitro Influences of TiO_2 Nanoparticles on Barley (Hordeum vulgare L.) tissue culture *Biol. Trace Elem. Res.* **150** 376 doi: 10.1007/s12011-012-9480-z

[7] Qin Y H, da Silva J A T, Bi J H, Zhang S L and Hu G B 2011 Response of in vitro strawberry to antibiotics. *Plant Growth Regul.* **65** 183 doi: 10.1007/s10725-011-9587-9

[8] Tambarussi E V, Rogalski M, Nogueira F T S, Brondani G E, de Martin V F and Carrer H 2015 Influence of antibiotics on indirect organogenesis of Teak *Ann. For. Res.* **58** 177 doi: 10.15287/afr.2015.345

[9] Zhang M, Gao B, Chen J and Li Y C 2015 Effects of graphene on seed germination and seedling growth. *J. Nanopart. Res.* **17** 78 doi: 10.1007/s11051-015-2885-9.

[10] Chen J, Yang L, Li Sh and Ding W 2018 Various physiological response to graphene oxide and amine-functionalized graphene oxide in wheat (*Triticum aestivum*). *Molecules* **23** 5 1104 doi: 10.3390/molecules23051104.

[11] Begum P, Ikhtiari R and Fugetsu B 2011 Graphene phytotoxicity in the seedling stage of cabbage, tomato, red spinach, and lettuce. *Carbon* **49,**12 3907 doi: 10.1016/j.carbon.2011.05.029

[12] Begum P and Fugetsu B 2013 Induction of cell death by graphene in Arabidopsis thaliana (*Columbia ecotype*) T87 cell suspensions. *J. Hazard. Mater.* **260** 1032 doi: 10.1016/j.jhazmat.2013.06.063.

[13] Hu X, Kang J, Lu K, Zhou R, Mu L and Zhou Q 2014 Graphene oxide amplifies the phytotoxicity of arsenic in wheat. *Sci. Rep.* **4** 6122 doi: 10.1038/srep06122

[14] Hummers W S and Offeman R E 1958 Preparation of graphitic oxide. *J. Am. Chem. Soc.* 1339 doi:10.1021/ja01539a017.

[15] Murashige T and Skoog F 1962 A revised medium for rapid growth and bio-assays with tobacco tissue cultures. *Physiol. Plant* **15** 473

[16] Trigiano R N and Gray D J 2005 Plant development and biotechnology *CRC Press LLC*. 358 p.
[17] Lloyd G and McCown B 1980 Commercially feasible micropropagation of mountain laurel, Kalmia latifolia, by shoot tip culture. *Combined Proceedings, International Plant Propagators' Society* **30** 421
[18] Gaetani G F, Ferraris A M, Rolfo M, Mangerini R, Arena S and Kirkman H N 1996 Predominant role of catalase in the disposal of hydrogen peroxide within human erythrocytes *Blood.* **87** 4 1595
[19] Genty B, Briantais J M and Baker N R 1989 The relationship between the quantum yield of photosynthetic electron transport and quenching of chlorophyll fluorescence. *BBA-Gen Subjects* **990** 87 doi.org/10.1016/S0304-4165(89)80016-9
[20] Akhavan O and Ghaderi E 2010 Toxicity of graphene and graphene oxide nanowalls against bacteria. *ACS Nano* **4** 10 5731 doi: 10.1021/nn101390x.
[21] Chen J, Peng H, Wang X, Shao F, Yuan Z and Han H 2014 Graphene oxide exhibits broad spectrum antimicrobial activity against bacterial phytopathogens and fungal conidia by intertwining and membrane perturbation. *Nanoscale* **6** 3 1879 doi: 10.1039/c3nr04941h
[22] Anjum N A, Singh N, Singh M K, Sayeed I, Duarte A C, Pereira E and Ahmad I 2014 Single-bilayer graphene oxide sheet impacts and underlying potential mechanism assessment in germinating faba bean (*Vicia faba* L.). *Sci. Total Environ.* **15** 472 834 doi: 10.1016/j.scitotenv.2013.11.018.
[23] Zhao G, Li J, Ren X, Chen Ch and Wang X 2011 Few-layered graphene oxide nanosheets as superior sorbents for heavy metal ion pollution management. *Environ. Sci. Technol.* **45** 24 10454 doi: 10.1021/es203439v.

IOP Conf. Series: Materials Science and Engineering **693** (2019) 012038 doi:10.1088/1757-899X/693/1/012038

Influence of GO-Ag and GO-CuO composites on micropropagated shoots during adaptation to greenhouse environment

O Zakharova[1,2], E Kolesnikova[3], E Kolesnikov[2], P Baranchikov[1], N Strekalova[1] and A Gusev[1,2*]

[1]*Derzhavin Tambov State University*, 33 Internatsionalnaya Street, Tambov 392000, Russia

[2]*National University of Science and Technology "MISIS"*, 4 Lenisnky Avenue, Moscow 119991, Russia

[3]*All-Russian Scientific Research Institute of Sugar Beet and Sugar named after A L Mazlumov*, 86 VNIISS, Ramon district, Voronezh region 396030, Russia

[*]E-mail: nanosecurity@mail.ru

Abstract. This paper focuses on the effects produced by GO-Ag and GO-CuO composite materials on microclones of hybrid white poplar and aspen (*Populus alba x Populus tremula*), hairy birch (*Betula pubescens*), crack willow (*Salix fragilis*), red oak (*Quercus rubra*) and scots pine (*Pinus sylvestris*) during their adaptation to greenhouse environment. The study revealed a positive effect of microclones treatment with GO-CuO and GO-Ag solutions at a concentration of 3 μg/l. The number of surviving and adapted microclones increased considerably, as did such parameters as the height of the plantlets, the number of leaves and the plantlets condition on a 1 to 5 scale. Photosynthetic activity was high in the white poplar and aspen hybrid, in birch, willow and pine. At the same time, in oak, the maximal stimulation of biomorphological parameters was accompanied by the decrease in photosynthetic activity. GO-Ag solution at a concentration of 4.5 μg/l produced an inhibitory effect on all the studied species except red oak.

1. Introduction

In the current economic situation, the microclonal propagation method seems to be the most promising and effective one as it quickly produces many plants that are clones of each other.

Tissue culture technique for micro propagation includes several stages: 1) initiation and establishment of culture; 2) multiplication in the aseptic environment; 3) rooting of shoots; 4) hardening of plantlets moved to greenhouse conditions.

The establishment of plantlets in soil with their further adaptation to the field environment is the final and the most critical stage. The success of the whole micropropagation process often depends on it [1].

Adaptation of microclones to *ex vitro* conditions, also known as hardening, is the most cost- and labor-consuming process. The industrial microclonal propagation suffers heavy losses when plantlets are transferred to non-sterile conditions. As very often their development is arrested, the plantlets lose leaves and die. In some cases, less than 50% of the plantlets survive the transferal.

If the microclones survive the adaptation stress, then they take root in the non-sterile conditions. Successful *ex vitro* establishing of the plantlets allows for the industrial microclonal propagation.

The success of the plant adaptation is largely determined by the selection of media for the rooting stage. The *in vitro* plantlets often do not develop fully functional root hairs, cuticle and stomata; considerable losses may occur when the cloned shoots are transplanted to non-sterile environment. Thus, the soil composition and structure should be carefully selected for each specific plant species and variety [2]. Hardening of the microcloned plantlets can have a positive effect on their further growth and development. As a result, the hardened and adapted plants can be established in the field [3].

According to a well-established opinion, the industrial microclonal propagation is viable only when the stage of transferal of the plantlets to non-sterile conditions is successful.

Application of nanoparticles and nanomaterials as antimicrobial agents is one of the key directions of the nanobiotechnology development [4].

Silver and copper oxide nanoparticles are well known for their antibacterial activity [5, 6]. However, the researchers have described both positive and adverse effects of these nanomaterials on plants [7-10]. Graphene oxide also possesses antibacterial properties [11-13]. Moreover, hydrophilicity of graphene oxide makes it a useful material for stabilizing hydrophobic particles in colloidal systems [14]. It should also be noted that copper is an essential micronutrient, which is incorporated in many proteins and enzymes, therefore, playing a significant role in plant health and nutrition [15].

Based on the aforesaid, the aim of our research is to elucidate the effects of nanopreparations based on graphene oxide and metallic nanoparticles on the microclonal shoots of such woody plants as white poplar and aspen hybrid (*Populus alba* x *Populus tremula*), hairy birch (*Betula pubescens*), crack willow (*Salix fragilis*), red oak (*Quercus rubra*) and scots pine (*Pinus sylvestris*) in the period of their adaptation to greenhouse environment when the plantlets previously cultivated in the aseptic conditions are exposed to severe contamination with phytopathogenic microorganisms.

2. Methods and materials

The graphene oxide (GO) nanoparticles were fabricated by the chemical exfoliation method, i.e. by Hummers method [16], consisting in graphite oxidation with sulfuric acid and potassium permanganate. Average size of the flakes varies from 0.1 to 3 μm while their mean thickness is less than 1 nm.

The silver nanoparticles were obtained by electric explosion of conductive wire in the atmosphere of inert gases (Advanced Powder Technologies, LLC, Tomsk, Russia). Particles are predominantly spherical in shape. The arithmetic average particle size is about 100 nm. Individual particles, on average, have sizes of 70-150 nanometers and form microagglomerates.

The copper oxide nanoparticles were obtained by the chemical deposition method with subsequent reduction of the analytically pure $Cu(NO_3)_2 \cdot 6H_2O$ solution in the alkaline medium [17]. Copper oxide particles are flakes (as thick as 100 nm) united in aggregates about 5 μm in size.

The nanocomposite materials were produced by mixing the aqueous colloidal solutions of the components with subsequent ultrasound treatment, while the ratio of graphene oxide to silver/copper oxide was 1:10.

The analysis of the obtained composite materials was carried out on the high-resolution scanning electron microscope "Merlin" (Carl Zeiss, Germany) with an energy dispersive analyzer "10mm^2 SDD Detector - X-Act" (figure 1).

SEM and EDX analysis showed the presence of silver and copper oxide nanoparticles, concentrated on graphene oxide flakes. This indicates the successful formation of nanocomposite materials.

The study of the nanomaterials effects on microclones of hybrid white poplar and aspen (*Populus alba* x *Populus tremula*), hairy birch (*Betula pubescens*), crack willow (*Salix fragilis*), red oak (*Quercus rubra*) and scots pine (*Pinus sylvestris*) was carried out during their adaptation to greenhouse environment. Plantlets of the uniform height with 4-5 leaves and at least 2 cm-long roots were transferred to the non-sterile conditions of the greenhouse. The plantlets with a well-developed stem and root system were removed from the tubes with long forceps or with a special hook. The roots were rinsed from the remaining agar with 1 % potassium permanganate solution. Then, the microclones were planted into technological cassettes filled with moist cultivation soil and placed into a greenhouse propagating frame.

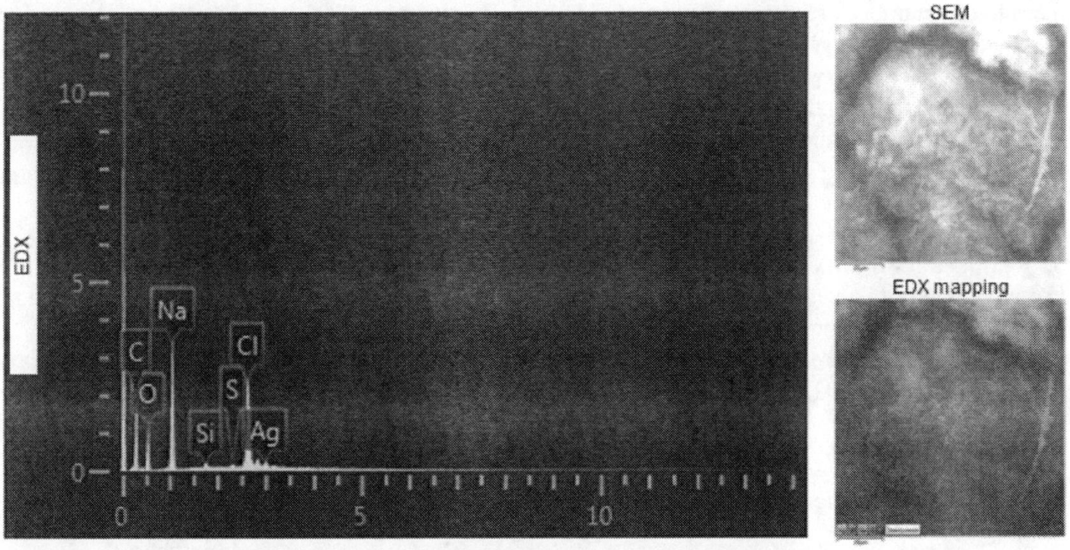

(a)

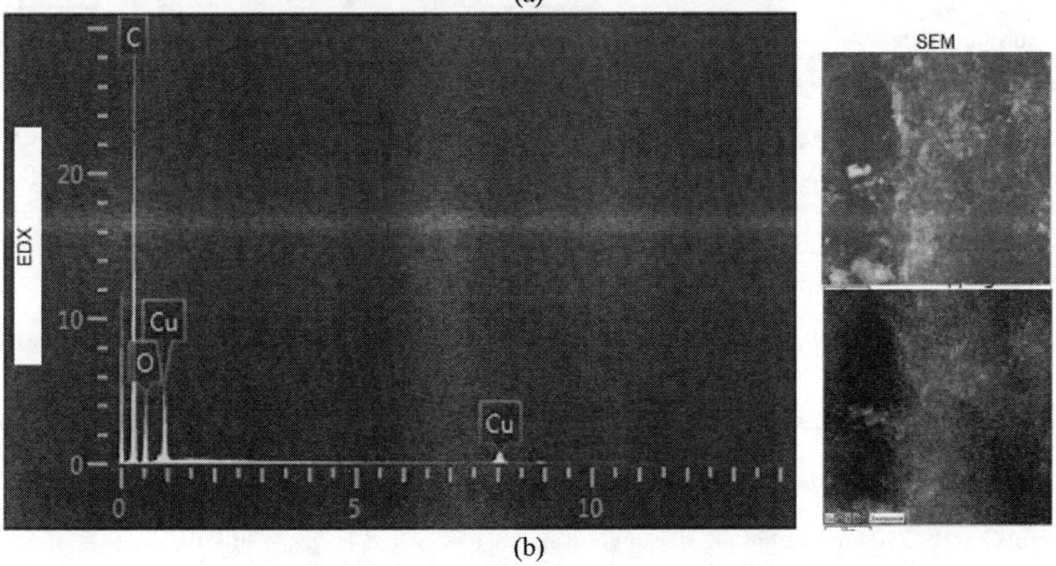

(b)

Figure 1. Results of SEM and EDX analysis of the samples of composite materials: (a) GO-Ag, (b) GO-CuO.

The soil consisted of pH-neutral peat and perlite at the ratio of 3:1. Perlite improves soil air permeability due to its considerable oxygen-retaining and wicking abilities. On the other hand, this material has low absorptive capacity, and plants cultivated on perlite or on soils with a high perlite content require frequent watering. The peat contained the following elements: nitrogen (N), phosphorus (P_2O_5), potassium (K_2O). The pH value of the peat aqueous extract was 6.7. The soil was treated with fungicidal agent 24 hours prior the plantlets relocation.

After the transferal, the microclones were treated with aqueous dispersions containing GO-Ag and GO-CuO preparations at 3 or 4.5 μg/l concentrations. The reference growth regulator 2,4-epibrassinolide at a concentration of 667 μg/l was used for positive control.

The pots with the plantlets were placed in a greenhouse and cultivated at 22-24 °C and humidity of 85-90 %.

In order to assess the root-taking and morphometric characteristics, the following parameters were measured: a number of surviving regenerated plants, %; plant height in cm; a number of leaves, pcs; a number of wilted leaves, pcs; a number of axillary shoots, pcs; a number of adapted regenerated plants, %; the microclones condition on a 1 to 5 scale.

Biochemical evaluation was carried out by measuring the photosynthesis activity which serves as the integral diagnostic indicator of plant resistance against stressors. The photosynthesis activity in the chlorophyll-containing tissues was determined by means of a fluorimetric indicator of physiological state IFSR-2 (Genty et al. method) [18].

3. Results and discussion

For adaptation of rooted microclones of the white poplar x aspen hybrid, the 3 μg/l concentrations of the two studied solutions produced better results than the same substances at 4.5 μg/l.

The best results were observed after foliage application of GO-CuO (3 μg/l). The number of surviving microclones reached 92%, thus exceeding the control by 32%. The mean plantlets height after 3 weeks of the experiment was 9.2 cm, while the mean number of leaves was 11 pcs. The microclones condition reached 5 points, while 60% of the plantlets were fully adapted to the new conditions (table 1).

Table 1. Adaptation effectiveness indicators for the white poplar x aspen hybrid microclones.

Active substances in the solution	Number of surviving microclones, %	Plant height, cm	Number of leaves, pcs	Number of wilted leaves, pcs	Number of adapted regenerated plants, %	Number of axillary shoots, pcs	Microclones condition on a 1 to 5 scale
Control	60.0±5	8.0±0.8	8±0	2±0	50.0±3	0	4±0
2,4-epibrassinolide	70.0±5	8.5±0.4	10±1	2±1	60.0±2	0	5±0
GO-CuO 3 μg/l	92.0±3	9.2±0.4	11±1	0±0	60.0±2	0	5±0
GO-Ag 3 μg/l	72.0±3	7.6±1.1	8±0	2±0	63.0±3	0	5±0
GO-CuO 4.5 μg/l	60.0±8	8.5±0.7	6±0	3±0	35.0±4	0	4±1
GO-Ag 4.5 μg/l	48.0±7	7.0±0.6	7±1	4±1	36.0±4	0	3±1

The lowest values were noted after the treatment with the solution containing 4.5 μg/l of GO-Ag. Only 48% of the plantlets survived; this value falls short of the control by 12%. Also, this group displayed the minimal height and the highest number of wilted leaves, while the number of adapted microclones did not exceed 36% compared with 50% of adapted plants in the control group.

For hairy birch rooted microclones adaptation, the concentration of 3 μg/l of all the nanomaterials also produced better results than the concentration of 4.5 μg/l.

The best results were observed after foliage application of GO-CuO and GO-Ag at a concentration of 3 μg/l. The number of surviving microclones reached 91% upon GO-CuO application and 90% after treatment with GO-Ag, thus exceeding the control by 23% and 24%, respectively. The mean plantlets height after 3 weeks of the experiment was 5.8 cm and 5.3 cm, with the mean number of leaves equaling 6 pcs. The microclones condition reached 5 points on a 1 to 5 scale, while 70% of the plantlets were fully adapted to the new conditions with only 1 wilted leaf (table 2).

The lowest values were noted after the treatment with the solution containing 4.5 μg/l of GO-Ag. The number of surviving plantlets fell short of the control by 15%, and the number of adapted plants decreased by 20% compared to the control. The microclones condition reached 3 points on a 1 to 5 scale.

Table 2. Adaptation effectiveness indicators for hairy birch microclones.

Active substances in the solution	Number of surviving microclones, %	Plant height, cm	Number of leaves, pcs	Number of wilted leaves, pcs	Number of adapted regenerated plants, %	Number of axillary shoots, pcs	Microclones condition on a 1 to 5 scale
Control	65.0±5	4.0±0.2	5±1	4±0	50.0±5	0	4±0
2,4-epibrassinolide	70.0±3	4.5±0.5	6±1	2±0	60.0±2	0	5±0
GO-CuO 3 μg/l	91.0±4	5.8±0.3	6±0	1±0	70.0±2	0	5±0
GO-Ag 3 μg/l	90.0±2.3	5.3±0.2	6±1	2±0	65.0±3	0	5±0
GO-CuO 4.5 μg/l	60.0±6	4.0±0.4	5±1	3±0	35.0±4	0	4±0
GO-Ag 4.5 μg/l	50.0±6	4.1±0.2	5±0	4±0	30.0±4	0	3±0

The same tendencies were noted in crack willow. The best results were observed after foliage application of GO-CuO solution at a concentration of 3 μg/l. The number of surviving microclones reached 100%, thus exceeding the control by 20%. The mean plantlets height after 3 weeks of the experiment was 5.8 cm with the mean number of leaves equaling 6 pcs. The microclones condition reached 5 points on a 1 to 5 scale, while 70% of the plantlets were fully adapted to the new conditions with only 1 wilted leaf.

The lowest values of the adaptation effectiveness parameters were recorded upon the treatment with GO-Ag at a 4.5 μg/l concentration (table 3).

Table 3. Adaptation effectiveness indicators for crack willow microclones.

Active substances in the solution	Number of surviving microclones, %	Plant height, cm	Number of leaves, pcs	Number of wilted leaves, pcs	Number of adapted regenerated plants, %	Number of axillary shoots, pcs	Microclones condition on a 1 to 5 scale
Control	80.0±3	4.8±0.2	5±1	3±0	50.0±2	0	4±1
2,4-epibrassinolide	85.0±2	5.0±0.2	6±1	2±0	60.0±2	0	5±0
GO-CuO 3 μg/l	100.0±0	6.2±0.1	6±0	1±1	70.0±1	0	5±0
GO-Ag 3 μg/l	100.0±0	5.6±0.2	5±0	2±0	65.0±3	0	5±0
GO-CuO 4.5 μg/l	71.0±2	4.0±0.1	5±1	2±0	35.0±4	0	4±0
GO-Ag 4.5 μg/l	70.0±3	2.9±0.3	5±1	4±0	30.0±4	0	3±1

The analysis of the red oak growth dynamics showed that the treatment with 3 μg/l solution of GO-Ag increased the number of surviving plantlets by more than 2 times compared with the control; the maximal values for plant height and for adapted microclones number were observed in the same group. At the same time, the maximal number of leaves (6 pcs per plant) was in the group treated with GO-CuO at 3 μg/l, similar to the number of leaves in the control group. Also, in the group treated with GO-CuO at 3 μg/l, we observed the least number of wilted leaves; here 50% of microclones survived, compared with 30% of surviving plantlets in the control. The number of adapted plantlets was three times higher than that in the control (table 4). One should note that every variant of treatment increased the number of surviving plantlets. GO-CuO and GO-Ag solutions applied to the soil at a concentration of 4.5 μg/l, reduced the plantlets height by about 1 cm; the values of all the other parameters remained close to that of the control.

Table 4. Adaptation effectiveness indicators for red oak microclones.

Active substances in the solution	Number of surviving microclones, %	Plant height, cm	Number of leaves, pcs	Number of wilted leaves, pcs	Number of adapted regenerated plants, %	Number of axillary shoots, pcs	Microclones condition on a 1 to 5 scale
Control	30.0±5	3.5±0.2	4±1	2±0	10.0±2	0	3±0
2,4-epibrassinolide	45.0±3	3.9±0.1	6±1	2±0	20.0±1	0	3±0
GO-CuO 3 μg/l	50.0±2	3.6±0.1	6±0	1±1	30.0±1	0	4±0
GO-Ag 3 μg/l	70.0±2	4.2±0.2	4±0	2±0	50.0±1	0	4±0
GO-CuO 4.5 μg/l	40.0±6	2.5±0.3	2±1	2±0	10.0±2	0	4±1
GO-Ag 4.5 μg/l	50.0±7	2.6±0.3	2±1	3±0	10.0±3	0	4±0

In case of scots pine microclones, GO-Ag solution at a concentration of 3 μg/l produced the best results. The number of surviving microclones reached 70%, thus exceeding the control by 10%. The mean plantlets height reached its maximal value of 4.8 cm, as did the mean number of leaves – 12 leaves per plant. The maximal number of adapted plants, namely, 70% of the plantlets, was observed in the same group (table 5).

Table 5. Adaptation effectiveness indicators for scots pine microclones.

Active substances in the solution	Number of surviving microclones, %	Plant height, cm	Number of leaves, pcs	Number of wilted leaves, pcs	Number of adapted regenerated plants, %	Number of axillary shoots, pcs	Microclones condition on a 1 to 5 scale
Control	60.0±3	3.0±0.2	8±1	0±0	60.0±2.3	0	5±1
2,4-epibrassinolide	60.0±2	3.5±0.2	10±1	0±0	60.0±2	0	5±0
GO-CuO 3 μg/l	60.0±4.5	4.6±0.1	10±0	0±0	60.0±3	0	4±0
GO-Ag 3 μg/l	70.0±2.5	4.8±0.3	12±1	0±0	70.0±4	0	5±0
GO-CuO 4.5 μg/l	60.0±2.0	3.4±0.5	6±0	1±0	30.0±2.2	0	4±1
GO-Ag 4.5 μg/l	30.0±3.0	3.3±0.6	6±0	1±0	30.0±5.0	0	3±1

The treatment with GO-CuO at 3 μg/l also produced good results. The lowest values were noted in the variant of GO-Ag at 4.5 μg/l, when the number of surviving and adapted plantlets decreased by two times in comparison with the control. The microclones were evaluated at 3 points on a 1 to 5 scale, this being the worst value for the overall plant condition.

Biochemical analysis was carried out for the variants with the most positive response to the treatment with nanomaterials. Photosynthetic activity measurements in the white poplar x aspen hybrid cultivated in the medium treated with GO-CuO at a concentration of 3 μg/l showed (figure 2) that they were characterized by a higher level of photosynthetic activity compared with the control. At the same time, the parameter variability within each individual plant was lower, thus prompting a suggestion that the treatment not only increases the resistance potential in plants but also improves stability of their condition.

The analysis of the effect produced by composite materials on photosynthetic activity in hairy birch plantlets showed that the highest results were produced by GO-Ag at 3 μg/l (figure 3). The obtained data, as well as the results of biomorphological research, allow one to assume that GO-Ag nanoparticles solutions at 3 μg/l can increase resistance potential in hairy birch microclones if applied during *in vivo* adaptation.

IOP Conf. Series: Materials Science and Engineering **693** (2019) 012038 doi:10.1088/1757-899X/693/1/012038

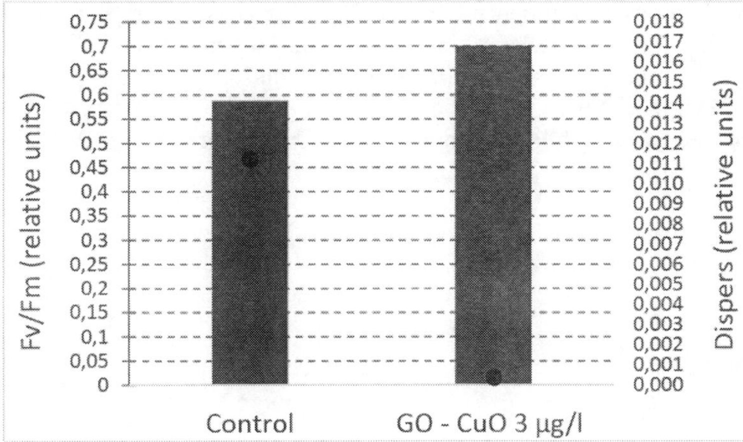

Figure 2. Photosynthetic activity of the white poplar x aspen hybrid microclones.

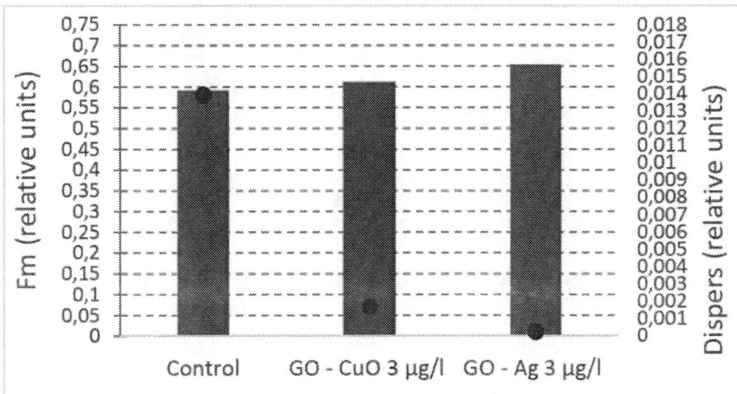

Figure 3. Photosynthetic activity of the hairy birch microclones.

Photosynthetic activity measurements in crack willow microclones showed that the treatment with GO-CuO at a concentration of 3 μg/l increased the studied parameter (figure 4); the plants from this group combined a higher level of photosynthetic activity with reduced parameter variability within each individual plant.

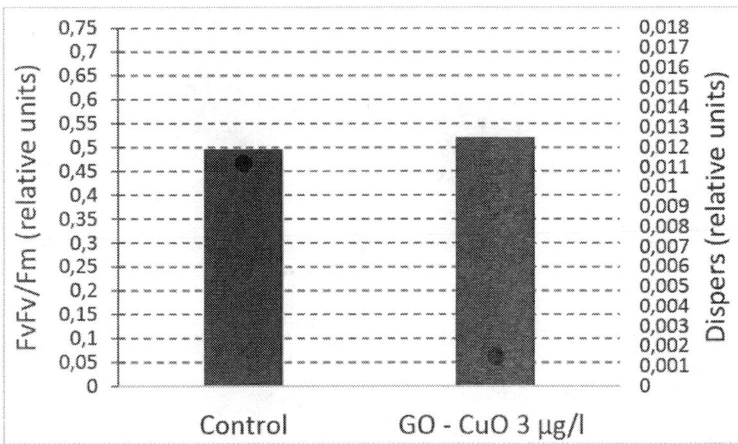

Figure 4. Photosynthetic activity of the crack willow microclones.

Unlike in the above described species, where increase in photosynthetic activity happened in the same treatment variants that produced the strongest stimulating impact on the biomorphological parameters, in the red oak microclones, inverse relationship was detected. The treatment with GO-Ag solution at 3 μg/l had the strongest stimulating effect on the adaptation processes while photosynthetic activity was below the activity observed in the control group (figure 5). At the same time, it should be noted that the parameter variability within each individual plant in the treated group was considerably below the level of variability observed in the control group.

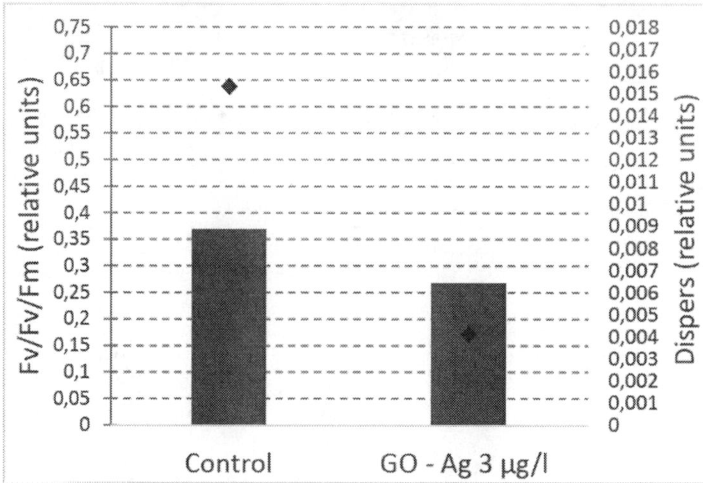

Figure 5. Photosynthetic activity of the red oak microclones.

For the scots pine plantlets, the highest values of adaptation parameters were detected when treated with GO-Ag solution at 3 μg/l. When we carried out the photosynthetic activity analysis in this group, the results showed that in the treated plants, photosynthesis processes were more than twice as active as in the control (figure 6).

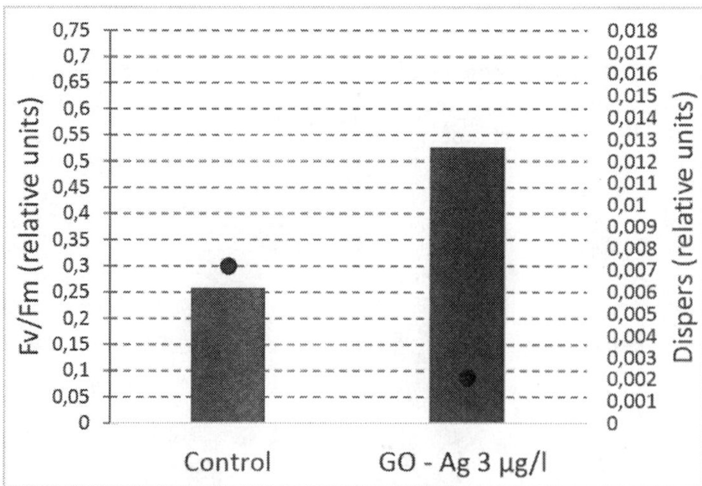

Figure 6. Photosynthetic activity of the scots pine microclones.

4. Conclusion

In order to summarize the results, we would like to note that the plant treatment with GO-CuO and GO-Ag solutions at a concentration of 3 μg/l had a pronounced positive effect on the plant adaptation process. The application of these solutions considerably increased the number of surviving and adapted microclones as well as plant height, the number of leaves and overall plantlet condition on a 1 to 5 scale. We assume that antibacterial activity of the nanoparticles contributed the most to this improvement [12]. Besides that, the nanoparticles may have influenced expression of the genes responsible for the synthesis of the growth-influencing phytohormones [19].

The analysis of photosynthetic activity in the white poplar x aspen hybrid, birch, willow, and pine plantlets revealed the increase in the values of the parameters and the reduction in their variability within an individual plant, which can point at a high resistance potential and stability of the condition of these microclones. Stimulation of photosynthetic activity by copper nanoparticles was described by Nekrasova et al. in their paper [20].

At the same time, the red oak microclones displayed a "non-standard" reaction to the treatment with nanopreparations. Namely, the treatment with GO-Ag solution at 3 μg/l produced the maximal level of stimulation of biomorphological parameters, while the photosynthetic activity fell below the values observed in the control. The decrease in photosynthetic activity may be attributed to the toxic action of silver nanoparticles on the photosynthetic system of the studied oak plants [21, 22].

The solution of GO-Ag at a concentration of 4.5 μg/l had a considerable toxic impact on all the plants except red oak. Phytotoxic effects of silver and graphene oxide nanoparticles have been described in a number of papers [23-26]. The researchers have observed that the toxic effects are produced mainly by high concentrations of nanoparticles; this agrees with the results of our research. The noted resistance of the oak plants against the adverse effects of nanoparticles may be attributed to a high content of tannin and other phenolic compounds in the tissues. It is known that phenolic compounds perform protective functions preventing mechanical damage to the cell walls as well as reducing the amount of reactive oxygen intermediates [27], i.e. they effectively block the main assumed phytotoxicity mechanisms of nanoparticles [28, 29]. However, the mechanisms of the effects of GO-Ag solutions on oak plants require further study.

Thus, composite materials based on graphene oxide and silver and copper oxide nanoparticles can be considered promising components of sterilizing products, taking into account their potential phytotoxicity at high concentrations.

5. Acknowledgments

This study was supported by the Ministry of Education and Science of the Russian Federation, project no. RFMEFI57417X0159.

References

[1] Turk B A, Swartz H J, Zimmerman R H 1984 Adventitious shoot regeneration in vitro–cultured leaves of Rubus genotypes. *Plant Cell. Tiss. Org. Cult.* **38** 11

[2] Pierik R L M 1987 In vitro culture of higher plants. *Martinus Nijhoff Publishers* 344

[3] Preece J E 2002 The most tricky part of micropropagation: establishing plants in greenhouses and fields *Comb. Proc. Intern. Plant Propagators' Soc.* **51** 300

[4] Ahmed F, Arshi N, Kumar S, Gill S S, Gill R, Tuteja N 2013 *Nanobiotechnology: scope and potential for crop improvement, in Crop Improvement Under Adverse Conditions*, eds N. Tuteja and S. S. Gill (New York: Springer) pp 245–269 doi:10.1007/978-1-4614-4699-0-11

[5] Cho K H, Park J E, Osaka T, Park S G 2005 The study of antimicrobial activity and preservative effects of nanosilver ingredient. *Electrochim Acta.* **51** 5 956 doi.org/10.1016/j.electacta.2005.04.071

[6] Yoon K Y, Byeon J H, Park J H, Hwang J 2007 Susceptibility constants of Escherichia coli and Bacillus subtilis to silver and copper nano particles. *NCBI.* **373** 2–3 572 doi:10.1016/j.scitotenv.2006.11.007

[7] Yin L, Colman B P, McGill B M, Wright J P, Bernhardt E S 2012 Effects of silver nanoparticle exposure on germination and early growth of eleven wetland plants. *PLoS ONE7* **7** 47674 doi: 10.1371/journal.pone.0047674

[8] Gusev A A, Kudrinsky A A, Zakharova O V, Klimov A I, Zherebin P M, Lisichkin G V, Vasyukova I A, Denisov A N, Krutyakov Y A 2016 Versatile synthesis of PHMB-stabilized silver nanoparticles and their significant stimulating effect on fodder beet (Beta vulgaris L.). *Mater. Sci. and Eng.* **62** 152 doi: 10.1016/j.msec.2016.01.040

[9] Rawat S, Pullagurala V L R, Hernandez-Molina M, Sun Y, Niu G, Hernandez-Viezcas J A, Peralta-Videa J R and Gardea-Torresdey J L 2018 Impacts of copper oxide nanoparticles on bell pepper (Capsicum annum L.) plants: a full life cycle study. *Environ. Sci.: Nano.* **5** 1 83 doi:10.1039/C7EN00697G

[10] Rahmani F, Peymani A, Daneshv E and Biparva P 2016 Impact of zinc oxide and copper oxide nano-particles on physiological and molecular processes in Brassica napus L. *Indian J. of Plant Physiology* **21** 2 122 doi:10.1007/s40502-016-0212-9

[11] Akhavan O, Ghaderi E 2010 Toxicity of graphene and graphene oxide nanowalls against bacteria. *ACS Nano.* **4** 10 5731 doi.org/10.1021/nn101390x

[12] Chen J, Peng H, Wang X, Shao F, Yuan Z, Han H 2014 Graphene oxide exhibits broad spectrum antimicrobial activity against bacterial phytopathogens and fungal conidia by intertwining and membrane perturbation. *Nanoscale.* **6** 1879 doi: 10.1039/c3nr04941h

[13] Gusev A, Zakharova O, Vasyukova I, Muratov D S, Rybkin I, Bratashov D, Lapanje A, Il'inikh I, Kolesnikov E, Kuznetsov D 2019 Effect of GO on bacterial cells: Role of the medium type and electrostatic interactions. *Mater. Sci. & Eng.* **99** pp 275–281 doi:10.1016/j.msec.2019.01.093

[14] Alves D C B, Silva R, Voiry D, Asefa T, Chhowalla M 2015 Copper nanoparticles stabilized by reduced graphene oxide for CO_2 reduction reaction. *Mater. for Renewable and Sustainable Energy.* **1** 4 doi.org/10.1007/s40243-015-0042-0

[15] Rastogi A, Zivcak M, Sytar O, Kalaji H M, IIc X, Mbarki S, Brestic M 2017 Impact of Metal and Metal Oxide Nanoparticles on Plant: A Critical Review. Frontiers in chemistry. *NCBI.* **5** 78 1 doi:10.3389/fchem.2017.00078

[16] Hummers W S, Offeman R E 1958 Preparation of graphitic oxide. *J. Am. Chem. Soc.* **60** 1339 doi.org/10.1021/ja01539a017

[17] Kundu M 2017 Negative electrode comprised of nanostructured CuO for advanced lithium ion batteries. *J. of Cluster Sci.* **28** pp. 1595 doi:10.1007/s10876-017-1170-8

[18] Genty B, Briantais J M, Baker N R 1989 The relationship between the quantum yield of photosynthetic electron transport and quenching of chlorophyll fluorescence. *Biochimica et Biophysica Acta.* **990** 1 87 doi.org/10.1016/S0304-4165(89)80016-9

[19] Siddiqui M H, Al-Whaibi M H, Mohammad F 2015 Nanotechnology and Plant Sciences: Nanoparticles and Their Impact on Plants. *Springer.* 1007 doi: 10.1007/978-3-319-14502-0

[20] Nekrasova G F, Ushakova O S, Ermakov A E, Uimin M A 2011 Effects of copper (II) ions and copper oxide nanoparticles on Elodea densa Planch. *Russ. J. Ecol.* **42** 458 doi:10.1134/S1067413611060117

[21] Cvjetko P, Zovko M, Stefanic P P, Biba R, Tkalec M, Domijan A M, Vrcek I V, Letofsky-Papst I, Sikic S, Balen B 2018 Phytotoxic effects of silver nanoparticles in tobacco plants. Environmental Science and Pollution Research *NCBI.* **25** 6 5590 doi: 10.1007/s11356-017-0928-8

[22] Matorin D N, Todorenko A, Kh N S, Zayadan B K, Rubin A B 2013 Effect of silver nanoparticles on the parameters of chlorophyll fluorescence and P700 reaction in the green alga Chlamydomonas reinhardtii. *Microbiology.* **82** 6 809

[23] Mazumdar H and Ahmed G U 2011 Phytotoxicity effect of Silver nanoparticles on Oryza sativa. *Int. J. of ChemTech Res.* **3** 1494

[24] Cui P, Zhang Y, Ma X, He Y, Li Y, Zhao Z Z 2014 Phytotoxicity of silver nanoparticles to cucumber (Cucumis sativus) and wheat (Triticum aestivum). *Zhejiang Univ-Sci A.* **15** 8 662 doi.org/10.1631/jzus.A1400114

[25] Hu X, Kang J, Lu K, Zhou R, Mu L 2014 Graphene oxide amplifies the phytotoxicity of arsenic in wheat. Zhou. *Scientific Reports 4.* **19** 6122 doi.org/10.1038/srep06122
[26] Chen J, Yang L, Li S and Ding W 2018 Various Physiological Response to Graphene Oxide and Amine-Functionalized Graphene Oxide in Wheat (Triticum aestivum). *Molecules.* **23** 5 doi:10.3390/molecules23051104
[27] Gershenzon J 2003 *Plant Defenses: Surface Protectants and Secondary Metabolites. In:Plant Physiology,* Eds. L Taiz and E Zeiger (Massachusetts: Sinauer Associates) pp 347–376 doi:10.1093/aob/mcg079
[28] Dietz K J and Herth S 2011 Plant nanotoxicology. *Trends Plant Sci.* **16** 11 582 doi:10.1016/j.tplants.2011.08.003
[29] Tripathi D K, Tripathi A, Shweta S, Singh Y, Singh K, Vishwakarma G, Yadav S, Sharma V K, Singh R K, Mishra R G, Upadhyay N K, Dubey Y L and Chauhan D K 2017 Uptake, Accumulation and Toxicity of Silver Nanoparticle in Autotrophic Plants, and Heterotrophic Microbes: A Concentric Review. *Front. Microbiol* doi:10.3389/fmicb.2017.00007

IOP Conf. Series: Materials Science and Engineering **693** (2019) 012039 doi:10.1088/1757-899X/693/1/012039

Kinetic studies on toluene removal from aqueous solutions using carbon nanotubes

A Kucherova[1*], I Shubin[1], A Yermakov[1], A Gerasimova[1], N Memetov[1] and A Popova[1]

[1]Department "Technology and Methods of Nanoproducts Manufacturing", *Tambov State Technical University,* 106 Sovetskaya Str, 392000 Tambov, Russian Federation

[*1]E-mail: anastasia.90k@mail.ru

Abstract. The effectiveness of Taunit-M carbon nanotubes in removing toluene from aqueous media was studied. Adsorption kinetic data were analyzed implementing the simple Weber-Morris diffusion model and very common phenomenological models (Lagergren and Ho-McKay) which simulate mass transfer processes through using formal equations for chemical kinetics. Investigating the correspondence of the experimental data to different models as regards adsorption kinetics, it can be should assumed that the toluene adsorption on the carbon nanotubes represents a two-stage process consisting of the phenomena of diffusive transfer and adsorption itself. The analysis of the temperature effect on the equilibrium sorption shows that the Gibbs free energy ($\Delta G°$) values vary in the range typical of adsorption interactions.

1. Introduction

Volatile organic compounds (VOCs) are widely used in the manufacturing of paints, synthetic rubber, agricultural chemicals, pharmaceuticals, as well as in the oil refining and other industries. Each year, large amounts of wastewater containing these organic compounds are discharged into aquatic media from industrial, transport and target sources of disposal. Since toluene is a flammable, toxic, carcinogenic and/or mutagenic substance [1], its availability in air, water and soil is dangerous even at low levels both for human health and the environment. Thus, purifying wastewater from organic compounds that threaten the well-being of the population represents an urgent task.

To date, lots of different methods have been developed to dispose of the organic compounds. They include membrane absorption [2], condensation [3], thermal combustion method [4], catalytic and photocatalytic oxidation technologies [5], non-thermal plasma [6], phytoremediation [7], and biotechnology [8]. However, each of these purification options has certain technological and economic limitations. Among them, adsorption methods are one of the most efficient ways for the removal of low (residual) contents of VOCs, since they are more technologically and economically productive in these cases [9, 10].

Carbon nanotubes (CNTs) are among the most widely investigated carbon nanomaterials. Their hollow layered structure and high specific surface area make it possible to consider them as a promising material for the adsorption of various types of organic compounds. Studies comparing the adsorption properties of CNTs with those of activated carbons prove that the former are much more effective in removing both organic [11] and inorganic [12] compounds from wastewater.

Analysis of the scientific literature shows that functionalized CNTs with different grafted groups are usually used in studies on the removal of organic pollutants from aqueous solutions. For instance, in the paper, [13] in order to improve the adsorption active sites for powerful connection between the adsorbent and the dye, surface modification of adsorbent was performed to create functional groups through surface oxidation of MWCNTs and zinc oxide (ZnO). At the same time, the adsorption properties of untreated (pristine) single-walled and multi-walled CNTs remain poorly investigated.

Considering the aforementioned, the purpose of this study was to study the adsorption behavior of unoxidized MWCNTs in purifying aqueous solutions from toluene.

2. Materials and methods

In the present study, Taunit-M CNTs, produced by NanoTechCenter Ltd (Tambov, Russia) were employed as an adsorbent. Before using, they were repeatedly washed with distilled water and then dried in a vacuum drying oven at 110 ° C for at least 2 h.

Toluene with (purity >99 %), purchased from Merck (Darmstadt, Germany) was used as an adsorbate. It was diluted with deionized water to the desired concentrations.

Experiments were carried out in 50- mL tubes containing the sorbent and toluene solutions. The tubes were shaken at 100 rpm on BioRS-24 rotators (Biosan, Riga, Latvia). After that, the mixtures were centrifuged using a 5810 R centrifuge (Eppendorf, Hamburg, Germany) at 10670 g for 10 min to separate the phases. The results of preliminary experiments showed that the adsorption equilibrium was achieved within 60 min. The concentration of toluene solutions was determined using a UV-1201 spectrophotometer (Shimadzu, Kyoto, Japan) at 264 nm. The effects of the following factors on the adsorption process (efficiency of toluene removal) were assessed: contact time (0-80 min), adsorbent dose (0.01-0.07 g), adsorbate concentration (25-145 mg/L), and solution temperature (25-45 °C).

All the tests were performed in triplicate and repeated three times, and only mean values are reported herein. The maximum deviation between parallel experiments did not exceed 3 %.

3. Results and discussion

3.1. Adsorption tests

3.1.1. Adsorbate concentration effect. The isotherm of the toluene adsorption on the Taunit-M CNTs (figure 1) belongs to the S-type, characteristic of the adsorption of sparingly water-soluble liquid compounds, according to the classification proposed by Giles and his colleagues [14]. It has a slight concave initial section, and a further increase in the adsorbate concentration leads to the formation of a plateau responsible for achieving the saturation of the adsorbent. In this case, the maximum adsorption capacity was found to be 121.7 mg/g under the following conditions: temperature – 25 °C, C_0 – 105 mg/L, and C_e – 31.98 mg/L). Therefore, the other tests were carried out under these conditions.

3.1.2. Adsorbent dose effect. The adsorbent dose is an important parameter, since it determines the adsorption capacity of the adsorbent at a given initial concentration of the adsorbate solution. The dependence of the toluene removal efficiency on the adsorbent dose (0.01-0.07 g) is presented in figure 2.

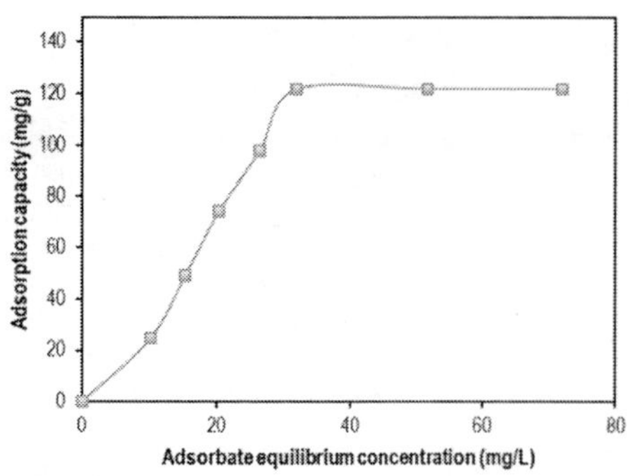

Figure 1. An isotherm of the toluene adsorption on the Taunit-M CNTs (temperature 25 °C).

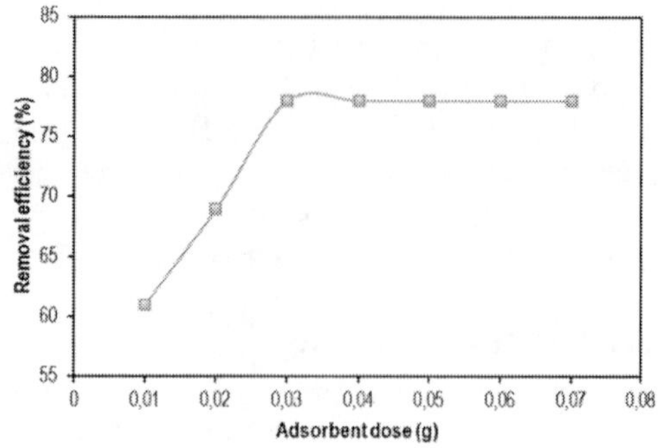

Figure 2. The Taunit-M CNTs dose effect on the toluene removal.

It can be seen that the maximum recovery is achieved with 0.03 g of the CNTs. The initial increase in the adsorption capacity is related to the specific surface area and the availability of active sites, as reported in [15]. However, a further increase in the dose of the adsorbent introduced does not lead to simbate changes in the toluene removal efficiency, which may be due to the adsorbent agglomeration as a result of high CNTs concentration, accompanied by partial blocking of the adsorption sites. Since no noticeable increase in the removal efficiency was observed after 0.03 g, further studies were conducted using this adsorbent dose.

3.1.3. Time effect. Kinetics plays a significant role in solute adsorption, and also represents adsorption efficiency of an adsorbent. The experimental results of studying the contact time effect on the toluene removal are shown in figure 3. From this figure, it follows that equilibrium is achieved within 60 min, and the removal rate is uniform throughout the entire period of time before the onset of equilibrium. Thus, for this study, the optimum length of time is 60 min.

3.1.4. Temperature effect. Since wastewater discharges come from the chemical industry at different temperatures, it is necessary to investigate the temperature effect on the toluene level reduction in solution using the adsorbent material considered herein. It was found that the adsorption capacity of the CNTs increases when raising the temperature from 25 to 35 °C. A further increase had the reverse impact on the adsorption capacity (figure 4). This confirms the results reported elsewhere [16].

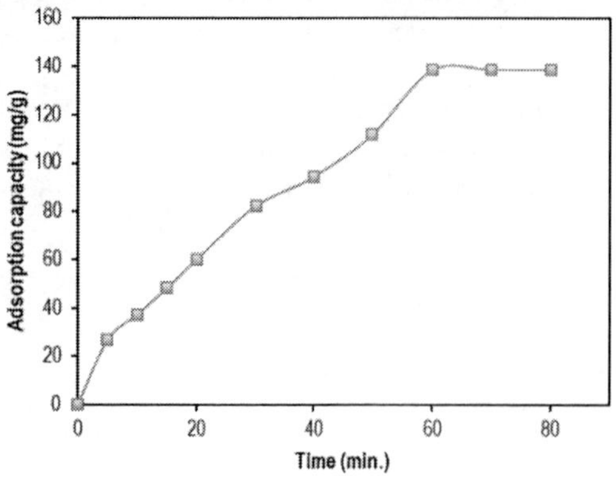

Figure 3. A kinetic curve obtained for the toluene adsorption on the Taunit-M CNTs (temperature 25 °C).

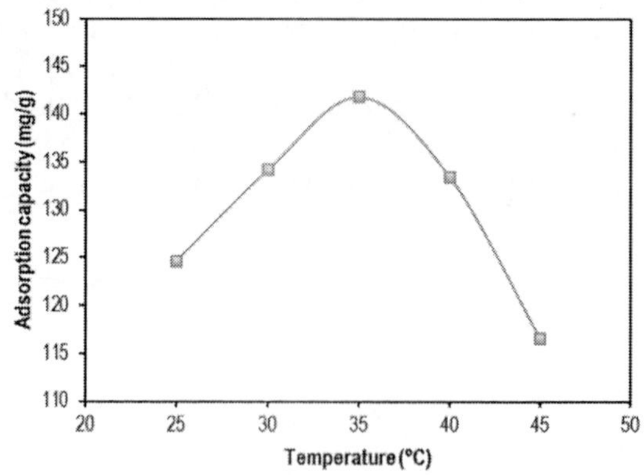

Figure 4. The temperature effect on the toluene adsorption capacity using the Taunit-M CNTs.

IOP Conf. Series: Materials Science and Engineering **693** (2019) 012039 doi:10.1088/1757-899X/693/1/012039

Favorable toluene adsorption (141.8 mg/g) at a rather low temperature (35 °C) is due to intermolecular interactions between the adsorbent and the adsorbate, which increase the affinity of the adsorption sites and, in turn, increase the diffusion of organic compounds toward the adsorbent. However, a decrease in the adsorption capacity is associated with a weakening of the interaction forces between the adsorbent and the adsorbate due to desorption at elevated temperatures.

3.2. Adsorption modeling

3.2.1. Kinetics. The adssorption at the solid-solution interface with a reaction localized in the adsorbent pores is usually described by four main stages [17]:

1) external mass transfer: from the bulk of the solution to the boundary layer – a liquid film on the adsorbent surface;
2) film diffusion: the adsorbate penetrates through the liquid layer of the adsorbent surface;
3) internal mass transfer: the adsorbate is transferred from the surface into the pores of the adsorbate;
4) adsorption: the interaction between the adsorbate with the active sites.

One (or more) of these stages can have a decisive role. The adsorption is usually the quickest step. The external diffusion is not often critical, especially when the experimental system is well mixed (agitated), as in the case considered herein. Thus, it is believed that only the intraparticle diffusion controls the adsorption kinetics [18, 19].

Reported in 1963, the Weber-Morris (intraparticle diffusion) model has been implemented many times to describe liquid-phase adsorption [20, 21]. This model was developed for adsorbents possessing a system of pores, the size of which allows adsorbate molecules to penetrate inside and settle there, and the adsorption process involves the rate of the internal mass transfer stage, *i.e.* adsorbate diffusion in the adsorbent pore system.

The model is based on earlier studies, especially the work [22], considering a theoretical approach developed for spherical adsorbent particles.

It can be assumed that the internal diffusion is the stage that limits the process rate, if the dependence is a straight line passing through the origin (C=0). However, when constructing the dependence of q_t *vs.* square root of t, a straight section, not passing through the origin, was obtained, which may be due to the difference in the mass transfer rate at the initial and final adsorption stages [23]. Moreover, such a deviation of the straight line from the origin shows that the diffusion inside the pores is not the only stage of rate control [24]. The results show that the mechanism of the toluene adsorption on the CNTs is complex, and both the surface adsorption and the intraparticle diffusion contribute to the actual process. The k_{id}value of the intraparticle diffusion model was found to be 17.9 mg/(g min$^{1/2}$). The correlation coefficient (R^2~0.99) clearly indicates that the adsorption obeys this model (table 1).

Taking into account the complexity of quantitative description of diffusion processes through simple models, one should also analyze the possibility of implementing very common phenomenological models that simulate mass transfer processes using formal equations for chemical kinetics. This approach most often involves the implementation of the pseudo-first- and pseudo-second-order models.

In 1898, Lagergren introduced the first-order rate equation for the adsorption of oxalic and malonic acids on carbon to explain adsorption kinetics on solid surfaces. The Lagergren (pseudo-first-order) equation is used to describe kinetic processes based on the solution concentration and the adsorption capacity of a solid [25]. It is the first rate equation developed to characterize adsorption in liquid/solid systems.

It should be noted that the pseudo-first-order kinetic model mathematically coincides with the equation characteristic of diffusion processes, but the adsorption rate constant is used as a fixed parameter. However, in the case of diffusion, the process rate depends on the size of adsorbent particles and the film thickness. If the chemical kinetics is the limiting stage, then the adsorption rate does not depend on the above factors, but depends only on the concentration and the temperature. Thus, when the adsorption kinetics is described by the pseudo-first-order kinetic model, the adsorption of the adsorbate is preceded by its diffusion.

The analysis of the kinetic curves constructed for the toluene adsorption on the Taunit-M CNTs in the coordinates of $log(q_e - q_t)$ *vs.* t showed that the dependence is linear (R^2>0.9). The toluene adsorption capacity predicted by the Lagergren model (q_L=139.6 mg/g) almost coincided with the experimentally obtained maximum adsorption capacity (q_e=141.8 mg/g). The pseudo-first-order adsorption rate constant was found to be 3.6 10^{-2} min^{-1} (table 1).

Table 1. Kinetic parameters of the toluene adsorption on the Taunit-M CNTs.

Adsorbate	*Kinetic models*							
	Weber-Morris		Lagergren			Ho-McKay		
	k_{id}	R^2	K_1	q_L	R^2	K_2	q_{H-M}	R^2
Toluene	17.9	0.99	$3.6 \cdot 10^{-2}$	139.63	0.97	$1.3 \cdot 10^{-4}$	192.31	0.92

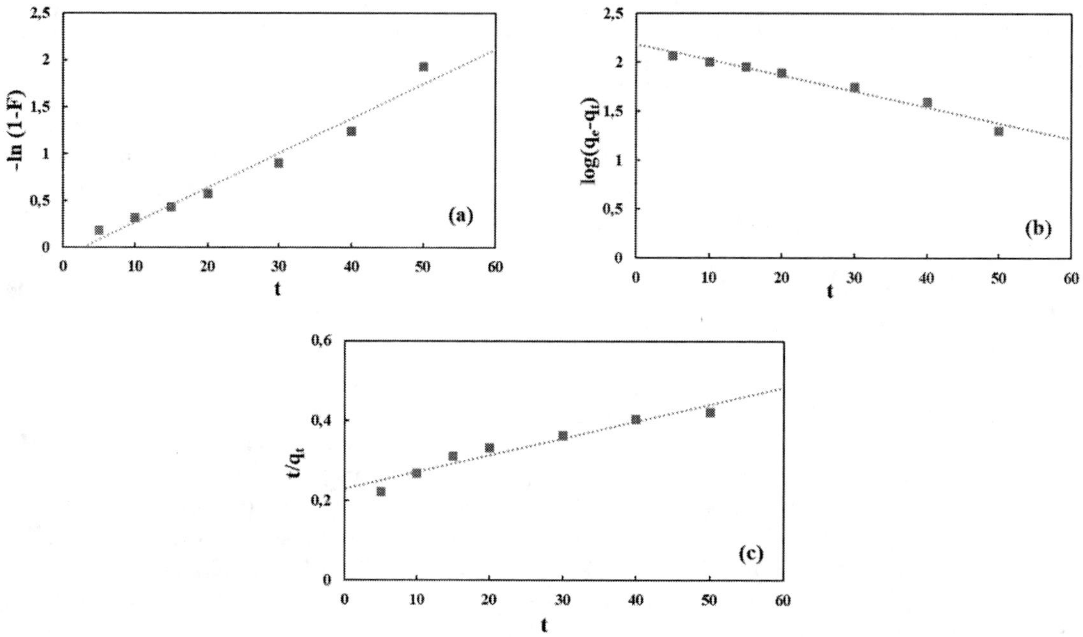

Figure 5. Kinetic plots: (a) Weber-Morris, (b) Lagergren, (c) Ho-McKay for toluene adsorption on the Taunit-M CNTs.

The mathematical form of the pseudo-second-order equation was first proposed in 1984 by Blanchard and his colleagues to describe the adsorption kinetics of heavy metals on natural zeolites [26]. Currently, this is the most commonly used mathematical expression used to process and correlate kinetic data for solid/solution adsorption systems [27]. The classical form of this model was presented by Ho and McKay.

The applicability of this model is usually accompanied by a conclusion about the occurrence of a chemical reaction that limits the adsorption kinetics. This is obviously due to the fact that every known theoretical rationale for the pseudo-second-order model is based on the fundamental theories of surface reactions. One of the earliest theoretical interpretations of this model was proposed in 2008 by

Azizian [28]; it is based on the classical Langmuir adsorption kinetics model, in which the adsorption is considered as a chemical reaction occurring on an energetically homogeneous solid surface. A similar approach was presented in 2008 by Liu and Shen [29]. Both these works proved that the Langmuir kinetics can be reduced to the equations of either the pseudo-first- or the pseudo-second-order rate. However, there exist papers, in which the pseudo-second-order model is considered as a fairly flexible mathematical formula capable of simulating the kinetics of the internal mass transfer (diffusion) adsorption for systems with flat and spherical particles [30].

Like in the previous case with the pseudo-first-order model, the analysis of the kinetic curves obtained for the toluene adsorption on the Taunit-M material in the coordinates of t/q_t *vs.* t showed that the dependence is also linear (R^2>0.9) (figure 5). However, the adsorption capacity predicted by the Ho-McKay model ($q_{H\text{-}M}$ =192.3 mg/g) differed from the experimentally obtained maximum adsorption capacity (q_e=141.8 mg/g). The pseudo-first-order adsorption rate constant was found to be $1.3 \cdot 10^{-4}$ g/(mg min) (table 1).

3.2.2. Thermodynamics. To study the temperature effect on the equilibrium adsorption, the following Gibbs equation was used:

$$\Delta G° = -RTlnK_c \qquad (1)$$

where $\Delta G°$ is the Gibbs free energy (kJ/mol), R is the universal gas constant equal to 8.314 J/(mol K), T is the absolute temperature (K), and K_c is the thermodynamic equilibrium constant, which can be estimated as follows:

$$\ln K_c = \frac{\Delta S°}{R} - \frac{\Delta H°}{RT} \qquad (2)$$

where $\Delta H°$ is the enthalpy (kJ/mol), and $\Delta S°$ is the entropy (J/mol K).

The analysis of the dependence of lnK_c vs. $1/T$ showed that it is linear (R^2>0.9), thereby making it possible to determine the ΔH° and ΔS° values, and then the $\Delta G°$ values at different temperatures (table 2).

Table 2. Thermodynamic parameters of the toluene adsorption on the Taunit-M CNTs.

Adsorbate	*Thermodynamic parameters*					R^2
	ΔG° at			*ΔS°*	*ΔH°*	
	298.15 K	303.15 K	308.15 K			
Toluene	-45.21	-45.98	-46.74	151.81	41.77	0.99

The negative Δ*G*° values, varying from -46.74 to -45.21 kJ/mol, indicate the spontaneity of the process and correspond to the range of values typical of physical adsorption. The standard enthalpy shows the energy change in the system. Its positive $\Delta H°$ value evidences the endothermic process. The positive $\Delta S°$ value stands for a change in the structure of the Taunit-M CNTs in the toluene adsorption process, leading to an increase in system disorder. The similar results were reported in the paper [31].

4. Conclusion

The adsorption efficiency of the Taunit-M CNTs was assessed regarding toluene. The maximum toluene adsorption capacity of this material was found to be 141.8 mg/g, which exceeds the adsorption capacity of analogous materials given in table 3.

Table 3. Comparison between the adsorption studies performed using different adsorbents.

Adsorbent	Toluene adsorption capacity (mg/g)	Conditions	Reference
NaOCl-oxidized single-walled CNTs	103.2	pH: 7, T: 25 °C, C_0: 20mg/L	1
HCl-oxidized multi-walled CNTs	60.0	pH: 7, T: 25°C, C_0: 20mg/L	32
H_2SO_4-oxidized multi-walled CNTs	36.7	pH: 7, T: 20 °C, C_0: 200 mg/L	32
NaOCl-oxidized multi-walled CNTs	-	-	33
NaOCl-oxidized multi-walled CNTs	99.9	pH: 7, T: 35 °C, C_0: 200 mg/L	34
Powdered activated carbon	40.0	pH: 6.4, T: 23 °C, C_0: 50 mg/L	35
HNO_3/30-% NaOCl-oxidized multi-walled CNTs	59.5	pH: 7, T: 20 °C, C_0: 200 mg/L	36
KOH-oxidized multi-walled CNTs	63.3	pH: 7, T: 25 °C, C_0: 20mg/L	37
Taunit-M CNTs	141.8	pH: 7, T: 35 °C, C_0: 25mg/L	Present work

The adsorption kinetic data were analysed using the simple Weber-Morris diffusion model and very common phenomenological models which simulate mass transfer processes using formal equations for chemical kinetics — the Lagergren and Ho-McKay models. As a result of processing the data, it was shown that the considered models are suitable, regardless of the adsorption process stage, *i.e.* intraparticle diffusion (Weber-Morris equation) or surface reaction (Lagergren equation). As noted in the literature, there is no consensus on the interpretation of formal equations for chemical kinetics.

Considering this, more studies are required to elucidate the toluene adsorption mechanisms. However, the analysis of the temperature effect on the equilibrium adsorption showed that the Gibbs free energy values vary in the range characteristic of physical sorption. Besides, these values were identified as negative, therefore the spontaneity of the adsorption process was established. The enthalpy and entropy values were found to be positive, thereby evidencing the endothermic process and showing an increase in randomness at the solid/solution interface after the toluene adsorption on the surface of the Taunit-M CNTs.

Thus, the results obtained can be useful in the development of adsorption technology for the purification of natural and waste water from toluene using CNTs.

Acknowledgment

The research was funded by the Ministry of Science and Higher Education of the Russian Federation under Project No. 16.1384.2017/PCh.

References

[1] Wibowo N, Setyadhi L, Wibowo D, Setiawan J and Ismadji S 2007 Adsorption of benzene and

toluene from aqueous solutions onto activated carbon and its acid and heat treated forms: influence of surface chemistry on adsorption *J. Hazard. Mater.* **146** 237

[2] Al-Gharabli S, Kujawski W, El-Rub Z A, Hamad E M and J. Kujawa 2018 Enhancing membrane performance in removal of hazardous VOCs from water by modified fluorinated PVDF porous material *J. Membr. Sci.* **556** 214

[3] Farrell E S and Pacey E S 2010 Rapid removal of selected volatile organic compounds from gaseous mixtures using a new dispersive vapor extraction technique: a feasibility study *Talanta* **82** 739

[4] Cloirec P Le, Pré P, Delage F and Giraudet S 2012 Visualization of the exothermal VOC adsorption in a fixed-bed activated carbon adsorber *Environ. Technol.* **33** 285

[5] environment air purification: the state-of-the-art *Appl. Catal. B* **203** 247

[6] Feng X, Liu H, He C, Shen Z and Wang T 2018 Synergistic effects and mechanism of a non-thermal plasma catalysis system in volatile organic compound removal: a review *Catal. Sci. Technol.* **8** 936

[7] Kim J, Khalekuzzaman M, Suh J N, Kim H J, Shagol C, Lagergren H – H and Kim H J Kim 2018 Phytoremediation of volatile organic compounds by indoor plants: a review *Hortic. Environ. Biotechnol.* **59** 143

[8] Cheng Y, He H, Yang C, Zeng G, Li X, Chen H and Yu G 2016 Challenges and solutions for biofiltration of hydrophobic volatile organic compounds *Biotechnol. Adv.* **34** 1091

[9] Aditya D, Rohan P and Suresh G 2011 Nano-adsorbents for wastewater treatment: a review *Res. J. Chem. Environ.* **15** 1033

[10] Mohmood I, Lopes C B, Lopes I, Ahmad I, Duarte A C and Pereira E 2013 Nanoscale materials and their use in water contaminants removal—a review *Environ Sci Pollut R.* **20** 1239

[11] Hosseini S J, Kokhdan S N, Ghaedi A M and Moosavian S S 2011 Comparison of multiwalled carbon nanotubes and activated carbon for efficient removal of methyl orange: kinetic and thermodynamic investigation *Fresenius Environ Bull.* **20** 219

[12] Li Y-H, Dinga J, Luan Z, Di Z, Zhua Y, Xua C, Wua D and Weic B 2013 Competitive adsorption of Pb^{2+}, Cu^{2+} and Cd^{2+} ions from aqueous solutions by multiwalled carbon nanotubes *Carbon* **41** 2787

[13] Seyed Arabi S M, Lalehloo R S, Olyai M R T B; Ali Gomaa AM, Sadegh, H 2019 Removal of congo red azo dye from aqueous solution by ZnO nanoparticles loaded on multiwall carbon nanotubes *Physica E.* **106** 150

[14] Rashid A, Bhatti H N, Iqbal M and Noreen S 2016 Fungal biomass composite with bentonite efficiency for nickel and zinc adsorption: A mechanistic study *Ecol. Eng.* **91** 459

[15] Giles C H, MacEwan T H, Nakhwa S N and Smith D 1960 Studies in adsorption: A system of classification of solution adsorption isotherms *J. Chem. Soc.* **3** 3973

[16] Dehghani M H, Mostofi M, Alimohammadi M, McKay G, Yetilmezsoy K, Albadarin A B, Heibati B, AlGhouti M,. Mubarak N M and Sahu J N 2016 High-performance removal of toxic phenol by single-walled and multi-walled carbon nanotubes: Kinetics, adsorption, mechanism and optimization studies *J. Ind. Eng. Chem.* **35** 63

[17] Yang J Y, Jiang X Y, Jiao F P and Yu J G 2017 The 2018 oxygen-rich pentaerythritol modified multi-walled carbon nanotube as an efficient adsorbent for aqueous removal of alizarin yellow R and alizarin red S *Appl. Surf. Sci.* **436** 198

[18] Knappe D and Rudiger U 1996 *Predicting the removal of atrazine by powdered and granular activated carbon*, PhD thesis, University of Illinois at Urbana-Champaign, 1996.

[19] McKay G 1998 Application of surface diffusion model to the adsorption of dyes on bagasse pith *Adsorption* **4** 361

[20] Cook D, Newcombe G and Sztajnbok P 2001 The application of powdered activated carbon for MIB and geosmin removal: predicting PAC doses in four raw waters *Water Res.* **35** 1325

[21] Schwaab M, Steffani E, Barbosa-Coutinho E and Júnior J B S 2017 Critical analysis of adsorption/diffusion modelling as a function of time square root *Chem. Eng. Sci.* **173** 179

[22] Pandey P K, Sharma S K and Sambi S S 2015 Removal of lead (II) from waste water on zeolite-NaX. *J. Environ. Eng.* **3** 2604

[23] Boyd G E, Adamson A W and Myres L S 1947 Kinetics of ionic exchange adsorption processes *J. Am. Chem.* **69** 2836

[24] Pandey K K, Prasad G and Singh V N 1986 Mixed adsorbent for Cu(II)removal from aqueous solutions *Environ Technol. Lett.* **50** 547

[25] Poots V J P, McKay G and Healy J J 1978 Removal of basic dye fromeffluent using wood as an adsorbent *J. Water Poll.* **50** 926

[26] Lagergren S 1898 About the theory of so-called adsorption of soluble substances *Kungliga Svenska Vetenskapsakademiens Handlingar* **24** 1

[27] Blanchard G, Maunaye M and Martin G 1984 Removal of heavy metals from waters by means of natural zeolites *Water Res.* **18** 1501

[28] Ho Y S and McKay G 1998 Sorption of dye from aqueous solution by peat *Chem. Eng. J.* **70** 115

[29] Azizian S 2008 Comments on "Biosorption isotherms, kinetics and thermodynamics" review *Sep. Purif. Technol.* **63** 249

[30] Liu Y and Shen L 2008 From Langmuir kinetics to first- and second-order rate equations for adsorption *Langmuir* **24** 11625

[31] Plazinski W, Dziuba J and Rudzinski W 2013 Modeling of sorption kinetics: the pseudo-second order equation and the sorbate intraparticle diffusivity *Adsorption* **19** 1055

[32] Lu C, Su F and Hu S 2008 Surface modification of carbon nanotubes for enhancing BTEX adsorption from aqueous solutions *Appl. Surf. Sci.* **254** 7035

[33] Chen W, Duan L and Zhu D 2007 Adsorption of polar and nonpolar organic chemicals to carbon nanotubes *Environ. Sci. Technol.* **41** 8295

[34] Lu C, Chung Y.L and Chang K F 2005 Adsorption of trihalomethanes from water with carbon nanotubes *Water Res.* **39** 1183

[35] Koh S-M and Dixon J B 2001 Preparation and application of organo-minerals as sorbents of phenol, benzene and toluene *Appl. Clay Sci.* **18** 111

[36] Pourzamani H, Hajizadeh Y and Fadaei S 2015 Efficiency enhancement of multi-walled carbon nanotubes by ozone for benzene removal from aqueous solution *Int. J. Environ. Health Eng.* **4** 29

[37] Abbas A, Abussaud B A, Al-Baghli N A, Redhwi H H and Zheng J 2017 Adsorption of toluene and paraxylene from aqueous solution using pure and iron oxide impregnated carbon nanotubes: kinetics and isotherms study *Bioinorg. Chem. Appl.* **2017** 2853925

Features of optimization synthesis of equipment for feeding nanodispersed materials

A Popova, I Shubin*, A Kucherova, N Memetov and D Tarov

Department "Technology and Methods of Nanoproducts Manufacturing", *Tambov State Technical University,* 106 Sovetskaya Street, 392000, Tambov, Russian Federation

*E-mail: i.shubin77@yandex.ru

Abstract. For the first time, in the analysis of the behavior of a nanodispersed material (*i.e.*, its outflow and batching from the hopper unit, as well as movement and mixing), a conventional approach was used to calculate the hardware and technological design of its functionalization. Besides, the features of the behavior of the nanodispersed layer and single nanoparticles as applied to the feeding process from the hopper under vibrations, its movement along the guide blade and the pouring of the flow into the receiving tank were considered, dependencies were formulated, and recommendations were given for designing the hopper assembly used to process the nanodispersed materials. The results were tested and showed high convergence in practical implementation.

1. Introduction

Today, nanotechnologies are smoothly moving from the category of basic and theoretical research, laboratory and demonstration samples to the category of pilot and mass production, which implies stable product quality - whether it be carbon nanomaterials (CNMs) directly, or equipment for their production or processing. At the same time, not the CNMs themselves, but the hardware and technological design of the production process comes out on top, providing guaranteed characteristics of the resulting construction or functional nanomaterial. Medicine and pharmacology, powder metallurgy, as well as sorbent, polymer and construction material production, represent just a small list of areas where knowledge of the physical-mechanical characteristics of nanodispersed materials (NDMs), consisting of a continuous gas or liquid medium and dispersed nanoparticles, and the features of process equipment under their application, considering scaling and the actual production conditions, is in demand [1,2].

A large number of works are devoted to studying the characteristics of NDMs. The authors mark a complete analogy in the properties of these materials and particulate bulk materials (including powders and industrial dusts) [3,4]. So, for example, the influence of the electrostatic properties of nanosized materials on the processing process was considered in the paper [6]. The general physical-mechanical characteristics of sinle nanoparticles and the laws of their movement were discussed in [10], and in [5], the hygroscopicity and density of the nanomaterials as applied to transistors and microelectronics were determined. However, only certain NDMs characteristics are considered in relation to laboratory research, and not all their parameters, as well as operating conditions and equipment calculation, are taken into account, *i.e.* there are no works devoted to the practical use of NDMs [5-11].

Thus, the goal of this work is to apply classical approaches to the calculation and design of technological equipment for the nanoindustry, as the most dynamically developing area of innovative materials and technologies.

2. Modeling NDMs in calculations of typical equipment

The authors of the present work examined the design of a feeding unit device for a nanodispersed system consisting of a bunker, a vibration bottom and an inclined blade, and carried out an optimization synthesis to combine single units, each of which provides a specific state of the nanoparticle layer. The following stages of the NDMs state were taken as assumptions: in the bunker - the layer is fixed (nanoparticles are in constant contact with each other); on the vibration bottom – the nanoparticles are in the vibro-fluidized bed (chaotic nanoparticle motion); on the inclined blade – the nanoparticles are in the rolling layer (ordered layered motion); the detachment from the blade edge and the nanoparticle fall in the fan represent a free single-particle motion (up to the logjam).

Let us consider in more detail the stages of the nanodispersed system motion on the example of the above-mentioned bunker device (figure 1).

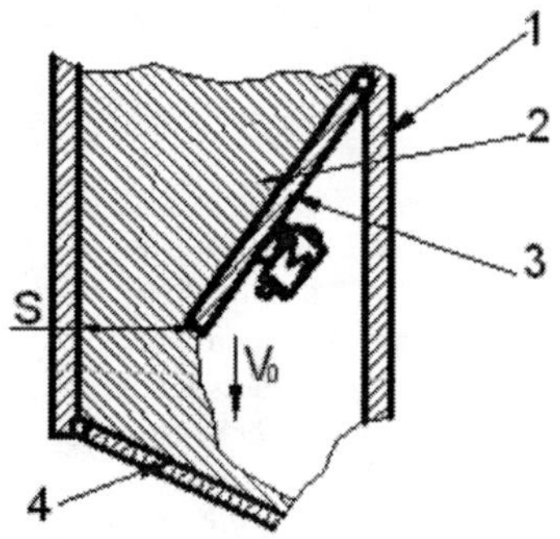

Figure 1. A pattern of the NDM motion at the outflow from the bunker device. 1 – NDM; 2 – bunker; 3 – vibration bottom; 4 – lead blade.

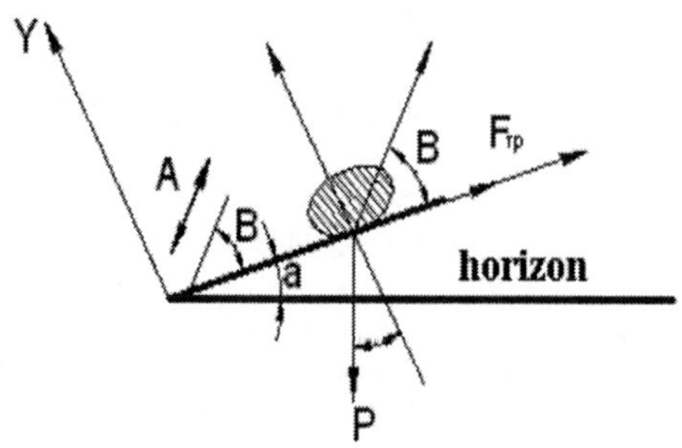

Figure 2. A diagram of forces acting on the nanoparticle when applying vibration.

At the first stage of feeding, the NDM 1 flows out of the bunker 2 onto the lead blade 4 along the vibration bottom 3, which, in our case, is used to equalize the flow rate and destruct arching.

Let us consider the vibration bottom as a plate (figure 2) oscillating with frequency ω and amplitude A in the direction forming angle β with the vertical axis [12,13]. The nanoparticle of mass m is acted upon by the gravitation force $P=mg$, the friction force F_{fr} along the plate surface, the normal surface reaction N, and the inertia force I. The nanoparticle motion relative to the vibration bottom is described by the following equations:

$$m\ddot{x} = m\mathrm{A}\omega^2 \cos\beta \sin\omega\tau - mg\sin\alpha - F_{TP} \quad (1)$$

$$m\ddot{y} = m\mathrm{A}\omega^2 \sin\beta \sin\omega\tau - mg\cos\alpha + \mathrm{N} \quad (2)$$

If the NDM is fed onto the vibration bottom, but does not come off from it, then the acceleration of the material relative to this surface is equal to $y = 0$. Then, from equation (2), it follows:

$$\mathrm{N} = mg\cos\alpha - m\mathrm{A}\omega^2 \sin\beta \sin\omega\tau \quad (3)$$

Numerous studies [13–16] have shown that during the vibration exposure, the properties of the nanodispersed system layer change significantly with the onset of the detachment of nanoparticles from each other and from the vibrating surface, on which they are located. Prior to the beginning of the separation, the NDM is liquefied on the vibrating surface.

After the onset of the detachment, with an increase in the intensity of oscillations, arbitrary motion of the NDM begins, the layer "boils up", *i.e.* its porosity increases. The material nanoparticles are detached from the vibration bottom, when the normal reaction is $N=0$. Then, according to equation (3), we get the following:

$$A_{kp}\omega_{cr}^2 = \frac{g}{\sin\omega\tau}\cdot\frac{\cos\alpha}{\sin\beta} \tag{4}$$

where ω_{cr} is the critical frequency. From equation (4), it can be seen that the minimum critical vibration acceleration of the vibration bottom, at which the nanoparticle separation occurs, will be at $\sin(\omega\tau)=\pm 1$. At these moments, the relative critical vibration acceleration is equal to:

$$\frac{A_{kp}\omega_{cr}^2 \sin\beta}{g\cos\alpha} = \pm 1 \tag{5}$$

As can be seen, for the nanodispersed system exposed to vibration, the transition of the layer to vibration liquefaction depends not on the nanoparticle mass, but on the vibration acceleration ($A\omega$) of the bottom. This makes it possible to control the vibration liquefaction and, accordingly, the porosity in the boundary layer of the material by adjusting the amplitude and frequency of the vibrator.

Thus, regarding the NDM mass effect, neglected when considering the motion of a nanoparticle along the vibration bottom, it is necessary to maintain a number of boundary conditions (assumptions):

1. The NDM is not detached from the vibrating surface (subject to the condition $N>0$);
2. The controlled liquefaction of the nanoparticle boundary layer occurs;
3. The destruction of arching occurs in the flow bulk;
4. The porosity remains constant in the NDM bulk, since its increase (nanoparticle liquefaction in the boundary layer along the vibration bottom) is compensated by compaction above the underlying layer due to vibration.

From the above, it is clear that the vibration will affect the flow indirectly, *i.e.* it will only "equalize" the flow rate, making the flow stable. Then, the nanoparticle rate v can be determined as follows:

$$v = \lambda\cdot\sqrt{3.2gR} \tag{6}$$

where R is the hydraulic radius of the hole (slit) of the outflow $R=F/L$; g is the gravitational acceleration; λ is the outflow coefficient; L is the perimeter of the hole (slit). The rectangular area F is expressed as follows:

$$F = (A-a')(B-a'); \tag{7}$$

where A and B are the rectangle sides; $a' = kd$; d is the particle diameter; κ=1.25-1.70 (coefficient).

Considering equation (7), R can be defined as follows:

$$R = \frac{(A-a')(B-a')}{2(A+B-2a')} \tag{8}$$

At the second stage of the motion, the NDM comes from the vibration bottom to the lead blade.
The equation expressing equilibrium for the NDM moving at angle α to the horizon:

$$\ddot{x}m + f_{Tp}mg\cos\alpha - mg\sin\alpha = 0 \tag{9}$$

where $\ddot{x}$ is the acceleration of the nanoparticle motion, which is determined as follows:

$$\ddot{x} = f_{Tp}mg\cos\alpha - mg\sin\alpha \tag{10}$$

Consistently integrating equation (10), we obtain the expression for determining the rate of the NDM motion and the distance it has traveled:

$$\dot{x} = (g\sin\alpha - f_{Tp}\cos\alpha)\tau + c_1 \quad (11)$$

under the following boundary conditions:

$$\dot{x} = v_0, \tau = 0, \Rightarrow c_1 = v_0 \quad (12)$$

$$\dot{x} = v_0 + g(\sin\alpha - f\cos\alpha)\tau \quad (13)$$

Next, we write the equation of the distance traveled along the guide blade by integrating equation (13):

$$x = v_0\tau + \frac{1}{2}g(\sin\alpha - f\cos\alpha)\tau^2 + c_2 \quad (14)$$

where $c_2 = 0$ is the integration constant under the initial conditions; τ=0; x=0.

Thus, equation (14) takes the following form:

$$x = v_0\tau + \frac{1}{2}g(\sin\alpha - f\cos\alpha)\tau^2 \quad (15)$$

Being in the rolling layer on the lead blade, the NDM increases the porosity. The degree of loosening can be characterized by the loosening coefficient *K*. It is an important parameter which determines the ratio of elemental volumes of the NDM in the bunker and in the flow moving along the lead blade. To obtain this value, it is necessary to find the average flow rate $\dot{x}_{av}$:

$$\dot{x}_{av} = \frac{\int_0^L \dot{x}\partial Q}{\sum Q} \quad (16)$$

where $\dot{x}$ is the function of the nanoparticle rate in the flow depending on the distance traveled; ∂Q is the material amount on the lead blade. The NDM volume on the lead blade V_{lb} can be determined by the following formula:

$$V_{lb} = L\int_S dS \quad (17)$$

where L is the length of the lead blade; S is the integration area, *i.e.* the surface area of the lead blade. Analyzing equation (16) for the average rate and simplifying it, we can write the following:

$$\dot{x}_{av} = v'_{av} = \frac{L_{iav}}{\tau_{iav}} \quad (18)$$

where L_{iav} is the average distance traveled by the ith nanoparticle along the blade; τ_{iav} is the average time spent by the ith nanoparticle for traveling. Whence it follows that the loosening coefficient for the NDM flow moving along the blade is equal to:

$$K = \frac{v'_{av} \cdot h_L}{S_\tau} \quad (19)$$

where h_L is the thickness of the flow at the edge of on the edge of falling off the blade; $S\tau$ is the area of the nanodispersed layer overlooking the falling-off line during traveling the length by the nanoparticles.

At the third stage of the NDM motion, free falling off the lead blade occurs. To analyze this motion, it is necessary to introduce the following assumptions confirmed in some papers [4,7,9]:

1. The nanoparticles fall off the blade as a flow consisting of a series of sublayers, with a thickness of each one determined by the average nanoparticle size.
2. At the time of the detachment from the blade, the nanoparticles of the same sublayer have the similar rate.
3. At the time of the detachment, the nanoparticle rate increases along the flow thickness, as the distance from the falling-off edge increases.

As a consequence of these assumptions, it follows that after the detachment from the blade, the nanoparticles do not collide with each other, so the "single-particle" approach can be employed to study the motion [17].

The motion of the nanoparticle in free fall after the detachment from the lead blade can be considered as a body motion in a gravitational field at some initial rate. The equation of the trajectory of the ith nanoparticle can be written as follows:

$$x = -h_i \sin\alpha - v_i \tau \cos\alpha \tag{20}$$

$$y = h_i \cos\alpha - v_i \tau \sin\alpha - \frac{g\tau^2}{2} \tag{21}$$

where x and y are the current coordinates of the moving nanoparticle; h_i is the distance between the nanoparticle and the falling-off edge of the lead blade at the time of the detachment, in the direction normal to the particle rate vector; α is the angle between the rate vector v_i and the horizoт; τ is the current time from the moment of particle detachment from the lead blade.

As noted before, the nanodispersed layer moving along the lead blade has a section close to rectangular. For a particle located at the lower boundary of the layer, $h_i = 0$ and $v_i = v_L$. The lower boundary of the nanoparticle incidence fan can be described by the following equations:

$$x = v_L \cdot \tau \cdot \cos\alpha \tag{22}$$

$$y = v_L \cdot \tau \cdot \sin\alpha + \frac{g\tau^2}{2} \tag{23}$$

For a particle moving along the upper boundary of the layer, $h_i = h$ and $v_i = v_U$. The upper boundary of the fan can be described by the following equations:

$$x = -h_i \sin\alpha - v_U \tau \cos\alpha \tag{24}$$

$$y = h_i \cos\alpha - v_i \tau \sin\alpha - \frac{g\tau^2}{2} \tag{25}$$

The fan surface S of the NDM falling off the lead blade is expressed as follows:

$$S = (L_U + L_L) L_W \tag{26}$$

where L_U is the length of the upper boundary of the fan from the falling-off edge of the lead blade up to the open surface of the logjam; L_L is the length of the lower boundary of the fan; L_W is the width of the blade.

The lengths L_U and L_L can be found from equations (22)-(25) through curvilinear integrals in the following form:

$$L_{(U,L)} = \int_L f(x; y) dL \tag{27}$$

The integral expressed by equation (27) can be transformed into an ordinary integral by substituting into the integrand:

$$x = x(\tau), y = y(\tau), dL = \sqrt{[\dot{x}(\tau)]^2 + [\dot{y}(\tau)]^2} d\tau,$$

after which you can take the integral in the measurement interval τ, which corresponds to the integration line.

The time of falling off the lead blade for the nanoparticles τ_f, is determined from the moment of their detachment from the blade up to the moment of contact with the open surface of the logjam. To define the numerical value of τ_f, it is required to combine the solution of the equation of the trajectory of the nanoparticle incidence (expressions (20) and (21)) and the equation of the open surface of the logjam.

The open surface of the logjam has a complex shape, and to date, there are no dependencies describing this boundary. The comparison between the experimental and calculated values, with an accuracy sufficient for engineering calculations, showed that when determining the time of the nanoparticle fall from the blade, the equation of the open surface of the logjam can be taken in the following form:

$$y = x \cdot \tan \alpha_f \tag{28}$$

where α_f is the angle between the open surface of the logjam and the horizon.

Solving equations (20), (21) and (28) jointly, we can obtain the following dependencies to determine the coordinates of the point of the intersection of the trajectory of the falling nanoparticle with the open surface of the logjam and the time of the nanoparticle fall.

$$x = -h_i \sin \alpha - v_i \cos \alpha \left(M + \sqrt{M^2 - N} \right) \tag{29}$$

$$y = h_i \cos \alpha - v_i \sin \alpha \left(M + \sqrt{M^2 - N} - \frac{g}{2} \left(M + \sqrt{M^2 - N} \right)^2 \right) \tag{30}$$

$$\tau = M + \sqrt{M^2 - N};$$

$$M = \frac{1}{g} \left(v_i \cos \alpha + \tan \alpha - v_i \sin \alpha \right) \tag{31}$$

$$N = \frac{2}{g} \left(h_i \cos \alpha - h_i \sin \alpha \cdot \tan \alpha \right)$$

$$\alpha = \arctan f_f$$

Substituting the values of v_i and h_i, obtained for the nanoparticles moving along the upper and lower boundary of the flow, into expression (31), we can find the upper integration limit for expression (27). In this particular case, the lower limit is equal to 0.

To determine the particle incidence rate, when they fall from the blade, it is necessary to differentiate equations (20) and (21) by time:

$$v_H = v \cos \alpha \tag{32}$$

$$v_V = v \sin \alpha + g\tau \tag{33}$$

where V_H and V_V are the horizontal and vertical components of the nanoparticle motion rate, respectively.

The absolute modulus v of the nanoparticle motion rate is defined as follows:

$$v = \sqrt{v_H + v_V^{\ 2}} \tag{34}$$

To assess the motion rate of the nanoparticle when it passes through the horizontal section 1-1, lagging behind the falling-off edge directed at distance H (figure 3), it is required to substitute the value $y=H$ into equation (22), find the value of τ, and determine the values of V_H and V_V using equations (32) and (33).

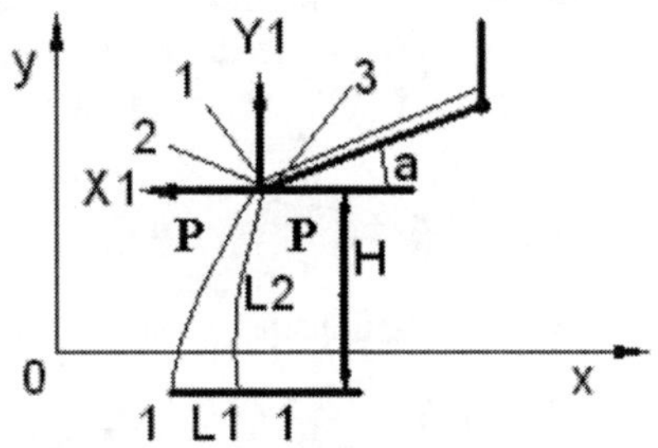

Figure 3. Determining the loosening coefficient for the NDM in the fan.

Earlier, it was noted that when falling off the lead blade, the NDM flow is loosened. By analogy with the loosening coefficient in the rolling layer, the loosening coefficient K for the NDM flow falling off the blade can be represented as a ratio of the volume V_g of the NDM amount Q in motion, to the volume V_P occupied by the lead blade.

To clarify the causes of the loosening of the incident NDM flow, we consider the loosening of the flow in the horizontal and vertical directions. When determining the loosening coefficient K_H for the horizontal direction, it is necessary to take into account the fall of nanoparticles 1 and 2 located in different rate sublayers on the blade.

The distance L_2 between the centers of gravity of the nanoparticles in the horizontal direction, when they pass through the section $P–P$ held horizontally through the falling-off edge of the lead blade, can be determined from the following expression:

$$L_2 = \frac{v_2^2 \cos\alpha \cdot \sin\alpha}{g} + v_2 \sqrt{\frac{v_2^2 \cdot \sin^2\alpha}{g^2} + \frac{2h \cdot \cos\alpha}{g}} \tag{35}$$

where V_1 and V_2 are the motion rates of nanoparticles 1 and 2, respectively, at the time of their detachment from the lead blade; α is the angle between the rate vectors and the horizon; h is the distance between the center of gravity of nanoparticle 2and the falling-off edge of the blade, in the direction normal to the nanoparticle motion rate vectors. The distance L_1 between the points of the intersection of the trajectories of the nanoparticle incidence with the horizontal section 1-1 can be determined as follows:

$$L_1 = \frac{v_2^2 \sin\alpha}{g} + v_2 \sqrt{\frac{v_2^2 \sin^2\alpha}{g^2} + \frac{2 \cdot (h\cos\alpha + H)}{g}} + \frac{v_1^2 \sin\alpha}{g} - v_1 \sqrt{\frac{v_1^2 \sin^2\alpha}{g^2} + \frac{2H}{g}} \tag{36}$$

Given that the loosening of the NDM at the falling-off edge of the blade is insignificant, the numerical value of the loosening coefficient K_H for the section 1-1 can be obtained from the following relation:

$$K_H = \frac{L_1}{L_2} \tag{37}$$

To determine the loosening coefficient K_V for the vertical direction, it is necessary to consider the fall of nanoparticles 2 and 3 detached from the blade one by one from the same sublayer. At the moment of the detachment from blade, the nanoparticles have the same motion rate, but nanoparticle 3 is detached later than nanoparticle 2. The lag time τ_{32} can be determined from the following expression:

$$\tau_{32} = \frac{L_{32}}{v_3} \tag{38}$$

where L_{32} is the distance between the centers of gravity of the nanoparticles in the direction of their motion at the moment of the detachment of nanoparticle 2.

For some time τ passed from the moment of the detachment of nanoparticle 3 from the lead blade, the distance between the centers of gravity in the horizontal direction can be determined as follows:

$$L_\tau = v_2 \cdot \tau_{32} + \frac{g\tau_{32}^2}{2} + g\tau\tau_{32} \quad (39)$$

From equations (37) and (39), it follows that the loosening of the NDM in the fan varies in width and height.

When carrying out practical calculations, it is more convenient to use the volume loosening coefficient K:

$$K = \frac{V_M}{V_R} \quad (40)$$

where V_M is the volume occupied by a certain amount of the moving material; V_R is the volume occupied by the same amount of the material at rest.

Let us consider in more detail the definition of the volume loosening coefficient. Let for some time τ, the NDM amount q falls off the blade. This amount, being on the lead blade, occupied the volume V_R according to equation (17):

$$V_R = L\int_S dS \quad (41)$$

When passing through the section 1-1, the volume V_M, occupied by the material amount q in motion, can be determined from the following expression

$$V_M = L_1 \cdot L \cdot v_{Vav} \cdot \tau \quad (42)$$

where L_1 is the width of the fan in the section 1-1; v_{Vav} is the vertical component of the average motion rate of the material when it passes through the section 1-1. After substituting equations (41) and (42), expression (40) takes the following form:

$$K = \frac{L_1 \cdot v_{Vav} \cdot \tau}{\int_S dS}. \quad (43)$$

3. Discussion

Analyzing the NDM motion from the bunker, first along the vibration bottom, then, along the lead blade, and finally, when falling in the fan, a number of conclusions and assumptions were made as for the properties of the moving layer of the material. In this regard, one can say that the nanoparticle motion rate along the height of the material flow fan will increase. The flow loosening coefficient (porosity of the nanoparticle flow) will be directly proportional to the height and rate of the nanoparticles when the NDM falls. The acceleration rate of the flow on the lead blade will be affected by the inclination angle, apart from the physical and mechanical properties of the NDM. The width of the opening of the NDM flow fan will change only to a certain height, and after that, it will remain almost constant; the area of the gap and the angle of the lead blade inclination will affect the performance of the feeder, whereas the effect of the NDM physical and mechanical properties will be reduced due to the vibration.

4. Conclusion

Thus, when analyzing the NDM behavior, a conventional approach was used for the first time to calculate the industrial technological equipment. Besides, the features of the behavior of a nanodispersed layer and single nanoparticles were considered as applied to single elements of the technological equipment. Moreover, again for the first time, dependencies were formulated and recommendations were given for designing a bunker assembly used for the functionalization of the NDMs. As a result, a general approach to the optimization synthesis of equipment for feeding nanodispersed materials based on a three-stage method is proposed. The features of the behavior of the nanodispersed layer at each stage when feeding the NDMs from the hopper to its subsequent processing are established.

References

[1] Tkachev A G, Popov A I and Shubin I N 2010 *Industrial Technology and Innovation. Equipment for Nanoindustry and its Manufacturing Technology* [in Russian] (Tambov: TSTU Publishing House) p 180
[2] Mishchenko S V and Tkachev A G 2008 *Carbon Nanomaterials. Production, Properties, Application* [in Russian] (Moscow: Mashinostroenie) p 320
[3] Ferapontov Yu A, Ulyanova M A, Bukatin A I, Tkachev A G and Shubin I N 2007 Determination of the particle size of carbon nanostructured materials obtained by the pyrolysis of a propane-butane mixture on a metal catalyst *Transactions of the TSTU* [in Russian] **13** 94
[4] Pershin V F, Alsayad T H K, Pasko T V and Pasko A A 2018 Determination of angles and coefficients of friction of carbon nanomaterials *Polzunov Bulletin* [in Russian] **4** 184
[5] Ghorai S 2013 *Chemical, physical and mechanical properties of nanomaterials and its applications*, PhD thesis, University of Iowa
[6] Čitaković N 2019 Physical properties of nanomaterials *Milit. Tech. Cour.* **67** 159 doi.org/10.5937/vojtehg67-18251
[7] Guo D, Xie G and Luo J 2014 Mechanical properties of nanoparticles: basics and applications *J. Phys. D* **47** 3001 doi.org/10.1088/0022-3727/47/1/013001
[8] Filippov A A and Fomin V M 2018 Determination of the mechanical characteristics of nanomaterials under tension and compression. *Journal of Physics: Conference Series* **991** 012020 doi :10.1088/1742-6596/991/1/012020
[9] Gatoo M A, Naseem S, Arfat M Y, Dar A M, Qasim Kh, and Zubair S 2014 Physicochemical properties of nanomaterials: implication in associated toxic manifestations *BioMed Research International* **2014** 498420 https://doi: 10.1155/2014/498420
[10] Jeon S K, Jang H S, Kwon O H and Nahm S H 2016 Mechanical test method and properties of a carbon nanomaterial with a high aspect ratio *Nano Convergence* **3** 29 doi.org/10.1186/s40580-016-0089-3
[11] Tarov D V, Gurova T V and Shubin I N 2015 Hardware design of the functionalization of carbon nanotubes with titanium stearate *Transactions of the TSTU* [in Russian] **21** 360
[12] Andrianov E I 1982 *Methods for Determining the Structural and Mechanical Characteristics of Powdered Materials* [in Russian] (Moscow: Khimiya) p 256
[13] Varsanofiev V D and Kolman-Ivanov E E 1985 *Vibration Technology in the Chemical Industry* [in Russian] (Moscow: Khimiya) p 240
[14] Makarov Yu I 1973 *Apparatus for Mixing Bulk Materials* (Moscow: Mashinostroenie) p 216
[15] D6128-00 2000 *Standard Test Method for Shear Testing of Bulk Solids Using the Jenike Shear Cell* (ASTM International, West Conshohocken, USA)
[16] Roessler T and Katterfeld A 2018 Scaling of the angle of repose test and its influence on the calibration of DEM parameters using upscaled particles *Powder Technol.* **330** 58
[17] Pershin V F, Odnolko V G and Pershina S V 2009 *Processing of Bulk Materials in Drum-Type Devices* [in Russian] (Moscow: Mashinostroenie) p 220

IOP Conf. Series: Materials Science and Engineering **693** (2019) 012041 doi:10.1088/1757-899X/693/1/012041

Kinetic of the dyes adsorption on a polyhydroquinone/graphene nanocomposite under dynamic conditions

E Mkrtchyan[1], E Neskoromnaya[1], I Burakova[1], D Kurnosov[1], A Burakov[1], A Tkachev[1] and I Ali[2]

[1]Department of Technology and Methods of Nanoproducts Manufacturing, *Tambov State Technical University,* 106 Sovetskaya Str., Tambov 392000, Russian Federation
[2]Department of Chemistry, *Jamia Millia Islamia (Central University)*, Jamia Nagar, New Delhi 110025, India

[1]E-mail: elina.mkrtchyan@yandex.ru

Abstract. Conducting adsorption experiments in a dynamic mode makes it possible to simulate process parameters as close to real conditions as possible. In the present work, the kinetics of the adsorption of some organic dyes (methylene blue, alizarin red S, and malachite green) on a novel polyhydroquinone/graphene nanocomposite, developed by the authors, were studied.It was established that the equilibrium of the adsorption system is achieved within 40, 120 and 57 min for methylene blue, alizarin red S and malachite green, respectively. Data processing using kinetic models showed that chemical interactions take place between the dye molecules and the active sites of the adsorbent during the adsorption. Besides, the contribution of the diffusion through the film on the adsorbent surface was also marked. The maximum adsorption capacity of the adsorbent was found to increase in the following sequence: malachite green (688.1 mg g^{-1}) < methylene blue (2.610 mg g^{-1}) < alizarin red S (2.923 mg g^{-1}).

1. Introduction

Contaminated water is one of the factors of human viral diseases [1]. Due to rapid urbanization growth rates, the process of environment all pollution is increasing fast [2]. Improper disposal of waste water coming from various enterprises is one of the fundamental causes of water contamination. The textile, rubber, plastic and pulp and paper industries use organic dyes in their production activities [3], thereby being the main sources of organic pollution.

Methylene blue (MB; $C_{16}H_{18}ClN_3S$; thiazine dye), alizarin red S (ARS; $C_{14}H_7NaO_7S$; anthraquinone dye) and malachite green (MG; $C_{23}H_{25}N_2$; N-methylated diamin-otriphenylmethane dye), used particularly for textile purposes, belong to synthetic (basic) dyes. They are persistent pollutants and can cause permanent damages to the eyes, skin and gastrointestinal tract of living beings [3].

At present, various methods of removing organic molecules from aquatic media are used – *e.g.*, adsorption, membrane filtration, oxidation, and biological treatment [1]. Among these waste water/water purification techniques, adsorption is considered to be one of the promising options from both economic and process instrumentation points of view.

Different researchers have employed adsorbents of various chemical nature, including carbon nanomaterials, to extract MB, ARS and MG from contaminated wastewater. The brief overview on using some of these materials for the adsorption of the above-mentioned species (conditions and adsorption capacity) is given in table 1.

Table 1. An overview of carbon materials used for the adsorption of organic molecules (ARS, MB, MG), their adsorption capacity, and experiment conditions.

Material	C_{in}*	t*	m*	T*	Q_e*
		ARS			
Oxidized pentaerythritol-modified MWCNTs** [4]	50-400	2	0.005	35	257.7
SWCNTs** [5]	100-1.000	1.08	0.03	45	312.5
MWCNTs[5]	100-1.000	1.67	0.03	45	135.2
AC** synthesized by carbonization and chemical activation of zinc [6]	100	9	0.005	25	74
Magnetic AC-based nanocomposite [7]	1000	4	0.01	-	108.7
		MB			
Reduced GO**-hybridized polydopamine-kaolin composite [1]	500	24	0.04	27	39.7
Magneticalginate-functionalized MWCNTs [8]	30-230	0-66	0.01	25	905.5
Regenerated cellulose/GO airgel [9]	20	0.5	0.05	25	68
Graphene-like porous carbon nanostructure [10]	100	0.25	0.05	50	469
MWCNTs filled Fe_2O_3 particles [11]	500	1	0.05	25	42.3
Mesoporous carbon capsules with magnetic nanoparticles [12]	50	0.16	0.005	20	608
A- COx/G nanofilamentous membrane [13]	20	-	-	-	95%
HDPy-clay [14]	150	1	0.1	25	142.5
HDTMA-clay [14]	150	1	0.1	25	142.22
		MG			
Oxidized mesoporous carbon [15]	700	6	0.01	40	963.1
Hyberbranched polyamine-functionalized MWCNTs [16]	40	2	0.005	-	78.7
AC/Fe_2O_4 composite [17]	15	-	0.002	20	20.13
Carbon nanoparticles with large pores [18]	500	-	0.01	25	1.892
GO [19]	200	7	0.05	-	476.2
Zeoliticimidazolate-modified GO/MWCNTs composite [20]	100	-	0.03	20	3,300/2.034
3D porous graphene hydrogel [21]	140	12	0.008	-	738.1
HDPy-clay [14]	150	1	0.1	25	145.5
HDTMA-clay [14]	150	1	0.1	25	147.99

* C_{in} – initial concentration of sorbate in solution, mg L^{-1}; t – adsorption time, h; m – adsorbent weight, g; T– the solution temperature, °C;Q_e – adsorbate concentration in the adsorbent at equilibrium, mg g^{-1}.
** MWCNTs – multi-walled carbon nanotubes; SWNTs – single-walled carbon nanotubes; AC – activated carbon; GO – graphene oxide.

As can be seen, only a few materials showed the high adsorption capacity the organics being considered. In this regard, there is a need for developing new adsorbents that could efficiently purify wastewater from such pollutants.

Considering the aforementioned, in the present study, a novel material, polyhydroquinone/graphene (PHQ/G) nanocomposite, was used for the removal of MB, ARS and MG. Besides, the kinetics of the dye adsorption processes was studied. The studies were carried out using the original dynamic setup based on the spectrophotometric method for determining concentrations in solution. Adsorption studies were carried out for a number of organic dyes (methylene blue, alizarin red, malachite green) of various chemical nature and, accordingly, the nature of the adsorption interaction with the studied adsorbent.

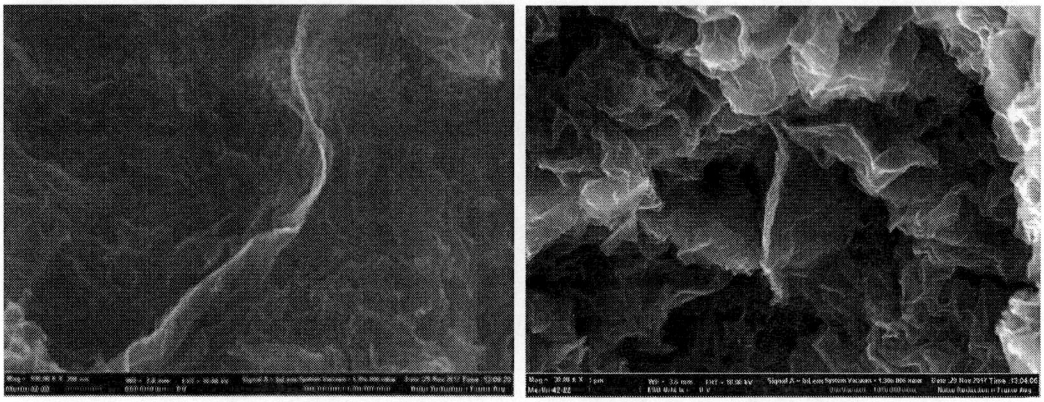
a b

Figure 1. SEM images of the surface structure of the PHQ/G nanocomposite. Magnification:100.00 KX (**a**) and 30.00 KX (**b**).

The morphology of the developed PHQ/G nanocomposite is shown in figure 1. At various image magnifications (100.00 and 30.00 KX), one can observe a graphene-like structure of the material in the form of flakes of carbon layers.

2. Experimental part

2.1. Material and reagents
The PHQ/G nanocomposite, herein employed as adsorbent, represents an organic ally modified graphene oxide. Its production technology is described in [22]. The material was developed at the Research Department "Technology and Methods of Nanoproducts Manufacturing" (Tambov State Technical University, Tambov, Russia). Solutions of MB, ARS and MG all acquired from Laverna Lab Ltd., Moscow, Russia, were use dasad sorbates.

2.2. Kinetic study
Experiments, the scheme of which is presented in figure 2, were performed in a dynamic mode, *i.e.* aqueous solutions of the organic dyes continuously passed through an adsorption cell (figure 2). In real conditions, the adsorption proceeds dynamically, since this is more acceptable for automation of the production process and its continuity.

100 mL of 1500-mgL^{-1} initial solutions of the MB, ARS and MG were separately fed from a container R to an adsorption cell (filled with 0.03 g of the adsorbent) along a loop 1.1 using a peristaltic pump, and then returned to the container R along a feedback loop 1.2.

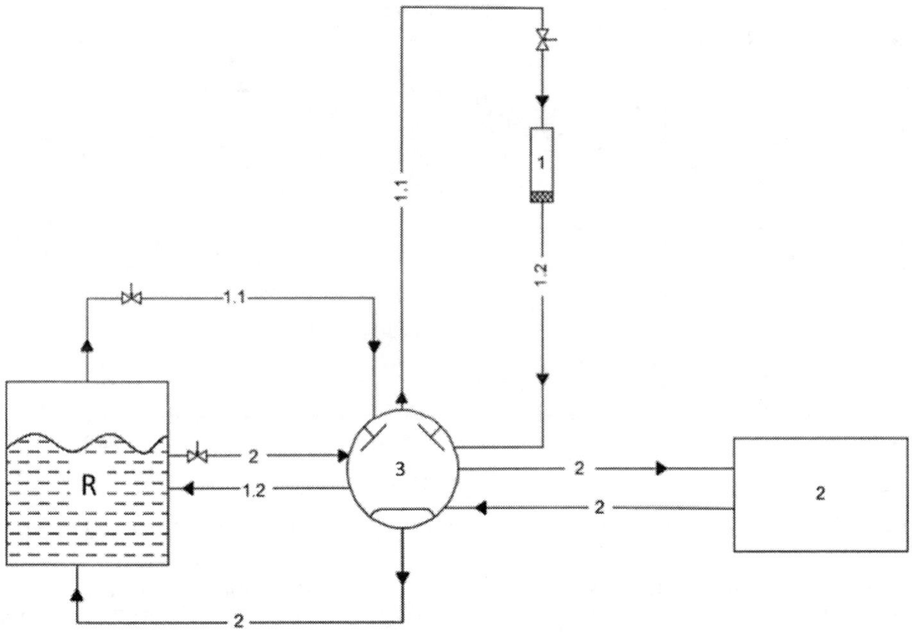

Figure 2. A scheme of experiments on studying the dye adsorption in a dynamic mode (R-container with the initial solution, 1- adsorption cell, 2 - spectrophotometer, and 3 -peristaltic pump).

The parallel loop 2 was intendedfor feeding the solutions to a cuvette located in a spectrophotometer 2 (PE 5400V instrument, Ekros, St. Petersburg, Russia). The optical density of the solutions was recorded every 180 s at the wavelengths of 570, 660 and 710 nm (for the MB, ARS and MG, respectively).

3. Results and discussion

The integral kinetic curves obtained for the adsorption of the MB, ARS and MG on the PHQ/G nanocomposite are presented in figure 3. It can be seen that the adsorbent material is characterized by a high adsorption rate during the initial period (up to 25 min). With a further increase in the contact between the adsorbent and the solutions, the removal rate monotonously decreases.The adsorption equilibrium of the systems is achieved within 40 (MB), 120 (ARS) and 57 (MG) min. The maximum adsorption capacities (Q_e^{exp}) were found to be as follows: 2.610 mg g^{-1} (MB), 2.923 mg g^{-1} (ARS), and 688.1 mg g^{-1} (MG). Comparing these values with those ones reported for the materials used as adsorbents by the other researchers (Q_e, table 1), it can be observed that in the case of the MB and ARS, the adsorption capacity of the PHQ/G nanocomposite is the highest, whereas in the case of the MG, the capacity is lower than that of the other graphene-based materials and mesoporous carbon.

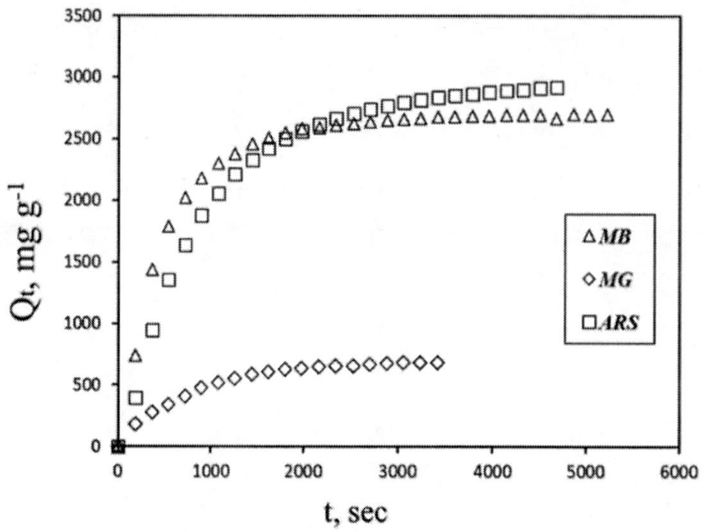

Figure 3. Integral kinetic curves obtained for the dye adsorption on the PHQ/G adsorbent.

The experimental data obtained were adjusted to equations of the known kinetic and diffusion models (pseudo-firs-t and pseudo-second order, Elovich, and intraparticle diffusion) (table 2). The results are presented in figure 4 and tables 3 and 4.

Table 2. Kinetic model equations implemented [23].

Model	Equation*
Pseudo-first-order	$lg(Q_e - Q_t) = lg(Q_e) - \frac{k_1 t}{2{,}303}$
Pseudo-second-order	$\frac{t}{Q_t} = \frac{1}{k_2 \cdot Q_e^2} + \frac{1}{Q_e} \cdot t$
Elovich	$Q_t = \frac{1}{\beta} ln(\alpha\beta) + \frac{1}{\beta} ln(t)$
Intraparticle diffusion	$Q_t = k_{id} \cdot t^{0,5} + c$

* t - adsorption time, min; Q_e - adsorption capacity at equilibrium, mg g^{-1}; Q_t-adsorption capacity at time t, mg g^{-1}; k_1 - pseudo-first-order adsorption rate constant (min^{-1}); k_2 -pseudo-second-order adsorption rate constant, g mg^{-1}min^{-1}; k_{id} - intraparticle diffusion coefficient, mg g^{-1} min$^{-0.5}$; c – boundary layer thickness constant, mg g^{-1}; α - initial flow rate, min^{-1}mg g^{-1}; β- desorption constant (degree of surface coverage and activation energy of chemisorption), g mg^{-1}.

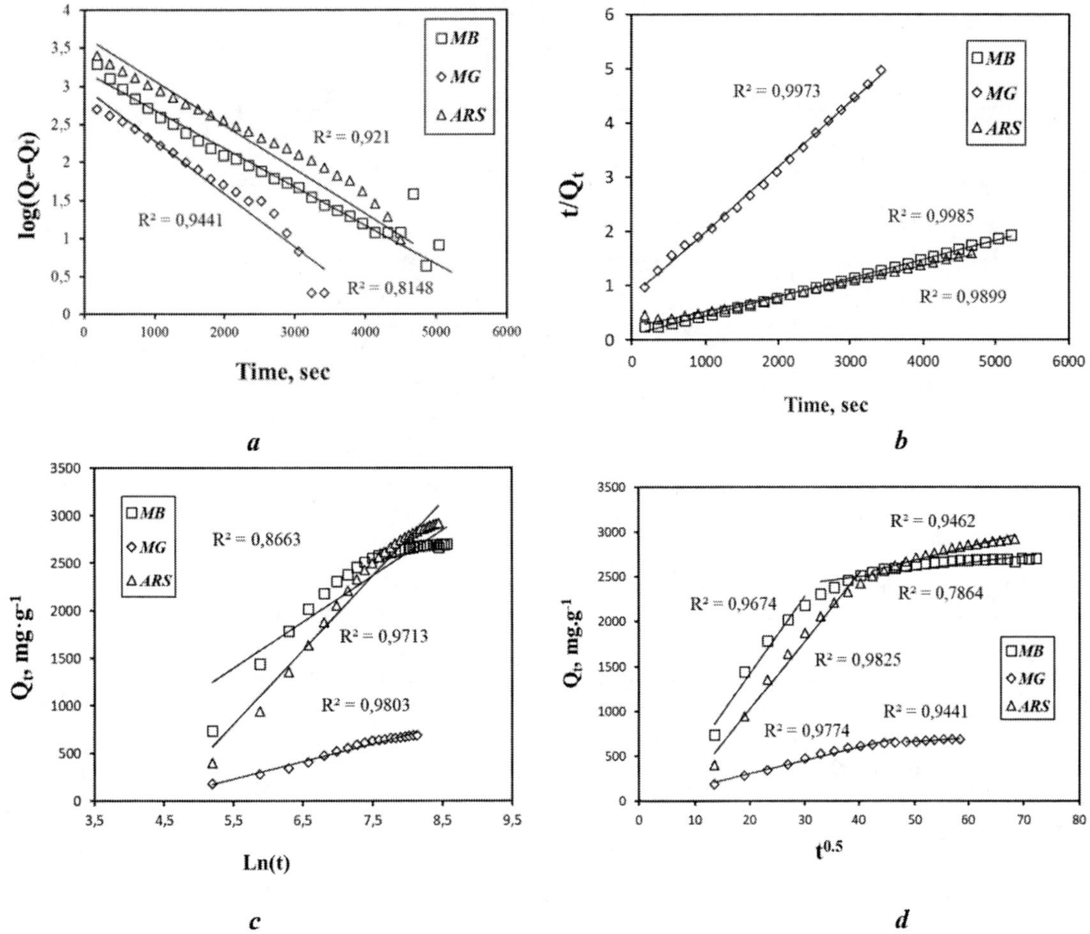

Figure 4. Kineticand diffusion models fitted to the experimental data obtained for the dye adsorption on the PHQ/G nanocomposite: pseudo-first-order (**a**), pseudo-second-order (**b**), Elovich model (**c**), and intraparticle diffusion (**d**).

Table 3. Parametersof the kinetic models fitted to the data on the dye adsorption on the PHQ/G nanocomposite.

Dye		Pseudo-first-order			Pseudo-second-order			Elovich		
	Q_e^{exp}, mg g^{-1}	Q_e, mg g^{-1}	k_1, min^{-1}	R^2	Q_e, mg g^{-1}	k$_2$, g mg^{-1} min^{-1}	R^2	α, min^{-1} mg g^{-1}	β, g mg^{-1}	R^2
MB	2.610	1.558	0.0012	0.921	3.333	$9 \cdot 10^{-7}$	0.9985	35.71	0.0020	0.8663
MG	688.1	948.7	0.0016	0.944	840.3	$1.78 \cdot 10^{-6}$	0.9970	2.63	0.0054	0.9803
ARS	2.923	2.826	0.0626	0.996	3.333	$2.25 \cdot 10^{-5}$	0.9899	516.9	0.0013	0.9713

Table 4. Parameters of the intraparticle diffusion model fitted to the data on the dye adsorption on the PHQ/G nanocomposite.

Dye	Intraparticle diffusion					
	Section 1			Section 2		
	$k_{id,}$ mg $g^{-1}min^{-0.5}$	c, mg g^{-1}	R^2	$k_{id,}$ mg $g^{-1}min^{-0.5}$	c, mg g^{-1}	R^2
MB	86.25	304.40	0.9674	7.74	2.191	0.7864
MG	14.55	15.50	0.9774	3.43	491.3	0.9441
ARS	75.37	476.08	0.9825	15.32	1.917	0.9462

It can be seen (judging by the highest determination coefficients R^2) that, in general all the models adequately describe the dye adsorption kinetics, but the best correlation to the experimental data is provided by the pseudo-second-order model (figure 4b, table 3): R^2=0.9985 (MB), 0.9899 (ARS) and 0.9973 (MG).

Based on the pseudo-first- and pseudo-second-order models (figures 4a and 4b), it can be assumed that chemical interactions occur between the adsorbent functional groups and the dye molecules. Theoretically, these interactions are stoichiometric, *i.e.*a single molecule occupies a single position on the adsorbent surface.

The Elovich model (figure 4c) implies heterogeneity of adsorption sites available on the material surface, meaning that these sites possess different binding energies and chemical affinities for the adsorbates.

The dependence describing the intraparticle diffusion (figure 4d) represents a multilinear curveconsisting of two linear sections. Since the approximating lines do not pass through the origin, the effect of the boundary diffusion layer takes place; the diffusion is not a limiting stage, and the superficial adsorption stage is superimposed on the diffusion process.

4. Conclusion

As a result of the kinetic studies conducted in the dynamic mode, a comparative analysis was made for the adsorption capacity of the PHQ/G nanocomposite material regarding the removal of the dye organic molecules – MB, ARS, and MG. The following adsorption capacity values were found: 2.923, 2.610 and 688.1 mg g^{-1} for the ARS, MB and MG, respectively. In the case of the ARS and MB adsorption, the novel material demonstrated the highest adsorption capacity among the adsorbents presented in table 1.

Besides, it was established that the adsorption equilibrium is achieved within 40, 120 and 57 min for the MB, ARS and MG, respectively. Moreover, the kinetic and diffusion models implemented to fit the experimental data adequately describe the dye adsorption processes. Based on those models, it was elucidated that in the course of the adsorption, chemical interactions take place between the dye molecules and the active sites of the material surface, and the diffusion through the film on the adsorbent surface also occur.

Thus, the PHQ/G nanocomposite proposed herein can be efficiently used in purifying wastewater from some dyes. However, further studies are required to investigate the dye adsorption based on adsorption isotherm construction.

Acknowledgment

The research was supported by the Ministry of Science and Higher Education of the Russian Federation in the framework of Project No. 16.1384.2017/PCh.

References

[1] He K, Zeng G, Chen A, Huang Zh, Peng M, Huang T and Chen G 2019 Graphene hybridized polydopamine-kaolin composite as effective adsorbent for methylene blue removal *Composites B* **161** 141

[2] Nagpal M and Kakkar R 2019 Use of metal oxides for the adsorptive removal of toxic organic pollutants *Sep. Purif. Technol.* **211** 522

[3] Yagub M T, Sen T K, Afroze S and Ang H M 2014 Dye and its removal from aqueous solution by adsorption: a review *Ang. Adv. Colloid Interface Sci.* **209** 172

[4] Yang J-Y, Jiang X-Y, Jiao F-P and Yu J-G 2018 The oxygen-rich pentaerythritol modified multi-walled carbon nanotube as an efficient adsorbent for aqueous removal of alizarin yellow R and alizarin red S *Appl. Surf. Sci.***436** 198

[5] Machado F M, Carmalin S A, Lima E C, Dias S L P, Prola L D T, Saucier C, Jauris I M, Zanella I and Fagan S B 2016 Adsorption of Alizarin Red S dye by carbon nanotubes: An experimental and theoretical investigation *J. Phys. Chem. C* **120** 18296

[6] Wanassi B, Hariz I B, Ghimbeu C M, Vaulot C, Hassen M B and Jeguirim M 2017 Carbonaceous adsorbents derived from textile cotton waste for the removal of Alizarin S dye from aqueous effluent: kinetic and equilibrium studies *Environ. Sci. Pollut. Res.* **24** 10041

[7] Fayazi M, Ghanei-Motlagh M and Taher M A 2015 The adsorption of basic dye (Alizarin red S) from aqueous solution onto activated carbon/γ-Fe_2O_3 nano-composite: kinetic and equilibrium studies *Mater. Sci. Semicond. Process.* **40** 35

[8] Boukhaifa N, Boutahala M, Djebri N and Idris A 2019 Maghemite/alginate/functionalized multiwalled carbon nanotubes beads for methylene blue removal: Adsorption and desorption studies *J. Mol. Liq.* **275** 431

[9] Ren F, Li Z, Tan W Z, Liu X H, Sun Z F, Ren P G and Yan D X 2018 Facile preparation of 3D regenerated cellulose/graphene oxide composite aerogel with high-efficiency adsorption towards methylene blue *J. Colloid Interface Sci.* **532** 58

[10] Gupta K, Gupta D and Khatri Om P 2019 Graphene-like porous carbon nanostructure from Bengal gram bean husk and its application for fast and efficient adsorption of organic dyes *Appl. Surf. Sci.* **476** 647

[11] Qu S, Huang F, Yu Sh, Chen G and Kong J 2008 Magnetic removal of dyes from aqueous solution using multi-walled carbon nanotubes filled with Fe_2O_3 particles *J. Hazard. Mater.* **160** 643

[12] Zhang Yu, Xu S, Luo Yu, Pan Sh, Ding H and Li G 2011 Synthesis of mesoporous carbon capsules encapsulated with magnetite nanoparticles and their application in wastewater treatment *J. Mater. Chem.* **21** 3664

[13] Homaeigohar Sh, Strunskus Th, Strobel J, Kienle L and Elbahri M 2018 A flexible oxygenated carbographite nanofilamentous buckypaper as an amphiphilic membrane *Adv. Mater. Interfaces* **5** 180000

[14] Ullah H, Nafees M, Iqbal F, Awan M S, Shah A and Waseem A 2017 Adsorption kinetics of malachite green and methylene blue from aqueous solutions using surfactant-modified organoclays *Acta Chim. Slov.* **64** 449

[15] Zhang X, Lin Q, Luo S, Ruan K and Peng K 2018 Preparation of novel oxidized mesoporous carbon with excellent adsorption performance for removal of malachite green and lead ion *Appl. Surf. Sci.* **442** 322

[16] Hu L, Yang Z, Wang Y, Li Y, Fan D, Wu D, Wei Q and Du B 2017 Facile preparation of water-soluble hyperbranched polyamine functionalized multiwalled carbon nanotubes for high-efficiency organic dye removal from aqueous solution *Sci. Rep.* **7** 3611

[17] Datta D, KerkezKuyumcu O, Bayazit S S and Abdel Salam M 2017 Removal of rhodamine B from aqueous solution using palm shell-based activated carbon: Adsorption and kinetic studies *J. Disp. Sci. Tech.* **38** 1556

[18] Tripathi P K, Gan L and Liu M 2016 Synthesis of large pore carbon nanoparticles for removal of malachite green *J. Nanosci. Nanotech.* **16** 892
[19] Gupta K and Khatri O P 2017 Reduced graphene oxide as an effective adsorbent for removal of malachite green dye: Plausible adsorption pathways *J. Coll. Int. Sci.* **501** 11
[20] Abdi J, Vossoughi M, Mahmoodi N M and Alemzadeh I 2017 Synthesis of metal-organic framework hybrid nanocomposites based on GO and CNT with high adsorption capacity for dye removal *Chem. Eng. J.* **326** 1145
[21] Shi Y C, Wang A J, Wu X L, Chen J R and Feng J J 2016 Green-assembly of three-dimensional porous graphene hydrogels for efficient removal of organic dyes *J. Coll. Int. Sci.* **484** 254
[22] Babkin A V, Melezhik A V, Kurnosov D A, Mkrtchyan E S, Burakova I V, Burakov A E and Galunin E V 2018 Kinetics of the Adsorption of Synthetic Dyes on a Polyhydroquinone/Graphene Carbon Nanocomposite *J. Phys.: Conf. Ser.* **1124** 081030
[23] Ho Y S and McKay G 1999 Pseudo-second order model for sorption processes *Process Biochem.* **34** 451

IOP Conf. Series: Materials Science and Engineering **693** (2019) 012042 doi:10.1088/1757-899X/693/1/012042

Studying the fire hazard properties of multi-walled carbon nanotubes by the method of synchronous thermal analysis

V Dan[1], N Barbin[1,2], O Bezzaponnaya[1], D Terentiev[1] and S Alexeev[3]

[1]*Ural Institute of State Fire Service of Emercom of Russia*, 22 Mira Street, 620062 Ekaterinburg Russian Federation

[2]*Urals State Agrarian University*, 42 Karl Liebknecht Street, 620075 Ekaterinburg Russian Federation

[3]*Ural Research Institute of All-Russian Voluntary Fire Organization*, 32 Uchiteley Street, 620137 Ekaterinburg Russian Federation

[1]E-mail: DanVP_92@mail.ru, NMBarbin@mail.ru

Abstract. The present paper considers oxidation of multi-walled carbon nanotubes at different heating rates and temperatures to ensure the process in the kinetic region. The nanotubes were analyzed by the method of synchronous thermal analysis. Experiments were carried out at different heating rates: 5, 10 and 20 °C/min. As a result, the character of thermogravimetric and differential thermogravimetric curves was determined, and thermograms were constructed, thereby allowing to define weight loss stages and establish exact weight loss values at a programmed temperature increase. Besides, the temperature characterized by the maximum weight loss rate was established. Differential scanning calorimetry revealed the maximum temperature of endothermic and exothermic peaks, as well as changes in thermal effects (enthalpy changes). The most important indicators of fire hazard of the multi-walled carbon nanotubes – ignition and self-ignition temperature – were determined. The ignition temperature was approximately equal to the temperature of the beginning of the thermal distortion of the sample in the air – 553.7, 578.6 and 584.3 °C for the heating rates of 5, 10 and 20 °C/min, respectively. The self-ignition temperature corresponded to the temperature of the maximum yield of volatiles – 627.5, 651.9 and 670.8 °C for the heating rates of 5, 10 and 20 °C/min, respectively.

1. Introduction

Due to economic reasons, multi-walled carbon nanotubes (MWCNTs) are the most relevant from the point of view of carbon materials applications. Currently, the total global production of carbon materials is about 1 million tons per year [1]. The increase in the productive capacity inevitably entails an increase in the area of accommodation of the finished product. In this regard, the problem of combustion and explosion of the carbon materials is of particular relevance.

When considering the explosion and fire hazard of dust, special attention is usually paid to its explosive properties [2-16], which are known since the 18th century [2-4]. Explosion and fire hazard of dust (dust cloud) and a dust layer characterize the following indicators: maximum explosion (over)pressure, maximum rate of overpressure rise, index of fire and explosion hazard (the *Kst* parameter, or deflagration index), minimum explosive concentration (low explosion limit), normal flame propagation speed, self-ignition and ignition (fire) temperatures , smoldering and layer ignition, heat of combustion, minimum ignition energy, limiting oxygen concentration, burning time, safe experimental maximum clearance, flame emissivity, minimum phlegmatizing concentration of gaseous deterrents, toxicity of combustion products, ability to burn when in contact with water, oxygen and other substances and to spontaneous combustion and exothermic decomposition, and combustibility group [2-18].

The requirements for determining the indicators of explosion hazard of substances and materials are set out in the paper [19]. Kinetics of thermal oxidation of the carbon nanomaterials was studied by differential thermal analysis in air [20]. However, the fire hazard indicators are not considered in these works.

Considering the aforementioned, the aim of the present work was to study the oxidation of MWCNTs in the air atmosphere, and determine ignition and self-ignition temperatures by a thermal analysis method.

2. Experimental part

"HW NANO" MWCNTs (Xuzhou, Jiangsu, China), with a purity of 98%, average diameter of 0.83 nm and length of 1-2 um were emploued in experiments. A corundum crucible with a volume of 85 μL was used, in which 3 to 4 mg of the MWCNTs was loaded.

The method of synchronous thermal analysis makes it possible to measure the dependence of weight changes in the test material and the inert sample (for comparison) and heat flow on temperature (time). It is successfully used to study the heat resistance of flame retardants, as well as the indicators of fire hazard of substances and materials. In the present research, this analysis was implemented in an air medium at the heating rates of 5 (experiment 1), 10 (experiment 2) and 20 °C/min (experiment 3). The measurement chamber was purged with a mixture of oxygen and nitrogen to create an environment as close to air as possible (20 % / 80 %). The MWCNTs were placed in an alumina crucible without lid.

A series of experiments were carried out using an STA 449 F5 Jupiter synchronous thermal analyzer (NETZSCH, Selb, Germany). This instrument is included into the State Register of Measuring Instruments and has a certificate of approval, as well as a valid certificate of verification. It represents a measuring complex that combines the functions of high-precision analytical scales and a differential scanning calorimeter.

The thermogravimetry (TG) analysis allows to determine the stages of weight losses based on TG curves constructed, and to establish the exact weight loss values at a programmed temperature increase. From differential TG (DTG) curves, it is possible to establish temperature characterized by a maximum weight loss rate. The data presented by differential scanning calorimetry (DSC) curves allow to reveal the maximum temperatures of endothermic and exothermic peaks, as well as changes of thermal effects (enthalpy changes).

The test results were obtained on certain graphics, the processing of which was performed using the Proteus Thermal Analysis software (NETZSCH).

3. Results and discussion

The graphical results (TG and DTG curves) of the synchronous thermal analysis are presented in figures 1-3.

Up to 110 °C, the weight loss areas are associated with the evaporation of adsorbed water. The weight loss in the temperature range of 110-500 °C is negligible. This range represents a section of the thermal stability of the MWCNTs. In the temperature range of 500-650 °C for experiment 1, 500-720 °C for experiment 2 and 500-810 °C for experiment 3, the weight loss caused directly the process of carbon oxidation. The most intense losses can be observed in the temperature ranges of 550-650 °C (experiment 1), 550-730 °C (experiment 2) and 550-810 °C (experiment 3), corresponding to the MWCNTs combustion process.

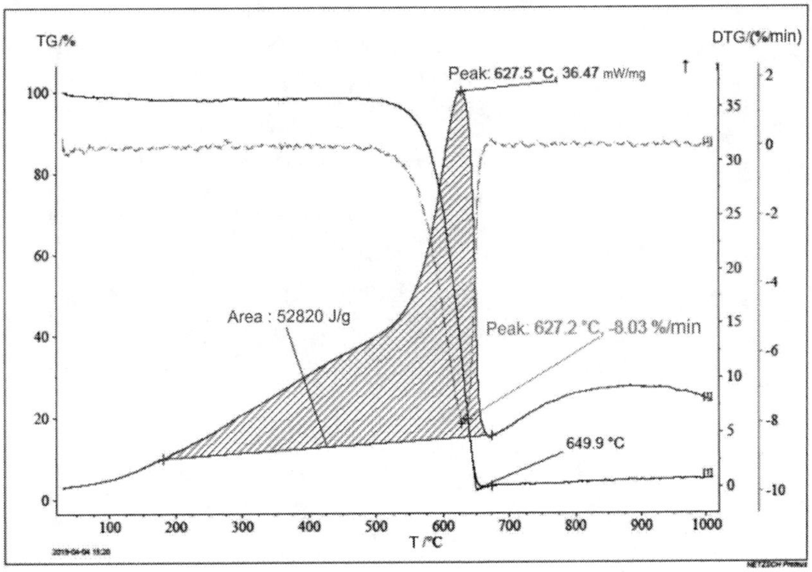

Figure 1. Graphical results of the synchronous thermal analysis at 5 °C/min.

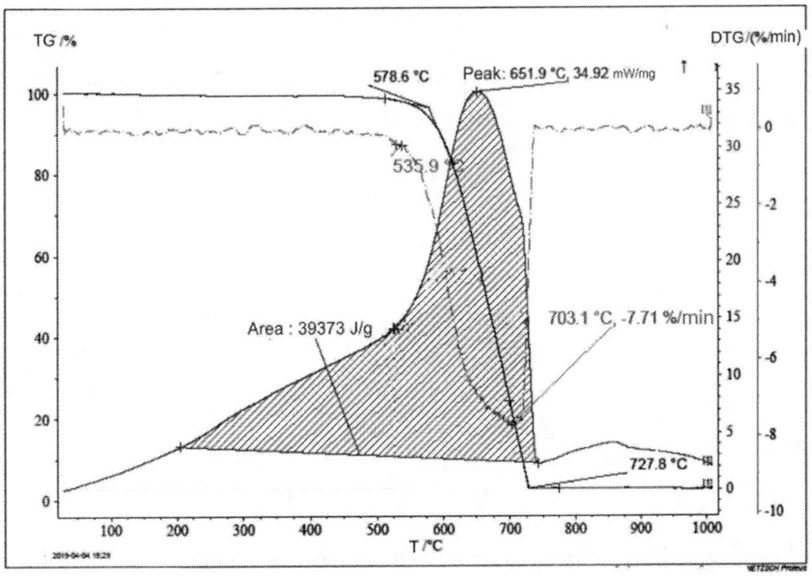

Figure 2. Graphical results of the synchronous thermal analysis at 10 °C/min.

Exothermic reactions occur in the temperature range of 110-649.9 °C for experiment 1, 110-727.8 °C for experiment 2 and 110-810.8°C for experiment 3. The graphs of the temperature dependence of heat have irregular shape with many peaks: at 550, 630 and 650 °C for experiment 1, at 550, 650 and 730 °C for experiment 2, and 550, 670 and 810°C for experiment 3. The process is accompanied by significant weight losses in the samples. The most intense losses take place in the temperature range of 500-650 °C for experiment 1, 500-720 °C for experiment 2, and 500-810 °C for experiment 3.

The enthalpy of the MWCNTs was found to be 52,820, 39,373 and 25,920 J/g for experiment 1, 2 and 3, respectively. Thus, the lower the heating rate, the higher the heat of nanotubes combustion.

The parameters of the TG, DTG and DSC curves are given in Table 1.

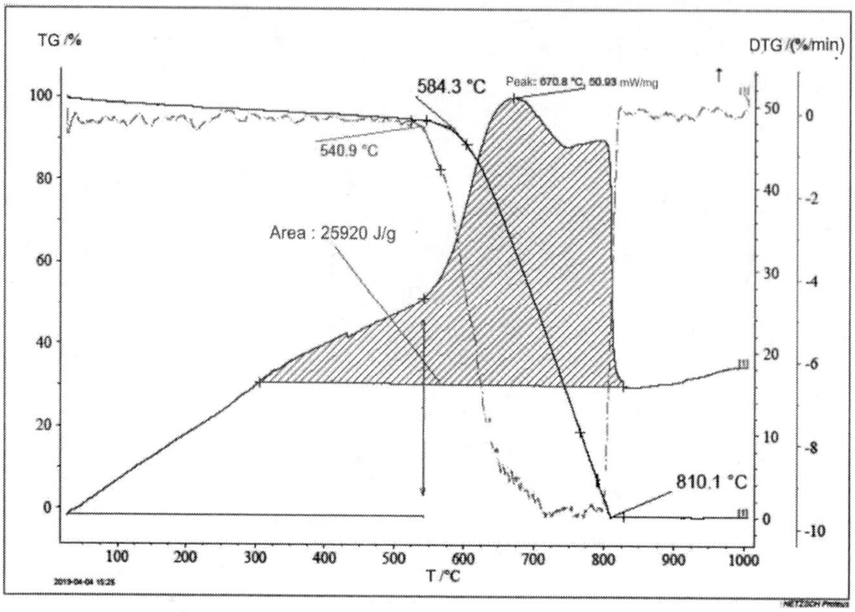

Figure 3. Graphical results of the synchronous thermal analysis at 20 °C/min.

Table 1. Parameters of the TG, DTG and DSC curves of experiments 1-3.

No.	Characteristic	Experiment 1	Experiment 2	Experiment 3
		TG		
1	Temperature of weight loss beginning	553.7 °C	578.6 °C	584.3 °C
2	Temperature of weight loss ending	649.9 °C	727.8 °C	810.1°C
		DTG		
3	Weight loss rate	8.03 %/min	7.71 %/min	7.38 %/min
4	Peak weight loss	627.2 °C	703.1 °C	780.0 °C
		DSC		
5	Enthalpy	52,820 J/g	39,373 J/g	25,920 J/g
6	Peak	627.5 °C	651.9 °C	670.8 °C

As can be seen from the table, the heating rate affects the temperature of the beginning and the ending of the weight loss: an increase in the heating rate leads to an increase in the temperature boundaries of this process.

The ignition temperature is the lowest temperature of the substance, at which vapors above the surface of the combustible substance are released at such a rate that leads to the ignition by the influence of an ignition source.

The self-ignition temperature is the lowest temperature of the combustible substance; while heating, there is a sharp increase of rate of exothermic volumetric reactions, leading to the appearance of a fiery combustion or explosion.

These indicators are the most important indicators of fire and explosion hazard. The studies carried out by the method of synchronous thermal analysis, made it possible to determine these indicators for the MWCNTs. The ignition temperature was found to be approximately equal to the temperature of the beginning of the thermal distortion of the sample in the air. The temperature of self-ignition corresponds to the temperature of the maximum volatile yield. The data obtained on these temperatures are presented in Table 2.

Table 2. Dependences of the ignition and self-ignition temperatures of the MWCNT on the heating rates.

Heating rate, °C/min	Ignition temperature, °C	Self-ignition temperature, °C
5	553.7	627.5
10	578.6	651.9
20	584.3	670.8

4. Conclusion

The oxidation of the MWCNTs at different heating rates was studied by the synchronous thermal analysis method.

As a result of the experiments, it was found that this method is applicable not only to determine the technological properties of the MWCNTs, but also some indicators of fire and explosion hazard. Thus, the ignition and self-ignition temperatures of the MWCNTs determined at different heating rates were respectively found to be 553.7 and 627.5 °C for experiment 1, 578.6 and 651.9 °C for experiment 2, and 584.3 and 670.8°C for experiment 3. The average self-ignition temperature of the MWCNTs in experiments 1-3 was 572.2 °C, whereas their average self-ignition temperature was 650.1 °C.

The temperature limits of the beginning and ending of the samples weight loss were also different: from 553.7 to 649.9°C for experiment 1, from 578.6 to 727.8°C for experiment 2, and from 584.3 to 810.1°C for experiment 3. Furthermore, the enthalpy of the MWCNTs was found to decrease with increasing the heating rate.

Thus, it can be concluded that an increase in the heating rate leads to an increase in the ignition and self-ignition temperature, an increase in the weight loss temperature range, and a decrease in enthalpy.

References

[1] Chesnokov N V, Mikova N M, Ivanov I P and Kuznetsov B N 2014 Obtaining carbon sorbents through chemical modification of fossil coals and plant biomass. *J. of Syberian Federal Univ. Chemistry* [Zhurnal Sibirskogo Federalnogo Universiteta. Khimiya – in Russian] **7** 42

[2] Eckhoff R K 2003 *Dust Explosions in the Process Industries* (Amsterdam: Gulf Professional Publishing) chapters 1 and 2 pp 1–198

[3] Eckhoff R K 2016 *Explosion Hazards in the Process Industries* (Amsterdam: Gulf Professional Publishing) chapter 7 pp 253–383

[4] Amyotte P 2013 *An Introduction to Dust Explosions. Understanding the Myths and Realities of Dust Explosions for a Safer Workplace* (Oxford: Butterworth-Heinemann) p 280

[5] Abbasi T and Abbasi S A 2007 Dust explosions – cases, causes, consequences, and control *J. Hazard. Mater.* **140** 7 https://doi.org/10.1016/j.jhazmat.2006.11.007

[6] Barton J 2002 *Dust Explosion. Prevention and Protection. A Practical Guide* (Rugby: Institution of Chemical Engineers) pp 8–25

[7] AIChE 2017 *Guidelines for Combustible Dust Hazard Analysis* (Hoboken: American Institute of Chemical Engineers) pp 15–24

[8] Pritchard D K 2004 *Report HSL/2004/12* (Buxton: Health & Safety Laboratory) chapter 3 pp 9–14

[9] Denkevitz A 2005 *Report on EFDA Subtask TW-TSS-SEA5.2* (Karlsruhe: Forschungszentrum Karlsruhe GmbH) pp 1–69

[10] Cheremisinoff N P 2014 *Dust Explosion and Fire Prevention Handbook. A Guide to Good Industry Practices* (Beverly: Scrivener Publishing) chapters 2 and 3 pp 29–92

[11] NFPA 652: 2019. *Standard on the Fundamentals of Combustible Dust* (Quincy, USA) p 89

[12] ASTM E1226–12a 2012 *Standard Test Method for Explosibility of Dust Clouds* (ASTM, West Conshohocken, USA) p 13
[13] ASTM E1515–14 2014 *Standard Test Method for Minimum Explosible Concentration of Combustible Dusts* (ASTM, West Conshohocken, USA) p 9
[14] ASTM E789–95(2001) 2007 *Standard Test Method for Dust Explosions in a 1.2-Litre Closed Cylindrical Vessel* (ASTM, West Conshohocken, USA) p 18
[15] UN 2009 *Recommrndations on the Transport of Dagerous Goods. Manual of Tests and Criteria* (UN, Geneva, Switzerland) subsection 33.2.1 pp 353–357
[16] Drysdale D 2011 *An Introduction to Fire Dynamics* (Chichester: John Wiley & Sons) subchapters 5.2.3 and 8.1.3 pp 221-222 and 325-329
[17] GOST (Russian National Standard) 12.1.044 2018 *System of Occupational Safety Standards. Fire and Explosion Hazard of Substances and Materials. Nomenclature of Indicators and Methods of their Determination* [in Russian – Sistema Standartov Bezopasnosti Truda. Pozharovzryvobezopasnost' Veshchestv i Materialov. Nomenklatura Pokazateley i Metody ikh Opredeleniya] (Moscow: Rosstandart) section 5 pp 6-8
[18] Federal Law of the Russian Federation №123-FZ (22 July 2008)) *Technical Regulations on Fire Safety Requirements* [in Russian – Tekhnicheskiy Reglament o Trebovaniyakh Pozharnoy Bezopasnosti] table 1 appendix 1, available at: http://www.consultant.ru/document/cons_doc_LAW_78699/8822583228e942acfd749911617579e9f9f065f4/
[19] Groenewoud W M 2001 *Characterisation of Polymers by Thermal Analysis* (Amsterdam/Boston/London/New York/Oxford/Paris/San Diego/San Francisco/Singapore/Sydney/Tokyo: Elsevier Science) p 396
[20] Glebova N V, Nechitailov A A, Kukushkina Y A and Sokolov V V 2011 *Tech. Phys. Lett.* **37** 438 https://doi.org/10.1134/S1063785011050063

IOP Conf. Series: Materials Science and Engineering **693** (2019) 012043 doi:10.1088/1757-899X/693/1/012043

Impact of plasma treatment in CH_4/N_2 on the properties of reduced graphene oxide

E P Neustroev[1*], A R Prokopiev[1], V B Timofeev[1], V I Popov[1], I I Kurkina[1], Z Y Davydova[1], A A Alekseev[1] and S O Semenov[2]

[1]Department of Radiophysics and Electronic Systems, *North-Eastern Federal University*, 58 Belinsky Street, Yakutsk, Republic of Sakha (Yakutia), 677000, Russian Federation

[2]*V.P. Larionov Institute of the Physical-Technical Problems of the North of the Siberian Branch of the RAS*, 1 October Street, Yakutsk, Republic of Sakha (Yakutia), 677890, Russian Federation

[*]E-mail: neustr@mail.ru

Abstract. The effect of the two-step process on the properties of graphene oxide (GO) is studied. At the first stage, CH_4 (or CH_2+N_2) plasma treatment was performed. The second one was heat treated at a temperature of 650° C. The formation of thin polycrystalline carbon films on the surface of graphene oxide was detected. The ratio of carbon to oxygen concentration corresponds to graphene oxide. Smooth surfaces of the films were detected by an atomic force microscope. Films have lower sheet resistance and higher carrier mobility than reduced GO (rGO) during the same heat treatment. An analysis of the Raman spectra shows that the formed films have larger graphene domains than the rGO. The presence of nitrogen in methane plasma leads to a decrease in the resistance and mobility of carriers in the carbon film.

1. Introduction

Graphene oxide (GO) belongs to the family of two-dimensional graphene-like materials. Due to its electrical, chemical and mechanical properties, reduced graphene oxide is of interest for use as optoelectronic devices [1, 2], bio- and gas sensors [3-9], electrodes [4, 9, 10] and fuel cells [4, 11, 12]. Graphene oxide can be given individual properties through functionalization [3, 8, 9]. Despite its advantages, GO needs to improve a stability of parameters and the quality of the crystal structure, eliminate oxygen functional groups and increase electrical conductivity. Effective methods for modifying a properties of GO through an addition of various functional groups are also required.

Methane is the main component in the chemical vapor deposition of graphene [10, 11]. By adding nitrogen to the methane stream, graphene can be doped [12]. The use of plasma in the CVD method significantly reduces the synthesis temperature [13, 14]. A modification of this method is the separation of plasma and heat treatment regions (remote PE-CVD) [15, 16]. This method allows to reduce defect formation and eliminate the vertical growth of graphene flakes during exposure in plasma [16]. In addition, plasma exposure to methane is used to functionalization and healing of lattice defects in graphene [17-19]. A feature of this work is the division of a single process in remote PE-CVD into two separate stages. At the first stage, carbon is deposition in methane plasma at room temperature. At the second stage, heat treatment is performed. This makes it possible to completely eliminate defect formation due to the effect of plasma reactive elements on the synthesis of a carbon film. This technique allows to reduce defect formation during subsequent synthesis of the carbon film during heat treatment. At the same time, this reduces a cost of producing carbon films relative to the remote plasma processing method. Thus, the aim of this work was to study the combined effect of the plasma treatment with methane and nitrogen in a two-step process on the properties of reduced GO.

2. Materials and methods

2.1. *Sample preparation and reagents*

A graphene oxide was obtained by an improved Hummers method [20] and a detailed description of the production of GO and its properties was carried out in our previous work [21]. To obtain graphene oxide were used graphite powder (Sigma-Aldrich,USA), H_2SO_4 (Sigma Tek, Russia), $NaNO_3$ and $KMnO_4$ (Chemman Mosreactive, Russia). An aqueous solution of GO on the SiO_2 surface of the Si/SiO_2 structure was precipitated and evaporated at room temperature for a day. Methane and nitrogen gases with a purity of ≥0.999 % for plasma processing were used.

2.2. *Measurement methods*

The Ntegra Spectra atomic force microscope (AFM) was used to measure the surface topography of GO films in tapping mode. To determine the thickness of GO on its surface, scratches with a sharp metal rod were made. As a result of the formation of a scratch, a threshold formed between GO and SiO_2. The height of this threshold was measured by AFM. The threshold height was equal to the thickness of the GO film, since the SiO_2 surface remained intact upon the formation of a scratch. To measurements of a micro Raman spectra on the Ntegra Spectra platform, a laser wavelength of 473 nm was used. The study of the surface morphology was carried out using a scanning electron microscope (SEM) (JEOL JSM 7800F) and an AFM. Evaluation of the content of chemical elements in the samples was carried out using energy dispersive X-ray spectroscopy (EDS) Inca Energy. The mobility of the samples was measured using the Ecopia HMS-5000 apparatus using the Hall effect method. To electrical measurements were made point contacts based on silver paste. Contacts were previously dried at a temperature of 120° C in air for 15 min.

2.3. *Processing method*

Plasma treatments were performed in radio-frequency (13.56 MHz) inductively coupled plasma. Figure 1 shows schematic diagram of the microwave plasma source and the reaction chamber. The samples of GO in the reaction chamber were arranged so that a surfaces of the samples were turned down ("face down") and were not subjected to bombardment by the reaction products in plasma, as was described in our previous works [22, 23]. After placing the sample in the center of the reaction chamber a reactor was evacuated to a pressure of less than 10^{-3} mbar. Methane (30 sccm) was introduced in the reaction chamber and a plasma was ignited at a pressure of 0.04 mbar with a microwave power of 150 W and 200 W. The same powers were used for plasma processing in a mixture of methane (30 sccm) and nitrogen (10 sccm). Then, a pressure in the chamber was 700 mbar. The process parameters for plasma treatment of samples are shown in table 1. It should be noted that the time of one process was 3 min and a total processing time was 12 min. After plasma treatment, all samples were annealed at 650° C in an argon atmosphere for 30 min.

IOP Conf. Series: Materials Science and Engineering **693** (2019) 012043 doi:10.1088/1757-899X/693/1/012043

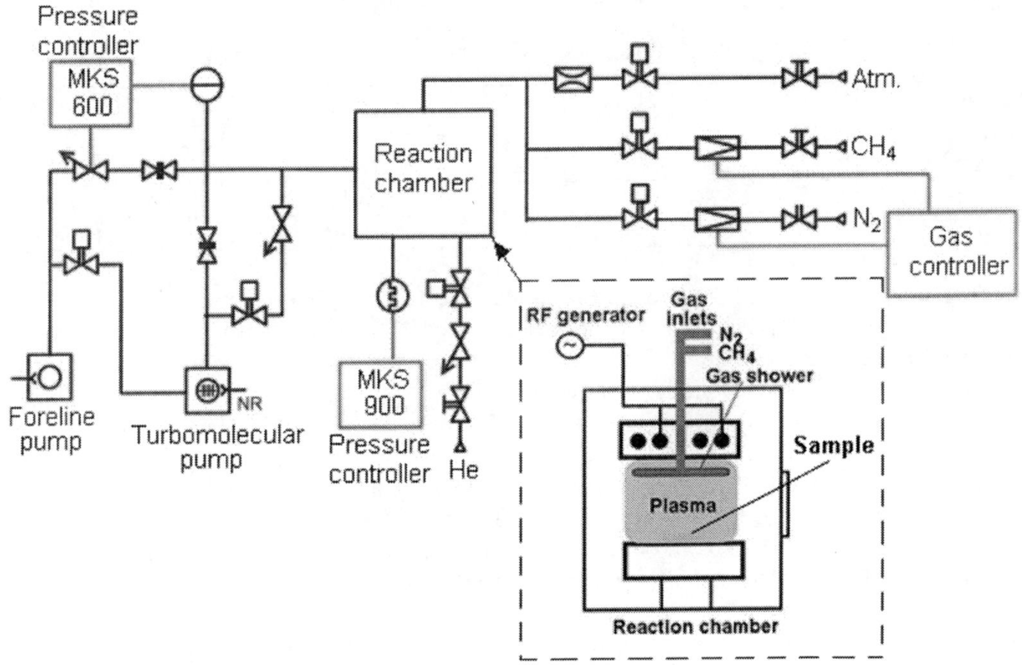

Figure 1. Schematic diagram of the microwave plasma source and the reaction chamber.

Table 1. Process parameters for plasma treatment of samples.

Symbol	CH_4-flow, sccm	N_2-flow, sccm	Power, W	Pressure, mbar
Process 1	30	-	150	400
Process 2	30	-	200	400
Process 3	30	10	150	700
Process 4	30	10	200	700

3. Results and discussion

Figure 2a presents an AFM image of the GO surface prior to plasma treatment. Measurements of the heights of the steps formed by the interface of GO with the surface of the substrate were carried out with a help of AFM, showed that the thickness of the film under study is in the range from 1 to 5 nm (Figure 2d). Figures 2b and 2f show the corresponding patterns of the sample surface and the profile of the step after plasma processing in methane with a power of P=200 W for 6 min. Figures 2c and 2f show similar images for a sample processed in a CH_4 + N_2 plasma with a power of P=200 W for 12 min. Table 2 presents the changes in thickness Δh_1 and Δh_2 after plasma treatment for 6 and 12 min, respectively. As can be seen, the deposition processes dominate the etching processes in plasma. These results can be explained by the deposition of carbon and the formation of an amorphized carbon (a:C) film as a result of plasma treatment [24]. For methane plasma, the deposition rate of an a:C film depends on the plasma power and increases with increasing power. The deposition rate is irregular and at the initial exposure times of the plasma reaches the highest value. The lowest deposition rate of a a:C film in plasma CH_4 + N_2 mixture was observed. Most likely, this is due to more intensive nitrogen etching of the deposited film. After annealing at 650° C, part of the amorphized carbon film was removed and part of the carbon atoms forms bonds with a GO film. As a result, a thickness of the initial GO film increased by 10–16 nm.

Table 2. The change in the thickness of the deposited film during plasma processing.

Change the thickness of the samples	CH_4, 150 W	CH_4, 200 W	CH_4+N_2, 200 W
Δh_1, nm	40	65-75	30
Δh_2, nm	60-70	~120	~(40-50)

In the inserts of figures 2a, 2b and 2c it is shown a surface roughness for the initial GO, after CH_4 (200 W, 6 min) and CH_4+N_2 plasma treatment obtained by AFM. Measurements showed that the surface roughness before and after plasma treatment decreased from ~ 2 nm to ~ 1 nm, respectively (Figures 2a and 2b). On the contrary, when using a mixture of methane + nitrogen plasma, the roughness increased to ~(4-5) nm (Figure 2 c). This confirms that etching occurs in a plasma involving nitrogen.

Figures 3a and 3b show SEM images of the surface of the initial GO and the surface of the sample after treatment in 200 W methane plasma for 6 min. As can be seen from the figure 3b, no wrinkles and corrugations typical for GO are observed due to the deposition and the formation of an a.C film. The surface roughness of the samples after heat treatment is not noticeably changed and remains the same.

Figure 4 shows the Raman spectra of the initial GO samples after plasma treatment in methane and in a mixture of CH_4 + N_2 for 12 min and subsequent annealing at T = 650° C. The Raman spectra of the initial samples have a typical form for GO, containing G and D peaks at ~ 1600 cm^{-1} and ~1350 cm^{-1}, respectively. The G peak characterizes the vibrations of the sp^2 carbon bond system in the planar structure of graphene [25, 26]. The D peak is related with sp^3 vibrations of hybridized carbon bonds and its activation requires the presence of structural disorders and/or edge effects [25, 26]. In the figure the intensity of the Raman peaks is normalized relative to the height of the G peak. After plasma treatment, a broad photoluminescence peak characteristic of amorphized carbon appears [24]. This band is most noticeable for samples treated in a CH_4 + N_2 gas mixture plasma (Figure 4b). The maximum of this band also shifted towards lower frequencies with increasing power and duration of the CH_4 + N_2 plasma exposure. During subsequent heat treatments (650° C, 30 min), the intensity of this peak becomes noticeably less and the D and G peaks become the most significant again. Table 3 shows the Raman parameters of the peaks of the initial GO samples, control sample (only annealing) and samples after plasma treatment and subsequent annealing at 650° C. Exposure of samples in plasma leads to a greater redshift of the maximum intensity of the G peak than in the control sample. The red shift of the G peak means an increase in the size of graphene domains or their number [27]. Estimates of the sizes of graphene domains were also performed according to the method proposed in [28].

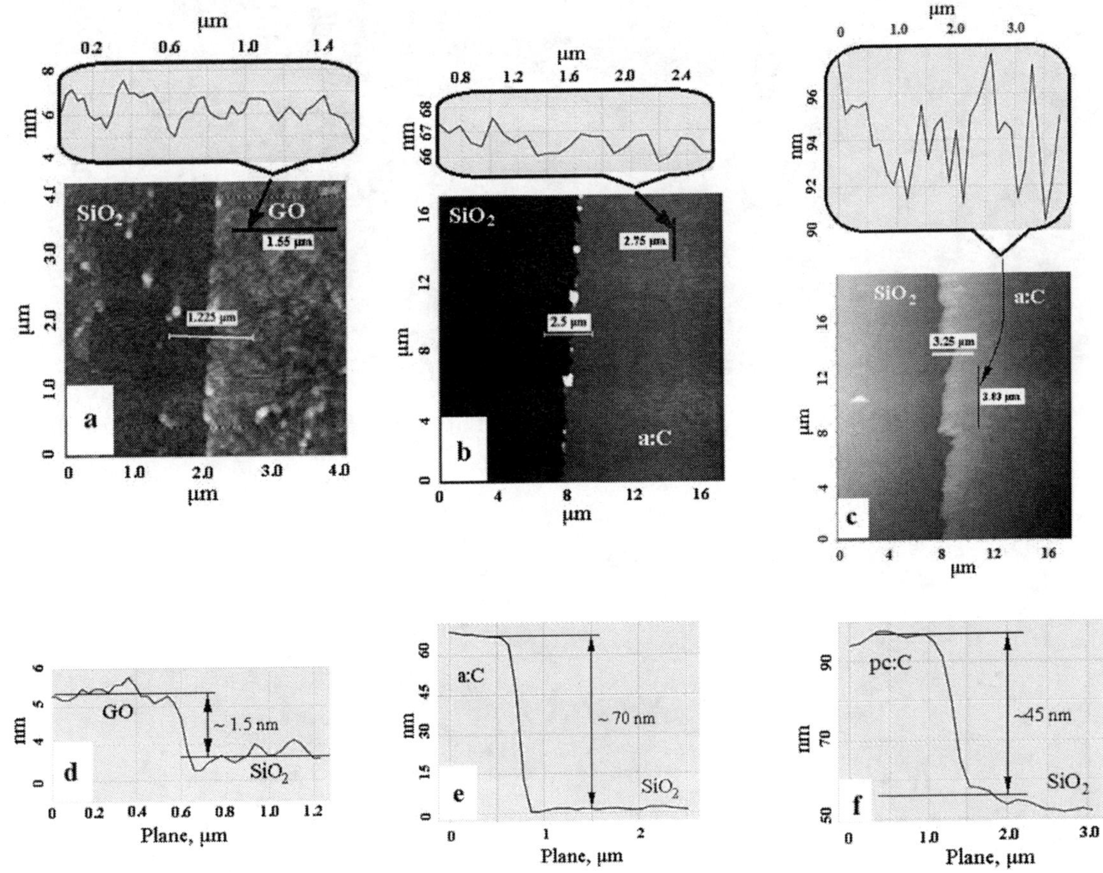

Figure 2. AFM image of the surface at the GO/SiO_2 interface for a) initial GO, b) after plasma processing in CH_4 (P= 200 W for 6 min) and c) after plasma processing in CH_4+N_2 (P= 200 W for 12 min). The insets of the figures show the corresponding surface roughness. Figures d, e and f show surface profiles along the white lines shown in figures a, b and c respectively.

For this was used the formula [25, 29, 30]:

$$L_a(nm) = (2.4\times10^{-10})\lambda_{laser}^4\left(\frac{I_D}{I_G}\right)^{-1},$$

where I_D/I_G is a ratio of the intensities D and G of the Raman peak of the spectra, λ_{laser} is the laser radiation wavelength. The evaluation showed that the size of the graphene domains after plasma treatment is approximately 14-16 nm. This is 1-2 nm more than for samples that are subject to thermal reduction only and 3-4 nm more than for the initial GO. An increase in the size of graphene domains is consistent with the results of work [17, 18, 31], in which treatment in the plasma of methane led to healing of defects in the graphene lattice. When using plasma power of 150 W or a shorter processing time (6 min), the size of the domains is not noticeably different from the size of the thermally reduced GO.

IOP Conf. Series: Materials Science and Engineering **693** (2019) 012043 doi:10.1088/1757-899X/693/1/012043

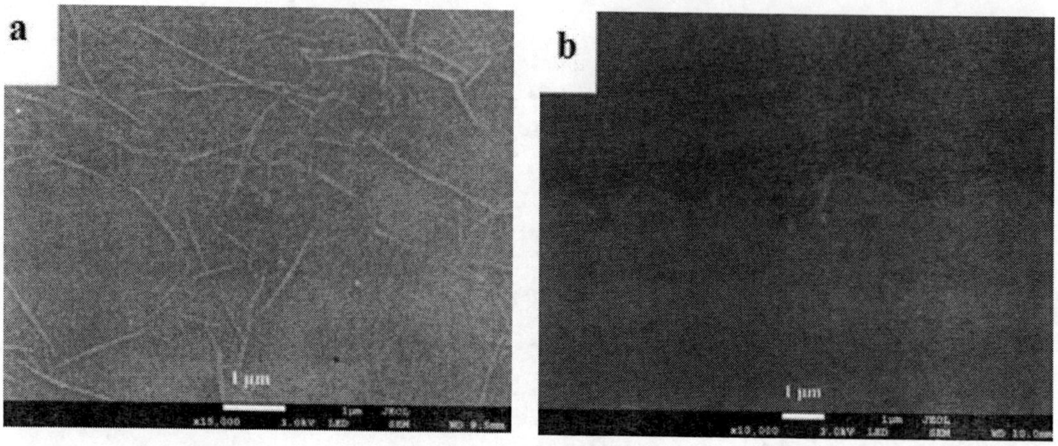

Figure 3. SEM images of a) the original GO surface and b) after exposure to 200 W methane plasma for 6 min.

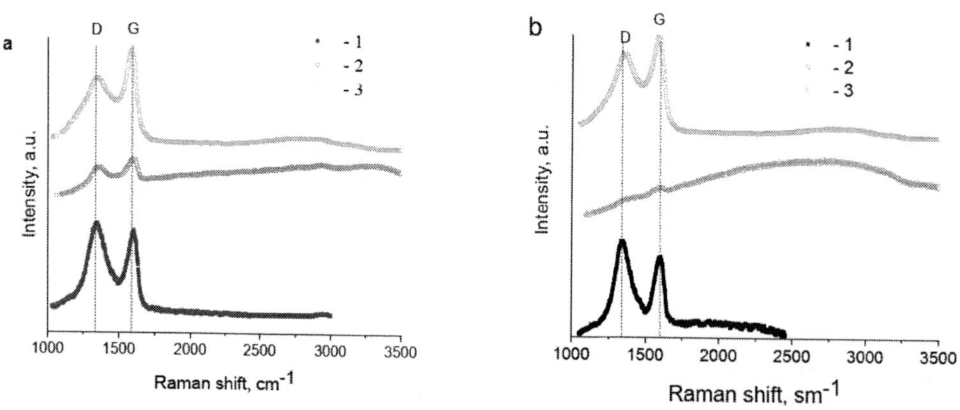

Figure 4. Raman spectra of samples of GO treated in plasma a) CH_4 and b) $CH_4 + N_2$ in accordance with processes 2 and 4, respectively. In the figures, the numbers indicate: 1 - initial GO, 2 - after exposure to plasma, 3 - after annealing at 650° C for 30 min.

Table 3. The parameters of the Raman spectra peaks for the initial samples of GO and after plasma treatment with CH_4 and $CH_4 + N_2$ and subsequent annealing.

Processing conditions	λ_G, sm^{-1}	I_D/I_G
Parameters Raman Spectra before processing (Initial GO)	1590	1.10
Parameters Raman Spectra of control sample (only annealing)	1584	0.88
Parameters Raman Spectra after processing 2	1581	0.76
Parameters Raman Spectra after processing 4	1580	0.81

Figure 5 shows the spectra of energy dispersive x-ray spectroscopy. As can be seen from the figure for the samples, plasma treatment of methane (process 2) and a mixture of methane + nitrogen (process 4) leads to an increase in the ratio of carbon to oxygen (C / O) on the surface of materials. It should be noted that the highest C / O values are observed upon exposure to methane plasma. The lower rates of formation of an amorphous carbon film during plasma treatment in a methane +

nitrogen mixture can decrease the C / O value. After annealing, the C / O ratio for two plasma-treated samples decreases. To the control sample of GO, as expected, the C / O ratio increases due to the removal of oxygen functional groups during heat treatment.

The reason for this may be associated with a decrease in the thickness of the amorphized carbon film at annealing. For thin films, the EDS results may be affected by the content of elements in the silicon substrate, in which the oxygen content is high. At the same time, a bound oxygen atoms of GO, protected by an amorphized film, will be removed less than in the control samples. As a result, a quantity of carbon atoms will decrease faster than oxygen and C / O will decrease. In contrast, plasma exposure to a $CH_4 + N_2$ mixture results in a higher C / O value after annealing. This can be explained, firstly, by the formation of a smaller thickness of the deposited carbon film. Secondly, by replacing oxygen with nitrogen under the action of plasma on the surface of GO. This is confirmed by the EDS data, which shows an increase in the number of nitrogen atoms relative to the initial content by ~ 5%.

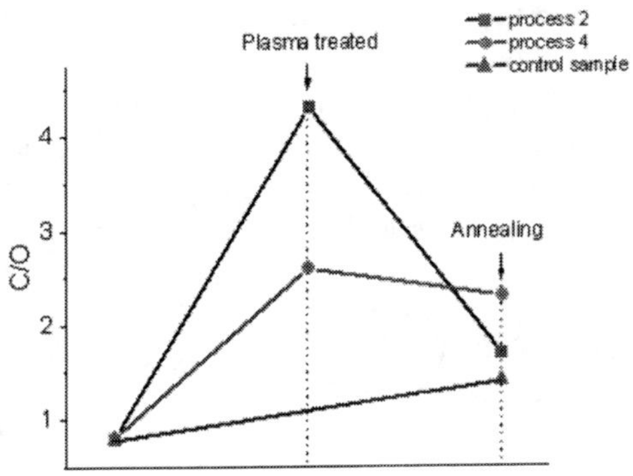

Figure 5. The change in the C/O ratio depending on the processing conditions.

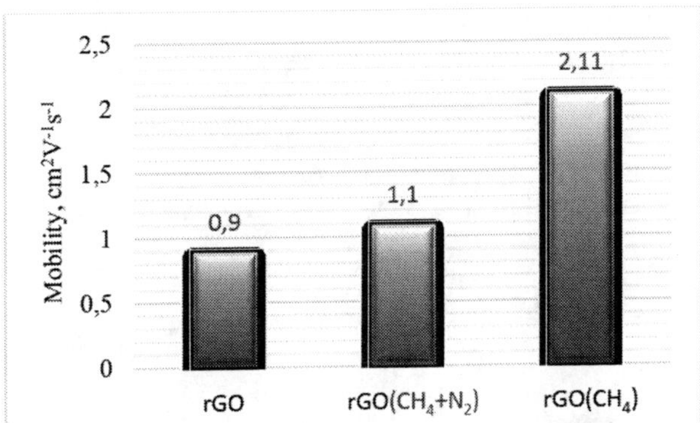

Figure 6. Carrier mobility in samples rGO (reduction GO), rGO (CH_4+N_2) and rGO (CH_4).

The sheet resistances of the samples were determined from the I-V characteristics measured at room temperature. After plasma treatment in methane and in a mixture of methane + nitrogen, the resistance of the samples after 12 min of treatment reaches hundreds of MΩ/sq. Heat treatment at T = 650° C reduces the resistance to several units of kΩ/sq. The ratio of the sheet resistances of the samples of the control sample and the samples treated according to methods 4 and 2 was 4:3:1, respectively. It should be noted that when nitrogen was added to plasma with lower power (process 3 in Table 1), a change in the type of conductivity from the p-type to the n-type was observed. It is possible that with a gentle plasma effect, the formation of nitrogen bonds with surface carbon atoms predominates over the etching processes, which leads to n-doping of graphene domains during subsequent heat treatment. Measurements of charge carrier mobility by the Hall effect method showed that, when using methane plasma (rGO(CH_4) in figure 6), the mobility was on average about two times higher than in the case of $CH_4 + N_2$ (rGO($CH_4 + N_2$) in figure 6) plasma treatment and reduction GO (rGO in figure 6). This may be due to the fact that heat treatment of amorphized carbon films deposited in plasma leads to the formation of films with a larger domain structure on the GO surface relative to the initial ones. Less resistance and mobility in samples exposed to $CH_4 + H_2$ plasma can be associated by a stronger surface damage during processing in such plasma leads to the formation of a smaller domain structure.

4. Conclusion

In this study, a simple two-step process was carried out for the treatment of GO in methane plasma (with or without the addition of nitrogen) and subsequent heat treatment. The experiment used radio frequency (13.56 MHz) inductively coupled plasma. Processing in plasma was carried out in the "face down" position to have a mild effect on the surface of GO. Heat treatment was carried out at T =

650 °C in an argon atmosphere for 30 min. Structural, optical, and electrical studies were carried out after plasma treatment and annealing. It follows from a results of investigations that, in a first stage of the process, plasma treatment in methane and in a mixture of CH_4 + N_2 leads to the deposition of amorphous carbon on the GO surface. The thickness of this film increases with enhancing processing time and plasma power. Sheet resistance of the films reaches hundreds of MΩ/sq. The presence of amorphous carbon is the cause of broadband photoluminescence in the Raman spectra in the range from 1000 to 3500 cm^{-1}. After the second stage of the process (heat treatment), a thickness of the carbon film decreases noticeably. Sheet resistance is reduced to several units of kΩ/sq. The intensity of broadband photoluminescence also decreases. In contrary, the intensities of the D and G peaks associated with the formation of ordered carbon structures significantly increase. It is most likely that the remaining carbon atoms form a crystalline structure with a GO substrate. As a result, an increase in the thickness of the relative initial GO film is observed (before plasma exposure). In addition, the red shift of the maximum intensity of the G peak of the Raman spectra and the decrease in the I_D/I_G ratio suggest that larger graphene domains are formed in comparison with initial samples. This is confirmed by the lower value of the layer resistance and the higher mobility of charge carriers in the samples treated in methane plasma. The surface roughness of this film is slightly less than that of the initial GO. Processing in a CH_4 + H_2 plasma increases the surface roughness, apparently due to more intense etching in the presence of nitrogen. As a result, such processing leads to the formation of a smaller domain structure. This is confirmed by lower mobility and higher layer resistance. The research results are supposed to be used to obtain GO-based carbon films with improved quality.

Acknowledgments
The reported study was funded by Ministry of Science and Higher Education of the Russian Federation (project FSRG-2017-0017) and RFBR project number 19-32-90133.

References

[1] Shelke N T, Karche B R 2017 Ultraviolet photosensor based on few layered reduced graphene oxide nanosheets *Applied Surface Science* **418** 374-379
[2] Fernandes G E, Kim J H, Oller D and Xu J 2015 Reduced graphene oxide mid-infrared photodetector at 300 K *Applied Physics Letters* **107** 111111
[3] Xu J, Wang Y and Hu S 2017 Nanocomposites of graphene and graphene oxides: synthesis, molecular functionalization and application in electrochemical sensors and biosensors *Microchimica Acta* **184** 1-44
[4] Singh R K, Kumar R and Singh D P 2016 Graphene oxide: strategies for synthesis, reduction and frontier applications *RSC Advances* **6** 64993-65011
[5] Yu X, Cheng H, Zhang M, Zhao Y, Qu L and Shi G 2017 *Materials* **2** 17046
[6] Lu G, Ocola L E and Chen J 2009 *Nanotechnology* **20** 445502
[7] Hu N, Yang Z, Wang Y, Zhang L., Wang Y, Huang X and Zhang Y 2013 *Nanotechnology* **25** 025502
[8] Chen D, Feng H and Li J 2012 *Chemical reviews* **112** 6027-6053
[9] Georgakilas V, Tiwari J N, Kemp K C, Perman J. A, Bourlinos A B, Kim K S and Zboril R 2016 *Chemical reviews* **116** 5464-5519
[10] Li X, Colombo L. and Ruoff R S 2016 *Advanced Materials* **28** 6247-6252
[11] Chen X, Zhang L and Chen S 2015 Large area CVD growth of graphene *Synthetic Metals* **210** 95-108
[12] Wei D, Liu Y, Wang Y, Zhang H, Huang L and Yu G 2009 *Nano letters* **9** 1752-1758
[13] Terasawa T O and Saiki K 2012 *Carbon* **50** 869-874
[14] Azam M A, Zulkapli N N, Dorah N, Seman R N A R, Ani M H, Sirat M S, Ismail E, Fauzi F B, Mohamed M A and Majlis B Y 2017 *ECS Journal of Solid State Science and Technology* **6** M3035-M3048
[15] Nandamuri G, Roumimov S, Solanki R 2010 *Applied Physics Letters* **96** 154101

[16] Cuxart M G, Šics I, Goñi A R, Pach E, Sauthier G, Paradinas M, Foerster M, Aballe L, Moreno Fernandez H, Carlino V and Pellegrin E 2017 *Carbon* **117** 331-342
[17] Baraket M, Walton S G, Wei Z, Lock E H, Robinson J T and Sheehan P 2010 s *Carbon* **48** 3382-3390
[18] Cheng M, Yang R, Zhang L, Shi Z, Yang W, Wang D, Xie G, Shi D and Zhang G 2012 Carbon **50** 2581–2587
[19] Zhu D, Pu H, Lv P, Zhu Z, Yang C, Zheng R, Wang Z, Liu C, Hu E, Zheng J, Yu K, Wei W, Chen L and Chen J 2017 Carbon **120** 274-280
[20] Hummers W S and Offeman R E 1958 J. Am. Chem. Soc. **80** 1339
[21] Neustroyev E P. Aleksandrov G N and Nogovitsyna M V 2015 Vestnik SVFU 50 **6** 87 (in Russia)
[22] Neustroev E P, Burtseva E K, Soloviev B D, Prokopiev A R, Popov V I and Timofeev V B 2018 *Nanotechnology* **29** 144002
[23] Neustroev E P, Nogovitsyna M V, Soloviev B D, Kurkina I I and Nikolaev D V 2018 *Journal of Structural Chemistry* **59** 793-798
[24] Hamelmann F, Aschentrup A, Brechling A, Heinzmann U, Abrashev M, Szekeres A and Gesheva K 2004 *Vacuum* **76** 139-142
[25] Beams R, Cançado L G and Novotny L 2015 *Journal of Physics: Condensed Matter* **27** 083002
[26] Kudin K N, Ozbas B, Schniepp H C, Prud'Homme R K, Aksay I A and Car R 2008 *Nano letters* **8** 36-41
[27] Jacobsohn L G, Franceschini D F, Maia da Costa, M E H and Freire Jr F L 2000 *Journal of Vacuum Science & Technology A: Vacuum, Surfaces, and Films* **18** 2230-2238
[28] Tuinstra F and Koenig J L 1970 *The Journal of Chemical Physics* **53** 1126-1130
[29] Pimenta M A, Dresselhaus G, Dresselhaus M S, Cancado L G, Jorio A and Saito R 2007 *Physical chemistry chemical physics* **9** 1276-1290
[30] Dresselhaus M S, Jorio A, Souza Filho A G and Saito R 2010 *Philosophical Transactions of the Royal Society A: Mathematical, Physical and Engineering Sciences* **368** 5355-5377
[31] Yang C, Gong J, Zeng P, Yang X, Liang R, Ou Q and Zhang S 2018 *Applied Surface Science* **452** 481-486

Electromagnetic characteristics of coatings based on graphene oxide- and multi-walled carbon nanotubes Taunit-M in a wide range of frequencies

V Suslyaev[1*], A Tkachev[2], E Korovin[1], R Stolyarov[2], K Dorozhkin[1], N Chapaksov[2], T Dyachkova[2], A Blokhin[2] and A Korshunov[1]

[1]Department of Radioelectronics, Faculty of Radiophysics, *National Research Tomsk State University*, Tomsk, 634050, Russia
[2] Department of Technology and Methods of Nanoproducts Manufacturing, *Tambov State Technical University*, Tambov 392000, Russia

[*1]E-mail: susl@mail.tsu.ru

Abstract. The frequency dependences of the reflection, transmission, and absorption coefficients of samples of materials based on graphene oxide and Taunit-M multi-walled carbon nanotubes were measured. Graphene oxide and carbon nanotubes are glued with acrylic adhesive to a double-sided sheet of plastic. The method of free space is used to measure in the frequency range 35 - 260 GHz. It is shown that the addition of MWCNT of 30 weight percent reduces the transmission coefficient by more than ten times.

1. Introduction

Carbon nanostructures (multi-walled and single-walled nanotubes, graphene, fullerenes) are highly conductive [1] and effectively interact with electromagnetic (EM) radiation [2, 3]. Composites based on them have a strong shielding effect in the microwave and radio frequency ranges [4, 5]. 2D structure - graphene oxide (GO) attracts interest from specialists in the field of development and application of radio absorbers in connection with the ability to control its dielectric properties. So, in [6, 7], a method is described for obtaining an effective absorber based on GO in the frequency range from 10 to 18 GHz with sample thicknesses of 3-4 mm.

Currently, the possibility of creating light thin protective screens is being actively studied. Screens should protect biological objects from the harmful effects of microwaves, reduce the radio visibility of military objects and solve the problems of electromagnetic compatibility. This can be done by developing multilayer multielement structures.

It should be noted that carbon nanostructures and carbon-filled composites are promising for detecting terahertz radiation [8, 9] and for creating passive elements (attenuators, polarizers, matched loads) in this frequency range [10-13].

In this paper, we study the high-frequency electromagnetic parameters of a thin screen based on GO obtained by the new technology. The degree of influence of MWNTs is studied.

2. Experimental part

2.1. Sampling technique

The samples were made from aqueous dispersions of MWCNTs Taunit-M (diameter 10-30 nm, length more than 2 μm) (figure 1a) and 0.5 wt.% Graphene oxide (GO (figure 1 b), (both manufactured by NanoTechCenter Ltd, Tambov, Russia). The plastic sheet was covered with an acrylic adhesive base 3M 468MP 200MP Adhesive (China). This is the first time this technology has been used. The elemental composition of the GO is presented in table 1.

After adding the MWCNTs, the solution was stirred on a mechanical stirrer (Witeg HT-50AX) for 5 min at 800 rpm. To improve the distribution of the nanotubes in the graphene oxide, the resulting mixture was processed on a three-roll machine (Exakt - 80E) in 2 stages: 1 - gaps between the rolls – 30-15 μm, rotation speed - 120 rpm, and 2 - gaps between the rollers –15-5 microns, rotation speed - 120 rpm. Next, the mixture was subjected to a 10-minute ultrasonic (US) processing (I-10 ultrasonic generator) with direct mixing at 300 rpm. The resulting composition was poured on a double-sided acrylic adhesive sheet and dried at room temperature.

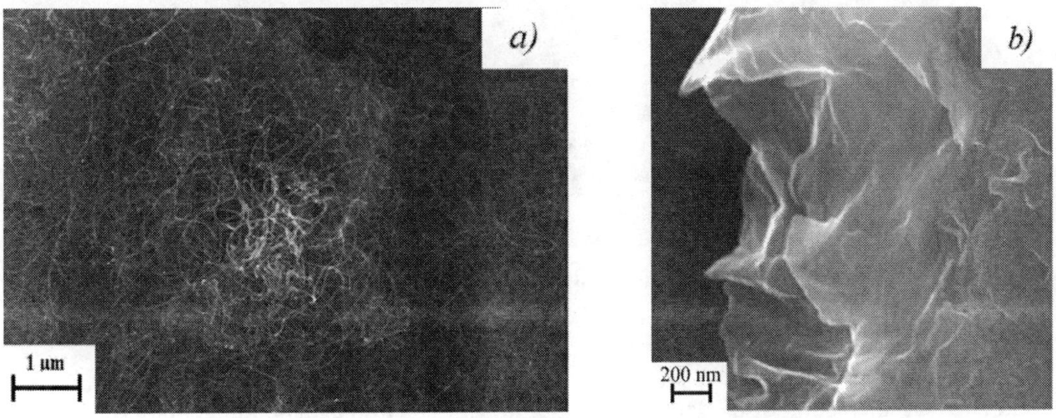

Figure 1. SEM images of the MWCNTs Taunit-M (a) and the GO (b).

Table 1. The elemental composition of the GO.

Element	Weight (%)	Atomic (%)
C	57.54	64.97
O	40.21	34.08
S	2.24	0.94
Sum	99.99	99.99

The MWCNTs were introduced into the GO aqueous solution at the concentrations shown in table 2.

Table 2. Characteristics of coatings.

Sample (No)	Composition (wt.%)	Size (mm)	Thickness (μm)
1	Coating base		123
2	GO		107
3	GO/MWCNTs 95/5	200x200	94
4	GO/MWCNTs 90/10		100
5	GO/MWCNTs 80/20		101
6	GO/MWCNTs 70/30		111

The structure of a slice of a nanocomposite of graphene oxide and carbon nanotubes is shown in figure 2. The design has a layered, porous structure with inclusions of carbon nanotubes. Carbon nanotubes are evenly distributed in the graphene oxide matrix. However, agglomerates are present in the structure. The size of the agglomerates corresponds to the size set between the rolls on a three-roll machine Exakt-80E.

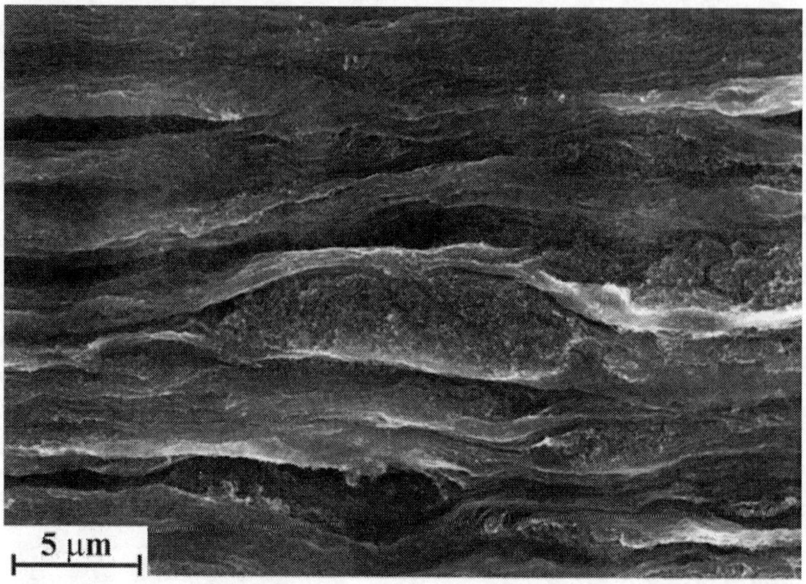

Figure 2. SEM image of the cross-sectional structure of a nanocomposite of carbon nanotubes and graphene oxide (CNT concentration of -30% mass).

2.2. Measurement of electromagnetic parameters
The efficiency of interaction of the fabricated material with electromagnetic radiation of the gigahertz and terahertz ranges was estimated by the reflection R and transmission T coefficients, which were obtained in the experiment, and the absorption coefficient A, which is $A = 1 - R - T$.

The coefficients R and T are mainly determined by the magnitudes of the magnetic and dielectric constant at a given frequency. The relationship is described by the following relationships for a plane wave that normally falls on a flat surface:

$$R = \rho(1 - e^{-2i\gamma d})/(1 - \rho^2 e^{-2i\gamma d}),$$

$$T = (1 - \rho^2)\, e^{-2i\gamma d}/(1 - \rho^2 e^{-2i\gamma d}),$$

where $\rho = (Z - 1)/(Z + 1)$ is the coefficient of reflection from the front surface of the sample with respect to the incident electromagnetic wave; $Z = (\mu / \varepsilon)^{1/2}$ is the wave impedance of the medium, $\gamma = k_0 (\mu\varepsilon)^{1/2}$ is the propagation constant, c is the speed of light, d is the thickness of the flat layer.

The free space method is used to experimentally determine the coefficients. This non-destructive method allows you to study samples of a large area. The measurements were carried out at three frequency sections: 30-56 GHz, 40 -120 GHz and 120-260 GHz.

The horn measurement method was used in the range of 30 - 56 GHz. This method modified to study the electromagnetic characteristics of materials with large losses in this range. The modification made it possible to increase the sensitivity of the horn method and preserve the plane-wave approximation, which is usually used to calculate the dielectric constant spectra [14]. The vector network analyzer PNA X N5247A (Agilent Technologies) was used for the generation, reception of electromagnetic radiation and the primary processing of measurement results.

The terahertz spectrometer of the Russian production of STD-21 was used to measure the electromagnetic response in the frequency range of 40 -120 and 120-260 GHz. The set of backward wave tubes was used to generate a continuous harmonic signal in these ranges. The spectrometer allows to measure the modules of the transmission and reflection coefficients and their phases [15].

The equipment of the Center for Collective Use "Center for Radio Measurement TSU" was used to measure the frequency dependencies of the listed coefficients [16]

3. Results

Coefficients *R* and *T* were measured on flat samples, the geometric characteristics of which are shown in table 2.

Figures 3-5 show the frequency dependences of the *R, T, A* coefficients of the samples (table 1). Electromagnetic energy almost completely passes through a layer of pure GO (figure 3), decreasing at high frequencies due to the increase in reflectivity (figure 4).

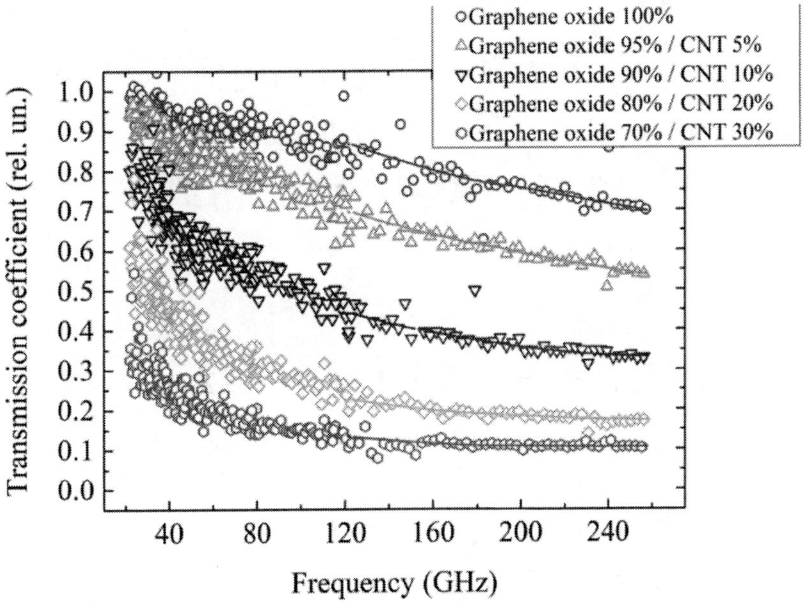

Figure 3. Frequency dependences of transmission coefficients.

IOP Conf. Series: Materials Science and Engineering **693** (2019) 012044 doi:10.1088/1757-899X/693/1/012044

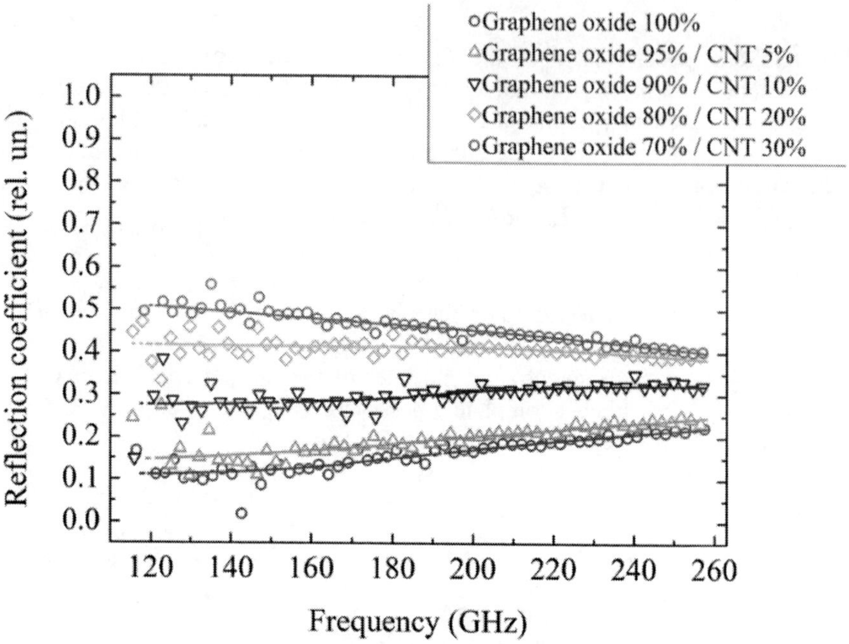

Figure 4. Frequency dependences of reflection coefficients.

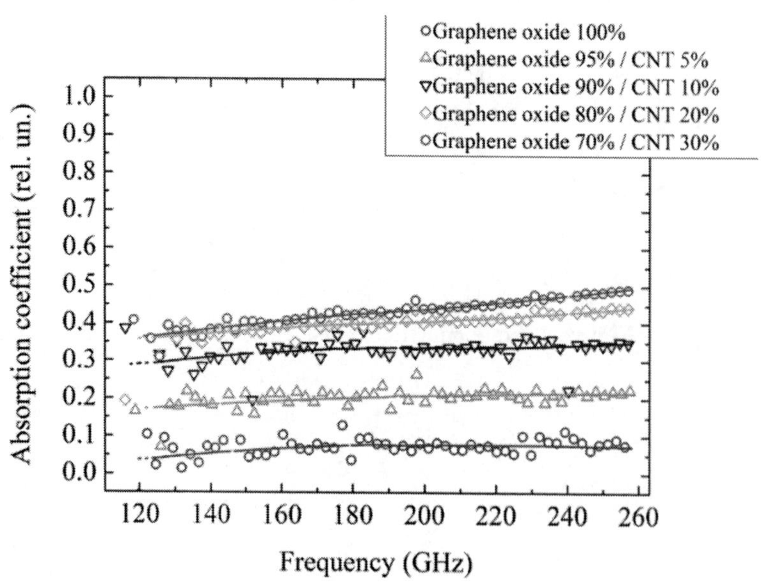

Figure 5. Frequency dependences of absorption coefficients.

An increase in the concentration of MWCNTs leads to an increase in the screening effect due to an increase in absorption (figure 5). The increase in coefficient A is caused by: 1- an increase in the hopping conductivity characteristic of MWCNTs; 2- by the layered structure that was formed during manufacture (figure 2), which provides polarization losses.

The sample (at 30 percent aqueous dispersion of MWCNTs) shields 90 percent of the incident energy, which is a high indicator for the selected thickness.

A high reflection coefficient is noted. It can be reduced by using an additional matching layer on the side of the incident electromagnetic wave.

The created structure surpasses the other composite in its shielding properties - epoxy resin+MWCNTs Taunit-M described in [17].

4. Conclusion

As a result of studying the electromagnetic response spectra (reflection coefficients R, transmittance T, absorption coefficient A) of graphene oxide material samples with different contents of Taunit-M MWCNTs, it is seen that an increase in MWCNT concentration leads to a significant increase in the protective qualities of the developed material, a relatively small increase reflection coefficient. A sample containing 30 wt.% aqueous dispersion of MWCNTs Taunit-M showed the greatest shielding property.

We can conclude that this material is promising for solving the problems of electromagnetic compatibility, protecting personnel from the harmful effects of electromagnetic radiation, and other practical applications.

Acknowledgment
This research was supported by "The Tomsk State University Competitiveness Improvement Programme".

References

[1] Burmistrov, N. Gorshkov, I. Ilinykh, D. Muratov. E. Kolesnikov, S. Anshin ,I. Mazov J-P Issi, D Kusnezov 2016 Improvement of carbon black based polymer composite electrical conductivity with additions of MWCNT *Composites Science and Technology* 129 79-85

[2] Shuba M V, Yuko D I, Kuzhir P P, Maksimenko S A, Crescenzi M D, Scarselli M 2018 Carbon nanotube sponges as tunable materials for electromagnetic applications *Nanotechnology* **29** 375202

[3] Zhang Q, Hároz E H, Jin Z, Ren L, Wang X, Arvidson R S, Lüttge A and Kono J 2013 Plasmonic nature of the terahertz conductivity peak in single-wall carbon nanotubes *Nano lett.* **13** 5991

[4] Bychanok D, Angelova P, Paddubskaya A, Meisak D, Shashkova L, Demidenko M, Plyushch A, Ivanov E, Krastev R, Kotsilkova R, Ogrin F Y and Kuzhirl P 2018 Terahertz absorption in graphite nanoplatelets/polylactic acid composites *J. Phys. D: P Appl. Phys.* **51** 145307

[5] Paddubskaya A, Demidenko M, Batrakov K, Valušis G, Kaplas T, Svirko Yu. and Kuzhir P 2019 Tunable Perfect THz Absorber Based on a Stretchable Ultrathin Carbon-Polymer Bilayer. *Materials* **12** 2

[6] Fan Wu, Qiao Zeng, Yilu Xia, Mengxiao Sun, Aming Xie 2018 The effects of annealing temperature on the permittivity and electromagnetic attenuation performance of reduced graphene oxide *Applied physics letters* **112** 192902

[7] Fang Ye , Changqing Song, Qian Zhou, Xiaowei Yin, Meikang Han, Xinliang Li , Litong Zhang and Laifei Cheng 2018 Broadband Microwave Absorbing Composites with a Multi-Scale Layered Structure Based on Reduced Graphene Oxide Film as the Frequency Selective Surface *Materials* **11** 1771

[8] Hartmann R R, Kono J and Portnoi M E 2014 Terahertz science and technology of carbon nanomaterials *Nanotechnology* **25** 322001

[9] Badin A V, Dorozhkin K V, Suslyaev V I, Berdyugin A I and Vigovskiy V Y 2017 Quasi-optical 2D system for non-contact non-destructive testing of defects in natural and artificial crystals *Proc. SPIE* **10466** 58771

[10] Valynets N I, Paddubskaya A G, Kuzhir P P, Sysoev V I, Bulusheva L G and Okotrub A V 2017 Fluorination as Effective Method for Tuning the Electromagnetic Response of Graphene. *Phys. Status Solidi B* **255** 1700226

IOP Conf. Series: Materials Science and Engineering **693** (2019) 012044 doi:10.1088/1757-899X/693/1/012044

[11] Shuba M V, Yuko D I, Kuzhir P P, S.A.Maksimenko, Kanygin M A, Okotru A V, Tenne R and Lambin P 2018 How effectively do carbon nanotube inclusions contribute to the electromagnetic performance of a composite material? Estimation criteria from microwave and terahertz measurements *Carbon* **129** 688

[12] Terukov E I, Babaev A A, Tkachev A G, Zhilina D V 2018 Radio-wave absorbing properties of polymer composites on the basis of shungite and carbon nanomaterial Taunit-M *Technical Physics* **63**(7) 1044–8b

[13] Kolanowska A, Janas D, Herman A P, Jedrysiak R G, Gizewski T and Boncel S 2018 From blackness to invisibility - Carbon nanotubes role in the attenuation of and shielding from radio waves for *stealth* technology. *Carbon* **126** 31

[14] Zhuravlev V A, Suslyaev V I and Zemlyanukhin Yu P 2011 Increase of the sensitivity and accuracy of quasioptical methods of measuring the electromagnetic response in the millimeter wavelength range *Russ. Phys. J.* **54** 828

[15] Dunaevskii G E, Suslyaev V I, Zhuravlev V A, Badin A V, Dorozhkin K V 2016 *Int. Conf. on Infrared, Millimeter, and Terahertz Waves (Copenhagen)* p 978-1-4673-8485-8

[16] http://ckp.tsu.ru/about/directions/radiophysics/

[17] Komarov F F, Tkachev A G, Milchanin O V, Parfimovich I D, Grinchenko M V, Parkhomenko I N, Bychenok D S 2017 A composite based on epoxy polymer and carbon nanotubes: Structure, optical properties and interaction with microwave radiation *J. Adv. Mater. & Technol.* **2** 19-25

Adsorption of Cu^{2+}, Zn^{2+} and Pb^{2+} ions on a novel graphene-containing nanocomposite: An isotherm study

A Babkin[1], I Burakova[1], A Burakov[1], D Kurnosov[1], E Galunin, A Tkachev[1] and I Ali[2]

[1]Department of Technology and Methods of Nanoproducts Manufacturing, *Tambov State Technical University,* 106 Sovetskaya Str., Tambov 392000 Russian Federation
[2]Department of Chemistry, *Jamia Millia Islamia (Central University)*, Jamia Nagar, New Delhi 110025 India

[1]E-mail: flex_trol@mail.ru

Abstract. The present research is dedicated to the problem of adsorption purification of aquatic media from toxic heavy metal ions. To extract some of these species (Cu^{2+}, Zn^{2+}, and Pb^{2+}) from aqueous solutions, a novel material - polyhydroquinone/graphene nanocomposite – was used as adsorbent. Its adsorption/desorption performance was assessed through isotherm studies. The following process parameters were determined: equilibrium metal concentrations in solution, metal distribution coefficients, percentage desorption, and adjusted metal distribution coefficients. Besides, based on experimental data, adsorption isotherms were constructed, from which it was possible to determine the maximum adsorption capacity of the nanocomposite regarding each heavy metal ion (Pb^{2+} - 1350 mg^{-1}; Cu^{2+} - 800 mg g^{-1}, and Zn^{2+} - 300 mg g^{-1}). In order to elucidate adsorption-desorption mechanisms occurring in the considered adsorbate/adsorbent systems and the process nature, the data were fitted to different isotherm models: empirical (Langmuir, Freundlich, Temkin, and Dubinin-Radushkevich), and combined (single- and dual-site Langmuir-Freundlich). All the models provided satisfactory fits to the data, thereby making it possible to assume heterogeneity of adsorption sites on the adsorption surface, and the irreversibility of the adsorption process.

1. Introduction

Graphene, a single graphite layer, is a very interesting material. It possesses unique physical, chemical and mechanical properties, due to which it can be used in many application fields [1, 2]. As for water/wastewater purification area, high specific surface area of graphene, together with its pore size and volume, makes it possible to employ this material for adsorption of various species, including heavy metal ions [3, 4].

In the literature, several graphene-based materials and composites are reported for the adsorption of heavy metals such as Cu^{2+}, Zn^{2+} and Pb^{2+} (table 1).

Table 1. Adsorption capacities of graphene oxide based materials.

Material	Metal ions	pH	$C_{init.}$	Q_{max},mg L^{-1}
Graphene Oxide(GO) [5]	Fe(III), Mn (II)	6	200 mgL^{-1}	Fe (III) – 24 Mn (II) – 21
Magnetic GO modified with a pyrrole-thiophene (mGO/SiO_2@coPPy-Th) [6]	Cu(II), Zn (II), Pb (II)	6	250	Cu(II) – 201, Zn (II) – 125, Pb (II) – 230
GO-TiO_2 [7]	Zn (II), Pb (II)	5.6	50	Zn (II) –89, Pb (II) – 65
GO [8]	Pb (II)	6	50-100	70
r-GO [9]	Pb (II)	4.4	150	93
GO/Fe_3O_4 [10]	Cd (II), Cu (II), Pb (II)	6	50-150	Cd (II) – 164, Cu (II) – 156, Pb (II) – 179
GO [11]	Cd (II)	6.5	~100	~900

Based on the tabular data, it can be established that graphene composites are actively used for the adsorption of heavy metals, exhibiting high sorption activity. It should be noted that the GO modification positively affects the adsorption capacity – it increases by 3-4 times.

However, the relevance and scientific novelty of activities on developing novel adsorption materials based on graphene and its derivatives are due to low elaboration and lack of systematics in studying processes of liquid-phase adsorption of heavy metals on the existing nanostructured graphene materials. In this regard, the development of directional methods for synthesizing these materials and the investigation of their physical and chemical characteristics is of undoubted scientific interest in the world community, since they provide the prospect of a significant improvement in the quality of fine adsorption purification of aquatic media [12].

Establishing patterns of the influence of the physical-chemical and structural parameters of graphene-based nanomaterials being developed on the selective adsorption of target pollutants (heavy metals) is an important trend. Besides, a comprehensive study on the mechanisms of extraction processes is required, which, in turn, will make it possible to predict the behavior of adsorption systems.

In this regard, studying the fundamentals of the heavy metal adsorption on graphene adsorbents based on empirical dependencies and mathematical models that take into account the availability of various adsorption sites on the adsorbent surface, as well as establishing the relationship between the structure of graphene adsorbent and adsorption mechanisms, appear to be important issues.

Considering the aforementioned, the aim of the present research was to propose a polyhydroquinone/graphene (PHQ/G) nanocomposite (synthesized by the authors of this paper) as a novel adsorbent for removing some heavy metal ions (Cu^{2+}, Zn^{2+} and Pb^{2+}), and study its adsorption and desorption behavior under batch conditions using an isotherm approach and employing empirical and combined isotherm models to fit experimental data.

2. Experimental part

2.1. Material and reagents

The PHQ/G nanocomposite was developed by the authors at the Research Department "Technology and Methods of Nanoproducts Manufacturing" (Tambov State Technical University, Tambov, Russia) and produced by NanoTechCenter Ltd. (Tambov, Russia). The technology of its manufacturing is described elsewhere [13].

Heavy metal salts: $Cu(NO_3)_2$, $Zn(NO_3)_2$ and $Pb(NO_3)_2$ (all – analytical grade) were purchased from Laverna Lab Ltd. (Moscow, Russia). Heavy metal stock solutions (1,000 mg L^{-1}) were obtained by diluting certain amounts of the corresponding salts in deionized water. Before using in tests, the solutions were diluted to the required concentrations: 12, 25, 50, 100, 200, 250, 450, 650, and 900 mg L^{-1}.

2.2. *Adsorption-desorption isotherm studies*

For adsorption studies, 0.01 g of the PHQ/G material and 30 mL of 12-900-mg L^{-1} metal solutions (pH=6.0) were separately placed into 50-mL conical tube sand then shaken at 100 rpm and ambient temperature for 1 h on BioRS-24 rotators (Biosan, Riga, Latvia). After that, the mixtures were centrifuged using a 5810 R centrifuge (Eppendorf, Hamburg, Germany) at 10 000 rpm for 10 min for phase separation, and supernatants were collected to measure the metal equilibrium concentrations on an MGA-915 MD atomic absorption spectrometer (Atompribor, Saint Petersburg, Russia).

For desorption tests, performed after the adsorption experiments, the solid phase (PHQ/G) was separated, dried at 45 °C for about 15 min to remove surface moisture, and placed to 50-mL conical tubes, to which 30 mL of metal-free buffering aqueous solutions (pH=6.0) were then separately added. Like in the adsorption tests, the tubes were also shaken at 100 rpm and ambient temperature for 1 h, the mixtures were centrifuged at 10 000 rpm for 10 min for phase separation, and supernatants were collected to measure the metal equilibrium desorption concentrations on the spectrometer.

2.3. *Experimental data processing*

The equilibrium adsorption capacity (Q_e, mg g^{-1}) of the material was determined according to the following expression:

$$Q_e = \frac{(C_{in}-C_e)V}{m}, \quad (1)$$

where C_{in} is the initial metal concentration in solution (before adsorption), mg L^{-1}; C_e is the equilibrium metal concentration in solution (after adsorption), mg L^{-1}; V is the solution volume, L; m is the adsorbent weight, g.

Correspondingly, the equilibrium adsorption capacity for the desorption case (Q_e^{des}, mg g^{-1}) was estimated as follows:

$$Q_e^{des} = \frac{(C_{in}-C_e-C_e^{des})V}{m}, \quad (2)$$

where C_e^{des} is the equilibrium metal concentration in solution (after desorption), mg L^{-1}.

Further, the coefficients of the distribution of the adsorbates between the phases were calculated for the adsorption (K_d, L g^{-1}) and desorption (K_d^{des}, L g^{-1}) cases according to the following equations:

$$K_d = \frac{(C_{in}-C_e)V}{C_e m} = Q_e\frac{V}{m}, \quad (3)$$

$$K_d^{des} = \frac{(C_{in}-C_e-C_e^{des})V}{C_e^{des} m} = Q_e^{des}\frac{V}{m}. \quad (4)$$

K_d^{des} provides information about the equilibrium displacement of the adsorbate between the adsorbent and the contact solution it is already incorporated into.

The percentage desorption (R^{des}, %), which shows the adsorbate amount percent leached back into the aqueous phase, and the adjusted distribution coefficient (K_d^{adj}, L g^{-1}), which corrects the adsorption (reversible)distribution coefficient K_d in the adsorbate fraction involved in the adsorption equilibrium, were assessed as follows:

$$R^{des} = \frac{100C_e^{des}}{(C_{in}-C_e)}, \quad (5)$$

$$K_d^{ADJ} = \frac{100K_d}{R^{des}}. \quad (6)$$

2.4. Experimental data fit

2.4.1. Empirical models. The metal adsorption data were fitted to the following empirical models: Langmuir, Freundlich, Temkin, and Dubinin-Radushkevich (table 2).

Table 2. Empirical isotherm models implemented [14].

Langmuir	Freundlich	Temkin	Dubinin-Radushkevich
$Q_e = Q_{max}\frac{K_L C_e}{1+K_L C_e}$ (7)	$Q_e = K_F C_e^n$ (8)	$Q_e = B_T ln(K_T C_e)$ (9)	$Q_e = Q_{max} exp(-K_{DR}\varepsilon^2)$ (10) $E = \frac{1}{\sqrt{2K_{DR}}}$ (10)

Q_{max}– maximum adsorption capacity, mg g^{-1}; K_L – Langmuir constant (adsorbate-adsorbent binding energy), L mg^{-1}; K_F – Freundlich constant (adsorbent distribution between the phases), L g^{-1}; n – indicator of heterogeneity of adsorption sites on the adsorbent surface, dimensionless; B_T – constant related to the heat of adsorption, J mol^{-1}; K_T –Temkin equilibrium constant (maximum binding energy), L g^{-1}; K_{DR}– Dubinin-Radushkevich isotherm constant related to the mean free energy of adsorption), mol^{2}kJ^{-2}; $\varepsilon = RT\ln(1+(1/C_e))$ – Polanyi potential, equal tkJ mol^{-1}; E- mean free energy of adsorption, kJ mol^{-1}.

2.4.2. Combined models. Two modifications of the combined Langmuir-Freundlich (Sips) model were implemented – single- (equation (11)) and dual-site (equation (12)):

$$Q_e = \frac{Q_{max}(K_{LF}C_e)^{n'}}{1+(K_{LF}C_e)^{n'}}, \quad (11)$$

$$Q_e = \frac{Q_{max1}(K_{LF1}C_e)^{n1'}}{1+(K_{LF1}C_e)^{n1'}} + \frac{Q_{max2}(K_{LF2}C_e)^{n2'}}{1+(K_{LF2}C_e)^{n2'}}, \quad (12)$$

where K_{LF}, KL_{F1}, K_{LF2} are the Langmuir-Freundlich constants (binding energy), L mg^{-1}; n', n_1', n_2'are indicators of heterogeneity of adsorption sites; subindexes "1" and "2" stand for adsorption sites possessing different affinity (binding energy) for the adsorbate.

3. Result and discussion

3.1. Adsorption isotherms

The isotherms constructed for the heavy metal adsorption on the PHQ/G nanocomposite based on the experimental data obtained are presented in figure 1.

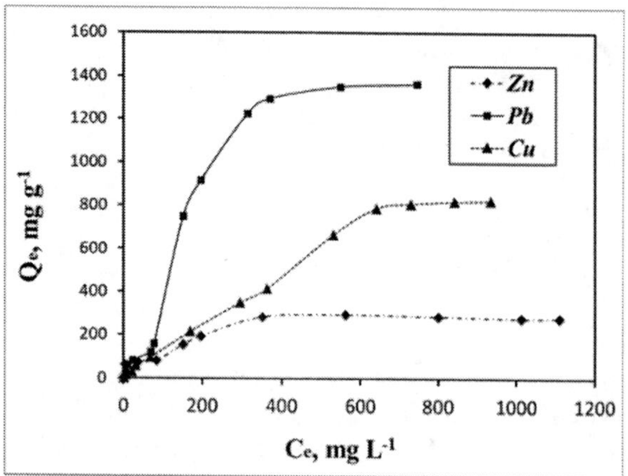

Figure 1. Isotherm constructed for the metal adsorption on the PHQ/G material. Conditions: adsorbent weight – 0.01 g, solution volume – 0.03 L, contact time – 1 h, temperature – 298 K, and pH 6.0.

As can be seen, all the isotherms may be related to the *L*-type by their shape, according to the classification introduced by Giles and its colleagues [15]. This means that the adsorbates possess a relatively high affinity for the adsorbent surface at lower concentrations. When the surface is saturated, with an increase in the adsorbate concentration, the number of vacant adsorption sites located on the adsorbent surface decreases, and the adsorbate amount decreases as well – the maximum adsorption capacity is reached (for Pb^{2+} - ~1350 mg g^{-1}, for Cu^{2+} - ~800 mg g^{-1}, and for Zn^{2+} - ~300 mg g^{-1}).

3.2. Adsorption and desorption parameters

Table 3 presents the data obtained for the adsorption and desorption of Pb^{2+} on the

PHQ/G nanocomposite (the data for Cu^{2+} and Zn^{2+} are not given herein, since all the ions showed the similar pattern; in this regard, Pb^{2+}, having the highest adsorption capacity, is taken as an example).

Table 3. Adsorption-desorption parameters obtained for the system Pb^{2+}-PHQ/G.

C_e, mg L^{-1}	Q_e, mg g^{-1}	K_d, L g^{-1}	C_e^{des}, mg L^{-1}	Q_e^{des}, mg g^{-1}	K_d^{des}, L g^{-1}	K_d^{des}/K_d	R^{des}, %	K_d^{adj}, L g^{-1}
4.2	23.4	5.57	1.41	19.17	13.60	2.44	18.08	30.82
10.2	59.4	5.82	1.59	54.63	34.36	5.90	8.03	72.52
23.3	80.1	3.44	2.58	72.36	28.05	8.16	9.66	35.58
70	120	1.71	2.62	112.14	42.80	24.97	6.55	26.17
77.67	156.99	2.02	2.97	148.08	49.86	24.67	5.68	35.61
151	747	4.95	6.30	728.1	115.57	23.36	2.53	195.53
195.5	913.5	4.67	6.17	894.99	145.06	31.04	2.03	230.60
312.7	1311.9	4.20	5.52	1295.34	234.66	55.93	1.26	332.36
369.2	1292.4	3.50	6.27	1273.59	203.12	58.03	1.46	240.52
550.7	1347.9	2.45	6.36	1328.82	208.93	85.36	1.42	172.91

It can be observed that when increasing the initial C_{in} (not shown) and, correspondingly, equilibrium C_e concentration of the adsorbate, adsorption capacity Q_e(equation (1)) progressively increases, the adsorption sites of the adsorbent surface become more and more saturated, and the coefficient of the adsorbate distribution between the solid and liquid phases K_d(equation (3)) decreases.

For the desorption case, the capacity Q_e^{des} (equation (2)) and the desorption distribution coefficient K_d^{des} (equation (4)), giving information about the equilibrium displacement of the adsorbate between the adsorbent and the contact solution it is already incorporated into, increase with increasing the equilibrium desorption concentration C_e^{des} (it directly depends on the equilibrium adsorption concentration C_e).

Comparing the K_d and K_d^{des} values, it can be seen that $K_d^{des}/K_d > 1$, thereby standing for the irreversibility of the Pb^{2+} adsorption, which is a good factor for efficient wastewater purification (the pollutant is poorly lixiviated from the adsorbent). The percentage desorption R^{des}(equation (5)) and the adjusted distribution coefficient K_d^{adj} (equation (6)) confirm this assumption. From figure 6 and table 1, one can see that the R^{des} decreases when increasing the Q_e^{des} (and, correspondingly, the C_e and C_e^{des}). The maximum percentage value was found to be only about 18.08 %, decreasing down to 1.42 %. Regarding the K_d^{adj}, which corrects the adsorption (reversible)distribution coefficient K_d in the adsorbate fraction involved in the adsorption equilibrium, it provides almost the same information as the K_d^{des} does; both the parameters were found to be satisfactorily correlated: $K_d^{adj} = 0.71\ K_d^{des} + 10.66$ ($R^2 = 0.85$).

3.3. Adsorption and desorption data fit implementing isotherm models

First, the metal adsorption data were fitted to the following empirical models: Langmuir, Freundlich, Temkin, and Dubinin-Radushkevich (table 4).

The Langmuir model (equation (7)) is based on the fact that the adsorbate is localized on the adsorbent surface by a single mechanism [16]. It is generally stated that a monomolecular layer of the adsorbate is formed on the adsorbent surface, and all the active sites possess equal energy and enthalpy.

The Freundlich model (equation (8)) is used to describe the adsorption process taking place on a heterogeneous surface. Since, according to this model, the adsorption sites have different energy values, the ones possessing maximum energies will be saturated first of all. However, the model does not consider maximum saturation.

According to the Temkin model (equation (9)), either the surface sites are not identical in activity and can be characterized by a certain distribution, or they are the same in activity, but the activity of each site decreases as a result of the appearance of chemisorbed molecules. The more the number of such molecules, the more the initial activity of this site changes, *i.e.* the activity of each site decreases when increasing the degree of surface coverage with the adsorbate [17].

To study the mechanisms of the adsorption process, the equilibrium data are processed using the Dubinin–Radushkevich isotherm model (equations (10)). Itis more general than the Langmuir isotherm, since it does not imply a homogeneous surface and a constant adsorption potential [18]. This model makes it possible to establish the nature (physical or chemical) of the adsorption process on the adsorbent surface, and can be used to calculate the mean free energy of adsorption E (equation (10')). The mechanism of the adsorbate binding is judged by the value of this parameter. If E <8 kJmol^{-1}, the adsorption process is physical, whereas if 8<E<16 kJ mol^{-1}, the process is governed by ion exchange, but if 16<E<40 kJ mol^{-1}, chemisorption takes place. The Q_{max} value determines the porosity of the adsorbent: the larger this value, the more developed the porosity, and the smaller the pore diameter [16-19].

Figures 2 and 3 demonstrate the experimental data fitted to the above-mentioned models, and table 3 presents the model parameters obtained.

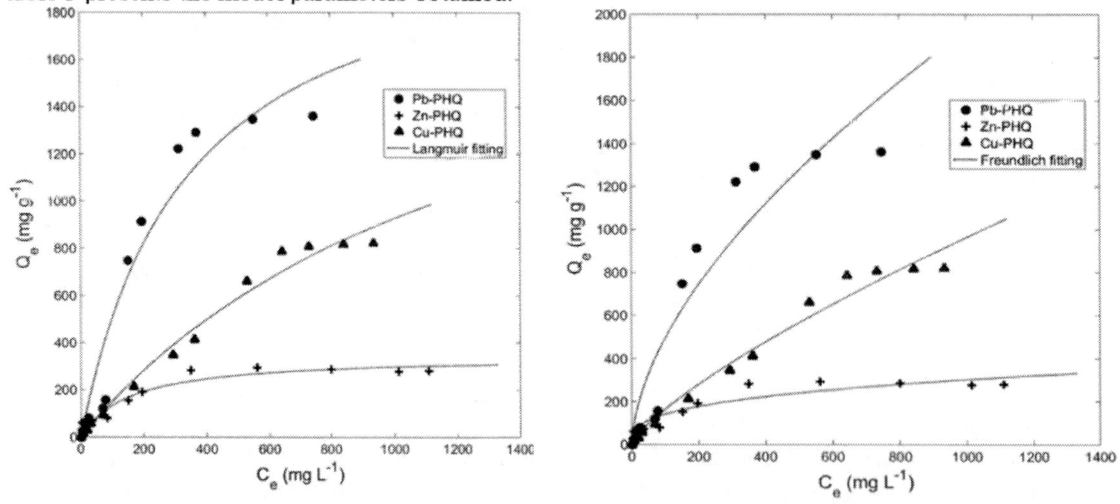

(a) **(b)**

Figure 2. Isotherms of the metal adsorption on the PHQ/G material: Langmuir **(a)**, and Freundlich **(b)**.

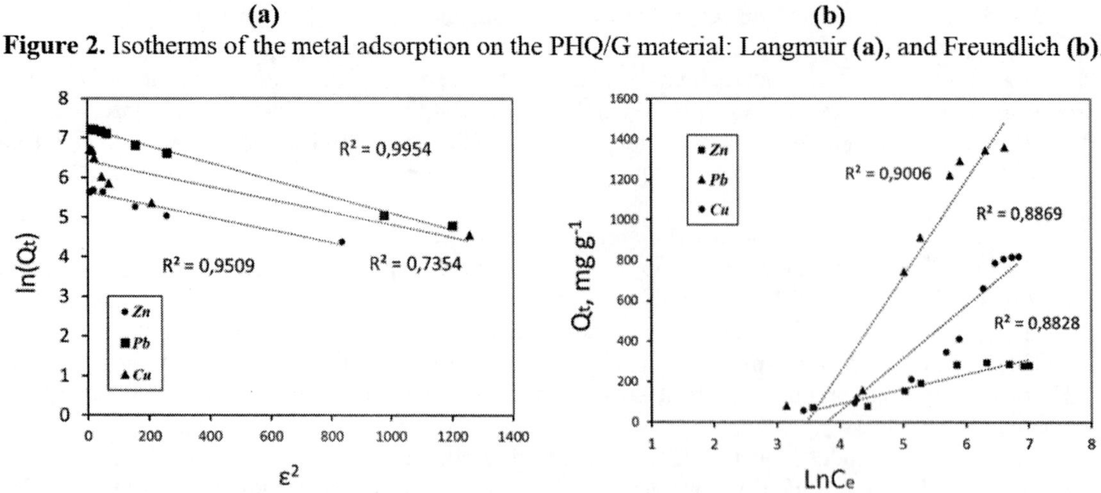

(a) **(b)**

Figure 3. Isotherms of the metal adsorption on the PHQ/G material: Dubinin-Radushkevich **(a)**, and Temkin **(b)**.

Table 4. Parameters of the adsorption isotherm models.

Model	Parameters			
		Cu^{2+}		
Langmuir	K_L	Q_{max}		R^2
	0.0007	2176		0.9848
Freundlich	n	K_F		R^2
	0.7663	4.85		0.9783
Temkin	K_T	B_T		R^2
	0.4633	257.23		0.8869
Dubinin-	K_{DR}	Q_{max}	E	R^2
Radushkevich	0.0016	609.2	17.68	0.7354
		Pb^{2+}		
Langmuir	K_L	Q_{max}		R^2
	0.0029	2211		0.9416
Freundlich	n	K_F		R^2
	0.5909	32.48		0.8945
Temkin	K_T	B_T		R^2
	0.6572	471.41		0.9006
Dubinin-	K_{DR}	Q_{max}	E	R^2
Radushkevich	0.0021	1364.4	15.43	0.9954
		Zn^{2+}		
Langmuir	K_L	Q_{max}		R^2
	0.0063	344		0.9329
Freundlich	n	K_F		R^2
	0.3285	31.36		0.8979
Temkin	K_T	B_T		R^2
	0.0651	71.433		0.8828
Dubinin-	K_{DR}	Q_{max}	E	R^2
Radushkevich	0.0016	279.85	17.68	0.9509

The graphic description shows that, in general, all these models are highly fitted to the experimental data (R^2~0.89-0.98). According to the Langmuir equation, Q_{max} varies in the following sequence: Zn^{2+} (344 mg g^{-1}) < Cu^{2+} (2176 mg g^{-1}) < Pb^{2+} (2211 mg g^{-1}). Comparing these values with those experimentally obtained (Zn^{2+} (300 mg g^{-1}) < Cu^{2+} (800 mg g^{-1}) < Pb^{2+} (1350 mg g^{-1})), it can be seen that the model overestimates the adsorption capacity more than 2 times for Cu^{2+} and Pb^{2+}, thereby implying the occurrence of the other process (*e.g.*, precipitation) parallel to the adsorption. For Zn^{2+}, the capacity values almost coincide, indicating the absence of precipitation.

The *E* values obtained from the Dubinin-Radushkevich model are as follows: 15.43 kJ mol^{-1}, and 17.68 kJ mol^{-1}, (both – Cu^{2+} and Zn^{2+}). This means that the ion exchange is the prevailing mechanism for the Pb^{2+} adsorption, whereas the chemisorption governs the Cu^{2+} and Zn^{2+} adsorption.

Judging by the *n* values of the Freundlich model, it can be seen that Zn^{2+} presents the highest heterogeneity of the adsorption sites on the adsorbent surface (0.3285), followed by Pb^{2+} (0.5909) and Cu^{2+}(0.7663). The assumption about the surface heterogeneity is confirmed by the Temkin model.

Using the combined sing- and dual-site Langmuir-Freundlich (or Sips) models (equations (11) and (12)) makes it possible to overcome some limitations of the Langmuir and Freundlich equations, and consider the effect of the heterogeneity of the adsorption sites on the adsorption process, since they contain more parameters and take into account energy changes from zero to infinity [20-22]. Moreover, the dual site model can separately describe the affinities of the sites possessing more or less energy and estimate their contribution to the process. The graphical interpretation of these models is given in figure 4, and their parameters are presented in table 5.

Both these models fit the adsorption data successfully (R^2~0.94-1.00), and thus confirm the assumptions of the Langmuir model (e.g., the existence of the precipitation process for Cu^{2+} and Pb^{2+}, and its absence for Zn^{2+}). Regarding the surface heterogeneity, based on the dual-site model, from the sum of $Q_{max1}+Q_{max2}$, the contribution of the adsorption sites with higher (H) and lower (L) binding energies (affinities) was determined (H/L, %): for Pb^{2+} - 24/76, for Zn^{2+} - 33/63, and for Cu^{2+} - 60/40. Thus, in the case of the Cu^{2+} adsorption, the process is controlled by higher-affinity surface sites responsible for complexation and chemical interaction, whereas in the case of the Pb^{2+} and Zn^{2+}, the lower-affinity sites responsible for electrostatic interactions and ion exchange prevail.

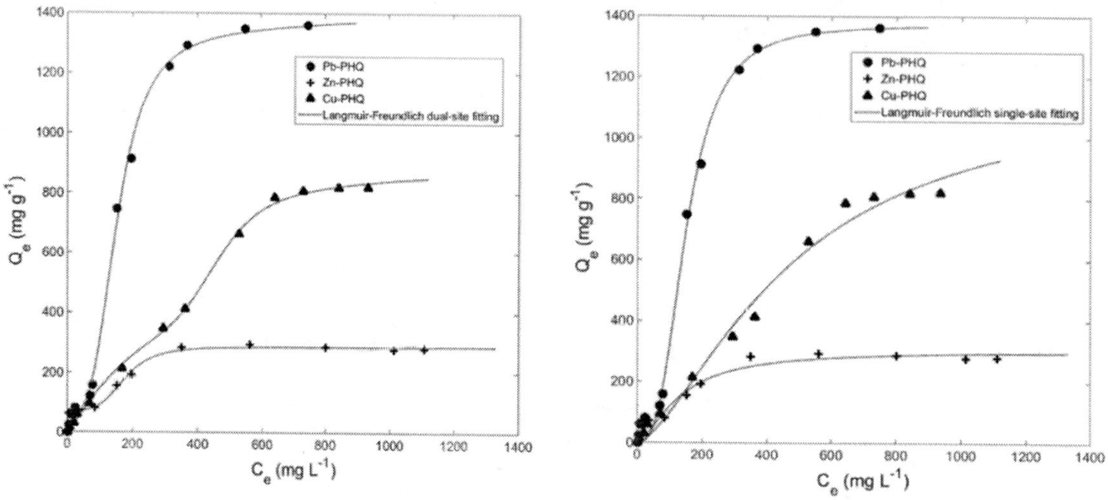

(a) (b)

Figure 4. Langmuir-Freundlich isotherms of the metal adsorption on the PHQ/G material: single-site **(a)**, and dual-site **(b)**.

Table 5. Parameters of the Langmuir-Freundlich models.

Adsorbate	Single-site Langmuir-Freundlich				Dual-site Langmuir-Freundlich						
	Q_{max}	K_{LF}	n'	R^2	Q_{max1}	K_{LF1}	n_1'	Q_{max2}	K_{LF2}	n_2'	R^2
Zn^{2+}	306	0.0075	1.563	0.9396	216	0.0056	4.020	71	0.952	1.546	0.9968
Pb^{2+}	1372	0.0067	2.936	0.9964	5078	$2.7\ 10^{-8}$	0.323	1214	0.007	3.473	0.9979
Cu^{2+}	1185	0.0021	1.525	0.9888	369	0.0022	6.994	545	0.005	1.252	0.9989

Finally, table 6 shows the parameters obtained after fitting the Pb^{2+} desorption data using the Langmuir, Freundlich and single- and dual-site Langmuir-Freundlich (Sips) models.

Table 6. Parameters of the isotherm models for the Pb^{2+}desorption from the PHQ/G material.

Langmuir			Freundlich		
Q_{max}	K_L	R^2	K_F	n	R^2
1.6 10^6	0.0001	0.7177	23.51	2.11	0.8406

Single-site Langmuir-Freundlich				Dual-site Langmuir-Freundlich						
Q_{max}	K_{LF}	n'	R^2	Q_{max1}	K_{LF1}	n_1'	Q_{max2}	K_{LF2}	n_2'	R^2
1107	0.2893	10.71	0.8936	1188	0.1437	2.37	600	0.2639	9.32	0.8822

It can be seen that all the models rather adequately describe the process (R^2~0.72-0.89). Besides, the percent contribution of the higher-affinity sites to the Pb^{2+} desorption) is about 34%. Moreover, in this system, the process is difficult to occur (n>1 in the Freundlich equation), which is additionally a favorable factor of the adsorption irreversibility confirmed above by the experimental data.

4. Conclusion

The novel material, PHQ/G nanocomposite, was proposed and tested as an adsorbent to remove Pb^{2+}, Cu^{2+} and Zn^{2+} from aqueous solutions. When using this adsorbent, rather high adsorption capacity was obtained, and the adsorption irreversibility was confirmed by the desorption studies and isotherm fitting (e.g., maximum 18 % of Pb^{2+} is desorbed from the adsorbent in to the aquatic medium). The isotherm studies were carried out implementing the Langmuir, Freundlich, Temkin, Dubinin-Radushkevich, as well as single- and dual-site Langmuir-Freundlich (Sips) models, all of which successfully fitted the experimental data. From the analysis of the results obtained, it was found that the adsorbent surface is heterogeneous – at least, two types of adsorption sites are available: higher-affinity (complexation and chemical interactions) and low-affinity (ion exchange and electrostatic interactions), as determined by the dual-site Langmuir-Freundlich equation. For the Cu^{2+} adsorption, the higher-affinity sites control the process, whereas for the other ions, the lower-affinity ones prevail. Thus, the PHQ/G material can be efficiently used to remove heavy metal ions for water/wastewater purification purposes.

Acknowledgment

This work was funded by the Ministry of Science and Higher Education of the Russian Federation under Project No. 16.1384.2017/PCh.

References

[1] Dreyer D R, Park S, Bielawski C W and Ruo R S 2010b The chemistry of graphene oxide *Chem. Soc. Rev.* **39** 228

[2] Norhayati H, Zuhailimuna M, Mohd Z H, Illyas M I, Azmi M, Azlan K, Suriani A B, Mazidah M and Adila M J 2016 A brief review on recent graphene oxide-based material nanocomposites: Synthesis and applications *J. Mater. Environ. Sci.*7 3225

[3] Gu D and Fein J B 2015 Adsorption of metals onto graphene oxide, surface complexation modeling and linear free energy relationships *Colloids Surf. A Physicochem. Eng. Aspects* **481** 319

[4] Duru I, Ege D and KamaliA R 2016 Graphene oxides for removal of heavy and precious metals from wastewater *J. Mater. Sci.***51** 6097

[5] Pourjavid M R, Sehat A A, Arabieh M, Yousefi S R, Hosseini M H and Rezaee M 2014 Column solid phase extraction and flame atomic absorption spectrometric determination of

manganese(II) and iron(III) ions in water, food and biological samples using 3-(1-methyl-1H-pyrrol-2-yl)-1H-pyrazole-5-carboxylic acid on synthesized graphene oxide *Mater. Sci. Eng. C* **35** 370

[6] Molaei K, Bagheri H, Asgharinezhad A A, Ebrahimzadeh H and Shamsipur M 2017 SiO_2-coated magnetic graphene oxide modified with polypyrrole–polythiophene: A novel and efficient nanocomposite for solid phase extraction of trace amounts of heavy metals *Talanta* **167** 607

[7] Lee Y-C and Yang J-W 2012 Self-assembled flower-like TiO2 on exfoliated graphite oxide for heavy metal removal *J. Ind. Eng. Chem.* **18** 1178

[8] Huang Z-H, Zheng X, Lv W, Wang M, Yang Q-H and Kang F 2011 Adsorption of Lead(II) Ions from Aqueous Solution on Low-Temperature Exfoliated Graphene Nanosheets *Langmuir* **27** 7558

[9] Zhang J, Xie X, Liang C, Zhu W and Meng X 2019 Characteristics and mechanism of Pb(II) adsorption/desorption on GO/r-GO under sulfide-reducing conditions *J. Ind. Eng. Chem.* In press https://doi.org/10.1016/j.jiec.2019.01.029

[10] Dahaghin Z, Mousavi H Z andSajjadi S M 2017 Trace amounts of Cd(II), Cu(II) and Pb(II) ions monitoring using Fe_3O_4@graphene oxide nanocomposite modified via 2-mercaptobenzothiazole as a novel and efficient nanosorbent *J. Mol. Liq.* **231** 386

[11] Jiang Y, Zhang X, Yin X, Sun H and Wang N 2018 Graphene oxide-facilitated transport of Pb^{2+} and Cd^{2+} in saturated porous media *Sci. Total Environ.* **631-632** 369

[12] AlamS N, Sharma N and Kumar L 2017 Synthesis of graphene oxide (GO) by modified Hummers method and its thermal reduction to obtain Reduced graphene oxide (rGO) *Graphene* **6** 1-18

[13] Burakov A E, Babkin A V, Burakova I V, Melezhik A V, Kurnosov D A, Mkrtchayn E S and Tkachev A G, Russian Federation Patent No. 2659285 (29 June 2018)

[14] Yao Y, Xu F, Chen M, Xu Z and Zhu Z 2010 Adsorption behavior of methylene blue on carbon nanotubes *Bioresour. Technol.* **101** 3040

[15] Giles C H, Smith D and Huitson A 1974 A general treatment and classification of the solute adsorption isotherm. I. Theoretical *J. Colloid Interface Sci.* **47** 755

[16] Langmuir I 1916 The constitution and fundamental properties of solids and liquids *J. Amer. Chem. Soc.* **38** 2221

[17] Tripathy S S and Raichur A M 2008 Abatement of fluoride from water using manganese dioxide-coated activated alumina *J. Hazard. Mater.* **153** 1043

[18] Dubinin M M 1965 The current state of the theory of volume filling of micropores during the adsorption of gases and vapors on carbon adsorbents *Zhurnal Fizicheskoy Khimii* [in Russian] **39** 1305

[19] Dada A O, Olalekan A P, Olatunya A and Dada A O 2012 Langmuir, Freundlich, Temkin and Dubinin-Radushkevich isotherms studies of equilibrium Sorption of Zn^{2+} unto phosphoric acid modified rice husk *J. Appl. Chem.* **3** 38

[20] Sips R 1950 On the structure of a catalyst surface II *J. Chem. Phys.* **18** 1024

[21] Koopal L K, van Riemsdijk W H, de Wit J C M and Benedetti M F 1994 Analytical isotherm equations for multicomponent adsorption to heterogeneous surfaces *J. Colloid Interface Sci.* **166** 51

[22] Sharma S and Agarwal G 2001 Interaction of proteins with immobilized metal ions: a comparative analysis using various isotherm models *Anal. Biochem.* **288** 126

IOP Conf. Series: Materials Science and Engineering **693** (2019) 012046 doi:10.1088/1757-899X/693/1/012046

The use of concrete and epoxy resin, modified with few-layer graphene for the production, repairs, and strengthening of concrete beams

V Pershin[1,*], K A Al-Shiblawi[2], A M R Al-Mashhadani[1], A Pasko[1] and D Melekhin[3]

[1]Department of Technology and Methods of Nanoproducts Manufacturing, *Tambov State Technical University,* 106 Sovetskaya Street, 392000, Tambov, Russia
[2]Ministry of Education, *School Buildings Department*, Najaf, St. Alkufah, Iraq
[3]Architecture Faculty, *Czech Technical University in Prague,* 1903/4 Zikova, 166 36 Prague 6 Czech Republic

[*]E-mail: pershin.home@mail.ru

Abstract. The paper deals with the modification of concrete and epoxy resin with a 2D carbon material (2DCM) obtained by liquid-phase shear exfoliation of graphite. It was experimentally established that the three-point bending strength of graphene-modified samples (0.05 wt.% of graphene in cement) in 1.5 times greater than that of pristine (unmodified) concrete. The water absorption of concrete modified with the 2DCM was found to decrease by 3 times. A method, differing from the known ones where few-layered graphene is obtained directly in the hardener through shear exfoliation of graphite, was developed for modifying the ED-20 epoxy resin using the 2DCM. It allowed us to increase the Charpy impact strength by 25-30 %. The tensile and bending strength were increased by 10 %. The test results of the restored beams after breaking at three-point bending, as well as on the adhesion of the reinforcing fabric to concrete, revealed a synergistic effect while using the concrete and epoxy resin modified with the 2DCM.

1. Introduction

Polymer composite fittings, in particular, fiberglass, are increasingly used in modern construction for internal and external reinforcement of beams. The results of experimental studies on a PBO-PRKM masonry reinforced composite are presented in [1,2]. Studies on the use of nanomaterials for modifying non-metallic reinforcement are of particular interest. In [3], the issues of modifying building composites with carbon Taunit nanotubes were considered. The article [4] describes the possibility of using complex polyfunctional additives for construction purposes based on carbon nanotubes and zeolites (synthetic and natural). The strength characteristics of composite reinforcement and, as a result, reinforced concrete products depend on the adhesion of reinforcement to concrete [5-7]. In fact, a binder is in contact with the concrete, which is used to form a fiber (fiberglass, basalt or carbon) rod. Thus, increasing the strength of the binder and adhesion of this binder with concrete is a reserve for improving the performance characteristics of composite reinforcement, when used in concrete products.

Improving the performance characteristics of concrete, such as compressive and tensile strength, crack resistance, water absorption, resistance in corrosive environments, etc. is a very urgent task. Many solutions have been developed to solve individual problems, but there is still no comprehensive solution. The recently introduced new material on the market, the so-called "nano-concrete", is not much different from ordinary concrete mixes. It also contains a mineral binder, aggregate, and water, only nanomaterials are used as modifiers: titanium dioxide nanoparticles [8], carbon nanoparticles [9], Melflux 1641 nanoparticles and plasticizer, carbon nanomaterials obtained as a by-product during plasma gasification of coal [10]. When these nano initiators interacting with cement crystallize and change its structure at the molecular level. In [11], issues of the influence of fillers containing nanosized particles on the formation of macro- and microstructures of concrete, the effect on the properties of concrete composites of early freezing of concrete mixtures with nano-modifiers were considered. It is revealed that frost resistance decreases, water resistance increases from 3 to 10 times.

In our opinion, the most promising direction in the use of nanomaterials for modifying concrete is the use of suspensions with 2D carbon materials (2DCM) in particular few-layers graphene [12], obtained by liquid-phase exfoliation of graphite into stator-rotor apparatuses [8].

In 2017, the International Organization for Standardization (ISO) published the first standard associated with the nomenclature of 2D materials [13]. The following classification has been proposed for graphene: graphene - single layer of carbon atoms; Bilayer graphene: two well-defined stacked graphene layers; few-layer graphene: consisting of three to ten well-defined stacked graphene layers; graphene nanoplatelet: thickness between 1 and 3 nm and lateral dimensions ranging from ≈100 nm to 100 μm. In the work [14], the authors note that these definitions are rather arbitrary from the physical point of view. Moreover, the term "graphene nanoparticles" is incorrect, since physically graphene represents essentially fine graphite. Given this circumstance, we use the term "2DCMs". The fact is that as a result of the liquid-phase shear exfoliation of graphite, a suspension is obtained, which simultaneously contains: graphene; double layer graphenc; a small layer of graphene; and graphene nanoplatelets. Depending on the process conditions, only the percentage of the indicated suspension components changes.

Considering the aforementioned, the present paper discusses options for modifying concrete epoxy resin with a 2DCM for the production, repair, and reinforcement of beams.

2. Concrete modification with few-layered graphene

To modify concrete, we used a 2DCM prepared in a rotor device with moving blades according to the technology described elsewhere [15]. After processing the suspension (5 L) for 45 min, it was centrifuged at accelerations in the sedimentation zone of 700 m/s^2. After removal of the precipitate, the concentration of few-layered graphene in the suspension ranged from 1.7 to 2.0 mg/mL. The aqueous suspension of this graphene was used as mixing water in the preparation of concrete lurries for the manufacture of samples. The suspension was diluted with pure water so that the few-layered graphene concentration ranged from 0.02 to 0.07 wt.% relative to cement. In the manufacture of control samples, the same ratio of cement, sand and pure water was used. During the experiments, the cement/sand ratios were changed in such a way as to obtain compressive strength of control samples from 5 to 30 MPa. To determine the strength characteristics of the cement, sample beams of 40×40×160 mm were made. The cement solutions and samples were prepared as described in [16]. The samples obtained were tested for compression, three-point bending, and water absorption. Figure 1 shows the characteristic dependences of the ultimate normal stresses under three-point bending at the moment of beam destruction, on the percentage of the few-layered graphene in concrete. The weight concentration was calculated for dry 2DCM with respect to cement.

As can be seen from the graph, the dependence has rather pronounced extremes of the maximum. The strength of the graphene-modified (graphene concentration of approximately 0.05 wt.% in relation to cement) was found to be about 1.5 - 1.8 times more than that of pristine (unmodified) concrete, whereas the water absorption of the 2DCM-modified concrete decreases by 3 times.

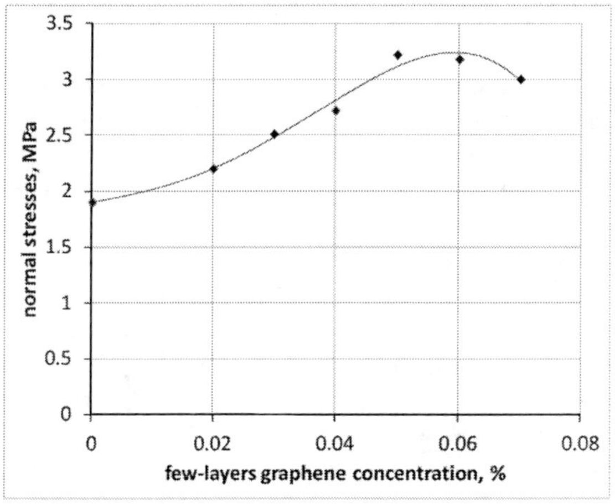

Figure 1. Dependence of ultimate normal bending stresses under three-point bending on the 2DCM concentration in concrete.

Figure 2. Concrete beam after failure at three-point bending.

3. Epoxy resin modification with few-layer graphene

In [17], the limiting stresses were experimentally determined for the transverse shear of glass composite reinforcement and the adhesion of reinforcement to concrete. It was established that the climatic conditions of operation of reinforced concrete products, in particular, cyclic freeze-thawing and soaking-drying, reduce the strength characteristics, especially the strength of adhesion of the reinforcement to the concrete, and these reductions for the ASC reinforcement are lower than for the metal reinforcement. It was especially noted that when testing the pulling of the reinforcement from concrete, quite often (approximately in 50 % of cases), there was a violation of the integrity in the binder material (epoxy resin), as well as the connection of the reinforcing fibers with the binder. Thus, one of the promising ways to improve the performance characteristics of the ASC is to increase the strength of the binder and its adhesion both with reinforcing fibers and with concrete.

A detailed analysis of methods for modifying epoxy resins with nanomaterials is given in [18, 19]. When modifying epoxy resins with graphene, a number of problems arise, in particular, in [20] attention is paid to the fact that the key point of polymer modification is the uniform distribution of the modifier in the matrix, which, in turn, largely depends on the affinity of the modifier polymer matrix. Usually, an interstitial fluid, usually acetone or isopropyl alcohol, is used to introduce graphene structures into the epoxy resin [20].

The results of our research have shown that both of these methods give unstable results and require a lot of labor and time.

We have developed a method for producing the 2DCM directly in an epoxy hardener. The exfoliation of graphite in polyethylene polyamines (PEPA) was carried out as follows. An initial suspension (approximately 1000 cm^3) was prepared with a graphite (GSM-2) concentration of about 45–55 mg/mL using a standard laboratory paddle mixer. After that, the suspension was weighed and poured into a cylindrical container with a diameter of 100 mm and a height of 150 mm. The rotary device was installed in the tank, and the graphite exfoliation was performed according to the procedure described above in Section 2. After 90 min processing the suspension, it was centrifuged at 5000 rpm (centrifugal accelerations in the sedimentation zone was 70 m/s^2) for 30 min, and then, the sediment was removed. The precipitate and graphene-containing hardener were weighed, and the graphene concentration in the hardener was calculated. Usually, about 2-4 g of graphite were converted to graphene, and the graphene concentration in the hardener was found to be about 0.2-0.4 wt.% Since the epoxy resin:hardener ratio is 10:1, this concentration in the hardener is not sufficient to obtain a concentration of 0.07 % in the cured resin as originally planned. Considering this circumstance, repeated centrifugation was carried out

IOP Conf. Series: Materials Science and Engineering **693** (2019) 012046 doi:10.1088/1757-899X/693/1/012046

at a speed of 5000 rpm until ¾ of the volume in the container, which was installed on the centrifuge, turned the color of a pure hardener. This part of the hardener was discharged, and its weighing and weighing of the residue were carried out. According to the weighing results, the concentration in the sediment was calculated. After that, the precipitate was mixed and used to prepare samples. The desired concentration of graphene in the epoxy resin was achieved by diluting the precipitate with the pure hardener in the required quantities. After the hardener was prepared with the required graphene concentration, it was mixed with ED-20 epoxy resin and samples were prepared for tensile, bending, and impact tests.

The forms and methods of making samples were described in [21]. A working hypothesis was formulated and experimentally confirmed that if the process of shear exfoliation is carried out in one of the epoxy resin components, then not only a good distribution will be ensured, but also good compatibility with the other components. A method of modifying the ED-20 epoxy resin with the 2DCM was developed, which, when the graphene concentration is 0.05 wt.%, increases the Charpy impact strength by 25-30 %. The method differs from the known ones, where few-layered graphene is obtained directly in the hardener by shear exfoliation of graphite. The tensile and flexural strength were increased less significantly, by only 10 %.

4. Use of modified epoxy to restore concrete beams

As shown by the results of the studies on three-point bend concrete beams with a size of 40*40*160 mm, when the beams break, practically no small fragments of concrete are formed. In other words, after the destruction, the two halves of the beam can be quite tightly interconnected, since the destruction surfaces are mirror reflections of each other (figure 2).

In the experiments, we used: 6 destroyed beams of standard concrete; 6 destroyed beams of modified concrete (0.05% 2DCM); ordinary (pristine) and modified epoxy resins. Three reference concrete beams were glued together with the ordinary epoxy resin and three beams – with the modified one. Similarly, glued beams made of modified concrete. After gluing, the beams were kept for 48 h at 22 °C. The results of the three-point bending tests are given in table 1.

Table 1. The values of the breaking load P for different samples.

Composition	Maximum load P, N			
	Sample 1	Sample 2	Sample 3	Average
Concrete	830	780	800	803
Concrete + epoxy resin	790	770	800	787
Concrete	840	790	750	803
Concrete + modified epoxy resin	890	840	800	843
Modified concrete	1090	1100	1020	1070
Modified concrete + epoxy resin	820	800	840	820
Modified concrete	1060	1040	1030	1043
Modified concrete + modified resin	1288	1368	1256	1304

It should be especially noted that the destruction of the repaired beams, i.e. glued together with the pristine epoxy resin, took place almost exactly along the plane of the foregoing destruction, and the epoxy resin was placed on two halves in separate sections. When using the modified epoxy resin, the newly formed surface displaced from the surface of the previous fracture towards one of the fixed supports. It has been concluded that at the restoration of beams from usual concrete by epoxy, bending strength decreases to 5%. Then, when restoring beams from ordinary concrete with epoxy resin modified with 2DCM, the flexural strength rises to 7%. Finally, when restoring beams from modified concrete with modified epoxy, bending strength increases by 20-30% (table 1).

The increase in the bending strength shows good compatibility of the modified concrete with the modified resin. It is possible that during the bonding process, the graphene structures in concrete are joined to the graphene structures located in the epoxy resin.

5. The study of adhesion of reinforcing fabric with concrete, using modified epoxy resin

As noted above, composite materials, in particular, fiberglass and epoxy resin are widely used to strengthen and repair concrete products, by external reinforcement. In this case, the main role in the effectiveness of the external reinforcement is played by the strength of the joint between the reinforcing fabric and concrete.

The experiments were carried out as follows: strips of fabric 40, 50, and 60 mm wide were glued to the beam in the form of a loop (figure 3, left). The sample was mounted on a tensile testing machine (figure 3, right). The tensile force was increased until the fabric was torn off from the concrete.

Figure 3. Tests for the separation of reinforcing fabric from concrete:source sample (left); setting the sample on a tensile testing machine (right).

The following combinations were used in the experiments: concrete + fabric + epoxy resin (BFC); concrete + fabric + modified epoxy resin (BTME); modified concrete + fabric + epoxy resin (MBTE); modified concrete + fabric + modified epoxy resin (MBTME).

To clarity that, in figure 4, the test results are presented in relative terms (the ratio of the effort of separation of a particular sample P to the force of separation of BFC – P_o). For all the samples, the cement:sand:water ratio, *i.e.* normal compressive stresses, were the same.

The deviations of the absolute values of the effort of separation from the average values did not exceed 5-7 %. For all the samples, in which the connection of the reinforcing fabric with concrete was carried out with the pristine, epoxy resin, the gap was practically observed on the resin layer adjacent to the concrete. The histogram does not show the relative tearing forces for the MBTE samples, since the efforts were equal, the efforts for BFC samples and destruction in all samples occurred on epoxy resin without damaging the concrete. In the BTME samples, as well as in the MBTM samples, the destruction occurred along the concrete, but in the MBTM samples, the zone of destruction was located farther from the outer surface. It is likely that when using the modified concrete and modified resin, a connection is made between the graphene structures that exist in the resin and in the concrete.

The diagram shows that when using both the concrete and epoxy resin modified with the graphene, the adhesion strength of the reinforcing fabric with the concrete increases, on average, by 30 % (figure 4).

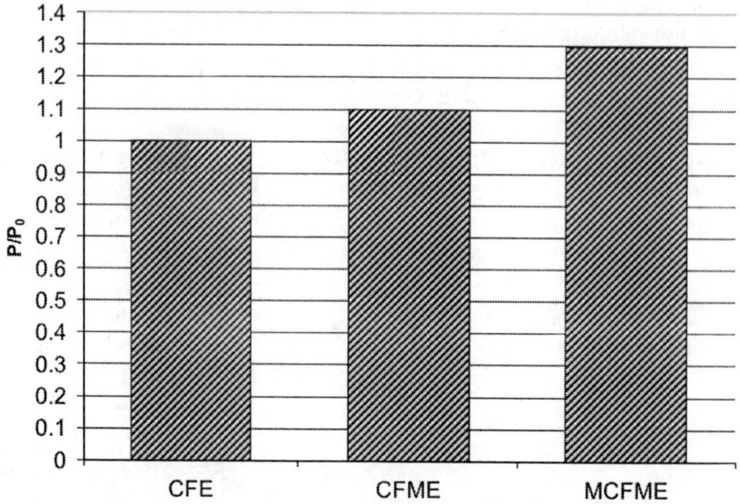

Figure 4. Histogram of the relative effort of separation of reinforcing the fabric from concrete.

Since the external reinforcement is mainly used when repairing damaged beams, comparing these results with the results presented in table 1, it can be concluded that with simultaneous use of concrete and epoxy resin modified with the graphene, there is a synergistic effect of increasing the strength of the repaired concrete products.

6. Conclusion

It has been experimentally established that the three-point bending strength of the graphene-modified samples (graphene concentration of about 0.05 wt.% with respect to cement) is about 1.5 times greater than that of the pristine concrete, whereas the water absorption of the 2DCM-modified concrete decreases by 3 times. Using the method for modifying the epoxy resin with the 2DCM the Charpy impact strength is increased by 25 %. The tensile and bending strength increase by 10%. The test results of the repaired beams after breaking at three-point bending, as well as on the adhesion of the reinforcing fabric to the concrete, revealed a synergistic effect while using the 2DCM-modified concrete and epoxy resin.

References

[1] Alecci V, Misseri G, Rovero L, Stipo G, De Stefano M, Feo L and Luciano R 2016 Experimental investigation on masonry arches strengthened with PBO-FRCM composite. *Composites B* **100** 228 http://doi.org/10.1016 / j.compositesb.2016.05.063

[2] Misseri G, Rovero L, Stipo G, Barducci S, Alecci V and De Stefano M 2018 Experimental and analytical investigations on sustainable and innovative strengthening systems for masonry arches. *Compos. Struct.* **210** 526 http://doi.org/10.1016/j.compstruct.2018.11.054

[3] Tolchkov Y N, Panina T I, Mikhaleva Z A, Galunin E V, Memetov N R and Tkachev A G 2017 Effect of surfactants on the distribution of carbon nanomaterials in aqueous dispersions with nano-modifying construction composites *Chemical Physics and Mesoscopy* [in Russian] **19** 292

[4] Panina T I, Tolchkov J N, Tkachev A G, Mikhaleva Z A, Galunin E V, Memetov N R and Popov A I 2016 The efficiency of application of a complex nano-modifying additive based on zeolites in building materials *Nanotechnology in Construction* [in Russian] **8** 116 http://doi.org/10.15828/2075-8545-2016-8-5-116-132

[5] Gil A I, Badalova E N and Lasouski Y D 2015 Fiberglass and carbon fiber reinforcement in construction: advantages, disadvantages, application prospects *Bulletin of Polotsk State University. Series F: Construction. Applied Science* [in Russian] **16** 48

[6] Benin A V and Semenov S G 2013 Experimental studies of the bonding of composite reinforcement with a flat winding with concrete *Industrial and Civil Engineering* [in Russian] **9** 74

[7] Khozin V G, Piskunov A A, Gizdatullin A R and Kuklin A N 2013 Coupling polymer composite reinforcement with cement concrete *News of the Kazan State University of Architecture and Civil Engineering* [Izvestiya KGASU – in Russian] **1** 214

[8] Falikman V R and Vainer A Ya 2015 New highly effective nano-additives for photocatalytic concrete: synthesis and research *Nanotechnology in Construction* [in Russian] **7** 8 http://doi.org/10.15828/2075-8545-2015-7-1-18-28

[9] Tolmachev S N and Belichenko E A 2014 Features of the effect of carbon nanoparticles on the rheological properties of cement paste and the technological properties of fine-grained concrete *Nanotechnology in Construction* [in Russian] **6** 13 http://doi.org/10.15828/2075-8545-2014-6-5-13-29

[10] Urkhanova S A, Lkhasaranov S A, Buyantuev S L and Kuznetsova A Yu 2016 On the effect of carbon nanomaterials on the properties of cement and concrete *Nanotechnology in Construction* [in Russian] **8** 16 http://doi.org/10.15828/2075-8545-2016-8-5-16-41

[11] Stenechkina K S 2017 Influence of hardening conditions on the properties of nano-modified concrete *Innovative Science* [in Russian] **5** 65

[12] Dimov D, Amit I, Gorrie O, Barnes M D, Townsend N J, Neves A I, Withers F, Russo S and Craciun M F 2018 Ultrahigh performance nanoengineered graphene–concrete composites for multifunctional applications *Adv. Funct. Mater.* **28** 1705183 https://doi.org/10.1002/adfm.201705183

[13] ISO/TS 80004-13 2017 *Nanotechnologies–Vocabulary–Part 13: Graphene and Related Two-dimensional (2D) Materials* (BSI Standards Publication, London, UK) https://www.iso.org/standard/64741.html

[14] Kauling A P, Seefeldt A T, Pisoni D P, Pradeep R C, Bentini R, Oliveira R V B, Novoselov K S and Neto A H C 2018 The worldwide graphene flake production. *Adv. Mater.* **30** 03784 http://doi.org/10.1002/adma.201803784

[15] Pershin V F, Al-Shiblawi K A, Baranov A A and Pasko T V 2019 Obtaining few-layer graphene by the method of liquid-phase shear exfoliation *Scientific and Technical Statements SPbPU. Natural and Engineering Sciences* [in Russian] **25** 143 http://doi-org/10.18721/JEST.25114

[16] Al-Shiblawi K A, Pershin V F and Pasko T V 2018 Cement modification with few-layer graphene *TSU Science Vector* [in Russian] **4** 6 doi.org/10.18323/2073-5073-2018-4-6-11

[17] Al-Shiblawy K A, Pershin V F and Yartsev V P 2018 Study of the properties of glass composite reinforcement and its adhesion to concrete *Building Materials, Equipment, Technologies of the 21st Century* [in Russian] **11-12** 42

[18] Al-Shiblawy K A, Pershin V F and Yartsev V P 2018 Modification of epoxy resins: Modern condition and prospects.. Part I. Modification of nanoparticles *Adv. Mater.Technol.* **2** 68 http://doi.org/10.17277/amt.2018.02.pp.068-078

[19] Al-Shiblawy K A, Pershin V F and Yartsev V P 2018 Modification of epoxy resins: Modern condition and prospects. Part II. Graphene and graphene oxide modification *Adv. Mater.Technol.* **4** 42 http://doi.org/10.17277/amt.2018.04.pp.042-053

[20] Amirova L, Surnova A, Balkaev D, Musin D, Amirov R and Dimiev A M 2017 Homogeneous liquid phase transfer of graphene oxide into epoxy resins *ACS Appl. Mater. Interface.* **9** 1 https://doi.org/10.1021/acsami.7b02243

[21] Al-Shiblawi K A, Pershin V F, Jarcev V P and Pasko T V 2018 Modification of epoxy resin by graphene *AIP Conf. Proc.* **2041** 020015 https://doi.org/10.1063/1.5079346

AUTHOR INDEX

AUTHOR INDEX